STOLEN TIME

STOLEN TIME

The History of Tempo Rubato

Richard Hudson

CLARENDON PRESS · OXFORD
1994

Oxford University Press, Walton Street, Oxford OX2 6DP
Oxford New York
Athens Auckland Bangkok Bombay
Calcutta Cape Town Dar es Salaam Delhi
Florence Hong Kong Istanbul Karachi
Kuala Lumpur Madras Madrid Melbourne
Mexico City Nairobi Paris Singapore
Taipei Tokyo Toronto
and associated companies in
Berlin Ibadan

Oxford is a trade mark of Oxford University Press

Published in the United States
by Oxford University Press Inc., New York

British Library Cataloguing in Publication Data
Data available

Library of Congress Cataloging in Publication Data
Data available
ISBN 0–19–816169–7

10 9 8 7 6 5 4 3 2 1

Typeset by Best-set Typesetter Ltd., Hong Kong
Printed in Great Britain on acid-free paper by Biddles Ltd., Guildford and King's Lynn

PREFACE

EVER since Pier Francesco Tosi first described a particular baroque performing technique in 1723 in terms of musical theft, a stream of writings has poured forth concerning *tempo rubato*. Some authors explain how, when, or where to use it. Some identify the purposes it serves or the effect it creates. Some describe it in prose; others depict it in musical notation. Some advocate its use; others complain of its abuse. Some attempt to distinguish various types. Some associate it with particular performers, composers, or periods in music history. The accumulated literature is vast, and yet questions remain unanswered.

There has recently been a crescendo of interest in tempo rubato, especially in its original form. Therefore the time seems right to consider the entire history of the device. The purpose of this book is to trace the development of rubato from its beginnings until the present day and to weave into a logical historical continuity the diverse and immense amount of information available from performers and composers, as well as from writers of all kinds—historians, biographers, theorists, critics, lexicographers, psychologists, and teachers. Beginning late in the last century, valuable information comes also, of course, from sound recordings.

Smaller parts of this history have been treated previously by others. New insight, however, comes to every part of the history when viewed from the perspective of the entire development. Inaccurate conclusions can be drawn from a limited study if one does not understand the preceding history. Consequently, certain early writers, such as C. P. E. Bach, have frequently been quoted out of context and in support of theories foreign to their concept of rubato. For this reason, I have included many complete quotations in this book, occasionally even those which have often been cited before. It is important that they be interpreted now within a broader view. In addition, I have included many plates and musical examples, hoping that they will provide clarity on a confusing subject and will also enable the reader, if he wishes, to draw his own conclusions. This book is not intended to be exhaustive nor to include every example ever written; it is, rather, a presentation of the broad historical development of tempo rubato, with the hope that it can provide a firm foundation for other scholars who will later study smaller parts of the history in greater depth.[1]

In the captions for the musical examples, as well as in the main text when not stated otherwise, dates of publication are in parentheses, other dates are not. Markings in the musical examples are in brackets when they actually occur earlier in the music. In prose quotations I have added accents in French and, for eighteenth-century English sources, have changed spelling, capitalization, and type style, when necessary, to conform to modern practice. Indications for rubato and associated terms, however, follow the scores themselves, for whether or not the word *rubato* is capitalized or in italics may in some cases indicate a difference in meaning, especially concerning the duration of the effect.

The preparation of this book has been funded for a number of years by grants from the University of California, Los Angeles. I owe a special debt of gratitude to my Research Assistant, Alyson McLamore, for her superb and creative work. I want to express my deep appreciation also to the staff of the UCLA Music Library for their enthusiastic support: to Shirley Thompson, who day after day cheerfully located materials and made them available to me; to Darwin Scott, who notified me of incoming scores and books relating to rubato; to Marsha Berman, who introduced me to the world of computerized data bases; and to Gordon Theil, who helped me locate audio sources. I appreciate also the patient and courteous service offered by the Interlibrary Loan Department of the University Research Library at UCLA.

I am grateful to the librarians throughout the world who made books and scores available to me, and to the Stanford Archive of Recorded Sound in Braun Music Center at Stanford University for copies of sound recordings. I thank my colleagues at UCLA who shared ideas with me: James Porter in regard to Bartók, Paul Humphries and Timothy Rice on ethnomusicology, Alden Ashforth on popular music, and Roger Bourland on the contemporary composer's point of view. For help with French and German translations, I thank Blair Sullivan and Marie Louise Göllner. I thank Sally Sanford for suggestions concerning vocal rubato and for her demonstrations of baroque singing styles. In addition, I am grateful for the patient help of Bruce Phillips, Music Books Editor at Oxford University Press, and the editorial staff.

---

[1] The original typescript for this book was so large it frightened both author and publisher. To bring it within manageable bounds, I deleted about 70 musical examples, 8 plates, 70 quotations, 46 minor composers, long lists of other rubatos by composers discussed, and considerable detail regarding the history of tempo fluctuation.

I want to express my special gratitude, finally, to Charles Rosen, for the insights I have gained from his many writings and recordings, and for his encouragement at critical moments in the preparation of this book. In 1988 he read my proposal for the book and an early version of the chapter on Chopin. In response he sent suggestions as well as several pages concerning rubato from the typescript of his Norton lectures at Harvard—lectures to be published eventually in a book entitled *The Romantic Generation, Music from the Death of Beethoven to the Death of Chopin*. In 1992 he read the original version of my completed typescript and again sent comments. I have often quoted in this book from his published works. With his kind permission, I have also included material from his Norton lectures as well as from the comments he sent me directly. His ideas, which range over an incredibly wide span of music history, have enriched this book immeasurably.

R.H.

Los Angeles
June 1993

ACKNOWLEDGEMENTS

THE author and publisher are grateful to the following for permission to reproduce material photographically: the British Library (for Plates I, IV, V, IX, X, XIV–XVIII, and XX); the Haags Gemeentemuseum: Music Department (Plate II), the Walter H. Rubsamen Music Library at the University of California, Los Angeles (Plates III and XI–XIII); the Musikabteilung of the Staatsbibliothek zu Berlin Preussischer Kulturbesitz (Plates VI–VIII); and the Warren D. Allen Music Library at Florida State University (Plate XIX).

They are also grateful for permission to reprint the following musical excerpts:

Ex. 1 from 'TEA FOR TWO' (Vincent Youmans, Irving Caesar) © 1924 (Renewed) WB MUSIC CORP., IRVING CAESAR MUSIC CORP. All rights on behalf of IRVING CAESAR MUSIC CORP. administered by WB MUSIC CORP. All Rights Reserved. Used by Permission.

Ex. 9.3 from Puccini, *Il tabarro*, reproduced by kind permission of G. Ricordi & C. S.p.a.

Ex. 10.1 from PIANO SONATA OP. 2, NO. 2, LAST MOVEMENT (from Beethoven's 32 Sonatas for the Pianoforte) Edited by Artur Schnabel. Copyright © 1935 (Renewed 1963) SIMON & SCHUSTER c/o EMI FEIST CATALOG INC. Reprinted by Permission of CPP/BELWIN, Inc., Miami, FL. All Rights Reserved.

Ex. 10.3 from D. Scarlatti/Bartók, Sonata K.332. Reprinted by permission of Editio Musica Budapest.

Ex. 10.4 from Stravinsky, PETROUCHKA. © Copyright by Edition Russe de Musique; Copyright assigned to Boosey & Hawkes, Inc. Revised Edition © Copyright 1947, 1948 by Boosey & Hawkes, Inc.; Copyright Renewed. Reprinted by permission.

Ex. 10.5 from Stravinsky, THE RITE OF SPRING. © Copyright 1921 by Edition Russe de Musique; Copyright Renewed. Copyright and Renewal assigned to Boosey & Hawkes, Inc. for all countries. Reprinted by permission.

Exx. 10.6 and 10.7 from 'Sonata For Violoncello and Piano' by Elliott Carter. Copyright © 1951, 1953 (Renewed) by Associated Music Publishers, Inc. (BMI). International Copyright Secured. All Rights Reserved. Reprinted by Permission.

CONTENTS

LIST OF PLATES

ABBREVIATIONS

Books, Series, and Periodicals

AMZ	*Allgemeine musikalische Zeitung* (repr., Amsterdam: N. Israel, Frits A. M. Knuf, 1964)
DJM	*Dwight's Journal of Music* (repr., New York and London: Johnson Reprint Corp.; New York: Arno Press, 1967)
DM	*Documenta musicologica, Erste Reihe* (Kassel: Bärenreiter-Verlag)
DTÖ	*Denkmäler der Tonkunst in Österreich*
EM	*Early Music*
HAM	*Historical Anthology of Music*, ed. Archibald T. Davison and Willi Apel, 2 vols. (Cambridge, Mass.: Harvard University Press, rev. edn. 1966)
JAMS	*Journal of the American Musicological Society*
JM	*Journal of Musicology*
LP	*Le Pupitre* (Paris: Heugel)
MD	*Musical Dictionaries*, Series I, microfiche from copies in The Library of Congress, unless stated otherwise (Washington, D.C.: Brookhaven Press, 1976)
MGG	*Die Musik in Geschichte und Gegenwart*
MMMLF	*Monuments of Music and Music Literature in Facsimile* (New York: Broude Brothers)
MMR	*Monthly Musical Record*
MOS	*Masterworks on Singing* (Champaign, Ill.: Pro Musica Press)
MQ	*Musical Quarterly*
MS	*Musicological Studies* (Brooklyn, NY: Institute of Mediaeval Music)
MTT	*Musical Theorists in Translation* (Brooklyn, NY, or Henryville, Pa.: Institute of Mediaeval Music)
New Grove DJ	*New Grove Dictionary of Jazz*, ed. Barry Kernfeld (London: Macmillan, 1988)
NHD	*New Harvard Dictionary of Music*, ed. Don Michael Randel (Cambridge, Mass.: Belknap Press of Harvard University Press, 1986)

PMMM	*Publications of Mediaeval Musical Manuscripts* (Brooklyn: Institute of Mediaeval Music)
PNM	*Perspectives of New Music*
PQ	*Piano Quarterly*
RILM	*Répertoire international de littérature musicale*
RR	*Recent Researches* [in the music of various periods] (Madison, Wis.: A–R Editions)
SM	*Studies in Musicology* (Ann Arbor, Mich.: UMI Research Press)
Stirling	*Frédéric Chopin, Oeuvres pour piano, Fac-similé de l'exemplaire de Jane W. Stirling* (Paris: Bibliothèque Nationale, 1982)

Libraries

BL	British Library, London
BN	Bibliothèque Nationale, Paris
GH	Gemeentemuseum, The Hague
LC	Library of Congress, Washington, D.C.
NYPL	New York Public Library at Lincoln Center
SML	Sibley Music Library, Eastman School of Music, University of Rochester
UCLA ML	University of California, Los Angeles, Walter H. Rubsamen Music Library
UCLA URL	University of California, Los Angeles, University Research Library

University Presses

CUP	Cambridge University Press (Cambridge)
HUP	Harvard University Press (Cambridge, Mass.)
IUP	Indiana University Press (Bloomington and Indianapolis, Ind.)
OUP	Oxford University Press (Oxford, London, New York)
PSUP	Pennsylvania State University Press (University Park, Pa.)
PUP	Princeton University Press (Princeton, NJ)
SUNYP	State University of New York Press (Albany, NY)
UCP	University of California Press (Berkeley and Los Angeles)
UChP	University of Chicago Press (Chicago)

UMP	University of Michigan Press (Ann Arbor, Mich.)
UNP	University of Nebraska Press (Lincoln, Neb.)
UTP	University of Texas Press (Austin, Tex.)
YUP	Yale University Press (New Haven, Conn.)

Other Publishers

AIM	American Institute of Musicology
AMP	Associated Music Publishers (New York)
Bärenreiter	Bärenreiter-Verlag (Kassel)
B & H	Boosey & Hawkes (London, New York)
Br. & H	Breitkopf und Härtel (Leipzig, Wiesbaden)
CNRS	Centre National de la Recherche Scientifique (Paris)
Da Capo	Da Capo Press (New York)
Ditson	Oliver Ditson (Boston)
Doubleday	Doubleday [Page] & Co. (Garden City or New York City, NY)
Dover	Dover Publications, Inc. (New York)
ERM	Édition Russe de Musique [Russischer Musikverlag] (Berlin)
Faber	Faber and Faber (London)
Greenwood	Greenwood Press (Westport, Conn.)
Gregg	Gregg International Publishers, Ltd. (Westmead, Farnborough, Hants, England)
Hawkes	Hawkes & Son (London)
Henle	G. Henle Verlag (Munich)
Knopf	Alfred A. Knopf (New York)
Marks	Edward B. Marks Music Corporation (New York)
Minkoff	Minkoff Reprint (Geneva)
Norton	W. W. Norton (New York)
Novello	Novello [Ewer] & Co. (London)
Olms	Georg Olms Verlag or Verlagsbuchhandlung (Hildesheim)
Prentice–Hall	Prentice–Hall (Englewood Cliffs, NJ)
Presser	Theodore Presser (Philadelphia or Bryn Mawr, Pa.)
Ricordi	G. Ricordi (Milan)
Schott	B. Schott's Söhne (Mainz)
S & S	Simon & Schuster (New York)
St Martin's	St. Martin's Press (New York)
UM	University Microfilms (Ann Arbor, Mich.)
UMI	UMI Research Press (Ann Arbor, Mich.)
Universal	Universal Edition (Vienna)

FALSTAFF. I am glad I am so acquit of this tinder-box: his thefts were too open; his filching was like an unskilful singer,—he kept not time.

NYM. The good humour is to steal at a minim's rest.

PISTOL. 'Convey' the wise it call. 'Steal'! foh! a fico for the phrase!

Shakespeare, *The Merry Wives of Windsor* (1602), Act I, sc. iii.

FALSTAFF. L'arte sta in questa massima:
    '*Rubar con garbo e a tempo.*'

Libretto by Arrigo Boito for Verdi's *Falstaff* (1893), Act I.

INTRODUCTION

THE Italian word *rubato* means robbed or stolen. *Tempo rubato* is stolen time. The expression appears during the first half of the eighteenth century to describe a practice in baroque vocal music: some note values within a melody are altered for expressive purposes while the accompaniment maintains strict rhythm. This type of rubato continues in vocal and violin music well into the nineteenth century. Keyboard music incorporates this earlier type of rubato during the second half of the eighteenth century. Eventually, however, the expression *tempo rubato* begins to refer to rhythmic alterations not only in the melody, but in the tempo of the entire musical substance. For at least the first half of the nineteenth century both types of rubato exist concurrently, but later in the century the earlier type disappears. It is the later type of rubato, finally, that continues to live in Western art music and is the type most familiar to us today.

Although other meanings, as we will see, occasionally occur for the expression *tempo rubato*, these are the two main types. I will refer to them as the *earlier* and *later* types of rubato, since these terms, it seems to me, will be easier for the reader to remember than the more descriptive names sometimes used by other authors.[1] The later rubato involves tempo flexibility, but usually of a more subtle and expressive nature than the retards and accelerations marked in a score. Although applied lavishly by late romantic performers, this type of rubato is generally employed today in a more restrained manner. It is one of the musical elements that we still understand and still employ in our art music today.

The modern music lover is often puzzled, on the other hand, when he reads a description of the earlier rubato, for he can recall no example from his classical listening experience. Earlier rubato has

---

[1] The earlier and later types are called *melodic* and *structural* by Howard Ferguson in *Style and Interpretation: An Anthology of 16th–19th Century Keyboard Music*, iv: *Romantic Piano Music* (OUP, 1964), 8–9, and in *Keyboard Interpretation from the 14th to the 19th Century: An Introduction* (OUP, 1975), 47–9. Robert Donington distinguishes between *borrowed* and *stolen* time in *The Interpretation of Early Music*, new edn. (Faber, 1975), 430–4, and Sandra P. Rosenblum, in *Performance Practices in Classic Piano Music* (IUP, 1988), 373–92, describes the two types as *contrametric* and *agogic*. In the 12th edn. of the *Riemann Musik Lexikon, Sachteil* (Schott, 1967), 945, they are *gebunden* and *frei*.

remained alive, however, in some of the popular performing styles of our century. Consider the music in Ex. 1 for the first four measures of 'Tea for Two'. The lower two staves present a piano accompaniment; the upper staff indicates the notes a singer might actually perform. The upper voice on the middle staff represents the melody as originally written. Comparing this melody to the sung version on the top staff reveals that the singer has altered the durational values of some of the notes. The E♭ on the word *just* in the second measure of the top staff has stolen time from the preceding note, so that it and the following G on *tea* come in too soon. In the fourth measure, the rest steals time from the G on *two*, and the E♭ on *for* extends itself, thus stealing time from the F on *tea*. At the end of the second measure two notes have anticipated their notated location by arriving too soon; in the fourth measure two notes are delayed and arrive too late. If the reader will snap his fingers (as the singer sometimes does) with four even beats per measure while singing the melody on the upper staff, he will, I believe, recall having heard this sort of effect in popular singing.[2]

The version on the top staff of Ex. 1 would not, of course, have ordinarily been written down in notation. The singer, if he needed to follow any score at all, would have been looking at the melody as notated on the middle staff. In most cases the actual performance would be so complex that it could be only approximately suggested by notation in any event. The alterations added by the singer are part of what can be described as performance practice, for they are so universally applied that they require no special indication in a musical score. Both the earlier and later types of tempo rubato are first applied in music history as a part of performance practice. After a period of time, the practice is described by writers and eventually the concept receives a name. Sometime later, composers actually indicate the device in musical scores by the term *rubato* or other words. At times, composers attempt to write out the effect in notes, either with or without verbal instructions. It will be the purpose of this book to trace the history of the two rubatos as they pass through these various phases. An attempt will also be made to distinguish between them as their paths of evolution intersect during the first half of the nineteenth century.

The word *rubato* appears in connection with vocal music in 1723, the expression *tempo rubato* in 1752. *Rubato* first occurs in sources for keyboard in 1755 and for violin in 1756. After 1789 the word

---

[2] See Henry Pleasants, *The Great American Popular Singers* (S & S, 1974), 27, where he refers to 'the popular singer's habit of body-swaying and finger-popping' as 'essential . . . in giving him an explicit beat to steal from (rubato)'.

Ex. 1 Alterations a singer might add to the refrain of 'Tea for Two' from *No, no, Nanette*, music by Vincent Youmans, words by Irving Caesar (New York: Harms, 1924)

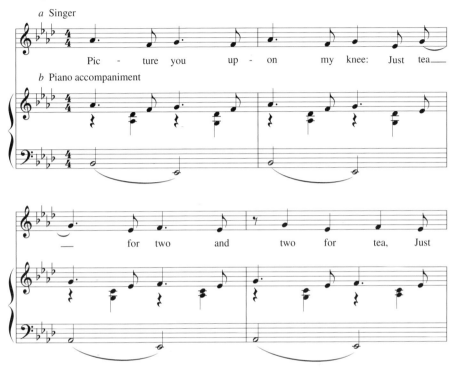

might also refer sometimes to a fluctuation of tempo. Long before this time, however, some of the techniques involved in both types of rubato actually existed. We can trace elements of the earlier type from the fourteenth century, the later type probably from the beginnings of music. The first two chapters of this book will deal with the background of rubato before the word itself appeared. We will turn first to the later type, since it involves a concept more familiar to the modern reader.[3]

---

[3] Some of the techniques described in the first two chapters are occasionally labelled *rubato* by modern writers, as I will indicate in the footnotes.

1 BACKGROUND OF THE LATER RUBATO

CHANGES of tempo during the course of continuous music occurred even before the word *rubato* was associated with it. Tempo markings first appeared during the Baroque period. Words for the modification of tempo, such as *ritardando, rallentando,* and occasionally *accelerando,* began to appear during the second half of the eighteenth century in the works Haydn, Mozart, and Beethoven. Before this time, however, flexible tempo sometimes occurred as a part of the performance practice of certain kinds of music. We find it in speech-like vocal music such as Gregorian chant, baroque monody, and recitative. We find it used to enhance the expression of text in the Renaissance madrigal as well as in baroque monody. We find it also employed for structural purposes to mark the beginning or end of a composition or a performance. Thus, it is involved with a retard or with an ornamented cadence or cadenza at the close of a piece. It is also involved with the preludial forms, which act as the introduction to the performance of other compositions. We shall now examine in more detail these tempo fluctuations employed for declamatory, expressive, and structural purposes.

Gregorian Chant

The main purpose of Gregorian chant is the effective communication of text. It is thus concerned with the clear pronunciation and proper accentuation of words and syllables, the rhythmic shaping of phrases, sentences, and other structural units, and the projection, on a very subtle level, of the meaning of the text. It is presumably sung, for the most part, with notes of equal value. Even when the music is confined to the repetition of a single pitch in a psalm tone, however, performance practice allows the slight lengthening or shortening of individual notes in order to accentuate the proper syllables or words.[1]

---

[1] *Liber usualis, with Introduction and Rubrics in English* (Tournai: Desclée, 1961), pp. xxxii–xxxiii.

Beginning in the tenth century, writers occasionally refer to a tempo change at cadence. They describe a gradual retard,[2] a slowing for phrases that end on certain important structural tones,[3] or the lengthening of the last two notes.[4] Acceleration is sometimes preferred for certain modes.[5] Some chants are sung quickly but slowed at the end and sometimes near the beginning; others are sung slowly, but hastened toward the end.[6]

In addition to tempo changes added by the performer, there are others designated by signs. In the ninth century, before the staff had been invented to identify pitches precisely, letters sometimes appeared above the neumes to indicate a special method of performance. The letter *c*, for example, refers to Latin words such as *cito* or *celeriter* which mean 'rapid', *t* to *trahere* or *tenere*, 'to drag out or hold'. There are letters also for different methods of pronunciation, and these, especially when they involve the consonants, may cause the lengthening or shortening of a syllable. Neumes marked *f*, for example, are performed with harsh or percussive attack.[7]

In addition, the shapes of the neumes could also indicate tempo changing.[8] A retard or lengthening of a note is shown by enlarging a portion of the neume figure, by separating notes that would otherwise have been joined together, or by adding a horizontal stroke.[9] Such a lengthening may occur not only to mark a cadence, but also to emphasize an important monosyllabic word or the accented syllable in a longer word. Similarly, abbreviated neume figures may indicate a

[2] Guido of Arezzo, *Micrologus* (*c*.1030), ed. Jos. Smits van Waesberghe in *Corpus scriptorum de musica*, iv (AIM, 1955), 175; trans. Warren Babb in *Hucbald, Guido, and John on Music: Three Medieval Treatises*, ed. Claude V. Palisca (YUP, 1978), 72.

[3] Engelbert of Admont, *De musica* (before 1325), ed. Martin Gerbert in *Scriptores ecclesiastici de musica sacra potissimum* (St Blasien, 1784; repr., Milan, 1931), ii, 368; see Joseph Dyer in 'A Thirteenth-Century Choirmaster: The *Scientia Artis Musicae* of Elias Salomon, *MQ* 66 (1980), 109–10.

[4] *Hieronymus de Moravia O.P.: Tractatus de musica*, ed. Simon M. Cserba in *Freiburger Studien zur Musikwissenschaft*, ii (Regensburg: Friedrich Pustet, 1935), pp. lxii–lxiii and 181–2; summarized by David Hiley in *Performance Practice: Music before 1600*, ed. Howard Mayer Brown and Stanley Sadie (Norton, 1989), 44.

[5] John Cotton, *De musica* (*c*.1100), trans. Warren Babb in *Hucbald, Guido, and John on Music*, 139.

[6] *Commemoratio brevis de tonis et psalmis modulandis* (anon., 10th cent.), ed. and trans. Terence Bailey in *Études médievales de l'Université d'Ottawa, Ottawa Medieval Texts and Studies*, iv (Ottawa: University of Ottawa Press, 1979), 103.

[7] *New Grove*, xiii, 132; J. Froger, 'L'Épître de Notker sur les "lettres significatives"', *Études grégoriennes*, 5 (1962), 23–72; and Eugene Cardine, *Gregorian Semiology*, trans. Robert M. Fowels (Solesmes, 1982), 224–7.

[8] Guido of Arezzo, *Aliae regulae* (*c*.1025), ed. Gerbert in *Scriptores*, ii, 37; see *Source Readings in Music History*, ed. Oliver Strunk (Norton, 1950), 120.

[9] For examples of such neume figures as well as the significative letters, see *Graduale triplex* (Solesmes, 1979) and Carl Parrish, *The Notation of Medieval Music* (Norton, 1957), 12 and Pl. II.

shortening of the notes, especially when they lead toward an important accented syllable.[10]

All of these changes of note value are slight nuances of expression which should be rendered in a warm and meaningful manner.[11] Gregorian chant is a comparatively solemn and restrained sort of music. Within this context, however, very small tempo changes may occasionally enhance the musical structure, the expression, or the declamation.

Madrigal, Monody, and Recitative

As the madrigal developed during the course of the sixteenth century, composers became more and more concerned with the expression of the meaning of the text. The literal and emotional suggestions of the text were mirrored in numerous ways in the notated score. Eventually performers applied expressive elements not included in the notation—elements such as dynamics and tempo changing. Around mid-century performers are described as singing fast and slow, as well as soft and loud, in response to the emotions in the text and music. This was likened to the practice of the orator, whose purpose was to affect the listener.[12] The intensity of expression increased as time progressed and led around the turn of the century to a decrease in the number of voices and the addition of an instrumental continuo. The continuo made it possible, finally, for a single solo voice to sing, and in a far more personal and intense manner. In some solo madrigals from the early seventeenth century the performer is specifically instructed to perform without a strict beat and according to the emotions.[13]

In the meantime the same desire for more intense expression gave rise shortly before 1600 to monody and early opera. Here the solo singer could change the tempo as he wished in order to achieve a speech-like eloquence.[14] Out of the essentially continuous flow of

[10] See Cardine, *Gregorian Semiology*, 24–31, 209, and 230.

[11] Joseph Gajard, *The Rhythm of Plainsong According to the Solesmes School*, trans. Aldhelm Dean (New York: J. Fischer & Bro., 1945), 44.

[12] Nicola Vicentino, *L'antica musica ridotta alla moderna prattica* (Rome, 1555), fac. in *DM* xvii (1959), fo. 94ᵛ (misnumbered 88); trans. Carol MacClintock, *Readings in the History of Music in Performance* (IUP, 1979), 78. See also Carl Dahlhaus, 'Über das Tempo in der Musik des späten 16. Jahrhunderts', *Musica*, 13 (1959), 767–9.

[13] Claudio Monteverdi, *Tutte le opere*, ed. G. Francesco Malipiero (Asolo, 1926–42), vii, 160 and 167; viii, 286. Fausto Razzi, in 'Polyphony of the *seconda prattica*: Performance Practice in Italian Vocal Music of the Mannerist Era', *EM* 8 (1980), 302, describes the flexibility in the latter piece as 'free rubato'.

[14] Giovanni de' Bardi, *Discourse on Ancient Music and Good Singing*, trans. Strunk, *Source Readings*, 299. See also Giulio Caccini, prefaces to *Le nuove musiche* (Florence, 1602) and

monody gradually crystallized the two contrasting styles of recitative and aria. During the fourth quarter of the seventeenth century each was incorporated finally in a separate movement. Italian recitative often retained some of the traits of monody, especially its free, declamatory rhythm.[15] At first it was accompanied only by continuo instruments, then around the end of the century it could sometimes be accompanied also by other instruments. In either case, the accompaniment generally followed precisely every rhythmic nuance initiated by the solo singer.

We find tempo changes, then, for expressive reasons in the madrigal and in early baroque monody. We find them for declamatory purposes in Italian monody and recitative. At the same time we occasionally find that when improvised ornaments are added by a singer, tempo fluctuation may occur even in metrically strict forms such as the aria,[16] and even in countries, such as France, which generally cultivated a more restrained singing style.[17]

Cadences and Cadenzas

Tempo changes may also articulate the beginning or end of a musical structure. We will consider in this section those changes that impart the feeling of conclusion. We have already noted tempo changes at the ends of phrases in Gregorian chant. The idea of retarding at cadences seems to have been transmitted also to polyphonic music. There are indications that a ritardando was made at the end of the medieval motet and conductus.[18] Late in the sixteenth century the cadence could be emphasized, even in types of music otherwise in strict rhythm,

Nuove musiche e nuova maniera di scriverle (Florence, 1614), both repr. (Florence: Studio per Edizione Scelte, 1983) and both ed. by H. Wiley Hitchcock in *RR in the Music of the Baroque Era*, ix (1970), 44–5 n. 10 and 54–5; xxviii (1978), Pl. IV and following page. Caccini uses the word *sprezzatura* (graceful neglect), which, according to Nigel Fortune in *New Grove*, xviii, 27–8, denotes 'concepts of expressiveness and rubato'.

[15] Sébastien de Brossard, *Dictionaire de musique* (Paris, 1703), repr. in *MD*, trans. and ed. Albion Gruber in *MTT* xii (1982), articles on *battuta*, *largo*, and *tempo*, as well as *recitativo*.

[16] Pier Francesco Tosi, *Opinioni de' cantori antichi e moderni, o sieno Osservazioni sopra il canto figurato* (Bologna, 1723), 63–8; trans. by John Ernest Galliard as *Observations on the Florid Song*, 2nd edn. (London, 1743; repr., London: William Reeves, 1926), 99–107.

[17] Jean-Jacques Rousseau, *Lettre sur la musique française* (Paris, 1753), trans. Strunk, *Source Readings*, 650. See also Newman Wilson Powell, 'Rhythmic Freedom in the Performance of French Music from 1650 to 1735', Ph.D. diss. (Stanford Univ., 1958; UM 59–3,719).

[18] Christopher Page, 'The Performance of Ars Antiqua Motets', *EM* 16 (1988), 154–5, and his chapter on 'Polyphony before 1400' in Brown and Sadie, *Performance Practice: Music before 1600*, 87–8. See also *The Conductus Collections of MS Wolfenbüttel 1099*, ed. Ethel Thurston in *RR in the Music of the Middle Ages and Early Renaissance*, xi–xiii (1980).

by extending the penultimate note or chord. In baroque practice, the usual dominant chord could thus be followed by a pause,[19] prolonged by a trill,[20] or simply held longer than indicated.[21] Such cadential retards can be observed in pieces recorded on contemporary mechanical instruments.[22] Modern listeners, however, often find them unexpectedly slight.[23] This should remind us that the degree of tempo changing at any given time in history is purely relative. To those who generally heard their music without any retards at all, a very slight slowing would have been noticeable. If we heard the same retard today, however, we might not consciously perceive it at all. It will be important to keep this in mind as we consider tempo changing from time to time during the course of this book.

The practice of adding ornamentation over the penultimate chord led during the early years of the eighteenth century to the cadenza. With the cadenza, however, we are no longer dealing merely with an embellishment of the penultimate chord in a cadence, but rather with a new section, sometimes of considerable length, to be inserted into a composition—such as an aria for solo voice, or a sonata or concerto for solo instrument—which otherwise observes strict metre. A motoristic sort of rhythm developed during the late Baroque period for almost all music except the cadenzas, the recitatives, and the preludial forms. It has a powerful forward drive and a forceful emphasis on a strict, even beat. The vigorous persistence of this beat creates an energy during the course of a piece that is difficult to stop. During the cadenza, however, the accompaniment ceases, leaving the soloist free to improvise in a style that deliberately contrasts with the regular rhythm of the main piece.[24]

---

[19] Thoinot Arbeau, *Orchésographie* (Langres, 1588), repr. of 2nd edn. of 1589 (Bologna: Forni, 1969), 48–9; trans. Mary Stewart Evans in *Orchesography* (Dover, 1967), 92. See also Georg Quitschreiber, *De canendi, elegantia, octodecim praecepta, musicae studiosis necessaria*, trans. Thurston Dart in 'How They Sang in Jena in 1598', *The Musical Times*, 108 (1967), 317.

[20] Carl Philipp Emanuel Bach, *Versuch über die wahre Art das Clavier zu spielen* (Berlin, 1753 and 1762; repr. Br. & H, 1969), ii, 254–5; trans. by William J. Mitchell as *Essay on the True Art of Playing Keyboard Instruments* (Norton, 1949), 375–6.

[21] Michael Praetorius, *Syntagma musicum*, iii, 2nd edn. (Wolfenbüttel, 1619), fac. in DM xv (1958), 80 (trans. in MacClintock, *Readings*, 151), and *Polyhymnia caduceatrix et panegyrica* (Wolfenbüttel, 1619), ed. Wilibald Gurlitt in *Gesamtausgabe der musikalischen Werke von Michael Praetorius*, xvii (Wolfenbüttel–Berlin: Georg Kallmeyer Verlag, 1930), 403.

[22] Concerning an example from 1617, see Albert Protz, *Mechanische Musikinstrumente* (Bärenreiter, *Vorwort* dated 1939), 46, and *Notenbeilage*, 6. For later notation, see Hans-Peter Schmitz, *New Grove*, vi, 202–3, and *Die Tontechnik des Père Engramelle* (Bärenreiter, 1953), 9, 24, and tables III and VII.

[23] George Houle, *Meter in Music, 1600–1800: Performance, Perception, and Notation* (IUP, 1987), 117 and 121–2.

[24] Daniel Gottlob Türk, *Klavierschule, oder Anweisung zum Klavierspielen für Lehrer und Lernende* (Leipzig and Halle, 1789), repr. in DM xxiii (1962), 308–13; trans. by Raymond H. Haggh as *School of Clavier Playing* (UNP, 1982), 297–301.

Occasionally one finds over a final cadence the word *adagio*, a term which means literally 'at ease' (*ad agio*) or 'leisurely'. Thus the word may sometimes imply a free rhythm as well as a slow tempo.[25] In addition, one sometimes also finds in the works of late baroque composers a tiny cadence which appears after the main cadence of an internal movement and moves through secondary chords to V or III. It is actually a half cadence and acts as a brief transition to the next movement. It is likewise often marked *adagio* or *adagio e piano*.[26] It was in the late Classic period, however, that the cadence became a dramatic event within a concept of tonal architecture. This was the same time, as we will see, that words such as *ritardando* gradually began to appear in scores.

Preludial Forms

The preludial forms for instruments were intended originally to precede the performance of a main composition, which itself could be vocal or instrumental, secular or sacred. They bear names such as *fantasia*, *ricercar*, *toccata*, *capriccio*, or *intonazione*, as well as *prelude*, *praeambulum*, or *preludium*. They provide a stylized representation of the process of warming up the fingers or checking the tuning (for those instruments tuned by the performer immediately before playing), or setting the pitch and mode for the listener or for other musicians involved in the main piece, or generally attracting the attention of the audience. They needed to have musical characteristics different from the main piece in order not to compete with it. Such music was no doubt originally improvised and in many cases this practice continued well into the nineteenth century. Numerous examples from the fifteenth century on, however, were written down, and their musical style reveals the traits associated with improvisation: a relative lack of organized structure and a free sense of rhythm.[27]

Rhythmic freedom occurs as early as 1536 in the vihuela fantasia, where chords are to be played slowly, ornaments or diminutions

---

[25] *New Grove*, i, 88–9. For late baroque examples, see Handel's Concerto Op. 6 No. 2 (2nd, 3rd, and 4th movts.) and J. S. Bach's Toccata and Fugue in D Minor for organ. Even earlier Froberger marks *à* or *avec discrétion* near the end of three of his gigues; see *10 Suittes de clavessin* (Amsterdam, *c*.1710), fac. in *17th Century Keyboard Music*, iv, with intro. by Robert Hill (New York: Garland, 1988), 4, 30, and 38.

[26] Such a cadence moves to III at the end of the opening movt. of Bach's Third Brandenburg Concerto, and to V at the end of the 1st movt. of Handel's Organ Concerto Op. 4 No. 1.

[27] Johann Mattheson, *Der vollkommene Capellmeister* (Hamburg, 1739), repr. in *DM* v (1954), 87–8; trans. Ernest C. Harriss in *SM* xxi (1981), 217.

rapidly, and a high point preceded by a slight pause.[28] Early in the seventeenth century tempo changing in the keyboard toccata was compared to that in the madrigal; one should play slowly at the beginning and at cadence, and when performing syncopations, trills, runs, and expressive passages.[29] Around mid-century the word *discrétion* occasionally refers to such rhythmic freedom in laments or *tombeaux*, some of which act as allemandes to introduce a set of movements in a suite.[30] In France this sort of rhythmic flexibility was depicted visually in the scores for unmeasured preludes. Beginning around 1630 in music for lute and continuing later in pieces for viol and finally clavecin, this notation involved a succession of slurred and unbarred notes to indicate arpeggiated chords as well as brief melodic passages.[31] Even after this type of notation was abandoned in the eighteenth century, French as well as English preludes were expected to be performed in the same rhythmically free style.[32]

Most of the longer examples from the seventeenth century are multi-sectional, with the sections themselves usually contrasting in texture, metre, note values, or general musical style, and hence presumably to be performed with different tempos or with different degrees of rhythmic flexibility. This was part of the baroque concertato style, and the changing of tempo between sections was essentially different from tempo flexibility within a section while music was in

[28] Luis de Milán, *Libro de música de vihuela de mano intitulado El maestro* (Valencia, 1536; repr. Minkoff, 1975), fos. Diii^r, E^r, F^r, Nv^r, and Qv^v; trans. Charles Jacobs (PSUP, 1971), 298–300. In *Tempo Notation in Renaissance Spain*, MS viii (1964), 8, Jacobs states that Milan's expression *tañer de gala* (graceful playing) 'seems to refer to a veritable *tempo rubato*'.

[29] Girolamo Frescobaldi, *Toccate e partite d'intavolatura di cimbalo, Libro primo* (Rome, 1616); *Il primo libro di Capricci fatti sopra diversi soggetti* (Rome, 1624); and *Fiori musicali* (Venice, 1635), prefaces repr. in Claudio Sartori, *Bibliografia della musica strumentale italiana stampata in Italia fino al 1700* (Florence: Leo S. Olschki, 1952), i, 219, 295–6, and 345; portions trans. in MacClintock, *Readings*, 133–6.

[30] See my book *The Allemande, the Balletto, and the Tanz* (CUP, 1986), i, 158–9; Mattheson, *Der vollkommene Capellmeister*, 89, Harriss's trans., 219; and examples by Froberger in *DTÖ* xxi, Jg. x/2, ed. Guido Adler, 110, 114–17. See Henning Siedentopf, *Johann Jakob Froberger: Leben und Werk* (Stuttgart: Verlagskontor, 1977), 67–8; and Avo Somer, 'The Keyboard Music of Johann Jakob Froberger', Ph.D. diss. (Univ. of Michigan, 1962; UM 63–5,019), 8, 27–8, and 454–5.

[31] Davitt Moroney, *New Grove*, xv, 212–14, and 'The Performance of Unmeasured Harpsichord Pieces', *EM* 4 (1976), 143–51; also Paul Prévost, *Le Prélude non mesuré pour clavecin (France 1650–1700)*, in *Collection d'études musicologiques*, lxxv (Baden-Baden and Bouxwiller: Éditions Valentin Koerner, 1987), especially 275–83 on the use of *rubato* or *discrétion*. For examples for viol see *Publications de la Société Française de Musicologie, Première série*, xx: *Concerts a deux violes esgales du Sieur de Sainte-Colombe*, ed. Paul Hooreman, 184 and 197–8. Concerning the general style of French lute music of the period, see André Souris' intro. to *Œuvres du vieux Gautier* (CNRS, 1966), p. xxxvi, section entitled 'Rubato virtuel et rubato écrit'.

[32] François Couperin, *L'Art de toucher le clavecin*, 2nd edn. (Paris, 1717), modern edn. by Anna Linde, trans. Mevanwy Roberts (Br. & H, 1933), 33.

progress. It was very late in the Baroque period, finally, that the multi-sectional preludial forms become multi-movement works. Thus with J. S. Bach a previously multi-sectional form may become a toccata, fantasy, or prelude and fugue in two separate movements, or even a toccata in three or four distinct and contrasting movements.

The Adagio movement in a violin sonata, a form deriving from the multi-sectional canzona, often takes on a preludial role in relation to a succeeding Allegro. As in the aria, the solo performer often added embellishment to the written score. Some performers apparently also took liberties with the tempo, as one would in a recitative. Roger North, however, decried the insertion of ornamentation that caused the measure of time to be broken.[33] The violin, as the main solo instrument of the Baroque period, did, however, imitate the solo vocal forms, and it was no doubt appropriate in certain sections or movements to employ the sort of rhythmic freedom used by singers in recitatives. This influence is evident also in keyboard music, as we can see when J. S. Bach uses the word *Recitativ* for an internal section of his Chromatic Fantasy and Fugue.

Rhythmic flexibility, which robs the tempo of its regular beat, occurs, then, in a variety of musical forms and for a variety of purposes. In one sense a certain amount of flexibility appears in the performance of any piece, for modern scientists have demonstrated that human beings are simply incapable of performing in an absolutely strict rhythm.[34] Furthermore, music in general—from any period in history—is not to be performed mechanically and without regard for the sense of the sounds. On the other hand, flexibility of a somewhat higher degree has been recognized at particular times in history, as we have just seen, as being suitable for certain types of music in order to project certain effects. One can thus use music on occasion to declaim a text, to express an emotion, or to mark the beginning or ending of a structural unit. Often music of a flexible nature is deliberately contrasted with music of a stricter sort by setting them side by side: thus the psalm tone and its antiphon, the recitative and the aria, the prelude and the fugue, the Adagio and Allegro, and the prelude that introduces a series of dance movements in a suite.

All of these forms, except the psalm tone and antiphon, crystallized during the late Baroque period and at the very time when the word

---

[33] BL, MSS Add. 32531, fo. 18ʳ, and Add. 32533, fo. 144ᵛ. See *Roger North on Music*, ed. John Wilson (Novello, 1959), 184–5, especially n. 18.

[34] See Jonathan D. Kramer, *The Time of Music* (New York: Schirmer Books, 1988), 73–4; and Alf Gabrielsson, 'Interplay between Analysis and Synthesis in Studies of Music Performance and Music Experience', *Music Perception*, 3 (1985), 82.

rubato first appeared. The rhythmic flexibility traced in this chapter represents part of the background of the later type of rubato of the nineteenth century. Quite a different sort of robbery occurs, however, in the earlier type, for only some note values in a melody are stolen, while its accompaniment maintains strict rhythm. We shall turn our attention now to the background of this earlier type of rubato.

2 BACKGROUND OF THE EARLIER RUBATO

THE sort of rhythmic robbery involved in the earlier rubato occurs well before 1723 in a number of compositional devices and techniques of performance practice. It appears during the Middle Ages and Renaissance as a notated aspect of melodic variation in compositions for keyboard and plucked string instruments. During the second half of the sixteenth century a similar procedure occurs also in the improvised ornamentation practised by singers and players of melody instruments. After 1600 the technique is employed more and more for expressive purposes. At the same time, the detailed elements of the evolving variation technique are codified, especially in France, and depicted by signs and small notes. French sources also describe the effect caused by the expressive pronunciation of text. In addition, robbery similar to that in the earlier rubato sometimes occurs also in the rhythmic transformation of melodies.

Melodic Variation

Rhythmic alteration occurs as one component of a technique of melodic variation that began around 1320 and continued through the eighteenth century. The technique generally involves the addition to a melody of new pitches, or sometimes rests, in such a way that the original structure is not altered. The varied version thus corresponds with the original melody phrase by phrase and measure by measure. The technique involves the principle that the effect of a note can be prolonged by the addition of other pitches that act either to ornament the original tone or to connect it with the next one. Alteration of note values often occurs as a consequence of this procedure.

The technique first appears in instrumental intabulations of vocal music. Ex. 2.1 shows how the top voice of a motet (*a*) was altered (*b*) when intabulated for keyboard around 1320.[1] In the opening measure,

[1] *Robertsbridge Fragment* from BL, MS Add. 28550, opening measures of the upper voice from 'Tribum quem'. The motet 'Tribum quem/Quoniam secta/Merito' comes from the appendix

Ex. 2.1 Intabulation of a vocal melody for keyboard, Robertsbridge Fragment, *c*.1320

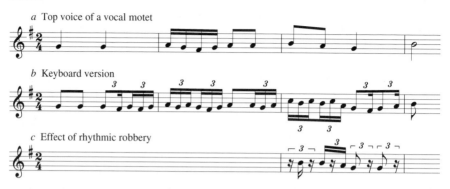

a Top voice of a vocal motet

b Keyboard version

c Effect of rhythmic robbery

the two original notes on G have been ornamented, first by simply repeating the pitch, then by alternation with the lower neighbour. At the end of the third measure, the note A in line (*b*) has been added to connect the original note G in (*a*) with the succeeding note B at the beginning of the fourth measure. This process produces a far more active melodic line, more suitable, presumably, for a keyboard instrument than the original slower vocal line.

When new pitches are added, however, they necessarily steal time from the original notes. In order to isolate this rhythmic robbery, I have substituted rests in Ex. 2.1*c* for the added pitches in the third measure. A comparison of the third bar in lines (*a*) and (*c*) shows that rhythmically the opening note B is initially delayed, then interrupted, and finally extended by stealing time from the second note A. This is not unlike the sort of robbery we saw in the fourth bar of 'Tea for Two' in Ex. 1.

The practice of arranging vocal works reached a high point in the sixteenth century.[2] An intabulation was performed as a separate and independent piece, and those who performed or listened to it would not necessarily be familiar with its vocal model. The effect of rhythmic robbery is considerably enhanced, however, when the listener has previously heard the robbed entity in its original or unrobbed form. This can happen in dance music when a section is immediately repeated

added around 1316 to the *Roman de Fauvel* (BN, MS fr. 146). See *Corpus of Early Keyboard Music*, i (1963): *Keyboard Music of the Fourteenth and Fifteenth Centuries*, ed. Willi Apel, 6; and Willi Apel, *The History of Keyboard Music to 1700*, trans. and rev. Hans Tischler (IUP, 1972), 26–7.

[2] See my book *The Allemande, the Balletto, and the Tanz*, i, 24–31; also Howard Mayer Brown, 'Embellishment in Early Sixteenth-Century Italian Intabulations', *Proceedings of the Royal Musical Association*, 100 (1973/4), 49–83; and *Instrumental Music Printed before 1600* (HUP, 1965).

with variation.[3] It can also occur when new voices are added to a borrowed melody. A Lutheran chorale or an Elizabethan popular tune would already have been well-known to a listener. When such a melody was subjected to embellishment, the added pitches and the rhythmic alterations would therefore have been much more apparent. In a set of variations, moreover, the opening statement often presents the melody in a simple, unembellished form which can be compared with succeeding variations.

In addition to being incorporated into various types of composition, this process of melodic variation was also improvised in the sixteenth century by singers and by performers on melody instruments such as the recorder or viol. Instruction books on this art of improvised variation appeared as early as 1535 and became more numerous at the end of the century. Some of these sources present lists of intervals or brief successions of pitches, showing for each a series of possible ways in which they could be embellished. Ex. 2.2*a* shows a brief passage

Ex. 2.2 Embellishment of a brief passage (Ganassi, 1535)

from Ganassi's recorder book of 1535, together with four embellished versions. In Ex. 2.2*b* the opening note G from (*a*) has been robbed, resulting in the anticipation of the A. In (*c*), (*d*), and (*e*), the four quarter notes of (*b*) have been replaced by five, six, or seven notes,

---

[3] *The Allemande, the Balletto, and the Tanz*, i, 21, as well as numerous examples in vol. ii.

Ex. 2.3 Embellishment of a madrigal by Rore (Bovicelli, 1594)

a Vocal original

An - cor——— che co'l par - ti - - re

b Embellished vocal version

An - cor ——— che co'l— par - ti - - re ——

c Effect of rhythmic robbery

resulting in more complex robbery.[4] Notation involving an unusual number of notes such as five or seven was no doubt an attempt to suggest in notation the type of rhythmic freedom which performers actually employed. Later, a similar practice became associated with tempo rubato, as we shall see, in the works of C. P. E. Bach.

Some sources include examples of this embellishment technique applied to entire compositions. The beginning of a madrigal by Cipriano de Rore appears in Ex. 2.3*a*, and below it a vocal embellishment by Bovicelli in 1594.[5] Ex. 2.3*c* reveals that rhythmically the opening note has been delayed, the final note anticipated, and the penultimate note reduced to half its original duration. In addition to this displacement of notes, Bovicelli also indicates very clearly that two of the syllables of text are likewise to be shifted. The final syllable of *partire* is displaced simultaneously with its pitch (see the third measure of Ex. 2.3*b*). With the second syllable of *ancor*, however, the text is anticipated but not its pitch: '*cor*' appears in the original piece

---

[4] Sylvestro di Ganassi dal Fontego, *Opera intitulata Fontegara* (Venice, 1535; repr. Bologna: Forni, 1969 and 1980), fos. Ciii[v], Hiii[v], Niii[v], and Qii[v]; trans. by Dorothy Swainson as *A Treatise on the Art of Playing the Recorder and of Free Ornamentation*, ed. Hildemarie Peter (Berlin–Lichterfelde: Robert Lienau, 1959), 28, 48, 68, and 80. Frederick Neumann, in *Ornamentation in Baroque and Post-Baroque Music, with Special Emphasis on J. S. Bach*, 3rd impression with corrections (PUP, 1983), 21, describes these examples as 'a graphic suggestion of astounding rhythmic freedom and tantalizing rubato effects'. See also Lodovico Zacconi, *Prattica di musica* (Venice, 1592; repr. Bologna: Forni, 1967), fo. 58[r] and 62[v]; and Howard Mayer Brown, *Embellishing Sixteenth-Century Music* (OUP, 1976), 25.

[5] Giovanni Battista Bovicelli, *Regole, passaggi di musica, madrigali et motetti passeggiati* (Venice, 1594), fac. in DM xii (1957), 46. See *Italian Diminutions: The Pieces with More than One Diminution from 1553 to 1638*, ed. Richard Erig with Veronika Gutmann, *Prattica musicale*, i (Zurich: Amadeus Verlag, 1979), 187–210; and Brown, *Embellishing Sixteenth-Century Music*, 43–6. The technique is similar to that applied in our own century by jazz soloists.

on the G on the fourth beat (compare the opening measures of Ex. 2.3*a* and *b*).[6] We shall note again later the involvement of text in rhythmic robbery, especially in the works of French baroque theorists.

The vocal or instrumental performer of such embellished music was, of course, closely acquainted with the original work, since this was the version he saw in notation as he improvised his variations. Listeners to such a performance might or might not know the work in its notated form. With vocal music especially, however, it seems to me that a listener would in any event notice at least the more active ornamentation when it occurred, if only due to the faster note values and the different tone of voice employed. There is also some evidence that occasionally both original and embellished lines were performed simultaneously—a Western type of heterophony. Ortiz indicates that when the viol ornaments any voice of a composition except the upper one, the accompanying keyboard player should play all the original parts. He also describes a method of performance in which the viol embellishes a plainchant while a keyboard plays the original version with added chords above it.[7] Vicentino advocates that instruments should play the original lines when singers are embellishing pieces for four voices. There is also evidence that sometimes instruments embellished parts while singers sang the original lines.[8] When an embellished and unembellished melody are heard simultaneously, the effect of the added pitches and the rhythmic alterations they cause would, of course, have been immediately apparent.

Although sources occasionally mention that the type of ornamentation should match the sense of the text,[9] embellishment in the sixteenth century seems mainly to have been decorative or virtuosic. It was in the next century and the Baroque period in general that improvised as well as notated embellishment became closely associated with expression. The continuing practice of improvisation is sometimes illustrated by notated examples, as in 'Possente spirto' from Monteverdi's *Orfeo* (published in 1609) and in slow movements from the violin sonatas of Corelli (published with ornaments in 1710).[10] It

[6] Concerning text placement, see Bovicelli, *Regole*, 7–9; Imogene Horsley in *New Grove*, iii, 124; and Brown, *Embellishing Sixteenth-Century Music*, 67.

[7] Diego Ortiz, *Tratado de glosas sobre cláusulas y otros géneros de puntos en la música de violones* (Rome, 1553), modern edn. Max Schneider (Bärenreiter, 1967), 55 and 68; trans. in Robert Donington, *The Interpretation of Early Music*, new version (Faber, 1975), 161 and 164.

[8] Nicola Vicentino, *L'antica musica ridotta alla moderna prattica* (Rome, 1555), repr. in *DM* xvii (1959), fo. 94ʳ (incorrectly numbered 88); trans. in Carol MacClintock, *Readings in the History of Music in Performance* (IUP, 1979), 77. See Brown, *Embellishing Sixteenth-Century Music*, 56–7 and 59.

[9] Brown, *Embellishing Sixteenth-Century Music*, 66–7.

[10] *Tutte le opere di Claudio Monteverdi*, ed. G. Francesco Malipiero, xi (Asolo, 1930), 84–98; Arcangelo Corelli, *Sonate a violino e violone o cimbalo*, fac. in *Archivum musicum*,

is this sort of embellishment by a solo performer that forms the direct background for the earlier type of rubato. Notated examples of earlier rubato sometimes also include added pitches as well as altered note values and look much like the examples of embellishment we have already seen. During the Baroque period, however, some of the detailed figures applied to single notes became identified as specific ornaments. These also have a rhythmic effect that constitutes part of the background of the earlier rubato.

Ornaments

The sort of embellishment discussed above applies generally to an entire melody. The technique is indicated by various terms in different countries and at different times: in English by *ornamentation, embellishment, variation, diminution,* or *division*; in Italian *coloratura, fioritura,* or *passaggio*; in Spanish *diferencia* or *glosa*; in German *Koloratur*. A specific figure applied to a single note, on the other hand, is designated by the word *grace* or *ornament* in English, or by *agrément* in French.[11] One can actually identify some of the main baroque ornaments in the embellished melodies from previous centuries. In the third measure of Ex. 2.1*b* a mordent appears on the original note G and a trill on B. Furthermore, the trill commences with its upper note and acts, as it does in the Baroque period itself, as an appoggiatura or leaning note. Ex. 2.3*b* displays a slide to the opening note A, a mordent on the following G, an anticipation of the G on the third beat of the second measure, and an appoggiatura to the A in the second measure.

Around the end of the sixteenth century attention turned in Italy to the expressive rendition of text. For this purpose small groups of ornamental notes were favoured over the virtuosic embellishment of entire melodies. Sometimes a performer added them as a part of performance practice; at other times they were written out in regular notation. Later, especially in French sources, they were indicated by signs, and finally by small notes. In any event, an ornament depended for its existence on time stolen from either the preceding or the following note.

Collana di testi rari, xxi (Florence: Studio per Edizioni Scelte, 1979). See Ernest T. Ferand, *Improvisation in Nine Centuries of Western Music* in *Anthology of Music (Das Musikwerk),* xii (Cologne: Arno Volk Verlag Hans Gerig KG., 1961); and Hans-Peter Schmitz, *Die Kunst der Verzierung im 18. Jahrhundert* (Bärenreiter, 1955).

[11] Donington, *The Interpretation of Early Music,* 160.

Ex. 2.4 *Anticipatione della syllaba*, Bernhard, *c.*1650

Zacconi emphasizes this sort of rhythmic alteration when he states in 1592 that some ornaments are produced 'by the slowing down and sustaining of the voice, and realized by taking a particle from one note and adding it to another'. One of his examples involves two notes, presumably of equal value, ascending a third from A to C. During the time allotted in the notation to the second note, the performer should repeat the first note (or more likely, tie it over), insert the pitch B to fill in the interval, and finally reach C later than its notated location. He describes the process as delaying the A by taking time from the C. He cautions against exaggeration, and only after describing the rhythmic aspect does he mention the ornamental note B that is added.[12]

The standard ornaments finally emerge late in the Baroque period. Some examples of the appoggiatura (or *port de voix*) and the trill, as well as various types of arpeggiation, will serve here to demonstrate their involvement with rhythmic robbery. We shall consider first some of the various types of appoggiatura. Familiar from late baroque music is the long cadential type, in which the first degree of the scale (by itself or included as the upper note of a trill) leans upon and delays the third of a dominant chord and hence makes this chord more emphatic before it finally resolves to the tonic. In the process the appoggiatura steals as much as one half or two thirds of the value of the following main note. Before this practice finally crystallized, however, a number of related but somewhat different types existed.

Around 1650 Christoph Bernhard, a pupil of Schütz, describes ornaments from the Italian singing style. He presents *anticipatione della syllaba* as in Ex. 2.4, where the syllables '*da*', '*bo*', and '*Do*' are anticipated by stealing time from the preceding syllable. In the case of '*bo*' the added pitch B also steals time from the preceding pitch C. The

---

[12] *Prattica di musica*, fo. 56ʳ. See Neumann, *Ornamentation in Baroque and Post-Baroque Music*, 22 (trans. of quotation), 23 n. 8, and 97–8 Ex. 14.2a; he refers to this as 'rubato technique' or 'rubato delay'.

Ex. 2.5 *Cercar della nota*, Bernhard, *c*.1650

Ex. 2.6 Italian unaccented cadence (Peri, 1600)

use of this ornament creates a syncopated and lively syllabic rhythm. Ex. 2.5 illustrates *cercar della nota* ('the searching-out of notes'); here the upper neighbour is added in 'Domino' and the lower in 'mi Jesu'.[13] In both cases the third syllable is displaced by stealing time and pitch from the preceding syllable. Similar placement of a syllable on a short note slurred to a long one occurs also in the cadence of Ex. 2.6, where an unaccented syllable is set to an unaccented note.[14] Since most Italian words have an accent on the penultimate syllable, this type of cadence occurs frequently in early baroque vocal music.

The early appoggiatura-like ornaments in the music of Michel Lambert are explained by Bacilly in 1668. In the case of a *port de voix* on an accented syllable (Ex. 2.7), Bacilly explains that 'it is necessary to divide the quarter note into two eighth notes of which the first will be sung on the word "la" and the second on the word "mort" before starting the *coup de gossier* [a 'stroke of the throat', indicated by the accent sign]...by repeating the *sol* [C] and then sustaining it afterwards.... It is necessary not only to borrow an eighth note from the preceding syllable but also to borrow by means of an anticipation a little of the time value of the upper note, to add it to that which has already been borrowed'.[15] The borrowing or stealing of time in Ex.

---

[13] *Von der Singe-Kunst, oder Manier*, sections 19–24; see 'The Treatises of Christoph Bernhard Translated by Walter Hilse', *The Music Forum*, 3 (1973), 17–19.

[14] Jacopo Peri, *Le musiche sopra l'Euridice* (Florence, 1600), fac. edn. Enrico Magni Dufflocq (Rome: Reale Accademia d'Italia, 1934), 30. See also *The Allemande, the Balletto, and the Tanz*, 175–6; Neumann, *Ornamentation in Baroque and Post-Baroque Music*, 99–101; and Donington, *The Interpretation of Early Music*, 525.

[15] Bénigne de Bacilly, *Remarques curieuses sur l'art de bien chanter* (Paris, 1668; 2nd edn., 1679, repr. Minkoff, 1971), 141; trans. by Austin B. Caswell as *A Commentary upon the Art of*

Ex. 2. 7 *Port de voix* on accented syllable (Bacilly/Lambert, 1668)

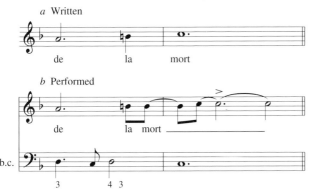

2.7 occurs without the addition of new pitches. In addition, the presence of a continuo part makes clear that the rhythmic alterations in the melody are accompanied by a conspicuous displacement between the final note C of the melody and the concluding C major triad in the accompaniment. There is also a displacement between the word *mort* and the final C major chord. Such displacements account in large measure for the expressive intensity of the ornament. For displacement to take place, of course, the continuo part must keep strict time while the singer alters the note values.

Although there is much controversy today concerning its correct execution, the trill involves rhythmic robbery in several ways.[16] The upper neighbour steals time from the main note as they alternate during the main body of the trill. If the trill commences with the upper note or a lengthier prefix on the beat, then the initial appearance of the main note is delayed, often in the manner of an appoggiatura. When there is a prefix that begins before the beat, it steals time from the preceding note. When the trill has a suffix, ordinarily the note below the main note and the main note itself, it steals time from the

Proper Singing (1668) in *MTT* vii (1968), 66. The outer 2 lines of the example come from *Les Airs de Monsieur Lambert* (Paris, 1668), transcribed in Austin B. Caswell, Jr., 'The Development of 17th-Century French Vocal Ornamentation and Its Influence upon Late Baroque Ornamentation-Practice', Ph.D. diss. (Univ. of Minnesota, 1964; UM 65–7,835), ii, final section, 5.

[16] Donington emphasizes the appoggiatura-like trill that begins *on* the beat with its upper note. Neumann presents impressive evidence that many other types of trill were also prevalent, some commencing before the beat and some beginning with the main note. One of the main purposes of his *Ornamentation in Baroque and Post-Baroque Music* is to demonstrate the large number of pre-beat ornaments of all types; concerning the trill, see 241–411. For Donington's point of view, see *The Interpretation of Early Music*, 236–59 and 620–40, and *New Grove*, xiii, 838–46.

end of the main note of the trill. When the alternation of the two notes in the body of the trill accelerates, another type of rhythmic freedom occurs. As we noted in Ex. 2.7, the accompaniment theoretically maintains strict rhythm in order for the rhythmic freedom of the melody to be evident. In an accelerating cadential trill, the accompaniment must likewise play strictly in time, or perhaps even with a ritardando, while the melody is accelerating. Composers occasionally depict this gradual acceleration by means of increasingly smaller note values.[17]

There are other baroque ornaments such as the turn, the mordent, the slide, and compound ornaments as well. All of them steal time from an adjacent main note. When the ornament commences on the beat, its time comes from the following note, thus delaying its entrance. When the ornament begins before the beat, it takes time from the end of the preceding note and serves to connect it to the next main note. In the latter case the initial appearance of the preceding main note is not affected. Hence, the most conspicuous robbery is caused by the on-beat graces, which necessarily delay the appearance of the following note.[18] In spite of exact notation in tables of signs, ornaments were meant to be performed with a sense of rhythmic freedom, and the performer was also generally free to add, omit, or change ornaments as he desired. It became customary to add more ornaments in the repetition of the opening section of a da capo aria; hence a listener might hear an ornament robbing time from the same note which he had earlier heard in an unornamented form. Occasionally ornamented and unornamented versions of the same note may be heard simultaneously. With pre-beat ornaments this causes no problem. With long appoggiaturas on the beat, however, and trills with a similar design, simultaneous sounding of the grace and its note of resolution causes a more noticeable clash. This sometimes occurs when a non-continuo performer executes the ornament and the continuo player realizes the unornamented note as part of a chord.[19] Sometimes such a clash is notated deliberately to produce an intensely emotional effect, as in the aria 'Erbarme dich' from J. S. Bach's St Matthew Passion.[20]

[17] See Giulio Caccini, *Le nuove musiche* (Florence, 1602; repr., Florence: Studio per Edizione Scelte, 1983), 4th unnumbered page of preface; modern edn. by H. Wiley Hitchcock in *RR in the Music of the Baroque Era*, ix (1970), 50. See also François Couperin, *Troisième livre de pièces de clavecin* (Paris, 1722), repr. in *MMMLF*, 1st ser., ix (1973), 11.

[18] Newman Wilson Powell, in 'Rhythmic Freedom in the Performance of French Music from 1650 to 1735', Ph.D. diss. (Stanford Univ., 1958; UM 59–3,719), 331, states that 'some features of the rhythmic performance of the *agréments* have the effect of creating an anticipation or delay in the rhythm—an effect which is analogous to the *tempo rubato* of the Italians and which may be regarded as the French counterpart of the Italian practice.'

[19] Donington, *The Interpretation of Early Music*, 315–19.

[20] *Neue Ausgabe sämtlicher Werke*, ii/v, 179–80, mm. 3 and 11: first the viola section, then the voice, sings C and B at the same time the solo violin plays appoggiaturas to these same notes.

Arpeggiation

During the Baroque period the arpeggio was also considered a type of ornament. It is such an important ancestor of early rubato, however, that it will be treated here in a separate section. It occurred often as part of performance practice, especially in music for lute, harpsichord, and clavichord. Eventually signs emerged to indicate it, most often the wavy line to the left of a vertical chord. A variety of methods were employed, separating the individual notes of a chord by commencing at the top, or at the bottom, or in some other order. The type that concerns us here is the one which lives on after the Baroque period and which moves from the lowest tone to the highest. As with the other ornaments, there is controversy today regarding its pre-beat or on-beat location.[21] When the arpeggio moves from the lowest to the highest pitch, and when it commences on the beat, it steals time from the upper note of the chord and thus delays a note that ordinarily functions as a part of the melody.[22] Since the speed at which the notes of a chord can be arpeggiated varies widely, a great diversity of effects can be achieved by this means. Long after the Baroque period the arpeggiated chord continued to be used on occasion for intensely expressive purposes.

A special type of arpeggiation occurs when only two notes are involved. Perrine describes in 1680 the *harpègement ou séparation* in two-note lute chords, indicated as in Ex. 2.8*b* by lines above the lower notes. Ex. 2.8*a* shows approximately how the melodic line is to be performed. He discusses only the ascending separation of notes and applies the technique also to three-note chords. It is perhaps important to note that the earlier *style brisé* in French lute music was also concerned with many successive single notes, but in that case they were all written as single notes in the tablature. Perrine, however, obviously wants the performer to see two vertical notes and to think of them as a vertical entity, but at the same time perform them separately.[23]

The same type of two-note arpeggiation occurs slightly later in

---

[21] Donington (*New Grove*, xiii, 853) stresses on-beat performance; Neumann expresses doubts (*Ornamentation in Baroque and Post-Baroque Music*, 492), but also adds (p. 494): 'Certainly there will be cases where an arpeggio in support of a melody note may be done on or astride the beat, but this would happen in the frame of an intended rubato delay.'

[22] Neumann states (ibid. 510) that 'the arpeggio has to be treated with a flexibility that permits its ready adaptation to any situation, including occasional rubato intents'.

[23] *Pièces de luth en musique* (Paris, 1680; repr. Minkoff, 1982), 6 and 6−7 of intro. Such a way of playing apparently became so common that some lute composers felt it necessary to use a special sign also when notes were to be played simultaneously; see Douglas Alton Smith and Peter Danner, '"How Beginners . . . should Proceed": The Lute Instructions of Lesage de Richée [c.1695]', *Journal of the Lute Society of America*, 9 (1976), 91.

Ex. 2. 8 *Harpègement* for lute (Perrine, 1680)

Ex. 2. 9 *Suspension* (F. Couperin, 1713)

clavecin music.[24] In 1713 Couperin defines an ornament called the
suspension (Ex. 2.9).[25] Significantly, he depicts it not as an isolated
note, but as an accentuated note to which three other notes lead. He
says that it is to be used in slow and tender pieces, that 'the duration
of the rest which precedes the note over which it is marked must be
left to the taste of the executant', and that 'in such cases where
stringed instruments would increase their volume of sound, the suspension (slight retardation) of the sounds on the harpsichord seems
(by a contrary effect) to produce on the ear the result expected and
desired.'[26] Couperin incorporates the *suspension* in his pieces as shown
in Ex. 2.10, where it occurs in (*a*) to accentuate the ascent to a long,
high note. Ex. 2.10*b* shows how this effect is even further intensified,
when the phrase recurs, by the addition of an arpeggiated octave in

---

[24] Michel de Saint-Lambert, *Les Principes du clavecin* (Paris, 1702; repr. Minkoff, 1972), 55;
trans. Rebecca Harris-Warrick, *Principles of the Harpsichord* (CUP, 1984), 94–5 and 124. See
also Gaspard le Roux, *Pièces de clavessin* (Paris, 1705; repr. Minkoff, 1982), table before p. 1.
[25] *Pièces de clavecin, Premier livre* (Paris, 1713), 75.
[26] *L'Art de toucher le clavecin*, 2nd edn. (Paris, 1717), modern edn. Anna Linde, English trans.
Mevanwy Roberts (Br. & H, 1933), 14–15. See also Donington's description of *broken notes* in
New Grove, xiii, 855. Thomas Sheridan, in *A Course of Lectures on Elocution* (London, 1762;
repr. Menston, England: Scolar Press, 1968), 77–8, says that similar pauses may be inserted by
orators 'before some very emphatical word' or 'where something extraordinary and new is
offered to the mind, which is likely to be attended to with an agreeable surprise'.

Ex. 2.10 *Suspension* in 'Les Nonètes' (F. Couperin, 1713)

the left hand.[27] The *suspension* continues later in works by Rameau[28] and is sometimes combined with other ornaments.[29]

This method of delaying notes for expressive purposes seems to have become almost a mannerism by mid-century. Foucquet states that 'in all pieces that require a gracious or tender execution, one ought to play the bass note before that of the melody, without altering the beat, which produces a suspension on each note of the melody.' In addition to this, he also says that 'when one encounters several notes in the bass, it is necessary to arpeggiate them, that is to commence with the lowest and so on, being careful for the sake of the melody to make the highest the last, which renders the touch mellow and graceful—indispensable for pieces of sentiment.'[30]

Jean-Baptiste-Antoine Forqueray published around 1747 some of his father's viol pieces arranged for clavecin. One of the sarabandes has this footnote: 'This piece must be played sensitively and with great taste; to show the proper interpretation I have added little crosses, which mean that the chords [or notes] in the left hand should be played before those in the right. In all other places the right hand should play first.' In addition, the notes of the piece are displaced

[27] *Pièces de clavecin*, *Premier livre*, 12. Sometimes Couperin indicates two-note arpeggiation by a wavy line, which, if it commences on the beat, would produce the same effect as the *suspension*.

[28] Jean-Philippe Rameau, *Pièces de clavecin* (Paris, 1724), modern edn. Kenneth Gilbert in *LP* lix (1979), 14.

[29] Louis-Claude Daquin, in *Premier livre de pièces de clavecin* (Paris, 1735; repr. Minkoff, 1982), combines it with an ascending appoggiatura and mordent; and Pierre-Claude Foucquet, *Pièces de clavecin, Livres I et II* (Paris, 1751; repr. Minkoff, 1982), i, 5, adds it to a trill or mordent.

[30] Foucquet, *Pièces de clavecin*, preface to *Second livre*: 'Dans toutes les Pièces d'éxécution gracieuse ou tendre, on doit toucher la note de basse, avant celle de dessus, sans altérer la mesure, ce qui opere une suspension sur chaque note du dessus. S'il se rencontre plusieurs notes dans la basse, il faut les harpeger, c'est à dire commencer par la plus basse et ainsi de suite, observant dans le dessus de faire entendre la plus haute, la dernière, ce qui rend le toucher moëlleux, gracieux et indispensable pour les Pièces de sentiments.'

vertically to show the correct order of playing the hands. In another sarabande, which is marked *tendrement*, a footnote reads: 'To play this piece in the way I should like it played, the performer should note how it is written [again, with right and left hand notes vertically aligned or displaced], the right hand being hardly ever quite together with the left.' In addition to extensive displacement between the two hands in these pieces, arpeggiation is often marked when either hand has two or more notes.[31]

Inequality

I use the English word *inequality* here to refer simply to rhythmic alterations resulting in either the unequal performance of equally notated pitches or the more unequal performance of already unequally notated pitches. The practice of performing a notated series of even notes by lengthening one note in each pair is described as early as the middle of the sixteenth century. In 1550 Bourgeois shows how to lengthen the first in each pair of notes.[32] Santa María gives examples in 1565 in which the first note, others in which the second note of a pair is lengthened.[33] Caccini says in 1602 that passages performed in such a manner 'have more grace'.[34] For Frescobaldi in 1616, the second of each pair of sixteenth notes should be 'somewhat dotted' when a passage of eighths and sixteenths are played together in both hands.[35] According to Couperin in 1713, the second of the two slurred eighth notes in a *coulé* should be prolonged.[36]

[31] A. Forqueray, *Pièces de clavecin*, ed. and trans. Colin Tilney in *LP* xvii (1970), 74, 94, and 104.

[32] Loys Bourgeois, *Le droict chemin de musique* (Geneva, 1550), repr. and trans. by Bernarr Rainbow as *The Direct Road to Music* in *Classic Texts in Music Education*, iv (Kilkenny, Ireland: Boethius Press, 1982), 120–3. See Frederick Neumann, 'The *Notes inégales* Revisited', *JM* 6 (1988), 140; repr. in Frederick Neumann, *New Essays on Performance Practice* in *SM* cviii (1989), 67–8.

[33] Tomás de Santa María, *Arte de tañer fantasia* (Valladolid, 1565; repr. Minkoff, 1973), *Primera parte, Capitulo* xix, fos. 45ᵛ–46ᵛ. See Donington, *The Interpretation of Early Music*, 454, for a trans. of the instructions. Neumann describes the rhythmic alterations in 'The *Notes inégales* Revisited', 142 (*New Essays*, 69), as 'rubato patterns' or 'rubato designs'.

[34] *Le nuove musiche*, 5th unnumbered page of the preface; modern edn. by Hitchcock, 51, and Strunk, *Source Readings*, 385. See Frederick Neumann, review of Anthony Newman's *Bach and the Baroque* in *The American Organist*, 21/4 (Apr. 1987), 40–3, repr. in *New Essays*, 235–41; on p. 238 he mentions 'Caccini's famous illustration of rubato singing . . ., whereby equally written notes are rhythmically manipulated in various manners, while remaining within the beat'.

[35] *Toccate e partite d'intavolatura di cimbalo, Libro primo*, 2nd edn. (Rome, 1616), item 7 in the preface; trans. by Dolmetsch in *Interpretation of the Music of the XVII and XVIII Centuries*, 6. See Claudio Sartori, *Bibliografia della musica strumentale italiana stampata in Italia fino al 1700* (Florence: Leo S. Olschki, 1952), i, 219.

[36] *Pièces de clavecin, Premier livre*, 75. See David Fuller in *NHD* 204.

The first note of a pair, on the other hand, is invariably the one lengthened in the French practice of *notes inégales*, which existed from the mid-seventeenth century to the end of the eighteenth. In this case, pairs of equally notated notes of a particular value in a given metre were performed with a slight lengthening of the first.[37] When a composer wished such notes to be performed equally as written, he needed to add special instructions or signs. There was also during the same period the different concept in all countries of accentuating 'good' notes in a series of even notes by slightly lengthening them and de-emphasizing the 'bad' notes by correspondingly shortening them. Thus, in a passage of eight sixteenth notes, for example, the first, third, fifth, and seventh are slightly longer than notated, the others slightly shorter. This procedure, unlike the *notes inégales*, could apply to notes of any value.[38]

In all the examples mentioned so far, the notes are written equally, but performed unequally. In some theoretical sources the method of performance is approximately indicated by the use of a dotted note. The lengthening in such cases, however, might actually be more or less than a dotted note, and in any event would vary. A different situation exists when dotted notes occur in a musical composition. Loulié explains in 1696 that in performing a dotted eighth and sixteenth the first note is 'held a bit longer' and the other 'passed through quickly'.[39] Whether such comments refer to a general practice of baroque over-dotting is the subject of present controversy.[40] Some scholars feel that

---

[37] There has also been much controversy in recent times concerning the *notes inégales*. See Donington, *The Interpretation of Early Music*, 452–63; David Fuller, *New Grove*, xiii, 420–7, and 'The Performer as Composer' in *Performance Practice: Music after 1600*, ed. Howard Mayer Brown and Stanley Sadie (Norton, 1989), 135–8; and George Houle, *Meter in Music, 1600–1800: Performance, Perception, and Notation* (IUP, 1987), 86–91 and 119–21. Frederick Neumann carefully separates the French *notes inégales* from the other situations in which one note in a pair is lengthened. See 'The French *Inégales*, Quantz, and Bach', *JAMS* 18 (1965), 313–58, repr. in *SM* lviii: *Essays in Performance Practice* (1982), 17–54 (see also 55–8 in reply to a critique by Donington); and 'The *Notes Inégales* Revisited', *JM* 6 (1988), 137–49 (*New Essays*, 65–76). On p. 138 of the latter (*New Essays*, 66), he writes: 'The *notes inégales* are a special case of rhythmic alteration that is related to agogic accents and to rubato.'

[38] See David Fuller, *New Grove*, xiii, 423. See also Grosvenor W. Cooper and Leonard B. Meyer, *The Rhythmic Structure of Music* (UChP, 1960), 8: 'in order to obtain the desired impression of grouping, the performer often slightly displaces unaccented beats in the temporal continuum so that they are closer in time to the accents with which they are to be grouped than if he had played them with rigid precision. (The rubato style of playing would seem to be an instance of such displacement.)'

[39] Étienne Loulié, *Éléments ou principes de musique mis dans un nouvel ordre* (Paris, 1696; repr. Minkoff, 1971), 16; trans. by Albert Cohen as *Elements or Principles of Music* in *MTT* vi (1965), 10.

[40] It is Neumann, again, who has questioned the general practice of over-dotting in baroque music; see the 5 articles repr. in *Essays in Performance Practice*, 73–182, and 2 others in *New Essays*, 77–90. See also David Fuller in *New Grove*, v, 581–3, and in Brown and Sadie, *Performance Practice: Music after 1600*, 130–5; Donington, *The Interpretation of Early Music*,

sharp over-dotting is appropriate in such styles as the opening duple section of a French overture or the dotted second beat of the French versions of the triple sarabande, folies d'Espagne, passacaille, and chaconne.[41] Loulié's instructions may, on the other hand, have been an attempt to describe the execution of a difficult rhythm. It is almost impossible to obtain the proper effect by counting four sixteenth beats commencing with the eighth note and then inserting the sixteenth note on the fourth count. This way the small note gets an incorrect accentuation. Occasionally the performer was instructed to insert a rest or silence of articulation between the dotted note and the sixteenth; this encouraged the concept that the small note was linked motivically with the following note, but without taking from it any of its accentuation.[42] Such silences of articulation, of course, also rob notes of a portion of their time. Since this sort of robbery does not affect the starting point of a note, however, it is usually perceived more as an effect of articulation.

In a few cases rhythmic alteration of the notated score involves more than two notes. Caccini illustrates the single and double *cascata* (Ex. 2.11), the fall through a descending scale from a high note. In both types the opening note steals time from the succeeding smaller notes so that the descent is more rapid when it does occur. In the single *cascata* the descent occurs in even notes that flow directly to the following cadence. In the two alternative methods for the double *cascata* in Ex. 2.11*b* and *c*, the final note of the descent is sustained longer and in (*c*) the notes increase in speed as they descend.[43]

As a final illustration of inequality, Ex. 2.12*b* shows a brief excerpt from a recitative in Lully's opera *Atys*. It is cited by Grimarest in 1707 as an example of discrepancy between text and music. He objects to the word *et* being set to a long note while the more important word *vous* appears on a very short one. He says that this can be remedied if

441–7; and Graham Pont, 'French Overtures at the Keyboard: How Handel Rendered the Playing of Them', *Musicology*, 6 (1980), 29–50 (see 39, 41, and 43–5, where he refers to the rhythmic alteration as 'tempo rubato'), and 'Handel's Overtures for Harpsichord or Organ: An Unrecognized Genre', *EM* 11 (1983), 309–22 (see 317–18 on over- and underdotting as tempo rubato).

[41] See my 4 volumes on *The Folia, the Saraband, the Passacaglia, and the Chaconne*, in *Musicological Studies and Documents*, xxxv (Neuhausen–Stuttgart: Hänssler-Verlag, 1982). A distant ancestor of baroque over-dotting may be the *modus non rectus* in 12th-cent. organum duplum, where the long values in modal rhythm are lengthened by the performer and the breves shortened. See William G. Waite, *The Rhythm of Twelfth-Century Polyphony* in *Yale Studies in the History of Music*, ii (YUP, 1954), 113–14. Waite shows how the quarter- and 8th-note pattern of the 1st rhythmic mode might be changed by a performer to a quarter note tied to a 16th plus a 16th note.

[42] Donington, *The Interpretation of Early Music*, 444–5.

[43] *Le nuove musiche*, 5th unnumbered page of preface; Hitchcock trans., 52, and Strunk, *Source Readings*, 385.

Ex. 2. 11 *Cascata doppia* (Caccini, 1602)

a Written

b Performed

c Performed

Ex 2. 12 Rhythmic alteration for the sake of text (Grimarest, 1707)

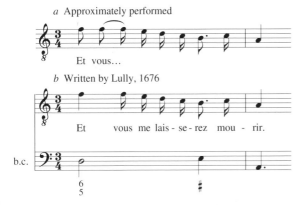

a Approximately performed

Et vous…

b Written by Lully, 1676

Et   vous me lais - se - rez mou - rir.

b.c.

the singer takes time from the note of the first syllable to put on the
second. I have shown in Ex. 2.12*a* approximately how one might
anticipate the word *vous* according to Grimarest's description, thus
making it last longer than *et*.[44]

Pronunciation

We have already seen how the placement of syllables can be affected
by rhythmic robbery. Bernhard in Exx. 2.4 and 2.5 shows how a

[44] Jean-Léonor de Grimarest, *Traité du récitatif, nouvelle édition* (The Hague, 1760; repr.,
New York: AMS Press, 1978), 135: 'celui qui chante prend de la note de la première silabe pour
mettre sur la seconde.' Grimarest does not show the music, which I have taken from the 2nd edn.
of Lully's *Atys* (Paris, 1709), copy at UCLA ML, 55 (Act I, sc. vi). See David Tunley, 'Grimarest's
Traité du récitatif: Glimpses of Performance Practice in Lully's Operas', *EM* 15 (1987), 363; and
Michael Alan Reid, 'Remarks and Reflections on French Recitative: An Inquiry into Performance
Practice Based on the Observations of Bénigne de Bacilly, Jean-Léonor de Grimarest, and Jean-
Baptiste Dubos', M.Mus. thesis (North Texas State Univ., 1985; UM 1326460 [1986]), 82.

syllable written on the beat can actually be sung a fraction of a beat too soon. Bacilly demonstrates the same procedure in the *port de voix* in Ex. 2.7. In these cases the displacement of syllables occurs in connection with the performance of ornaments. In Ex. 2.12 Grimarest shows that displacement can also act to lengthen and thus accentuate certain words of a powerful text.

Stolen time, however, may also be involved with the pronunciation of words. Ordinarily singers aim to commence the vowel sound of a syllable exactly when the time value of a note begins. Consonants are pronounced quickly and generally are not perceived to take any appreciable amount of time from the sung notes. Presumably, a consonant at the beginning of a syllable occurs before the note and takes its time from the preceding beat. Consonants act, like the silences of articulation, to insert brief rests at the very ends of notes. Thus, in Ex. 2.13*a* the written notation for this passage would, under ordinary circumstances, match fairly well the performed rendition.

In the types of music in which a powerful rhetorical delivery of text is significant, however, the execution of this passage can be altered on words of great passion. I have shown in Ex. 2.13*b* an anticipation of the final word and in (*c*) its delay, both representing the sort of robbery associated, as we will see, with the earlier type of rubato. Ex. 2.13*d*, *e*, and *f*, however, demonstrate a different type of robbery caused by the impassioned pronunciation of consonants.

In Ex. 2.13*d*, the '*l*' of *love* commences while the note D is still being sung and is sustained until the vowel occurs with the note G. The consonant '*l*' thus steals from the previous note its vowel sound but not its pitch. Other voiced consonants, such as *m*, *n*, *r*, *v*, or *z*, could act in a similar fashion in English.[45] In Ex. 2.13*e*, the '*s*' commences considerably before the note B, which coincides with the vowel of *sing*. The sustaining of '*s*', however, produces only the nonmusical sound of the exhalation of air, which interrupts the singing of any pitch. Therefore, the sustaining of '*s*' in this case robs the preceding pitch A of part of its pitch as well as part of its vowel sound. Ex. 2.13*f*, finally, shows a sustaining of '*s*' that begins on the beat. This time the note B is robbed of a portion of its pitch, which, in effect, then makes a delayed entrance. Other consonants in English which make, while they are being sustained, only various sounds of air passing from the mouth are *f*, *h*, *k*, *p*, *q*, and *t*, as well as *sh*, *ch*, *th*, and *wh*.

---

[45] See Neumann, *Ornamentation in Baroque and Post-Baroque Music*, 510 n. 36.

Ex. 2.13 The pronunciation of consonants

Four consonants that really cannot be prolonged in English the same way, I believe, are *b*, *d*, *g*, and *j*. They can, however, be pronounced, like the other consonants, with various degrees of vigour, depending upon the emotional content of the word. Other languages have different consonants that can be sustained: in French the '*j*' in *jamais*, for example, is voiced and can act like the '*l*' in Ex. 2.13*d*. Presumably consonants at the end of a syllable, such as the German '*ch*' in *ich*, could also, for expressive purposes, be sustained longer than usual. The singing of a syllable thus involves more than just prolonging a vowel sound on a particular pitch. It includes also all the percussive and breathing sounds required in the production of consonants. All of these nonmusical sounds can be produced with a wide range of effect—from the gentlest to the most vigorous expression.

They are therefore important to the actor and to the singer of highly emotional texts.[46]

During the Baroque period only a few French writers discuss the pronunciation of text. There was a close connection in France between acting and singing, and a tradition for the scholarly study of the French language. Bacilly's treatise from 1668, already mentioned above in connection with ornaments, is equally concerned with the relationship between text and music and includes several chapters on pronunciation. He defines two types of pronunciation: the simple type to make words understandable, and 'another type which has more force and energy in its delivery and which consists of giving weight or gravity to the words which one recites. This style is the great favorite of those associated with the theatre.'[47] In the chapter on the pronunciation of consonants he discusses the letter *r*. When an *r* is followed by another consonant, as in *pourquoy*, *pardon*, or *charmant*, it 'must be pronounced with a greater degree of force' as if it 'appeared as a double r and possibly with even more intensity according as the word demands greater or less expression'. When an *r* follows another consonant in the same syllable (*prendre* or *agréable*) it is not emphasized unless the meaning, as in *cruelle*, demands it.[48]

Bacilly's next chapter is entitled 'Suspending Consonants before Sounding the Following Vowels'. He says: 'There is a certain kind of pronunciation which is entirely peculiar to vocal music and declamation: when the singer (or speaker) wants to give a greater force of expression to a word, he emphasizes certain consonants before sounding the vowels which follow by prolonging or suspending them; this is called *gronder* (growling).' The letter '*m*' is most important and can be prolonged in words such as *mourir*, *malheureux*, and *misérable*. Other suitable consonants would be the '*f*' in *infidelle*, the '*n*' in *non*, the '*s*' in *sévère*, the '*j*' in *jamais*, and the '*v*' in *vous*. He gives musical examples with texts including these words, but cautions that *gronder* is used only when justified by an intense emotion.[49] Presumably this practice results in effects such as those I suggest in Ex. 2.13*d*, *e*, and *f*. In Ex. 2.12, then, the '*m*' in *mourir* could be prolonged by beginning it during the value of the dot on the preceding syllable, in the manner of Ex. 2.13*d*.

---

[46] See Fausto Razzi, 'Polyphony of the *seconda prattica*: Performance Practice in Italian Vocal Music of the Mannerist Era', *EM* 8 (1980), 302 and 305–8.

[47] *Remarques curieuses*, 248–9; Caswell trans., 129. See also Marilyn Feller Somville, 'Vowels and Consonants as Factors in Early Singing Style and Technique', Ph.D. diss. (Stanford Univ., 1967; UM 68–15,099).

[48] *Remarques curieuses*, 291–2; Caswell trans., 150. Rosen comments that the French *r* of this period was rolled more like the Italian than it is today.

[49] *Remarques curieuses*, 307–11; Caswell trans., 159–63.

In 1755 Bérard also discusses pronunciation. According to his theories, hard and soft pronunciation results from the strength with which the letters are produced, dark and clear pronunciation from the amount of air retained in the mouth or allowed free exit. He also speaks of the prolonging or 'doubling' of consonants by persons moved by passion. In that case, he says, 'there reigns . . . a certain disturbance in the organs, which causes the movements from which pronunciation of the letters results, to persevere too long a time'; this in turn causes 'the letters to be doubled strongly' and, as Blanchet adds in his second edition of 1756, 'during a considerable time'.[50]

Bérard gives some texts from Lully and Rameau operas in which he indicates the doubled consonants by writing each a second time above the word. Ex. 2.14 shows a passage from Lully's opera *Armide*;

Ex. 2.14 Doubled consonants in Lully's *Armide* (Bérard, 1755)

Bérard gives the melody at the end of his book, and I have added the instrumental accompaniment from another source. This accompanied recitative is sung by La Haine (Hate), who has been summoned by Armide in her struggle with Amour, who opposes the vengeance she has planned against Renaud. Bérard says that in this example 'one cannot devote oneself too much to strongly doubling the letters, and to pronouncing with much harshness and gloom'.[51] Hate sings: 'The

---

[50] Jean Antoine Bérard, *L'Art du chant* (Paris, 1755), repr. in *MMMLF*, 2nd ser., lxxv (1967), 93–4; trans. Sidney Murray (Milwaukee: Pro Musica Press, 1969), 97 (see also 86–90). Joseph Blanchet accused Bérard of plagiarism and published in 1756 his own *L'Art ou les principes philosophiques du chant*, copy at UCLA ML. In ch. 3 of the 2nd part, 'Art & utilité de doubler les Consonnes', he repeats and expands the passage I have quoted (53–4). Blanchet's examples are different, however, and he includes no music.

[51] Bérard, *L'Art du chant*, 96, and Murray's trans., 98; the melody is on p. 13 of the musical examples. Murray omits a doubled '*t*' which Bérard gives on p. 96 for the last syllable in the word *déteste*, but in the musical score itself Bérard doubles the '*s*' rather than the '*t*', as I show in

more one *understands love*, the *more* one *detests it*' (italicized words are those emphasized by doubled letters).[52] The sinister mood is projected by the tremolo figure in the bass as well as by the doubled consonants. A prolonged hiss on the '*s*' of *déteste* would certainly be effective, as would also, I would think, a doubling of the preceding '*t*' or the '*m*' in *amour*. The '*l*' in *plus* would have to be prolonged either on the pitch C of *et*, or on its own D, or both. In any case, the '*p*' interrupts any pitch, and one wonders whether most of the doubling could not more effectively occur on this letter.

In any event, the vigorous pronunciation of consonants described by Bacilly, Bérard, and Blanchet contributes substantially to the effective delivery of text. In order to lengthen a consonant by doubling or growling, it must necessarily steal more time than usual. This time may come from the previous note, as in Ex. 2.13*e*, from the note itself, as in Ex. 2.13*f*, or from some combination of the two.

Rhythmic Transformation

There is one other type of rhythmic robbery that should be considered as part of the background of the earlier rubato: the rhythmic transformation of a melody or motive. This technique occurs in various forms and styles of music from the twelfth century on. It is an important part of thematic transformation in the nineteenth century. In some cases a melody is so enormously altered rhythmically that it is difficult to determine which note steals from which other note and exactly how much has been stolen. At other times, the process is applied so simply and regularly that the robbery on single notes can be perceived as easily as with the intabulations, embellishments, ornaments, arpeggiation, and unequal notes discussed above. In the variation technique that we observed in the intabulations and examples of embellishment, there was in most cases a measure by measure correspondence between the original and the varied version, as well as a preservation of the metre, phrasing, and general rhythmic structure. Rhythmic transformation, on the other hand, can involve changes of metre, changes of accentuation, and changes in the basic rhythmic

---

Ex. 2.14. In addition, Bérard doubles the '*l*' in *le* when he gives the text alone, but not when he writes it below the melody. The 2 outer parts in Ex. 2.14 are from the copy in UCLA ML of the 2nd edn. of Lully's *Armide* (Paris, 1713), 107 (Act III, sc. iv).

[52] Trans. by Murray, 125 n. 19. See Betty Bang Mather, *Dance Rhythms of the French Baroque* (IUP, 1987), 161; also Patricia M. Ranum, 'Les "caractères" des danses françaises', *Recherches sur la musique française classique*, 23 (1985), 54–5.

design. There sometimes is a hazy line, however, between rhythmic variation and rhythmic transformation.

Most of the clausula and motet tenors of the Middle Ages were constructed as rhythmic transformations of Gregorian chant melismas. The chant was performed by itself, presumably, as a series of even notes, or at least in a manner rhythmically different from the motet tenors. Ex. 2.15*a* shows the initial notes from a borrowed passage

Ex. 2.15 Rhythmic transformation in motet tenors, School of Notre Dame, early 12th century

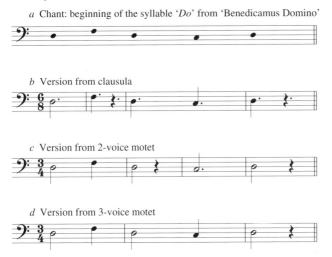

a Chant: beginning of the syllable '*Do*' from 'Benedicamus Domino'

b Version from clausula

c Version from 2-voice motet

d Version from 3-voice motet

from chant; below this are three rhythmically altered versions from a clausula or motet, showing in each case the rhythmic pattern which is repeated many times in the tenor during the course of the piece.[53] Rhythmic robbery becomes apparent as we compare each of the mensural versions with each other or with the chant itself. Comparing (*b*) and (*c*), we note a change of metre as well as a different internal structure: in (*b*) the second rhythmic unit begins with D, in (*c*) with C—notes that receive a special accentuation because of the rhythmic structure articulated by the preceding rest. Ex. 2.15*d* departs even

[53] See *HAM* i, 25: No. 28e, f, and g. Ex. 2.15*a* is from the *Liber usualis*, 124; (*b*) from Florence, Biblioteca Medicea-Laurenziana, MS Pluteo 29, I, fac. edn. Luther Dittmer in *PMMM* x, fo. 88ᵛ; (*c*) from Wolfenbüttel, Herzog August-Bibliothek, MS Helmstedt 1099 (1206), fac. edn. Luther Dittmer in *PMMM* ii (1960), fo. 179ᵛ, transcr. Gordon Athol Anderson in *MS* xxiv/2 (1976), 181; and (*d*) from Bamberg, Staatliche Bibliothek, MS Lit. 115 (*olim* Ed.IV.6), No. 31, see Pierre Aubry, *Cent motets du xiiiᵉ siècle* (Paris, 1908; repr., New York: Broude Bros., 1964), fac. in i, fo. 16ᵛ, transcr. in ii, 62.

further structurally, since the same notes which occupied four bars in
(b) and (c) now fill only three.

During the Renaissance a composer could employ the paraphrase
technique to transform a chant into a metrical line that displayed a
variety of note values. In the process the original rhythmic values were
altered and occasionally new pitches added. Ex. 2.16 shows how the

Ex. 2.16 Rhythmic transformation in paraphrase technique, 15th and 16th centuries

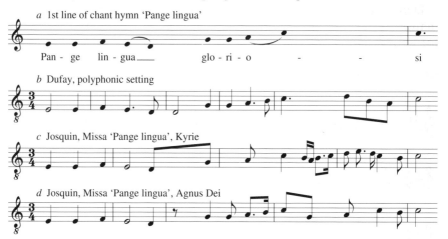

opening phrase of the chant hymn 'Pange lingua' in (a) could be
altered when set polyphonically by Dufay (b) or incorporated in a
paraphrase mass by Josquin (c and d).[54] Added pitches are most
numerous approaching the cadences. Although all the polyphonic
versions share the same metre, (b) reaches the final C in the sixth
measure, whereas (c) and (d) reach it in the fifth. Comparison of
certain specific pitches reveals how their role can change. The G in (a)
and (b), for example, becomes conspicuous because it is leapt to and
then repeated. Ex. 2.16d emphasizes it still further by a delay, but in
(c) it seems to play a totally different role as part of an anacrusic
sweep up to C.[55] Those who listened to these polyphonic settings were

[54] For (a) see *Liber usualis*, 957–9. (b) is in Guillaume Dufay, *Opera omnia*, v, ed. Heinrich
Besseler in *Corpus mensurabilis musicae*, i/5 (AIM, 1966), 53; (c) and (d) in *Josquin Desprez:
Werken*, ed. Albert Smijers (Amsterdam: G. Alsbach, 1952), xxxiii, 1 (tenor in mm. 1–5) and 22
(tenor in mm. 3–7 of the Agnus Dei), in both cases with note values halved (that is, quartered in
relation to the original notation).

[55] Accentuation of the G also changes with the text. In the original chant it coincides in the
first verse with the powerful word *gloriosi*. If transcribed metrically, Ex. 2.16a might begin with
a measure of 5/4 (2 + 3 on *Pange lingua*) followed by a measure of 3/2 (on *gloriosi*). The
polyphonic settings have different texts. Dufay's setting in (b) was to be used for verses 2, 4, and

no doubt already familiar with this well-known Gregorian hymn. Dufay's setting actually alternated with verses sung in chant, so in a complete performance of the hymn the listener heard both the original and the transformed version of the melody three times. Although the original chant is not heard in a performance of Josquin's mass, each movement is involved with different paraphrases of the melody and the listener can compare each with the others as well as with his memory of the chant.

Dance music of the Renaissance also utilized rhythmic transformation. The triple dance of a pair was often a simple metrical transformation of the preceding duple dance. Ex. 2.17 shows the opening

Ex. 2.17 Rhythmic transformation in 16th-century dance pairs

a 1st half of *Printzentantz*

b Nachtanz, method 1

c Nachtanz, method 2

section of a popular *Deutscher Tanz* along with two methods of constructing a *Nachtanz*.[56] The relationship is very simple and can be heard clearly, since one or two measures of each triple version correspond with each measure of the duple melody. Yet one cannot speak easily of one note being delayed or another one anticipated. One might consider that the half note and quarter in the opening measure of (*c*) were replacing two theoretical dotted quarter notes to correspond with the first two notes of (*a*). In this case, the G in (*c*) would have stolen an eighth beat from the F in the process of delaying its movement. With metre changing, it seems to me we are involved with a procedure essentially different from variation. Therefore, I am con-

6 of the strophic hymn. In Josquin's work (*c* and *d*), each movement, of course, uses the text of the mass, but text underlay, unfortunately, is not certain. Smijers, for example, places the opening syllable of *eleison* on the note G, whereas it is below the preceding note D in the transcription in Heinrich Besseler and Peter Gülke, *Schriftbild der mehrstimmigen Musik* in *Musikgeschichte in Bildern*, iii/5 (Leipzig: VEB Deutscher Verlag für Musik, 1973), 118 (original notation on p. 119).

[56] *The Allemande, the Balletto, and the Tanz*, i, 15–20, and numerous examples in ii.

sidering it in this book to belong to the technique of rhythmic transformation—probably the simplest and most obvious type, to be sure.

Around the turn of the seventeenth century the multi-sectional instrumental canzona began to employ rhythmic transformation. Composers achieved unity in the piece by repeating sections, or by basing one section on a rhythmic transformation of a theme from another. Ex. 2.18 gives melodies from a canzona by Trabaci in 1603.[57]

Ex. 2.18 Rhythmic transformation in the canzona (Trabaci, 1603)

a Theme of sections 1 and 5

b Theme of sections 2 and 4

c Theme of section 3

Fugal entries of (*a*) appear in the first section, which recurs, except for slight changes and an extended cadence, as the final section of the piece. The second and fourth sections are based on (*b*), a metrical transformation of (*a*): two measures of (*a*) become, in effect, four measures of (*b*), but there is not a direct correspondence between the first measure of (*a*), for example, and the first two of (*b*). The middle section employs the theme in (*c*), which is a different duple version of the pitches in (*a*). The effect of such rhythmic robbery can be seen by comparing the accentuation and general rhythmic sense of the note C, for example, in each version. In each it has a different value in relation to the neighbouring notes. Without any doubt, the composer expected and wanted the listener to relate the themes of the various sections and to sense thereby the unity they imparted to the composition as a whole. As some of the notes in Ex. 2.18*b* and *c* steal time from others in relation to their values in (*a*), they create new rhythmic configurations with the same pitches.

---

[57] *Canzona franzesa sesta* from *Ricercate, canzone franzese, capricci . . . Libro primo* (Naples, 1603); modern edn. in *Giovanni Maria Trabaci, Composizioni per organo e cembalo*, ed. Oscar Mischiati in *Monumenti di musica italiana*, 1st ser., iv/2 (Bärenreiter, 1969), 16–19, also in *HAM* ii, 16–17: No. 191.

Ex. 2.19 Rhythmic transformation in J. S. Bach's *Musical Offering* (1751)

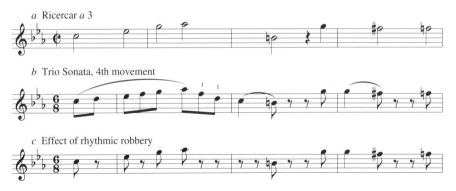

a Ricercar *a* 3

b Trio Sonata, 4th movement

c Effect of rhythmic robbery

The work of J. S. Bach in Ex. 2.19 illustrates transformation and variation at the same time. The main theme of the *Musical Offering*, which begins as in (*a*), commences in the last movement of the trio sonata as in (*b*). The added pitches are part of variation technique and theoretically could also have been added to the duple version in (*a*). In (*c*) I have replaced the added pitches with rests, as I did in earlier examples, in order to clarify the effect of the rhythmic robbery. There is not only a metre change, but a melody of four bars has been changed to three plus a brief anacrusis, and the rhythmic role of individual notes has drastically changed (consider the opening C, for example). Thus the change depicted by (*c*) represents a fairly complex example of rhythmic transformation, even though it still concludes the opening phrase on B♮ and is fairly obvious to the listener.[58]

At the end of the seventeenth century rhythmic robbery of various types was fairly extensive. In the last chapter we saw that the tempo was sometimes robbed of its strict beat in the recitative and the preludial forms as well as in the cadenza and the final cadence. In the present chapter we have seen notes in a melody steal time from other notes through the techniques of melodic variation and rhythmic transformation. In the late Baroque period especially, the same type of robbery resulted also from the effect of ornaments, arpeggiation, inequality, and the affective pronunciation of text.

The two types of robbery thus thrived and coexisted at the very time when the word *rubato* first appeared. For almost a century,

<hr />

[58] *Neue Bach-Ausgabe*, VIII/i, 14 and 43. See also the variation and transformation in the different versions of the *Art of Fugue* theme.

however, the term was applied exclusively to the type discussed in this chapter. So, it is to this earlier rubato that we will next turn—first considering it in music for solo voice or a melody instrument, later in music for keyboard.

3 THE EARLIER RUBATO IN VOCAL MUSIC

THE idea of robbery was first applied by Tosi in 1723. He was a popular castrato singer, also a composer and teacher of singing, who lived at different times in various cities of Italy, Germany, and England. He admired most the singing of arias in what he called the 'pathetic' style, a style which was out of fashion by 1723 and which was cultivated probably thirty or forty years earlier—thus at the end of the seventeenth century. It was in the pathetic arias that he recommended the 'stealing of time'.

The last decade or so of the seventeenth century was precisely the time at which a number of previously developing musical elements finally crystallized in Italian music and united into an integrated style. In the preceding two chapters we have mentioned in passing some of these elements. The solo performer dominated: in instrumental music the violinist, in vocal music the solo singer. The multi-sectional canzona had recently become a multi-movement violin or trio sonata. The recitative and aria had similarly become self-contained units within a numbers opera. The recitative, with its special type of free rhythm, was useful for narration and conversation and generally included no repetition of text. The aria, on the other hand, represented the prolongation of a single static affection at a moment of great emotion. By 1680 it included much repetition of text, a da capo structure, a clear tonal design, and a motoristic sense of recurring beat emphasized vigorously by the basso continuo. In contrast, the recitative had no closed formal structure, used tonality in a deliberately random fashion, and included in the continuo only slow-moving chords that followed the flexible declamation of the soloist.

At the end of the seventeenth century, major—minor tonality had just crystallized and brought with it harmonic sequences, modulations to nearly-related keys, a new significance for dissonances, a new importance for the cadence, and a new element for the design of formal structures. The new insistent and propulsive beat, derived from the rhythm of earlier ritornellos, dances, and more folk-like songs, provided a sense of vitality and urgency to the aria and served to distinguish its rhythm from that of the recitative. Much of the new rhythmic power was projected by the basso continuo, now organized into recurring patterned rhythms or simply 'walking' or 'running' in a series of even notes. The bass-line was like a second melody in this texture of continuo-homophony. Motives were increasingly associated with specific affections, which could thus be portrayed and prolonged during the course of a da capo aria.

The opera orchestra in Italy at the end of the seventeenth century usually included, in addition to the continuo instruments, only a chamber group of four or five solo strings, mainly for playing ritornellos and sinfonias. At the time when Tosi was presumably at the high point of his career as a singer, the solo voice in both recitative and aria was usually accompanied only by the continuo instruments. Therefore, every small nuance of the voice became conspicuous and expressive. It was totally free in recitative, where the accompaniment followed every rhythmic change. In the aria, it was constrained only by the strict rhythm of the continuo instruments. When the singer, under this rhythmic restraint, came to a place in the aria where he wished to intensify the expression beyond that allowed by the written notes themselves, he stole time from one note and gave it to another. In this way, the singer could project an intensified emotion, the accompaniment could continue its persistent momentum, and the listener could perceive the tension between the two. Tosi was clearly more interested in such sensitive nuances of singing than in a virtuosic display of ornamentation.

There are five essential aspects of the earlier rubato (the reader may wish to refer to Ex. 1 in the Introduction):

1. Some notes in a melody steal time from other notes. Sometimes the *lengthening* of notes is emphasized, with the shortening of others occurring simply as a consequence.

2. The accompaniment keeps strict time.

3. The steady bass imposes compensation on the alterations in the melody, so that the amount some notes are lengthened must exactly equal the amount the other notes are shortened. The shortening and lengthening are sometimes described as acceleration and retard.

4. The notated starting point of a note may be anticipated or delayed.

5. Notes in the melody that are written in the score in vertical alignment with notes in the accompaniment are, in fact, displaced and do not sound simultaneously. The amount of anticipation or delay is a measure of this displacement and dissonances often result.

In addition, for vocal music the pronunciation and placement of syllables of text may also be involved. Different sources that describe the technique sometimes emphasize one of these aspects over the others. It will be significant, as we trace the development of rubato in music for voice, then for violin, and finally for keyboard instruments, to note which of the aspects receives priority.

The Baroque Period

In the last chapter we saw that this process of robbery extended far back into history. What is special about Tosi's rubato is the use of the technique specifically for intensely expressive purposes within the context of the late seventeenth-century pathetic aria.

Tosi's *Opinioni de' cantori antichi e moderni, o sieno Osservazioni sopra il canto figurato* appeared in Bologna in 1723, but without any musical examples. J. Alençon published an abridged Dutch translation in 1731 at Leyden. John Ernest Galliard, a German composer and oboe player who lived in London from 1706 on, published an English translation of the book in 1742 as *Observations on the Florid Song* and included a few examples not connected directly with rubato. Roger North was an amateur English musician who retired from law and politics in 1688 and during the rest of his long life wrote extensively about the science, theory, and performance of music. In a manuscript from around 1726 he indicates that he had heard Tosi sing and knew about his theories. In an earlier manuscript from around 1695 he explains Tosi's rubato and gives numerous musical examples. Tosi came to London in 1692 and for the next two years gave concerts and taught singing. He returned again to London shortly after his book was published and remained there until at least 1727. North probably met Tosi during his first visit, and both North and Galliard could have seen him during the 1720s.[1]

Tosi mentions the stealing of time on several occasions throughout his book. First it appears in a list of the qualities of good song: 'going from one note to another', according to Galliard's translation, 'with singular and unexpected surprises, and stealing the time [*rubamento di tempo*] exactly on the true motion of the bass'.[2] Later he mentions cadences which the old singers invented 'without injuring the time' and hopes that the modern student will imitate this process of 'some-

---

[1] See *Roger North on Music*, ed. John Wilson (Novello, 1959), 151 n. 3, and John Hawkins, *A General History of the Science and Practice of Music* (London, 1776), fac. of 1853 edn. with intro. by Charles Cudworth (Dover, 1963), ii, 653, 764, and 823–4. See also the *New Grove* articles on Tosi, Galliard, and North.

[2] John Ernest Galliard, *Observations on the Florid Song*, 2nd edn. (London: J. Wilcox, 1743; repr., London: William Reeves, 1926), 129; corresponding to Pier Francesco Tosi, *Opinioni de' cantori antichi e moderni, o sieno Osservazioni sopra il canto figurato* (Bologna, 1723), 82.

what anticipating the time [*rubare un pò di tempo anticipato*]'.[3] In a chapter entitled 'Observations for a Singer' he gives more detail: 'Whoever does not know how to steal the time [*rubare il tempo*] in singing, knows not how to compose, nor to accompany himself, and is destitute of the best taste and greatest knowledge. The stealing of time [*il rubamento di tempo*] in the *pathetic* is an honorable theft in one that sings better than others, provided he makes a restitution with ingenuity.' Here Galliard adds this footnote:

Our author has often mentioned time; the regard to it, the strictness of it, and how much it is neglected and unobserved. In this place speaking of stealing the time, it regards particularly the vocal, or the performance on a single instrument in the *pathetic* and *tender*; when the bass goes an exactly regular pace, the other part retards or anticipates in a singular manner, for the sake of expression, but after that returns to its exactness, to be guided by the bass. Experience and taste must teach it. A mechanical method of going on with the bass will easily distinguish the merit of the other manner.[4]

In another of his observations Tosi mentions 'that art which teaches to anticipate the time, knowing where to lose it again; and, which is still more charming, to know how to lose it, in order to recover it again'.[5] Finally, one of the many instructions concerning passages or graces is 'that they be stolen on the time [*rubato sul tempo*], to captivate the soul'.[6] Thus Tosi's *rubamento di tempo*, and expressions such as *rubare di tempo* and *rubato sul tempo*, refer to an expressive manner of performing a passage of great passion—a manner in which altered note values in a melody act, when measured against a strict rhythm in the bass, as an anticipation—or, even more effectively, as a delay—in the time.

According to North, Tosi speaks earlier of 'breaking' rather than 'stealing' time. North therefore calls this early type of rubato 'the breaking and yet keeping of time', thereby describing the action of both the melody and the bass. In *An Essay of Musicall Ayre*, written probably sometime between 1715 and 1720, he explains how a performer can produce dissonant suspensions by delaying notes during a cadence. This is accomplished by 'dwelling upon some notes too long and coming off others too soon; that is, breaking time and keeping it, which Sig^r Tosi said was the chief art of a performer, and he showed it most luculently by his voice'.[7] In an early version of *The Musical*

[3] Galliard, 138; Tosi, 88.
[4] Galliard, 156, Tosi, 99.
[5] Galliard, 165; Tosi, 105.
[6] Galliard, 177; Tosi, 113.
[7] BL, MS Add. 32536, fo. 39^v (in section 66: 'The Performers give the Artificial Mixtures by Gracing'); *Roger North on Music*, 151.

Grammarian from around 1726, he considers this manner of performance 'absolutely necessary to a good hand or voice', referring, I
presume, to the instrumental performer as well as the singer. He
explains by referring to a brief two-bar melody:

when these notes are sounded, no one of them shall come on or off in its due
time, and yet the common measure shall be strictly observed. It would be a
vanity to attempt a description of this manner, but it is easily shown and
made understood by the demonstration of example (when an artist, as I
remember one Sig^r Tosi, an eunuch, was so obliging distinctly) to show it;
and at the same time he applauded the manner, calling it the breaking and yet
keeping time. And I am sensible most persons that perform well do the same
thing more or less, but *incogitanter*, and by habit rather than design. But it is
apparent enough that the reason of such elegance is the intermixture of
harshnesses that, like a poignancy or spice in sauces, relisheth the mess.[8]

In an untitled work written much earlier (around 1695 and thus
shortly after Tosi's first London visit), North includes many musical
examples, but without mentioning Tosi by name. The examples are
preceded by these comments:

Now for smooth and sliding graces [such as appoggiaturas], the great
secret is to break and yet keep the time. Our ordinary scholars think there is
such virtue in graces, that they will (for want of readiness) dwell too long,
and so break the step or time of the consort, which in walking music is
intolerable, and no grace can make amends for it. But there is a way of
breaking the time, and coming in again at [the] proper place, and affording
the just sound to answer the harmony, but by this error of time put in certain
discord-elegance inexpressible, and to be known only by observation, as in
the following examples.

The first example is shown in Ex. 3.1, the others in Plate I—and these
could be multiplied 'so to infinity, according to the skill, invention,
and practice of the master'.[9] North writes each example two ways:
first *plain*—as written in a score—then to the right of this a version
that is *grac't*—an example of the improvised embellishment added by
a performer. Below the top line of Plate I thus appear the words: *the
plaine, the grace, plaine, grac't, or thus* (showing a second graced
version). The *graces* included the breaking yet keeping of time as well
as the usual ornaments.

In all these examples North gives a melody and its bass, but no text.

---

[8] BL, MS Add. 32533, fos. 113^v–114^r (ch. 20: 'Of Gracing', section 123[2]: 'Of Tempering
of Notes, and of Breaking and Keeping Time'); *Roger North on Music*, 151–2.

[9] BL, MS Add. 32532, fo. 7^r, in section entitled 'Graces in Playing'. Ex. 3.1 is at the bottom of
fo. 7^r, Pl. I on fo. 7^v. See *Roger North on Music*, 152–3, for a selection from the musical
examples.

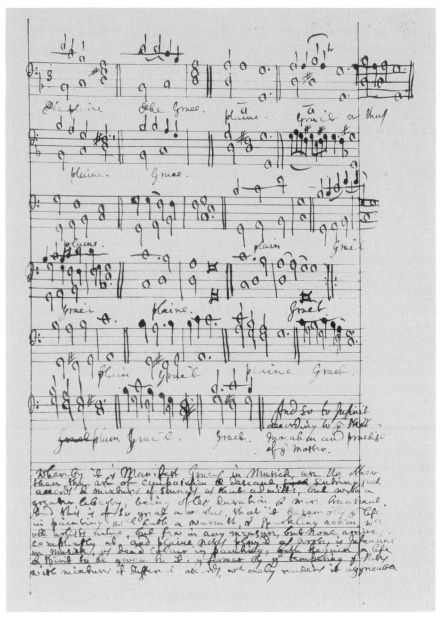

PLATE I. Roger North, 'Breaking and yet Keeping Time', BL, MS Add. 32532, c.1695, fo. 7ᵛ

The bass of each plain version remains essentially unaltered when the melody is graced (the omission of the first two bass notes in the second example on the second line is no doubt an oversight). He presents twelve groups, two of which contain two different varied versions. Of the fourteen graced versions, half include added notes—usually only one or two, but sometimes many, as in Ex. 3.1 and the second example on the second line of Plate I. Some, like the latter example, are clearly cadences, whereas others, such as Ex. 3.1, are not. In Ex. 3.2 I have written the last measure on Plate I, with the

Ex. 3.1 Graced passage with an anticipation and added notes, North, *c.*1695

Ex. 3.2 Graced passage with 2 delays, North, *c.*1695

presumed ungraced version, which North omitted, directly above it in order to show the two delayed notes more clearly. Similarly, I have aligned in Ex. 3.3 the three versions that commence in the middle of the third line on Plate I and continue on the fourth. Ex. 3.3*b* shows delays of the last three notes, whereas (*c*) displays an anticipation of the second and a delay of the last, as well as the addition of two notes and a third voice. The seven graced melodies with added notes contain three anticipations and six delays. The seven versions with no added notes display one anticipation and twelve delays; of the latter, three examples delay a single note, three others delay two notes, and one delays three. Clearly North, like Tosi, shows a preference for delaying.

North includes a few more examples in *An Essay of Musicall Ayre*.

Ex. 3.3 Passage with 2 graced versions, North, c.1695

Immediately following his comments quoted above concerning Tosi, he offers in the same paragraph several syncopated passages, presumably as examples of liberties that involve breaking yet keeping time. 'For notes falling away in sixths [see Ex. 3.4] the upper lags a

Ex. 3.4 Gracing of parallel 6ths and 3rds, North, c.1695

little behind, and in falling by thirds the upper will be (for the time) too soon upon the back of the lower, which making its way sets all right.' I have added in Ex. 3.4a the same passages as they would presumably be written. He makes clear elsewhere that the expressive effect comes from the dissonances produced by syncopation: in the parallel sixths the interval of the seventh, in the thirds the interval of the second. Further delaying of notes occurs in Ex. 3.5, to which I have again added in (a) the presumed ungraced version. This sort of

Ex. 3.5 Gracing of a passage, North, c.1695

a Presumably written

b Performed

syncopation 'lies within the liberty of a performer to fill in, whenever he hears the bass invite, as that movement egregiously doth'.[10]

North complains in *The Musical Grammarian* of the disturbance caused when members of an ensemble break time for one of these reasons:

> One is when any of the company for infirmity break time, for that discomposeth the rest that attend. And the other is for too much ability, when any out of peevishness or pride affect that which is called 'breaking and yet keeping time' ... and thereby disorder the rest, who for their measures are apt to depend on them; ...[11]

Finally, North describes the general effect of rubato in his autobiography from around 1695 entitled *Notes of Me*:

> And there is no greater grace than breaking the time ... and still holding it punctually upon the main, to conserve the grand beat or measure. For this sprinkling of discord or error is like damask ... or any unaccountable variegation of colours that renders a thing agreeable; and yet we discern not the distinction of parts, but only a pretty sparkling ... And a plain sound not thus set off is like a dull plain colour, or as a bad copy of a good picture, that wants the spirit and life which a sparkling touch gives it. Thus a life and warmth in the colouring of a picture is well resembled to graces in music, that are not the body but the soul that enlivens it, or as the animal spirits that cannot be seen or felt, but yet make that grand difference between a living and a dead corpse.[12]

Tosi describes rubato mainly in terms of anticipation and delay, and North provides examples, such as Ex. 3.2, in which a fixed bass

[10] BL, MS Add. 32536, fo. 39ᵛ. In Exx. 3.4 and 3.5 I have added a missing tie between the B naturals. See *Roger North on Music*, 153 and 27 n. 35.

[11] BL, MS Add. 32533, fo. 58ᵛ, in section on 'Timekeeping in Consort'; *Roger North on Music*, 105.

[12] BL, MS Add. 32506, in section from fo. 69ʳ to 87ᵛ; quotation from *Roger North on Music*, 27–8.

clearly depicts this aspect by showing the resulting displacement between the melody and its accompaniment. In his essay from around 1715–20 North does mention, as we have seen, that delaying is achieved by prolonging certain notes and shortening others. When Alençon summarizes Tosi's conception of rubato in his Dutch translation of 1731, he substitutes for anticipation and delay the idea of the increase and decrease of note values:

> He who wants to sing well, must try to acquire a sweet and pleasant management of the voice, and sometimes to prolong one note somewhat and diminish another note somewhat without injuring the time, which in Italian is called *Rubare il tempo* and gives the singing much charm.[13]

The alteration of note values involves the constraint of compensation: thus in Ex. 3.2. the lengthening of the first note by one quarter beat is matched exactly by the amount of time lost by the last note. On the other hand, anticipation and delay, since they depend only upon the starting point of a note, do not require such a balance: an anticipation does not need to be accompanied by a delay, nor a delay by an anticipation. In Ex. 3.2 the second and third notes are both delayed, yet none of the notes is anticipated; the second note even retains its full duration in spite of being shifted in time. In Ex. 3.1 the second note is anticipated, but this causes neither delay nor anticipation of the first note, since the stolen time comes from the end and not the beginning of its duration. All of this points out the considerable difference in these two ways of looking at tempo rubato. Tosi mentions only in passing the restitution involved in stealing time, although it must indeed have been the alteration of note values that suggested the image of robbery in the first place.

There is also a device in the pathetic style, called *strascino* by Tosi or *drag* by Galliard, which is somewhat related to rubato. Agricola, who published a German translation of Tosi's book in 1757, was puzzled by Tosi's explanation.[14] Galliard, however, provides two musical examples, which I include as Exx. 3.6c and 3.7b. Tosi describes dragging, in Galliard's translation, as follows:

> on an even and regular movement of a bass, which proceeds slowly, a singer begins with a high note, dragging it gently down to a low one, with the forte

[13] J. Alençon, *Korte aanmerkingen over de zangkonst, getrokken uit een italiaansch boek, betyteld Osservazioni sopra il canto figurato di Pier Francesco Tosi* (Leyden, 1731), copy in GH, 43: 'Die wel wil zingen moet zig eene zoete en aangenaame Leidinge van Stem zien te verkrygen; en somtyds de eene *Note* wat verlangen en de andere *Note* wat afneemen, zonder de Maat te krenken, 't welk in't Italiaansch genaamd wordt *Rubare il tempo*, en den Zang veel aardigheid byzet.'

[14] Johann Friedrich Agricola, *Anleitung zur Singkunst* (Berlin, 1757), repr. ed. by Erwin R. Jacobi (Celle: Hermann Moeck, 1966), 233–6, especially footnote (d).

and piano, almost gradually, with inequality of motion, that is to say, stopping a little more on some notes in the middle, than on those that begin or end the *strascino* or *drag*. Every good musician takes it for granted that in the art of singing there is no invention superior or execution more apt to touch the heart than this, provided, however, it be done with judgement and with putting forth of the voice [*portamento di voce*] in a just time on the bass [*sul tempo, e sul basso*].[15]

Galliard's first example is in Ex. 3.6*c*. I have added in (*a*) the melody as it would probably have been notated, and in (*b*) I show a

Ex. 3.6 *Strascino* or *drag* (Galliard, 1743)

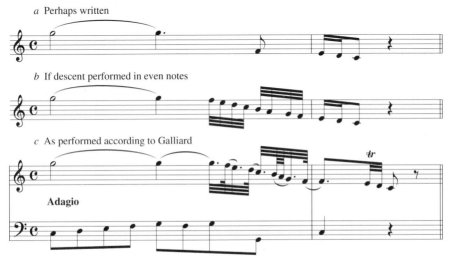

descent performed in even note values. Comparison of (*b*) with the upper staff of (*c*) shows the stealing of time and hence the relationship of dragging to rubato. Ex. 3.7*b* is Galliard's other example, and I have left his original note values, even though they are not mathematically correct, since they do depict the sense of Tosi's verbal description. Again in Ex. 3.7*a* I have added a possible written version. Galliard marks Ex. 3.6*c Adagio*, and in Ex. 3.7*b* adds an appropriate text.[16]

Bayley later connects dragging with Tosi's stealing of time and defines it as 'much the same motion as that of gliding, only with inequality, hanging as it were upon some notes descending, and hastening

[15] Galliard, 178–9; Tosi, 114.
[16] Galliard's musical examples are nos. 8 and 9 on pl. VI at the end of his book. In no. 8 the notes are not carefully aligned vertically, and those in the descent have twice the value I show in Ex. 3.6*c*.

Ex. 3.7 Another *drag* (Galliard, 1743)

the others so as to preserve the time in the whole bar'. After quoting Tosi, he then suggests two works by Jeremiah Clarke as especially suitable for dragging. Both begin by repeating a descending phrase, as though Clarke wanted to give the singer an opportunity to add more intense expression the second time.[17] Singers seem to have a special interest in embellishing a descending leap, as in the *cascata* of Caccini (Ex. 2.11) and later in the *portamento*, which is sometimes imitated on the keyboard.

When new pitches are added, the technique involved in examples of the earlier type of rubato looks very much like the melodic variation we saw in the last chapter: compare Ex. 3.1*b*, for example, with the intabulation in Ex. 2.1 or the vocal ornamentation in Ex. 2.3. The earlier rubato thrives in those periods in which improvised embellishment is most prevalent. The revival of virtuosic singing in the late seventeenth century brought with it a more thorough examination of its component parts, one of which was the stealing of time. Examples illustrating this particular aspect, however, often contain added notes, even, as we will see, during later periods.

In spite of the great importance of text, writers of this period do not, as far as I know, discuss the robbery caused by expressive pronunciation. North includes no text at all with his examples. Galliard gives a text for only one of his examples of dragging (Ex. 3.7*b*), as well as for a number of cadences. The latter show the final unaccented syllable of words such as *morte* and *amante* falling on a strong beat, rather than on an unaccented beat as they did earlier. It was the French writers, as we noted in the last chapter, who described

---

[17] Anselm Bayley, *A Practical Treatise on Singing and Playing* (London, 1771), copy at Bologna, Civico Museo Bibliografico Musicale, 44.

in greatest detail the affect of text. In 1740 Dubos writes that 'the most exact and most skilful composer of declamation left room for good actors to display their talents, and to show, not only in their gesture, but likewise in the pronunciation, their superiority over indifferent artists. 'Tis impossible to note all the accents, rests, softenings, inflections, shakes and breakings of the voice.' Referring specifically to the singing of a Lully opera, he defines the earlier type of rubato by linking the changes in note value with an acceleration or retard:

The good actor [opera singer] who enters into the spirit of what he sings, accelerates or slackens seasonably some notes, borrowing from one to lend to the other; he throws out or retains his voice; he dwells upon some places . . . Each actor supplies out of his fund, and in proportion to his capacity, the want of what could not be written in notes.[18]

Some baroque singers were noted for their ability to employ tempo rubato. Mancini praises Antonio Pasi of Bologna (1697 or 1710– 1770), a contralto castrato and pupil of Pistocchi, for his graceful *passaggi, trilli, mordenti e rubbamento di tempo*, 'all done to perfection and in their appropriate places, the whole making an individual and arresting style'.[19] Galliard reports that Francesco Antonio Mamiliano Pistocchi (1659–1726) was popular at the end of the seventeenth century; Tosi considered him the greatest of all singers.[20]

Famous about the same time as Pasi were the two great Italian sopranos Francesca Cuzzoni (*c*.1698–1770) and Faustina Bordoni (1700–81, wife of the composer Johann Adolf Hasse), whose rivalry during the production of Handel operas became a public scandal in London during the late 1720s. The two singers were compared in every detail of operatic singing. According to Tosi, Cuzzoni excelled in the pathetic, cantabile style, Faustina in the brilliant allegro.[21] Burney describes Cuzzoni as follows:

In a cantabile air, though the notes she added were few, she never lost a favourable opportunity of enriching the cantilena with all the refinements and embellishments of the time. Her shake was perfect, she had a creative fancy,

---

[18] Jean-Baptiste Dubos, *Réflexions critiques sur la poésie et sur la peinture* (Paris, 1740; original edn., 1719), iii, 316–17; English trans. from 5th edn. by Thomas Nugent in *Critical Reflections on Poetry, Painting and Music* (London, 1748), iii, 239–40. A copy of each work is in UCLA URL. The original French reads in part: 'Le bon Acteur . . . presse, ou bien rallentit à propos quelques notes, il emprunte de l'un pour prêter à l'autre; . . . il appuïe sur certains endroits.'

[19] Giambattista Mancini, *Riflessioni pratiche sul canto figurato*, 3rd edn. (Milan, 1777; repr. Forni, 1970), 22–3; trans. from *Practical Reflections on Figured Singing by Giambattista Mancini: The Editions of 1774 and 1777 Compared, Translated and Edited* by Edward Foreman in *MOS* vii (1967), 8.

[20] Tosi, 65; Galliard, 101–2.

[21] Tosi, 109; Galliard, 170–1.

and the power of occasionally accelerating and retarding the measure in the most artificial and able manner, by what the Italians call *tempo rubato*.[22]

Burney also quotes and elaborates on Quantz's description of the two singers from his autobiography of 1754. Quantz, according to Burney, had been present in London during 1727 at performances of Handel's *Admeto* in which both singers appeared. Faustina, according to Burney's translation:

sang *adagios* with great passion and expression, but was not equally success-ful, if such deep sorrow were to be impressed on the hearer, as might require dragging, sliding, or notes of syncopation, and *tempo rubato*.

The last two elements, 'notes of syncopation' and '*tempo rubato*', were added by Burney to Quantz's list.[23] In the same source, however, Quantz does mention that he heard 'the so-called tempo rubato' performed for the first time by the Italian soprano Lotti. He refers to Santa Scarabelli Stella or Santini, who was famous early in the cen-tury, married the composer Antonio Lotti, and was favourably men-tioned by Tosi, Galliard, and Burney.[24]

The period of these great singers marks the high point of baroque opera—an opera in which the solo singer, mainly through the da capo aria, was the uncontested centre of attraction. The only competition came from a few non-continuo instruments which usually played along with the voice in the arias and in a few accompanied recitatives. All sorts of embellishments could be improvised during the repetition of the opening section of an aria or during cadenzas. A singer was therefore almost totally free to display his powers of expression or of virtuosity. For those who performed in the expressive style, the device of rubato provided a striking method of delivery.[25]

---

[22] Charles Burney, *A General History of Music*, 2nd edn. (London, 1789), copy at UCLA ML, iv, 307; quoted in Carol MacClintock, *Readings in the History of Music in Performance* (IUP, 1979), 262–3, and in *New Grove*, v, 110.

[23] Charles Burney, *The Present State of Music in Germany, the Netherlands, and the United Provinces*, 2nd edn. (London, 1775), fac. in *MMMLF*, 2nd ser., cxvii (1969), ii, 189; and *A General History of Music*, 2nd edn. (London, 1789), iv, 319; quoted also in MacClintock, *Readings*, 265–6, and *New Grove*, iii, 47. See 'Herrn Johann Joachim Quantzens Lebenslauf, von ihm selbst entworfen' in Friedrich Wilhelm Marpurg, *Historisch-Kritisch Beyträge zur Aufnahme der Musik*, i (Berlin, 1754), repr. in *Facsimiles of Early Biographies*, v: *Selbstbio-graphien deutscher Musiker des XVIII. Jahrhunderts*, ed. Willi Kahl (Cologne, 1948; repr., Amsterdam: Frits Knuf, 1972), 241; trans. Paul Nettl in *Forgotten Musicians* (New York: Philosophical Library, 1951), 313.

[24] See 'Quantzens Lebenslauf', 213–14, and Nettl's trans., 292; Tosi, 66, and Galliard, 103–4 and footnote; Burney, *Present State of Music in Germany*, 176.

[25] For a recent survey of baroque vocal rubato, as well as tempo flexibility, see Ellen T. Harris's chapter in Brown and Sadie, *Performance Practice: Music after 1600*, 108–10.

The Classic Period

Tosi speaks of 'stealing the time', North of 'breaking yet keeping time', and Dubos of 'borrowing and lending'. It is the image of robbery, however, that finally becomes permanently attached to the device. In his autobiography of 1754 Quantz, as we have seen, reported hearing tempo rubato. It is in the chapter entitled 'Of the Manner of Playing the Adagio' in his flute book from two years earlier, however, that the expression *tempo rubato*, as far as I know, first appears. He illustrates *eine Art vom Tempo rubato* as shown in Ex. 3.8, where I have vertically aligned (*b*) and (*c*) with (*a*), which

Ex. 3.8 Explanation of *tempo rubato* (Quantz, 1752)

appears on a different page in his book. He describes the notes in (*b*) as follows: E is strong (meaning greater stress), F weak and crescendo, G and A the same, and C weak. In (*c*) E is weak with a crescendo to the dot, F and G weak and crescendo, A and C weak. Although he is speaking specifically about tonguing and bowing on the flute and violin, the same instructions would apply presumably to the voice. 'In the first example', he writes, 'the fourth against the bass is anticipated, replacing the third, and in the second the ninth is held in place of the third, and resolved to it.'[26]

[26] *Versuch einer Anweisung die Flöte traversiere zu spielen* (Berlin, 1752), modern edn. Arnold Schering (Leipzig: C. F. Kahnt, 1926), 61 (for Ex. 3.8*a*), 67 (for Ex. 3.8*b* and *c*), and 100

We have already noted both the syncopated notes and the concern with dissonance in North's examples of parallel sixths and thirds from around 1695 (Ex. 3.4). It was no doubt Quantz, however, who established the tradition of depicting the two aspects of rubato—the anticipation and the delay—in two separate syncopated versions of a single melody. North's examples, however, showed that anticipations and delays could both occur in the same brief passage (Ex. 3.3c). Furthermore, all the examples of North, Galliard, and Quantz considered so far have attempted to depict in notation a very subtle improvised device. The actual amount of time stolen can no doubt vary considerably from one performance to another and, in any event, can scarcely be notated with any precision at all. Quantz's examples are therefore abstractions to aid in the theoretical study of two different aspects of rubato. This representation itself, however, later begins a life of its own as one of the meanings of rubato and, as we will see, gives birth to still other definitions of the word.

Although Quantz's book primarily concerns flute playing, its method of describing rubato strongly influenced those who later wrote about the voice. These include Agricola, who was with Quantz at the court of Frederick the Great in Berlin, as well as Marpurg, Kirnberger, and Schulz, who were closely associated with Quantz or his ideas. Also at the Berlin court were C. P. E. Bach, who writes about keyboard rubato, and Franz Benda, famous for his violin rubato. At Leipzig, but also obviously influenced by the Berlin School, was Johann Adam Hiller. Most of the sources concerning vocal rubato during the Classic period come from Berlin or Leipzig and appear in theoretical works or in journal articles complaining of its abuse.

Agricola, who among other accomplishments was also a singer and a teacher of singing, translates *rubamento* and *rubare* in his translation and commentary on Tosi's book in 1757 as *Verziehung* and *verziehen* (to distort or alter). Rubato then becomes 'the alterations of the value of the notes . . . which, however, must be exactly fitted to the movement of the bass'.[27] He uses the image of robbery—'ein wenig Zeit . . . voraus wegzustehlen'—when Tosi speaks of 'anticipating the time' at a cadence.[28] To Tosi's main reference to rubato Agricola adds this footnote: 'To alter the notes [*rubare il tempo*] means actually to

---

(the prose explanation); repr. of 3rd edn. (Berlin, 1789) in *DM* ii (1953), 146, Table VIII Fig. 4, and Table X Fig. 4e and f; trans. by Edward R. Reilly as *Johann Joachim Quantz, On Playing the Flute*, 2nd edn. (New York: Schirmer Books, 1985), 174 and the examples on 137 and 142 (see also 172–3 for the explanation of the words *strong*, *weak*, and *crescendo*).

[27] *Anleitung zur Singkunst*, 196: 'Verziehungen der Geltung der Noten (*rubamento di Tempo*) welche aber der Bewegung des Basses genau angemessen seyn müssen.'

[28] Ibid. 202.

take away from a written note something of its value in order to add it
to another, and vice versa.' He illustrates with examples almost identical to those of Quantz in Ex. 3.8, except that they are all printed on
a single page, and in his last example (see Ex. 3.8c) he delays the
opening note with an eighth rest.[29] He cautions that if the singer
dislocates the notes too much the melody will not be intelligible, and
that it is best to learn the proper execution of rhythmic robbery
(*Zeitraub*) from excellent performers.[30] However, he does not mention anticipation, delay, or the dissonances they cause. He seems to
conceive of rubato mainly as changes in note values, and thus is
puzzled by Tosi's observation that losing time is even more charming
than anticipation: 'These rather enigmatic words of the author all
refer to the so-called *tempo rubato*, which was stressed almost too
much at that time.'[31]

Succeeding German writers, on the other hand, tend to describe
the technique in terms of anticipation, delay, and the resulting
displacement between melody and bass. In 1763 Marpurg defines
Tonverziehung as delaying or anticipating a note without the par-
ticipation of the harmony, but always with melodic displacements
(*Rückungen*). He had previously illustrated *Rückung* or *Syncopation*
with a series of suspensions, either sustained or restruck. His examples
of *Tonverziehung* are much like those of Quantz and Agricola (see Ex.
5.2b and c in Chapter 5). He also includes *Tonverbeissung*, the sup-
pression of notes, which occurs when each of the four notes in his examples are delayed by an eighth rest (Ex. 5.2d). The Italians, he says,
use the single expression *tempo rubato* to include both *Tonverziehung*
and *Tonverbeissung*.[32] We will later note some of Marpurg's other
ideas when we examine his books on keyboard performance. Hiller,
in a work from 1774, follows Marpurg's presentation, illustrating
Tonverziehung with the customary examples, but placing *Tonverbeissung* in a different category. In the second edition of the book in
1798, however, he states that *Tonverbeissung* really amounts basically
to the same thing as syncopation or tempo rubato.[33]

[29] Ibid. 219: 'Die Noten verziehen, (*rubare il tempo*) heisst eigentlich einer vorgeschriebenen
Note etwas von ihrer Geltung abnehmen, um es einer andern zuzulegen, und umgekehrt.'
[30] Ibid. 220.
[31] Ibid. 225: 'Diese ziemlich räthselhaften Worte des Verfassers zielen alle auf das sogenannte
Tempo rubato: aus welchem man, zu seinen Zeiten, fast zu viel Wunder machete.'
[32] Friedrich Wilhelm Marpurg, *Anleitung zur Musik überhaupt, und zur Singkunst* (Berlin,
1763), copy at UCLA ML, 148–9; Ex. 5.2 comes from p. 150, where (b), (c), and (d) are Figs.
3a, 3b and 4.
[33] Johann Adam Hiller, *Anweisung zum musikalisch-richtigen Gesange*, 2nd edn. (Leipzig,
1798), copy in GH, 177. See Suzanne Julia Beicken, 'Johann Adam Hiller's *Anweisung zum
musikalisch-zierlichen Gesange, 1780*: A Translation and Commentary', Ph.D. diss. (Stanford
Univ., 1980; UM 80–24,623), 221.

Kirnberger and his pupil Schulz both write primarily about musical composition, but occasionally mention performance. Kirnberger discusses syncopation in 1771 and depicts *anticipatio* and *retardatio* with examples much like those of Quantz in Ex. 3.8. He mentions certain progressions in which syncopation would produce unacceptable dissonances. When he refers to a melody and bass moving in thirds or tenths (as they are in Ex. 3.8), he adds this footnote:

When ... the two voices ... are separated by a third ..., syncopation can very easily be employed by the singer, even when it is not indicated by the composer. However, the voice part must not be accompanied by flutes or violins playing with it in unison. These would progress simultaneously with the bass, and the singer would then be blamed for hurrying or falling behind. A singer who wants to make such alterations must look carefully at the score so as not to make the changes at the wrong time or contrary to the nature of the accompaniment.

Although he does not use the expression *tempo rubato*, he explains the matter further in another footnote:

That the principal harmony and simple melody must be made perceptible in all embellished arias, particularly in adagios, is most clearly evident from the fact that the best arias are those in which the accompanying parts have only the main notes of the harmony; most slow arias by the famous Hasse are composed in this way. This has the advantage of leaving the singer freedom to add his embellishment at will and to anticipate and retard as he pleases.[34]

Schulz briefly summarizes the same ideas in 1774 by stating that singers and players apply *retardatio* (*Verzögerung*) and *anticipatio* (*Voreilung*) to good effect if their knowledge of harmony prevents them from violating the rules.[35] An anonymous treatise from Frankfurt in 1779 also mentions that true virtuosos know how to use *Tonverziehung* to strengthen the expression without hurting the harmony. The author cites the arias of Graun and Hasse as especially suited for rubato, but warns that as soon as the harmony changes the rubato should end as well.[36]

---

[34] Johann Philipp Kirnberger, *Die Kunst des reinen Satzes in der Musik*, i (Berlin and Königsberg, 1771; repr. Olms, 1968), 217–25 nn. 80 and 83; see also 192–3. Trans. by David Beach and Jurgen Thym as *The Art of Strict Musical Composition* (YUP, 1982), 228–34, 208–9.

[35] Johann Georg Sulzer, *Allgemeine Theorie der schönen Künste*, ii (Leipzig, 1774), copy in UCLA URL, 1236–7 (on *Verzögerung*); see also the article on *Verrükung* on 1218–19. Schulz wrote most of the music articles from S to Z in Sulzer's work.

[36] *Wahrheiten die Musik betreffend gerade herausgesagt von einem teutschen Biedermann* (Franfurt am Main, 1779), 117 ff.; quoted by Boris Bruck in *Wandlungen des Begriffes Tempo rubato* (Berlin: Paul Funk, [1928]), 16.

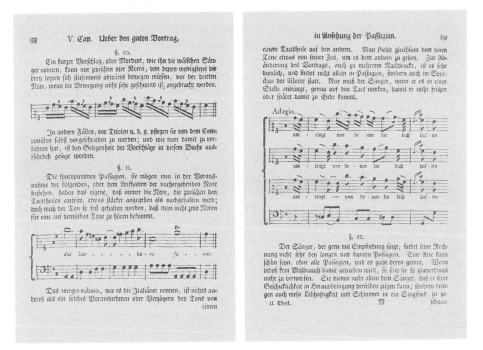

PLATE II. Johann Adam Hiller, *Anweisung zum musikalisch-zierlichen Gesange* (Leipzig, 1780), 88–9

In 1780 Hiller adds a new point of view, especially concerning text. He says (in paragraph 11 on Plate II) that 'syncopated passages, which may consist of the anticipation [*Vorausnahme*] of the following or retardation [*Aufhalten*] of the previous note, have the special characteristic that always the note that comes between the beats is performed somewhat stronger as it is held out; yet the tone must be held so firmly that one does not hear two notes for one on the same tone.' These instructions seem to contradict those of Quantz, who, as we have seen, stated that the syncopated notes (in Ex. 3.8) should be performed 'weak and crescendo', thus possibly sounding like the 'two notes for one on the same tone' to which Hiller objects. Hiller illustrates syncopation with the series of anticipations in an example with Italian text (on page 88 of Plate II). This was presumably notated by the composer in response to the powerful text.

The same sort of syncopations, he continues, may also be added by the singer:

Tempo rubato, as the Italians call it, is nothing more than such an anticipation [*Vorausnehmen*] or delaying [*Verzögern*] of a tone from one beat to the other. One steals, as it were, some time from one note, in order to give it to the other. This is very useful for varying the performance and for greater emphasis, and it occurs not only in *passaggi* [improvised embellishments] but also in the delivery of the words [*sprechen der Worte*]. But the singer, when using it, must exactly observe the beat, in order that he does not come to the end too soon or too late.[37]

Hiller then inserts an adagio example with German text (page 89 on Plate II), which shows the written melody on the top staff, a version with a series of anticipations on the second, and another version with several retardations on the third. To show that rubato occurs 'also in the delivery of the words', he places the text very carefully in this example, so that its displacement in relation to the top line is very clear. In most cases the displacement of a pitch and its syllable takes place simultaneously. In the third line, however, the word *dich* is anticipated, but not the pitch.

Unlike most of the earlier theorists, Hiller had a special interest in text and includes Italian or German words with many of his examples.[38] Earlier in the book he suggests shortening certain unaccented syllables that occur on melismas by moving them to a later note. Thus in the text 'ein Freund zum Trost erscheint' he moves the syllable '*er*' from the first to the second of two eighth notes, and in 'Alma mia, ritornero' delays '*ma*' and '*ne*' until later notes. The effect is thus to shorten the unaccented syllable and consequently to lengthen the preceding accented syllable—and this is accomplished by stealing time from the former to give to the latter. In this procedure, however, the pitches are not altered rhythmically. A rubato effect also occurs in syllabic music, where the singer should occasionally dot an accented syllable, as in '*gan*' in the phrase 'O Gott, mein ganzes Leben'. Time stolen from '*zes*' is then taken by '*gan*'. In this case, of course, both the pitch and the syllable are displaced.[39]

In the last chapter of his book, devoted to improvised variation in the aria, Hiller includes displacement of the tempo—'Verrückung des

---

[37] *Anweisung zum musikalisch-zierlichen Gesange* (Leipzig, 1780; repr., Leipzig: Edition Peters, 1976), 88–9. See also Beicken's trans., 105–6; as well as Sally Allis Sanford, 'Seventeenth and Eighteenth Century Vocal Style and Technique', D.M.A. diss. (Stanford Univ., 1979; UM 80–2,045), 212, and Carl LeRoy Blake, 'Tempo Rubato in the Eighteenth Century', D.M.A. thesis (Cornell Univ., 1988; UM 88–4,554), 38–9.

[38] Agricola does add a footnote on pp. 137–40 of *Anleitung zur Singkunst* in which he mentions Bérard's book as well as the importance of strong pronunciation, especially for the consonants.

[39] *Anweisung zum musikalisch-zierlichen Gesange*, 30–2 and 35–8, and Beicken's trans., 53–6 and 58–60.

Zeitmaasses (tempo rubato)'—as one of the methods of variation. He concludes the book with two embellished arias, one with German text (from which the excerpt on page 89 on Plate II comes), the other with Italian. Both display numerous examples of rubato.[40]

Lasser was a singer, violinist, and composer who worked mainly at Munich. According to his book of 1798, *rubare il tempo* means to displace (*verziehen*) the notes, a process involving 'Anticipation und Retardation, Vorausnahme und Verzögerung'. He feels that 'this tempo rubato makes an extraordinary effect when applied at the proper place', and gives two groups of examples, both with text. In one group the words *a respirar* are set to a melody almost identical to that in Ex. 3.8. Now, however, in both the *anticipation* and *retardation* there is also a rubato in the syllabic rhythm, with '*re*' on the first note of the opening bar stealing time from '*spi*' on the last.[41]

The concepts of Tosi were thus transmitted by German authors to the end of the century. The idea of stealing time refers specifically to an act which occurs in the melody. This terminology, in the form of the expression *tempo rubato*, continued to be used in the works of all the theorists above except Schulz and Kirnberger. It was the notion of displacement between a melody tone and the bass note written vertically below it, however, that seems to have attracted their attention even more. This must have been the most conspicuous aspect of rubato for those for whom the device was a novelty in a foreign style or for those who were not performing artists themselves. Quantz depicted this displacement, as we have seen, in syncopated examples that separated anticipation from delay. German nomenclature reflects this central concern, with *Verziehen*, *Rückung*, or *Verrückung* consisting of both *Vorausnehmen* and *Verzögern* or *Aufhalten*. Those more directly concerned with singing, such as Hiller and Lasser, show an increasing interest in the effect of text.

After the turn of the century, however, the activity in the melody seems to be emphasized more than displacement with the bass. Several works published in London illustrate this trend. In *The Singers Preceptor* of 1810 Domenico Corri includes no schematic abstractions such as Ex. 3.8. In his prose description (see Plate III) *Tempo rubato* is now the sole term for the device, which he describes in terms of

---

[40] *Anweisung zum musikalisch-zierlichen Gesange*, 129 for the quotation, 135–40 for the German aria (the text on 89 recurs on 135, 137, and 139), and 141–52 for the Italian; Beicken's trans., 143, 149–54, and 155–66.

[41] Johann Baptist Lasser, *Vollständige Anleitung zur Singkunst*, 2nd edn. (Munich, 1805), copy at LC, 154: 'Dieses *Tempo rubato* am rechten Orte angebracht macht ungemeine Wirkung'. Examples are on p. 158. Sandra P. Rosenblum, in *Performance Practices in Classic Piano Music* (IUP, 1988), 483 n. 65, states that this publication of 1805 is identical to the 1st edn. of 1798.

6   *Dialogue between Master and Scholar.*

Master. Towards accomplishing that end, you are already informed, that many qualities and acquirements **must** combine; we will proceed to the mention of others, which are also of great importance.

————RHYTHM OF TIME.————

It is an old adage "Hours (that is, a rigid observance of time,) were made for slaves."—This is no less true in the musical than in the moral principle. If we are to credit the wonderful effects produced by music amongst the ancients, as related by historians, we shall find no reason to presume they knew any thing of *time*. The rhythm of time appears, therefore, to be an invention of modern date, and from hence it has arisen, that melody being shackled and restrained within its strict limits, the energy or pathos of singing, and the accent of words, have become as it were cramped and fettered.

Scholar. I have always heard, that the acquisition of perfect time is a most essential requisite in musicians of every denomination.

Master. I must again request you will not misapprehend me, but attend to the spirit of the maxim. Time is indispensably necessary in music where many parts are combined, and consequently to be executed by many performers:— We are, therefore, under great obligation to this invention; but to meliorate the rigour of its laws in melody, eminent singers have assumed a licence, of deviating from the strict time, by introducing the Tempo Rubato.

Scholar. What is this Italian Term?

————TEMPO RUBATO————

Master. Is a detraction of part of the time from one note, and restoring it by increasing the length of another, or vice versâ; so that, whilst a singer is, in some measure, singing ad libitum, the orchestra, which accompanies him, keeps the time firmly and regularly. Composers seem to have arranged their works in such a manner as to admit of this liberty, without offending the laws of harmony: one caution, however, becomes highly necessary; namely, that this grace, or licence, is to be used with moderation and discretion, in order to avoid confusion; for too frequent a use of Tempo Rubato, may produce *Tempo indiavolato.*

Scholar. Although I should not be able to attain the perfection of the *Tempo Rubato,* I hope if I sing in perfect time I shall not offend refined ears.

Master. You give me a gentle critique on my own remarks;—in every profession, or art, the refinement, or finish, is the most difficult to attain; there is a certain progress in every study, which human ingenuity may soon reach; but there are many hard steps to surmount, before it can arrive at perfection;—observe that, before you begin the practice of the Tempo Rubato, you ought to be a proficient in the knowledge of perfect time.

This Italian licence of Tempo Rubato, may be used in any species of music where there is a leading or predominant melody, and the management of it must be left to the skill and prudence of the performer, on account of the various characters and meaning peculiar to different compositions, which the performer must carefully discriminate in order to know where this alteration of the time will produce happy effect; and in using this licence, he should artfully manage the lengthening and shortening of notes, to restore to the next what he has stolen from the preceding, by which means the laws of harmony may be preserved.

————QUICKENING OR RETARDING OF TIME.————

Another improvement, by deviation from strict time, is to be made by the singer delivering some phrases or passages in quicker or slower time than he began with, in order to give emphasis, energy, or pathos, to particular words; and I cannot illustrate what I would inculcate on this head more forcibly, than by reciting an instance of the effect produced by this kind of expression, by my much esteemed friend, Mr. Braham, in my song of "Victory," in the Opera of the Travellers." When I composed this song, in arranging the music to the first Stanza—

"He was fam'd for deeds of Arms;   }   I could not imagine that the same melody would be   {   "Battle now with fury glows,
"She a maid of envied charms."   }   suitable to the words of the second Verse.   {   "Hostile blood in torrents flows."
&c. &c.   }   {   &c. &c.

and accordingly varied the music.—At the first rehearsal, the melody of the first verse having produced some effect, Mr. Braham advised me not to change it, but repeat the same melody to the second verse, to which I reluctantly agreed, fearing to be criticised by the connoisseurs on two points; first, for having expressed two passions so contrary to each other with the same melody; and, secondly, as that melody had not any change of modulation, consequently might seem monotonous;—however, Mr. Braham thought that accelerating the time at the second verse, and adding a fuller accompaniment, would produce the change in point of sense and expression, the effect fully justified his advice and opinion, and to Mr. Braham's eminent talents I feel indebted for the popularity of this song.

Scholar. The more I hear from you, the more I am alarmed, lest my abilities will not be adequate to the attainment of so difficult an art.

Master. Remember that in music, as in all other arts, when you have passed through the gradation of the necessary rules, your mind, assimilating the various parts of this acquired knowledge, creates, as it were, a new perception, that renders all your further progress easy and pleasant. Our next subject in order is the

————PHRASE.————

Scholar. What is meant by a phrase in music?

PLATE III. Domenico Corri, *The Singers Preceptor* (London, 1810), 6

altered note values and a strict accompaniment. Expressive singing and the expressive projection of text, as he explains in the section 'Rhythm of Time', require deviation from strict time. Later in his book Corri describes the cantabile style, which 'comprehends all soft, slow movements, where all the charms of vocal music may be combined—the Messa di Voce, the Portamento, Tempo Rubato, etc. here are used to their full extent, united with an elegant and noble delivery of the words'. In addition, tempo rubato is effective also in 'the Allegro Agitato (of distress)', which, like recitative, 'should be uttered nearly as speaking in musical notes'.[42] Corri was born in Italy, studied with Porpora in Naples, moved to England in 1771, and later formed a music publishing company with the composer Jan Ladislav Dussek.

Richard Bacon, in his *Elements of Vocal Science* (London, 1824), defines tempo rubato as 'liberties with single passages or notes' or 'taking a portion of the duration from one note and giving it to another'. He considers it 'one of the greatest helps to powerful elocution in singing'. To illustrate the way a singer can use rubato to 'throw great force upon a word of importance, which would otherwise be deprived of its meaning', he cites Handel's setting of the word *repentance* on three eighth notes: 'Common feeling dictates the propriety of shortening the first and last [notes], and allowing the time thus taken, to the middle note' (for example, a sixteenth note, quarter note, and sixteenth). He also suggests abbreviating the opening syllable of the words *revenge* and *without* in works of Handel and Haydn in order to strengthen the effect of the first and correct the 'false accentuation' of the second.[43] Such alterations are similar to those suggested by Grimarest more than a century earlier (see Ex. 2.12).

Isaac Nathan, a student and assistant of Domenico Corri, includes similar 'corrections' in his *Musurgia vocalis* (second edition 1836), but does not label them *tempo rubato*. He facetiously includes rubato in the table of contents as 'an honest theft', and later defines it as 'borrowing from or adding to a note, a little duration of sound, and making up the time by contracting, or detracting from the sound of another in the same bar; or, in other words, accelerating the time of one part of a bar, and retarding the other part, and *vice versa*'. His reference to note alteration, from the singer's point of view, as 'accel-

[42] *The Singers Preceptor* (London: Chappell & Co., 1810), repr. in MOS iii: *The Porpora Tradition*, ed. Edward Foreman (1968), 6, 69, and 70.

[43] Richard Mackenzie Bacon, *Elements of Vocal Science; being a Philosophical Enquiry into some of the Principles of Singing* (London, 1824), modern edn. by Edward Foreman in MOS i (1966), 31–2 and 84–5 (see also p. 50). His alteration of *without* in the duet 'Graceful Consort' from an English trans. of Haydn's *Creation*, however, is risky, since the rhythmic motive involved, which fits the phrase perfectly in German, is sometimes doubled by the bass line.

erating' and 'retarding' is unusual and reminds us of a similar description by Dubos in 1740. Nathan warns that one must 'steal discreetly',
however, for 'caught bungling in the fact, not even the restoration of
the stolen property to its neighbour will compensate for the offence',
and the injudicious use of this larceny, especially in 'part-music', can
cause discord and confusion.[44]

Similarly, Corri cautions (on Plate III) that rubato should be used
with 'moderation and discretion' lest its too frequent use produce
devilish rather than *stolen* time. According to Bacon, students should
'carefully abstain from all such indulgences'. Several articles in the
Allgemeine musikalische Zeitung from Leipzig early in the nineteenth
century also complain about excessive rubato. An anonymous writer
in 1802, in an article 'On the Abuse of Tempo Rubato', describes the
technique as 'the most intolerable metamorphosis that one can undertake with the simple movement of a melodic phrase'.[45] Another author
in the same journal in 1813 complains about all sorts of affectations
which singers were using for increased expression, including audible
breathing (apparently fashionable at that time among actors as well as
singers) and the 'so-called *rubamento di tempo*'. He feels that the
latter, although effective in rare cases, has become a bad habit of most
Italian singers and has led to rhythmic capriciousness and dissolution.[46]
In 1822 Haeser describes '*rubamento di tempo* oder *tempo rubato*' as
both syncopation and the alteration of note values within a single
measure or occasionally two consecutive measures. The device, he
feels, is apt to become monotonous when used too often and to lead
to metrical chaos when exaggerated. He notes that such abuses of
rubato unfortunately mar the performance of many otherwise distinguished singers and instrumentalists, since 'it has become almost
fashionable to sing and play far more *a piacere* than *a tempo*'.[47]

[44] *Musurgia vocales, An Essay on the History and Theory of Music, and on the Qualities, Capabilities, and Management of the Human Voice*, 2nd edn. (London: Fentum, 1836), repr. in MOS iii: *The Porpora Tradition*, ed. Edward Foreman (1968), p. vii in the contents and 290–1 (see pp. 190–1 for some note alterations to correct the 'defective' accentuation of words). See also G. G. Ferrari, *A Concise Treatise on Italian Singing*, i (London, preface dated 1825), copy at UCLA ML, 10: 'When a performer steals a composer's time, in proper places, he will not be condemned for the theft by the indulging laws of music, because what he takes from one note he gives to another. The *Tempo Rubato* likewise produces effective accelerations and retardations, particularly when they heighten the expression of emphatic words.'

[45] 'Über den Misbrauch des Tempo Rubato', *AMZ* 5/9 (24 Nov. 1802), 147: 'die aller-misslichste Metamorphose, die man mit dem schlichten Gange eines melodischen Satzes vorneh-men kann'.

[46] 'Mittheilungen über Gesang u. Gesangsmethode (Beschluss aus der 10ten No.), V. Ruba-mento di tempo. Taktwillkühr. Hörbares Athemnehmen. Verzierung des Gesanges im Allge-meinen', *AMZ* 15/11 (17 Mar. 1813), 181–2.

[47] August Ferdinand Haeser, 'Übersicht der Gesangslehre, Drittes Kapitel: Vom Vortrage', *AMZ* 24/8 (20 Feb. 1822), 128: 'es beynahe zur Mode geworden ist, weit mehr a piacere als a tempo zu singen und zu spielen'. See Blake , 'Tempo Rubato in the Eighteenth Century', 105–6.

During the Classic period the solo singer became considerably less conspicuous in opera. He had to compete with an orchestra which finally assumed symphonic proportions and which played an increasingly important role even in arias and recitatives. He competed sometimes with a chorus, ensemble, or dance, and the aria, which had been the soloist's special domain in the numbers opera, either merged with recitative or assumed a simpler form. *Opera seria* competed with *opera buffa*, and separate numbers were increasingly replaced by multi-sectional scenes such as the three-part solo *scena* and the ensemble finale. Such scenes dealt with changing emotions during the course of continuous music, rather than the single static emotion of the baroque aria. The operatic reforms by Gluck and by Italians such as Jommelli, Hasse, and Traetta attempted to remove purely virtuosic display in order to achieve a more truthful representation.

It was the general desire of most later classic composers that the singer follow the written score except in improvised cadenzas. There was a considerable decrease in the number and complexity of ornaments indicated by signs. In spite of the composers' desire for control and for classic simplicity, improvised vocal embellishment never seems to disappear. Hiller's book of 1780, in addition to its description of rubato, also presents examples and rules for improvised embellishment. During the early nineteenth century there seems to be an increase in such improvisation, especially in Italian opera. In reaction to this, Rossini eventually writes out in full exactly the notes to be sung. Improvised ornamentation and expressive devices such as tempo rubato, however, continue to be practised, and by some of the greatest singers, until well into the nineteenth century.[48]

The Romantic Period

One of the most important sources for evidence of the continuation of embellishment and, at the same time, one of the richest sources from any period for a detailed description of the practice of vocal rubato is the second part of the *Traité complet de l'art du chant* first published at Paris in 1847 by Manuel Patricio Rodríguez García (1805–1906). Each part of the treatise was published also in later editions; both were combined and condensed in an English translation of 1857 as well as a French and German version in 1859. At least in the sections dealing with rubato, the abridgement consists mainly of eliminating

---

[48] See Eva Badura-Skoda's comments in *New Grove*, ix, 43–8. For notated examples of ornamentation, see *Embellished Opera Arias*, ed. Austin B. Caswell in *RR in the Music of the Nineteenth and Early Twentieth Centuries*, vii and viii (1989).

some of the numerous musical examples of the original work. In addition, García published another book, which was translated as *Hints on Singing* in 1894.[49]

His father was the famous Spanish tenor Manuel del Pópulo Vicente Rodríguez García (1775–1832), for whom Rossini wrote the part of Almaviva in *The Barber of Seville*. His sisters were the two well-known mezzo-sopranos Maria Malibran and Pauline Viardot. He retired from singing in 1829 and devoted himself to scientific studies on the voice and to teaching, first at the Paris Conservatoire (until 1848), then at the Royal Academy of Music in London (1848–95). Among his pupils were Jenny Lind and Julius Stockhausen. He was a sensitive observer of the great singers of his time and had the ability to analyse the technical details of their art. One of these details was *temps dérobé* or tempo rubato, which he describes at length and with many specific examples.

In the first part of the *Traité* García mentions a device called *temps d'arrêt* or, in the English translation, *prolongation*: 'In passages formed of equal notes, increase of value can be given to any one of them in order to heighten effect, or to support the voice on those parts of a bar which might otherwise be passed over.' He gives the illustrations in Ex. 3.9, where one can make a *prolongation* on the notes marked with

Ex. 3.9 *Temps d'arrêt* or *prolongation* (García, 1847)

a cross, each passage in (*b*) to occupy exactly the same amount of time as the written version in (*a*). This device, 'by giving support to the voice, permits it to render distinctly that which would lack clarity, and

[49] The following sources were available to me: (*a*) *Traité complet de l'art du chant*, repr. 2nd edn. of Part I and 1st edn. of Part II, both from 1847 (Minkoff, 1985); and copies in BL of the 3rd edn. of Part I (1851) and 1st edn. of Part II (1847). Trans. by Donald V. Paschke in *A Complete Treatise on the Art of Singing*, i: *The Editions of 1841 and 1872* (Da Capo, 1984), and ii: *The Editions of 1847 and 1872* (Da Capo, 1975). (*b*) *García's New Treatise on the Art of Singing* (London, [1857]), copy at BL. (*c*) *Nouveau traité sommaire de l'art du chant* or *Neue summarische Abhandlung über die Kunst des Gesanges* (Schott, [1859]), copy in UCLA ML. (*d*) *Hints on Singing* (London, [1894]), repr., with Introduction by Byron Cantrell (Canoga Park, Calif.: Summit Publishing Co., 1970).

the passages gain thereby much effect'. When he discusses rubato in Part II he adds this footnote in the French version: 'Le *temps d'arrêt* ... is the first element of tempo rubato.'[50]

The main description of rubato occurs in the English edition on pp. 50 and 51, which are reproduced in Plates IV and V. Like Corri, Bacon, and Nathan, he sees the alteration of melodic note values as the primary element in rubato. Unlike them, however, he emphasizes the lengthening of notes (as in Ex. 3.9), with the shortening of others resulting simply as a consequence (see the sentence following the heading 'Tempo Rubato' on Plate IV). This melodic process then becomes perceptible when the rhythm of the accompaniment is strictly maintained (opening sentence on Plate V). He emphasizes this point, in fact, when he describes (at the top of Plate IV) the strict rhythm appropriate in the music of the Classic composers, where tempo rubato is the only permissible manner of changing the note values.

One of the important aspects of his presentation is the identification of specific musical situations suitable for tempo rubato. He illustrates each with musical examples, and it is significant that he seldom includes the accompaniment or bass-line (as in the fourth example on Plate V). This reveals again that he conceives of tempo rubato, from a singer's point of view, as a purely melodic adjustment of note values, and he assumes that the orchestra will maintain its strict beat by itself. Since it is important for us, however, to know the nature of the accompaniment, I have added piano reductions of the orchestral parts to García's excerpts in Exx. 3.10 to 3.17. Most of his examples come from late classic and early romantic operas by composers such as Mozart, Rossini, Donizetti, and Bellini.

In the first situation García describes, rubato 'breaks the monotony of regular movements, and gives greater vehemence to bursts of passion'. He illustrates this with the excerpt at the top of Plate V from Donizetti's *Anna Bolena*—an excerpt in which Anna, after being reminded of the lover she rejected in order to marry Henry VIII, tells Giovanna not to let herself be seduced by the temptation of a royal throne. In Ex. 3.10 I have included a piano reduction of the orchestral accompaniment on the lower two staves and added the slurring and accentuation from modern editions to the written voice part on the second line.[51] In this example, as in most of the others, the accompaniment provides a decisive and insistent emphasis on each eighth-note

[50] García, *Traité*, i, 49, and ii, 24: 'Le temps d'arrêt ... est le premier élément du tempo rubato.' See *García's New Treatise*, 26. He actually writes the 2nd and 3rd passage in Ex. 3.9 a second time a step higher, as I show for the 1st one.

[51] García, *Traité*, ii, 24; Cavatina 'Come innocente giovane' from Act I, sc. iii, mm. 172–4: piano/vocal score (Ricordi, [1959]), 25; full score (New York: Edwin F. Kalmus, n.d.), 102.

50

Otello,—" L'ira d'avverso fato ;" the stretta finale of *Otello* ; and stretta finale of *Don Giovanni*. In such a case, a voice produces the effect of a percussive instrument, and proceeds in like manner by striking distinct blows.*

Time is of three different characters, viz., regular, free, and mixed. Time is regular when an air is characterized by a very decided rhythm, which rhythm—as we have said—is usually composed of notes of short duration. Warlike songs, or shouts of enthusiasm, especially require strongly-accented and regular measure (see Examples A). The compositions of Mozart, Cimarosa, Rossini, &c., demand great exactitude in their rhythmic movements. Every change introduced

into the value of the notes, should, without altering the movement of the time, be procured from adopting the *tempo rubato*.

Secondly, time is *free*, when, like discourse, it follows the impulse of passion and accents of prosody ; chanting and recitatives are examples of free-measure.

Thirdly, time is mixed when the feelings expressed in a piece exhibit frequent irregularities of movement, as is often the case in tender, melancholy sentiments. In such pieces, the value of the notes is generally long, and the rhythm but little perceptible. A singer should avoid marking the time too strongly, or giving it too regular and stiff a character (see Example B).

Irregularities in time are, *rallentando, accelerando, ad libitum, a piacere, col canto,* &c.

On Rallentando.

Rallentando expresses decrease of passion ; and consists in slackening the rapidity of a measure, in all its parts at once, in order to enhance its grace and elegance. It is also used as a preparation for the return of a theme or melody.

On Accelerando.†

Accelerando is the reverse of *rallentando*, as it increases the velocity of a movement, and adds greater spirit and vivacity to the effect.

On Ad-libitum.

In *ad-libitum* phrases, time is slackened ; but this kind of free movement must not be arbitrarily introduced. Consequently, whenever a singer intends risking it, he must not diminish the time

* This attack is effected by means of a stroke of the glottis, or stress on the consonant, according as a word begins with a vowel or consonant. If these notes were only feebly struck, the rhythmic element would be destroyed.

† In the quintett of *Beatrice di Tenda*, the forty-four last bars constantly increase in rapidity. Donizetti's music—and above all, Bellini's—contains a great number of passages, which, without indications either of *rallentando* or *accelerando*, require both to be employed.

throughout, but have recourse to the *tempo rubato*, which will be noticed immediately. Certain pieces admit of the voice and accompaniment being alternately free and in strict rhythm; when latitude is given to the vocal part, the time of the accompaniment must be well marked. (See above Example B.)

Suspensions and cadenza stop the accompaniment altogether, and leave the singer for some moments absolutely independent.

Tempo Rubato.

By *tempo rubato* is meant the momentary increase of value, which is given to one or several sounds, to the detriment of the rest, while the total length of the bar remains unaltered. This distribution of notes into long and short, breaks the monotony of regular movements, and gives greater vehemence to bursts of passion. Example :—

PLATE IV. Manuel García, *New Treatise on the Art of Singing* (London, 1857), 50

51

To make *tempo rubato* perceptible in singing, the accents and time of an accompaniment should be strictly maintained: upon this monotonous ground, all alterations introduced by a singer will stand out in relief, and change the character of certain phrases. *Accelerando* and *rallentando* movements require the voice and accompaniment to proceed in concert; whereas, *tempo rubato* allows liberty to the voice only. A serious error is therefore committed, when a singer, in order to give spirit to the final cadences of a piece, uses a *ritardando* at the last bar but one, instead of the *tempo rubato*; as while aiming at spirit and enthusiasm, he only becomes awkward and dull.

This prolongation is usually conceded to appoggiaturas, to notes placed on long syllables, and those which are naturally salient in the harmony. In all such cases, the time lost must be regained by accelerating other notes. This is a good method for giving colour and variety to melodies. Example:—

Two artists of a very different class—Garcia (the author's father) and Paganini—excelled in the use of the *tempo rubato*. While the time was regularly maintained by an orchestra, they would abandon themselves to their inspiration, till the instant a chord changed, or else to the very end of the phrase. An excellent perception of rhythm, and great self-possession on the part of a musician, however, are requisite for the adoption of this method, which should be resorted to only in passages where the harmony is stable, or only slightly varied —in any other case, it would appear singularly difficult, and give immense trouble to an executant. The annexed example illustrates our meaning*:—

The *tempo rubato*, again, is useful in preparing a shake, by permitting this preparation to take place on the preceding notes; thus:—

The *tempo rubato*, if used affectedly, or without discretion, destroys all balance, and so tortures the melody.

Of the Forte-piano, and Accents on single sounds.

Forte-piano, applied to isolated notes, is called accent. The most regular accents of song are founded on the emphasis of spoken language, and fall on the down-beats in a bar, and on long syllables in words. But as this arrangement would not be sufficient to give character to all kinds of rhythm, accents are also placed, when required on the weak parts or beats of a bar, in this way destroying the prosodic accent.† Example:—

* This passage presents an approximate example of the use which the author's late father made of the *tempo rubato*.

† Spaniards, much more frequently than Italians, make use of this liberty in their popular songs; and although the Spanish language has a prosody quite as much accentuated as the Italian, yet in popular tunes the accents of the music regulate those of the words,—a characteristic feature of their national music, perhaps not to be met with elsewhere.

PLATE V. Manuel García, *New Treatise on the Art of Singing* (London, 1857), 51

Ex. 3.10 Rubato in Donizetti's *Anna Bolena* (García, 1847)

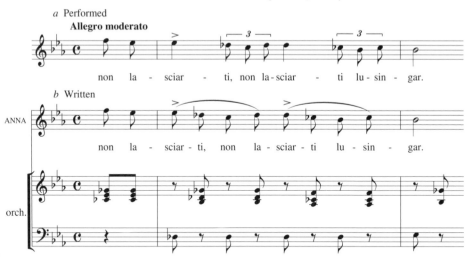

beat, thus constituting a strict rhythmic background against which the deviations of the voice become particularly conspicuous. The notes shown in the accompaniment are played by the strings, with winds sustaining the chords in half notes. Two measures before and after the excerpt in Ex. 3.10 the music is clearly in E flat major. The sudden and passionate harmonic movement is matched by a rubato in which the syllable '*sciar*' steals time twice from the following three syllables. He has thus applied here the *prolongation* which he depicted in Ex. 3.9.

At other times, tempo rubato 'aids the musical colouring, especially when repeating a phrase'. García illustrates this in the French version of the *Traité* with an excerpt from the Countess's aria 'Dove sono i bei momenti' from Mozart's *Marriage of Figaro* (see Ex. 3.11).[52] The Countess sadly wishes that her constancy would bring the hope *of changing the* count's *ungrateful heart*. Again, García shows only the voice parts, a portion of which I include on the upper two staves; I have added the accompaniment, played only by strings. Ex. 3.11 represents the second of two almost identical phrases, and García gives enough of the melody to make clear that the rubato is added only the second time. Robbery is here accompanied by an added pitch (the accentuated G on the opening beat) as well as a lengthening of the notes on the first and third beats. This bears remarkable similarity to the *strascino* or *drag* of Tosi and Galliard: see Ex. 3.7, where slower

[52] *Hints on Singing*, 62; *Traité*, ii, 24; Mozart, *Le nozze di Figaro*, Act III, sc. viii, *Neue Ausgabe sämtlicher Werke*, Serie II: 5/xvi/2, p. 411, mm. 58–60.

Ex. 3.11 Rubato in Mozart, *Le nozze di Figaro* (García, 1847)

notes occur at the middle of the descent and the fall commences with pitches higher than the presumed written note (see also Caccini's *cascata* in Ex. 2.11).

Tempo rubato can also 'give spirit' to final cadences (first paragraph below the upper two musical examples on Plate V). In the French edition García includes the excerpt on the top two staves of Ex. 3.12 from a duet in Rossini's *Barber of Seville*. After Figaro suggests that the Count gain entrance to Bartolo's house disguised as a soldier, they both sing 'what a wonderful *idea! truly great*'. García explains that 'one . . . commits a grave fault when, in order to render warmly the very animated cadences of the duet from the *Barber*, one suddenly uses the ritardando in place of the tempo rubato in the next-to-last measure'. This cadence actually occurs twice in the duet. The first time, the orchestra reaches the final chord on the first beat of the last measure along with the voices; the second time, as shown in Ex. 3.12, it delays the chord in a sort of notated tempo rubato, thus adding to the animation felt by the singers before the crashing dotted rhythm of the tonic chord finally appears. The full orchestra is playing and the actual pitches sung by the Count and Figaro in the penultimate measure are doubled in different octave levels by the violins (tremolo), flutes, oboes, and clarinets.[53] Therefore, when the voices use tempo

---

[53] *Traité*, ii. 24; trans. from Paschke's *Complete Treatise*, ii, 76. Duet 'All'idea di quel metallo' from Act I, sc. iv: piano/vocal score (Ricordi, 1973), 87; full score (Ricordi, 1969), 95. In both modern scores the passage is in G major, the 8th notes are staccato, García's 2 accent signs in (*b*) are missing, and Figaro sings 'bella, bella' while the Count has 'bravo, bravo'.

Ex. 3.12 Rubato in Rossini, *Il barbiere di Siviglia* (García, 1847)

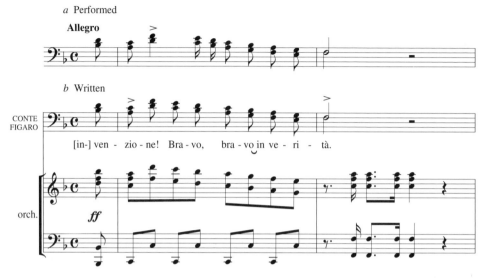

rubato, the melody is heard simultaneously in both its robbed and unrobbed forms. As we have seen, some eighteenth-century writers, such as Kirnberger and Schulz, warned against applying rubato when non-continuo instruments doubled the melody. This is no doubt good advice in general, although in Ex. 3.12 the lingering on the high notes at the expense of the succeeding two seems to work well enough. Tempo rubato, however, is ordinarily a device for the solo singer, for examples shown in notation are only a rough approximation of the actual time values, which would change from one performance to another depending on the intensity of motivation at the moment. For two singers to perform rubato simultaneously as in Ex. 3.12, it would require advance planning of a sort not usually associated with the device.

In the second paragraph below the upper examples on Plate V García gives a list of single notes that can be prolonged by tempo rubato in order to give 'colour and variety to melodies'. It is at this point in the French edition that he links the *temps d'arrêt* (Ex. 3.9) with rubato. In the example from Donizetti's *Lucia di Lammermoor* on Plate V he shows, by adding upward stems, those notes which should be lengthened slightly; the others, of course, are necessarily shortened. The orchestra plays a single chord on the opening beat and is then silent for the rest of the measure.[54] In the French edition he

[54] The cavatina from which this excerpt comes is missing from modern scores, replaced apparently by 'Regnava nel silenzio'. It is printed at the end of the piano/vocal score *Lucia di*

Ex. 3.13 Rubato in Donizetti, *Anna Bolena* (García, 1847)

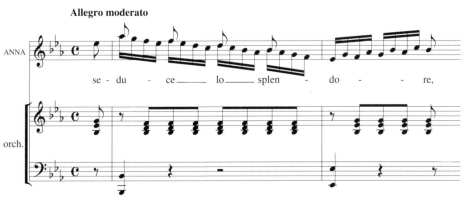

includes a second excerpt (on the top staff of Ex. 3.13), in which the
orchestral strings (which I include below the melody) provide a continuous eighth-note movement accompanied by a sustained chord in
the horns. He depicts the first note of each descending four-note group
as an eighth note, hence to be prolonged for about half the total time
of the group. Modern editions slur the four sixteenth notes together
and add an accent mark above the first note of each group. This
excerpt comes from the same portion of Donizetti's *Anna Bolena* as
Ex. 3.10; here Anna speaks to Giovanna: 'If ever the *splendour* of a
royal throne should *tempt* you, remember my anguish.'[55]

The next paragraph on Plate V describes a special type of tempo
rubato employed by García's father and by the violinist Paganini.
They, according to the French version, excelled in using tempo rubato
'by phrase' (*par phrase* in italics). The English translation in Plate V,
curiously, omits 'by phrase', words which define this particular situation in which rubato is applied over a far longer period of time and
involves a greater number of notes. In addition, new pitches have been
added in the third measure of this excerpt from Rossini's *Barber of
Seville*. García may include the bass-line here because of the complexity
of the rubato. The orchestra is actually silent on the opening anacrusis,

---

Lamermoor [sic], *rappresentata nel Teatro S. Carlo, a Napoli* (Paris: Bernard Latte, n.d.). The
piece, which is paged separately, is missing from the table of contents. It bears the title 'Perché
non ho del vento: Cavatina cantata dalla Signora [Fanny] Tacchinardi-Persiani' (who sang the
title role in the first performance of 1835). It occurs also at the end of the piano/vocal score
published by Ditson in Boston (with an introduction by J. Wrey Mould dated July 1854). Both
scores are at UCLA ML.

[55] *Traité*, ii, 25; 'Come innocente giovane' from Act I, sc. iii, mm. 213–14: piano/vocal score,
28; full score, 110.

and then the strings play chords above the bass notes. This example has particular significance, since it was García's father for whom Rossini wrote the part of Count Almaviva (see the footnote marked with an asterisk at the bottom of Plate V). The rubato in this case emphasizes a facetious comment by the Count: after Figaro says that the promise of gold inspires his mind to erupt, the Count replies, 'Then, let's see . . . *some singular prodigy from that volcano of your mind*'.[56]

The final rubato situation mentioned on Plate V is the preparation of a trill, to take place during the time of the preceding notes. This is illustrated by the example we have just discussed, as well as the more complex excerpt in Ex. 3.14 (see the last excerpt on Plate V, which

Ex. 3.14 Rubato in Rossini, *Il barbiere di Siviglia* (García, 1847)

should have the tenor rather than the treble clef). In the Count's serenade to Rossina in *The Barber of Seville* ('*the lovely dawn is breaking*'), a guitar plays the sixteenth-note triplets in the accompaniment, with strings on the eighth-note chords. The orchestra thus controls the rhythmic beat to a fine degree. As written, the voice and

[56] Act I, sc. iv, 'All'idea di quel metallo': piano/vocal score, 77–8; full score, 87. In *La Revue musicale*, 6 (1830), 228, Fétis describes a performance of this duet in 1829 in which García's father employed the technique of 'bending [*rompant*] the beat now and then with perfect grace, and arriving on time at the end of the phrase'; trans. James Vincent Radomski, 'The Life and Works of Manuel del Pópulo Vicente García (1775–1832): Italian, French, and Spanish Opera in Early Ninetenth-Century Romanticism', Ph.D. diss. (UCLA, 1992), p. 568.

guitar would produce a rhythm of two against three on the second half of each main beat. With the tempo rubato, however, this effect disappears and the syllable '*ta*' steals time from the following '*la*', and '*ro*' steals time, although García does not notate it exactly, from the preceding two syllables. Thus both the pitch and the syllable on the first '*la*' are delayed, whereas both pitches and the syllable '*ro*' on the trill are anticipated. A few new pitches also appear.[57]

We have already noted the intensifying effect of rubato when a passage is repeated (Ex. 3.11). In a later section of the book García discusses in more detail the variety one should give to the second of two identical or sequential passages: 'When the second section of a phrase is composed of the same values as the first, its coloring should be sometimes the *tempo rubato*, and sometimes the *piano* opposed to the *forte*.' In the French edition he illustrates the tempo rubato with a brief excerpt from Ferdinando Paer's opera *Griselda*. In this rare case he does include the accompaniment, perhaps because the singer is performing four sixteenth beats against a triplet in the orchestra. He explains further: 'When the identical thought is repeated several times in succession, as it is frequently with all composers, especially Mozart; or when the thought pursues an ascending or descending progression [that is, sequence], . . . each different development should be submitted, according to the sentiment of the phrase, to the *crescendo* or *diminuendo*—the *accelerando* or *ritardando*; in rarer instances, to isolated accents and the *tempo rubato*.'[58]

In an example from Rossini's *Otello* (Ex. 3.15), he uses tempo rubato to emphasize a change in dynamics when a two-note figure is repeated. He explains that 'when a pianissimo follows [a forte], it should be separated from the *forte* by a slight rest, striking the note an instant after the bass . . . This rest affords relief after loud notes, and prepares us for seizing all effects, however delicate, that follow,—especially if the first consonant that ensues after the rest is produced with vigour.' Below the excerpt he adds the note: 'Strike the C [presumably the letter '*c*' in the word *consolar*] after the bass.' In Ex. 3.15 I have added in the bass an $E^\flat$ in parentheses which is missing in the scores available to me, but which García must have had in mind. The text is part of Desdemona's prayer: 'Make my dearly beloved Otello *come to console me*.' The same music recurs to different text later in the aria, but García makes no comment on varying it.[59]

[57] *Traité*, ii, 25 (with tenor clef). Act I, sc. i, Cavatina 'Ecco ridente in cielo': piano/vocal score, 19; full score, 31–2.

[58] *New Treatise*, 55; *Traité*, ii, 32–3.

[59] *New Treatise*, 55; *Traité*, ii, 33. The accompaniment in Ex. 3.15 is from the piano/vocal score (Paris: Aulagnier, [1830]), copy in UCLA ML, 169 (from the *Preghiera* in Act III). The fac.

Ex. 3.15 Rubato in Rossini, *Otello* (García, 1847)

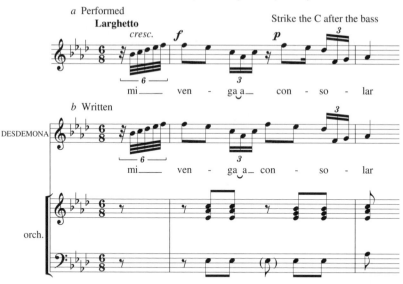

In another section concerning slurring, García includes an illustration of the 'accent used by Madame Pasta' when, in the title role of Bellini's *Norma*, she expresses, in an aside, her love for Pollione and declares that on his breast she would have 'life, *country, and heaven*'. Again, as in almost all of these rubato examples, the orchestra gives a strong, unhesitating sense of beat, in this case from the strings accompanied by sustained chords in half notes in the clarinets. In Ex. 3.16 the syllables '*tria*' and '*e*' have been delayed, but not their pitches; for '*lo*' and '*a*' both pitches and syllables have been delayed.[60] Giuditta Pasta was the very famous Italian soprano who sang the part of Norma in the first performance of this opera at Milan in 1831. In 1830 she similarly created the title role of Donizetti's *Anna Bolena*, from which Exx. 3.10 and 3.13 come; these examples of rubato may also reflect her method of performance. When Chorley in 1850 listed half a dozen examples that he could recall from the great artists who possessed the 'nicest sense of measurement of time', he included 'the

of Rossini's autograph MS, ed. by Philip Gossett in *Early Romantic Opera*, viii (New York and London: Garland, 1979), shows only rests in the wind ensemble parts on this beat, and the same for the other two appearances of the melody (see fo. 114^{r-v}).

[60] *Traité*, ii, 27, and *New Treatise*, 53; Act I, sc. iv, m. 178 ('Casta diva' begins in m. 55 of the same scene): piano/vocal score (B & H, n.d.), 45; full score (New York: Edwin F. Kalmus, 1970), 139.

Ex. 3.16 Rubato in Bellini, *Norma* (García, 1847)

tempo rubato of Madame Pasta'.[61] Pasta, as we will see later, was one of the great singers whom Chopin knew and admired.

Another excerpt which, like Ex. 3.16, prolongs a note and thereby delays the following descending notes, appears in the concluding section of the French edition where García includes several lengthy pieces along with a piano accompaniment and his suggested method of performance. He not only notates his embellished versions, but also makes comments from time to time. Above the second half of the first measure of Ex. 3.17 he writes 'ex. de temps dérobé' (example of tempo rubato), referring to the sustaining of the $B^\flat$ at the expense of the three following pitches and the syllables '*se*' and '*ta*'.[62] The large $B^\flat$ in Ex. 3.17a represents the grace-note appoggiatura in (*b*), which ordinarily would have taken half the value from the $A^\flat$ to which it is attached; this would have produced a descending group of four even eighth notes. In addition, the grace-note *port de voix* (slur of the voice) which precedes this $B^\flat$ in (*a*) is itself a type of tempo rubato, for it anticipates the pitch $B^\flat$, but not its syllable, by stealing time from the preceding F. To the right of Ex. 3.17a I have noted two ways of performing a *port de voix*: the first as customary in Italy and the second as they do it, according to García, 'too often in France'. His father, in a book entitled *Exercises and Method for Singing*, explains

[61] Henry F. Chorley, *Thirty Years' Musical Recollections* (London, 1862), 2nd rev. edn. by Ernest Newman (Knopf, 1926), 284.
[62] *Traité*, ii, 84. I have added a missing tie between the two large $B^\flat$s in Ex. 3.17a.

Ex. 3.17 Rubato in Cimarosa, *Sacrifice d'Abraham* (García, 1847)

that 'to sing in the Italian manner, the greatest attention is necessary to the carrying or a mode of slurring the voice, which is never executed with the syllable about to be taken, . . . which is the French manner, but with the syllable just pronounced . . . which is the true Italian method.'[63] We noted the French method in Chapter 2 in sources from two hundred years earlier: compare the French manner of execution in Ex. 3.17 with the opening two syllables of Bernhard's *anticipatione della syllaba* in Ex. 2.4. The same technique occurs in Bacilly's *port de voix* in Ex. 2.7. Although García does not connect this device with tempo rubato, it is clear that in the Italian method the pitch but not the syllable on '*for*' is anticipated, whereas in the French manner the syllable but not the pitch is anticipated—in each case by stealing time from the preceding note.

García concludes his discussion of tempo rubato with the warning (see Plate V) that 'if used affectedly, or without discretion', it 'destroys all balance, and so tortures the melody'. Presumably, it should be used only for special passages, and then only discreetly and with good taste. Fortunately for us, García identifies, as we have seen, some of the special occasions suitable for rubato: to vary a passage in even notes, to intensify passion, to emphasize a repeated phrase, to enliven a final cadence, to prolong appoggiaturas, notes on long syllables, and those important in the harmony, to embellish an entire phrase, and to

---

[63] *New Treatise*, 52–3, and *Traité*, ii, 27–8. His father's comments are in *Exercises and Method for Singing* (London: T. Boosey [1824]), copy in BL, 2 (Rule 8th).

prepare a trill. In all of these situations one or more significant notes are prolonged by stealing time either from the following notes or, in the case of the trill preparation, from the previous note. Even in Ex. 3.15 the important note, although retaining its duration, is prolonged past its original ending point. It should be emphasized once again that all but a few of his examples show only the voice part and thus convey a sense of melodic robbery, but not a sense of displacement between melody and accompaniment. The only time García comments on such displacement is the instruction to 'strike the C after the bass' in Ex. 3.15. The accompanimental parts which I have included show, however, that such displacement would have been a most conspicuous aspect to a listener. The type of music García chose for illustrative purposes almost always possesses an emphatic and mechanically recurring beat in the orchestra. With the strictness of the beat audibly secure, the singer was free to steal time in any manner he wished.

In spite of García's precise notation, however, the result, in order to be expressive, must *sound*, even to a listener who does not have a score to follow, as though the passion of the moment caused the notated rhythm to be uncontrollably altered. Aristide Farrenc, after quoting García's verbal description of rubato in the introduction to his collection of piano music in 1861, states:

As for the examples which accompany his precepts, I would say that they are insufficient, for there is in this device combinations and nuances of value which cannot be notated; one can only get an idea of it by hearing a great virtuoso . . .[64]

Although some of the texts of García's examples are of a comic nature, most are involved with love—especially the anguish of a betrayed, lost, difficult, or distant love. At the end of the treatise he categorizes the different singing styles. The *plain style* or *canto spianato* (smooth, even) refers to simple, relatively unadorned singing in the slower tempos. 'Its chief resources are—steadiness of voice, harmony, and delicacy of the *timbres*, swelled sounds of every variety, finest delicate shadings of the forte-piano, slurs, tempo rubato, and neatness of articulation.' By contrast, the *florid style* or *canto fiorito* 'is rich in ornament and coloring . . . In this, as in the *canto spianato*, an artist uses *mezza di voce* [swelled sounds], *tempo rubato*, *forte-piano*, slurs . . . ' He further subdivides this style into three types. The *canto di agilità* has brilliant and rapid embellishments and occurs in Allegro

---

[64] *Le Trésor des pianistes*, i (Paris, 1861; repr. Da Capo, 1977), 4: 'Quant aux exemples qui accompagnent ses préceptes, je dirai qu'ils sont insuffisants, car il y a dans cet artifice de l'exécution des combinaisons et des nuances de valeur qui ne peuvent se noter; il n'y a que l'audition d'un grand virtuose qui puisse en donner l'idée'.

movements. He mentions in a footnote that 'García [his father], Pellegrini, Tamburini, Mesdames Sontag and Damoreau, were remarkable for their excellence in this style.' The next subdivision of the florid style is *song of contrivance* or *canto di maniera* for singers with less power; it has briefer and more delicate and graceful passages and is represented by 'Mesdames Pasta, Persiani, and M. Velluti'. The final florid type is *bravura singing*, which is 'the *canto di agilità* with the addition of power and passion' and which is illustrated by the singing of 'Garcia, in [Rossini's] *Otello*, Mesdames Catalani, Malibran, and Grisi'.[65] All of these categories of plain and florid singing are suitable, then, for tempo rubato. I have included García's lists of singers because many of them are precisely the artists Chopin heard and described in his letters.

García discusses other styles, such as the declamatory and buffo, which do not ordinarily involve rubato. He does, however, describe some popular Spanish songs as follows:

> The Spaniard sprinkles his song with numerous mordents which attack the notes, and with frequent syncopations which displace the tonic accent, so as to add more piquancy to the effect by an unexpected rhythm. It is only at the end of the phrase that the voice coincides with the bass.[66]

This sounds very much like the tempo rubato *par phrase* which his father added to Count Almaviva's excerpt on Plate V.

He includes many other topics in his book as well. He discusses improvised embellishment and the various ornaments. In a section on expression he describes various ways to use the breath in performing a sob, a sigh, or a laugh. In his discussion of pronunciation he categorizes the different types of consonants and says they can be 'forcibly pronounced . . . to give strength to the expression of some sentiment' or 'to render words audible in large buildings'. He summarizes the role of consonants as follows:

> In music, the two elements of speech correspond with those of melody; vowels with sounds; consonants with time. Consonants serve to regulate or beat the time—to hurry or retard a passage, as well as to mark the rhythm; they indicate the moments at which an orchestra should blend with the voice, after an *ad libitum*, a cadence, or a pause. Finally, consonants impart spirit to the *stretta*, and concluding cadences. They should always be prepared beforehand, in order that they may fall precisely with the beat.[67]

---

[65] *New Treatise*, 72–4; *Traité*, ii, 66–70.
[66] *Traité*, ii, 70; trans. from Paschke's *Complete Treatise*, ii, 199–200.
[67] *New Treatise*, 43; *Traité*, ii, 7.

Manuel García was clearly a sensitive musician, a careful observer, and a patient teacher. He lived during a golden age of great singers. The many musical examples in his books preserve the various details of their style. Whether a singer could actually learn to execute tempo rubato by studying these notated examples, it is difficult to say, for it was a device to be used in a very personal manner in response to intense musical situations. For us today, however, García's examples reveal more precisely than any other source, I think, the true nature of the vocal rubato during the second quarter of the nineteenth century. It is fortunate for us that García was such a faithful chronicler of the singer's art.

In 1849 Laure Cinti-Damoreau published in Paris her *Méthode de chant*. She dedicated the book to her students at the Paris Conservatoire, where she taught singing from 1833 to 1856. García, who taught at the Conservatoire himself from 1847 to 1850, mentioned Damoreau, as noted above, as one of the great singers in the *canto di agilità* style. During the late 1820s and the 1830s leading operatic roles were created especially for her by Rossini, Auber, and Meyerbeer. The first part of her book consists almost entirely of exercises and compositions, all provided with vigorously rhythmic piano accompaniments much like the operatic piano reductions we have already seen in Exx. 3.10 and 3.15. Her students thus became accustomed to hearing a strictly maintained beat in the accompaniment. In the second part of the book she presents many examples of improvised cadenzas and variants of passages from the written scores of operas by Bellini, Rossini, Auber, Halévy, Meyerbeer, Adam, and others. These come no doubt from her own singing experience. She notates the variants, as García usually does, without accompaniment. Unlike García, however, she makes no analytical comments about her complex art of embellishment and hence does not mention by name any of the specific devices such as tempo rubato.

Some of the variants simply replace the composer's original passage with a completely new one that fits the same accompaniment. Others, however, remain close enough to the original melody for one to recognize the technique of melodic variation. It is in the latter type that one can sometimes find tempo rubato, as in Ex. 3.18 from Meyerbeer's *Robert le diable*. Cinti-Damoreau had herself created the role of Isabelle in the first performance of 1831. Ex. 3.18*a* shows her version, apparently, of the original melody, the second line her fifth variant; I have added the printed score in (*c*) from a contemporary piano reduction. Ex. 3.18*b* exhibits considerable anticipation of the first four notes of the descending arpeggio in (*c*). The top line, if it is indeed meant to be a variant, presents a most exciting augmentation

Ex. 3.18 Variation in Meyerbeer's *Robert le diable* (Cinti-Damoreau, 1859)

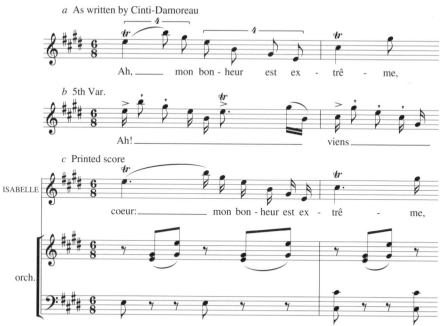

and anticipation of all the notes of the arpeggio in polymetre with the accompaniment. I have added the 4s and the time signature in this line, as well as the missing dot on the trilled note on the second line. Cinti-Damoreau's examples show that the art of vocal embellishment was still alive in French and Italian opera until at least the middle of the nineteenth century and that tempo rubato could be employed to delay notes or anticipate them.[68]

Although rubato is seldom mentioned in singing treatises after the mid-nineteenth century, it does appear in Faure's *La Voix et le chant* from around 1886. Faure performed major roles in French and Italian operas from his debut in 1852 until 1876, singing works by Mozart, Meyerbeer, Rossini, Verdi, Gounod, and Thomas. From 1857 to 1860 he taught at the Paris Conservatoire. In his treatise he defines *stentato* (laboured, halting) as 'with pain or trouble, stretched out with effort'. It involves both a special colour and 'an effect of broken rhythm without alteration of the metre, as in the *tempo rubato*'. He continues:

[68] Laure Cinthie Montalant, later Cinti-Damoreau, *Méthode de chant, composée pour ses classes du Conservatoire* (Paris, 1849), copy in BL, 97. Ex. 3.18c is from the piano/vocal score by Johann Peter Pixis (Paris: Maurice Schlesinger, 1831), copy in UCLA ML, 105–7. See Austin Caswell, 'Mme Cinti-Damoreau and the Embellishment of Italian Opera in Paris: 1820–1845', *JAMS* 28 (1975), 459–92.

Stentato indicates a retard in the phrase, but whereas the *ritardando* takes place at the will of the singer or the instrumentalist, in the *stentato* the performer seems to want to escape a mysterious pressure and surrenders only to the superior force of a feeling which imposes this retard on him; the notes are heavily stressed and even hammered, without ceasing to be bound together in the most rigorous manner.[69]

He gives two brief vocal lines in which the notes involved are all marked with accent signs by the composer. In the first, from Meyerbeer's *Les Huguenots*, Raoul, after discovering that Valentine loves him, sings in counterpoint to the latter's frightened exclamations: 'If my joy be a dream, *may I never come to waking*.' While Valentine holds a long note and the orchestra is suddenly silent, Raoul repeats the second half of the sentence with *stentato*. His passage is marked *trainez* [drag] *le mouvement* in the original edition of the opera and the other parts *suivez le chant*.[70] The other vocal excerpt comes from Verdi's *Ernani*, where five *stentato* notes emphasize the word *gloria* as the ensemble and chorus sing 'To Charles V both glory and honour'.[71]

In a later section on colour, Faure discusses the rhythmic effect called *anticipation*:

It is the process which consists in borrowing from a beat a little of its value in order to give it to the beat that follows. This is what the Italians call *tempo rubato*.

Strictly, one could write the *anticipations* as one does syncopations, with which they offer some analogy; but this would be giving it the letter and not the spirit. Employed with discernment, the *anticipations* give to the rhythm a greater freedom of movement and impart to the melody the stirring character of improvisation while preserving the feeling of the metre.[72]

He shows in an example how the Italian baritone Giorgio Ronconi (1810–90) interpreted the passage in the sextet from *Lucia di*

[69] Jean-Baptiste Faure, *La Voix et le chant: traité pratique* (Paris: Henri Heugel, date on portrait 1886), 171: 'Le *stentato* indique en effet un mouvement de retard dans la phrase, mais tandis que le *ritardando* s'opére par la volonté du chanteur ou de l'instrumentiste, dans le *stentato* l'exécutant semble vouloir échapper à une étreinte mystérieuse et ne céder qu'à la force supérieure d'un sentiment qui lui impose ce retard; les notes sont marquées lourdement et même martelées, sans cesser d'être liées entre elles de la façon la plus rigoureuse.'

[70] Full score (Paris: M. Schlesinger, 1836), copy at UCLA ML, Act IV, 751.

[71] *The Works of Giuseppe Verdi*, I/v (UChP and Ricordi, 1985), 368: Act III, concluding ensemble and chorus, m. 131.

[72] *La Voix et le chant*, 182: 'Parmi les variétés de rythme, il faut placer en première ligne les *anticipations*. C'est le procédé qui consiste à emprunter à un temps un peu de sa valeur, pour le reporter sur le temps qui suit. Ce que les Italiens appellent: le *tempo rubato*.

'À la rigueur, on pourrait écrire les *anticipations* comme on le fait pour les *syncopes*, avec lesquelles elles offrent quelque analogie; mais ce serait en donner la lettre et non l'esprit. Employées avec discernement, les *anticipations* laissent au rythme une plus grande liberté d'allure et communiquent au chant, tout en lui conservant le sentiment de la mesure, le caractère entraînant de l'improvisation.'

Ex. 3.19 Rubato and *stentato* in Donizetti's *Lucia di Lammermoor* (Faure, 1886)

Lammermoor where Enrico sings: 'She [Lucia] is my own flesh and blood and I *have betrayed her*! *She lies* between life and death.' In Ex. 3.19 I have shown the portion of the excerpt which includes his *anticipation* or *tempo rubato* on the syllable '*ta*'. Ronconi's 'sudden movement' from F to Eᵇ, on a closed then open vowel ('*o*' to '*a*'), 'threw on this phrase a light whose brilliance the skilled singer augmented even more by strengthening the following notes with the aid of *stentato*'. I have added a piano reduction of part of the orchestral accompaniment. The chorus and ensemble are singing at the same time, but do not interfere with the *anticipation*. The unnotated rhythmic alterations in the *stentato*, however, must create slight displacements with the same melody doubled by the flute, clarinet, and violins. The accompaniment provides additional rhythmic energy with its two beats against three in the melody.[73]

Later in the book Faure gives an example of rubato from *Les Huguenots*. Early in the opera Raoul describes a woman with whom he had fallen in love as 'plus blanche que la blanche hermine' (whiter than the white ermine). According to Faure one should sing the

---

[73] Ibid. 183: 'le passage subit du *fa naturel* fermé au *mi bémol* ouvert sur les paroles "*l'ho tradita*" jetait sur cette phrase une lumière dont l'habile chanteur augmentait encore l'éclat en renforçant les notes suivantes à l'aide du *stentato*'. I have drawn the piano reduction from both the fac. of a MS of the full score (New York: Edwin F. Kalmus, n.d.) and the piano/vocal score with English and Italian texts (New York: G. Schirmer, 1898 and 1926), finale to Act II. I have added an accent sign, which Faure omits, for the Aᵇ on the syllable '*di*'.

ascending eighth-note octave leap on the second *blanche* with 'une légère anticipation' of the high note—an alteration which would cause rhythmic displacement with the even eighth notes in the solo viola's accompaniment.[74] Although Faure mentions only anticipation in connection with tempo rubato, his *stentato* seems to involve both delay and anticipation if, indeed, the duration of the measure is not altered. His book, like García's later work of 1894, is one of the few examples of singing treatises from the late nineteenth century which still speak of rubato in the earlier sense.

German singers are also mentioned occasionally in connection with this sort of rhythmic freedom. García includes Henriette Sontag along with Cinti-Damoreau in his list of great singers of the *canto di agilità* style. In 1824 she sang in the first performances of Beethoven's Ninth Symphony and *Missa solemnis*. A contemporary critic was probably describing her intentional rubato when he wrote: 'it has been observed that in a division, say, of four groups of quadruplets, she would execute the first in exact time, the second and third would increase in rapidity so much that in the fourth she was compelled to decrease the speed perceptibly, in order to give the band the means of recovering the time she had gained'.[75] She also sang the title role in the first performance of Carl Maria von Weber's *Euryanthe* in 1823. In a letter the following year, Weber discusses some performance aspects of the opera and mentions a sort of natural vocal rubato:

The most difficult task of all, generally speaking, will always be to achieve a rhythmical relationship between the voice and the orchestra, so that they are perfectly blended and the voice is set off and carried along by the orchestra, which also enhances the emotional expressiveness of the vocal line—for the natures of the two are radically opposed to each other. Singing involves a certain fluctuation of the beat due to breathing and the articulation of words, which could perhaps be compared to the beating of waves on the shore. Instruments (especially stringed instruments) divide time into sharp segments like the beats of a pendulum. Truth of expression is only achieved when these contradictory characteristics are successfully blended.[76]

Friedrich Wieck, the father of Clara Schumann, refers to a singer's fluctuating rhythm in his *Klavier und Gesang* of 1853:

[74] *La Voix et le chant*, 200. See the full score (Paris: Schlesinger, 1836), copy in UCLA ML, Act I, 84–5; the melody, which is in the key of B flat here, is quoted by Faure in A major. In the full score the voice part is marked *très doux à demi voix*; the viola plays the accompaniment alone *avec délicatesse*.

[75] Quoted by George T. Ferris in *Great Singers* (New York: D. Appleton & Co., 1882), 202.

[76] Printed in the *Berliner allgemeine musikalische Zeitung*, 4/28 (11 July 1827), 218; and, with slight changes, in *AMZ* (from Leipzig), 50/8 (Feb. 1848), 126–7. Part of the trans. by Martin Cooper from *Carl Maria von Weber: Writings on Music*, ed. John Warrack (CUP, 1981), 305.

As for singing in time, particularly in German vocal music, . . . you might be a bit more indulgent. Of course, if it were not for breathing and the time-consuming consonants and many other matters! The greatest singers have their problems with German vocal music. Just reflect for a moment on the liberties that Lind [Swedish soprano who studied with the younger Manuel García] . . . took and had to take in order to do justice to the beauty of her voice.[77]

Another problem for the singer of German works may have been the nature of the orchestral score. When Weber speaks of the orchestra enhancing the expressive quality of the vocal part, he identifies an essential distinction between the German and the Italian–French styles. In the orchestral music for a German opera, as in the piano part for a Lied, the musical texture becomes, as the century progresses, increasingly more melodic and independently expressive. This, of course, attracts considerable attention away from the solo voice. The orchestral part of an Italian or French opera (see Exx. 3.16 and 3.18 by Bellini and Meyerbeer, for example), tends to confine itself exclusively to a strong statement of harmony and rhythm, against which every nuance of the voice can be clearly perceived. The orchestra here is an inconspicuous accompaniment for the singer, whereas in German works it provides an expressive and melodic counterpoint to the voice.

Occasionally German sources even use the expression *tempo rubato*. The pianist and conductor Ignaz Moscheles mentions hearing a singer named Rachel in *Les Horaces* in 1841: 'as she proceeds she seems to gather inspiration, the measured "Tempo" becomes a "Vivace", then a "Presto", and then a "Tempo rubato".'[78] The baritone Julius Stockhausen, who studied with García and who was later closely associated with Lieder singing and with Brahms, is described in 1911 by the conductor and composer Bernhard Scholz in the following anecdote regarding events in 1859:

It was a pleasure for me to accompany him with the orchestra or at the keyboard. At first I tried to follow every small inflection of his performance; then he requested that I remain peacefully and strictly in time even when he allowed himself small deviations here and there, for which he would later compensate. He moves himself with complete freedom, but on a firm rhythmic basis. . . . Through him the character of the 'Tempo rubato' first became completely clear to me: freedom of phrasing on a steady rhythmic foundation.[79]

---

[77] Trans. by Henry Pleasants from Friedrich Wieck's *Piano and Song (Didactic and Polemical)* in *Monographs in Musicology*, ix (Stuyvesant, New York: Pendragon Press, 1988), 155.

[78] *Recent Music and Musicians as Described in the Diaries and Correspondence of Ignatz Moscheles*, ed. by his wife (New York: Henry Holt, 1873; repr. Da Capo, 1970), 283–4.

[79] Bernhard Scholz, *Verklungene Weisen* (Mainz, 1911), 127. The passage is also included in *Julius Stockhausen: Der Sänger des deutschen Liedes*, ed. Julia Wirth geb. Stockhausen (Frankfurt am Main: Verlag Englert und Schlosser, 1927), 189. See also Edward F. Kravitt, 'Tempo as an Expressive Element in the Late Romantic Lied', *MQ* 59 (1973), 498.

For more than a century and a half, then—from Tosi to Fauré—vocal tempo rubato displayed a fairly consistent form in late baroque, classic, and earlier romantic music. As late as 1874, Lemaitre states in his French translation of Tosi's book:

The Italian singers have a manner of delaying the singing, or of losing the precision of the time at will, while the orchestra continues its prescribed movement, which has a great effect, when it is done with taste and when the singer knows how to regain his balance. One cannot give an example of this effect in singing; it is necessary to observe it in performance. This method may be called *vacillare*, which means to vacillate, hesitate, falter, waver, be in suspense.[80]

Fortunately for us, other writers did attempt to illustrate tempo rubato—each, however, from his own point of view. Examples from the Baroque and Classic periods depict the results of displacement between the voice and its accompaniment. North provided a great variety of examples of the sort that might have been extracted from real compositions (Plate I and Exx. 3.1–3.5). During the second half of the eighteenth century Quantz, followed then by Agricola, Marpurg, Kirnberger, and Schulz, portrayed the component aspects of the device in a series of three brief examples showing each of four melody notes occurring with, before, or after the corresponding bass note (Ex. 3.8). Agricola, who was, as far as I can tell, the only one of these writers ever seriously involved with singing as a performer or teacher, was also the only one to emphasize the melodic alteration of note values. North and the other German writers seem to perceive tempo rubato from the point of view of a listener rather than that of a performer. The listener, I believe, tends to hear rubato first of all—if he analyses the technical aspects of the device at all—as displacement between voice and accompaniment.

Those involved with singing, on the other hand, tend toward a different point of view. Tosi in 1723, Bayley in 1771, Corri in 1810, and later Nathan include no examples at all, but speak of rubato in terms of melodic alterations. Although Hiller gave Quantz-like examples in 1774, he changed in 1780 to examples with text drawn from actual compositions. Lasser in 1798 included brief progressions like Quantz, but added text. In 1824 Bacon described how to use rubato to emphasize the accented syllable of important words in familiar pieces. García, Cinti-Damoreau, and Fauré, finally, completed the break with the Quantz tradition and presented examples, almost

---

[80] Théophile Lemaitre (Professeur de chant, Paris), *L'Art du chant traduit de l'Italien* (Paris, 1874), 126; quoted in Bruck, *Wandlungen des Begriffes Tempo rubato*, 9–10; trans. Reginald Gatty, 'Tempo Rubato', *The Musical Times*, 53 (1912), 161. I have substituted *regain* for *preserve* as a trans. of *reprendre*, and Gatty gives the author's name as *Lemaire*.

exclusively, that included only the voice part. This is the natural way in which the performer would conceive of tempo rubato: as the stealing of time from one note to give to another. It was a singer, after all, who invented the analogy of rhythmic robbery in the first place. It was not a matter of the singer ignoring the accompaniment or its rhythmic effect, but of taking it for granted. The typical accompaniment in a French or Italian opera of the first half of the nineteenth century had just as precise and insistent a beat in its own way as the continuo in a late baroque aria.

The history we have just traced in vocal music represents the mainstream in the development of the earlier type of rubato, for it occurred earlier in this medium and persisted there longer. Although the expression *tempo rubato* could on rare occasions be used in vocal music for other meanings, it never referred, as far as I know, to the modification of tempo. It is obvious from Corri's work of 1810 in Plate III that he treats 'Quickening or Retarding of Time' as a topic separate from tempo rubato. Similarly, García, on Plate IV, has sections entitled 'On Rallentando' and 'On Accelerando' before he deals with tempo rubato. In Plate V, in fact, he makes the distinction very clear: '*Accelerando* and *rallentando* movements require the voice and accompaniment to proceed in concert; whereas, *tempo rubato* allows liberty to the voice only.' This precise meaning for tempo rubato continued in vocal music at least through the first half of the nineteenth century and, in some sources, such as the books of Faure and García in 1886 and 1894, until the end of the century.

Before 1850 the expression *tempo rubato* almost never appears in a vocal score, since the device was exclusively a part of improvised performance practice. It generally involved the delaying, anticipating, or prolonging of notes against a conspicuously rhythmic accompaniment. It was applied when certain musical situations coincided with an intense emotion in the text. It was meant to sound—even though the listener had never seen a score, and even if he had not previously heard an unrobbed version of the passage for comparison—as though some notated version had been altered, and as though the alterations were caused not by intellectual design but by the force of passion. It will be helpful to keep this relatively clear mainstream development in vocal music in mind as we turn to tempo rubato for the violin and keyboard instruments.

4 THE EARLIER RUBATO IN VIOLIN MUSIC

EARLY in the Baroque period the solo violin, a relatively new instrument at that time, became the instrumental equivalent of the solo voice. In canzonas, sonatas, and eventually in concertos, the violin imitated the styles and forms of vocal music. We have already noted the practice of improvised embellishment in Corelli's violin sonatas. Tempo rubato was therefore a part of this practice, in fact if not in name, as it had been in vocal music long before North and Tosi. The expression does not actually appear in violin sources, however, until the early Classic period, slightly after 1750.[1]

The Classic Period

In 1754 the violinist Giuseppe Tartini includes 'appoggiature, trilli, modi di tempo rubbato [*sic*], e protratto' in a list of techniques applied by a good singer.[2] Leopold Mozart, in his *Gründliche Violinschule* of 1756, speaks of the problems of accompanying a virtuoso performer (presumably either a singer or a violinist):

when a true virtuoso who is worthy of the title is to be accompanied, then one must not allow oneself to be beguiled by the postponing or anticipating of the notes . . . into hesitating or hurrying, but must continue to play throughout in the same manner; else the effect which the performer desired to build up would be demolished by the accompaniment.

[1] Referring to an even earlier date, Burney makes a curious comment about Francesco Geminiani, a virtuoso violinist and pupil of Corelli. In 1711 Geminiani became leader of the opera orchestra in Naples, but 'he was soon discovered to be so wild and unsteady a timist, that instead of regulating and conducting the band, he threw it into confusion, as none of the performers were able to follow him in his *tempo rubato*, and other unexpected accelerations and relaxations of measure.' If this story is true, Geminiani perhaps carried over into his conducting some of the devices he employed in his violin playing. See Charles Burney, *A General History of Music*, 2nd edn. (London, 1789), iv, 641; and *New Grove*, vii, 224.

[2] *Trattato di musica secondo la vera scienza dell'armonia* (Padua, 1754), repr. in MMMLF, 2nd ser., viii (1966), 149 (in ch. 5). See Fredric Johnson, 'Tartini's *Trattato* . . . : An Annotated Translation with Commentary', Ph.D. diss. (Indiana Univ., 1985; UM 86–17,820), 380; and Rodolfo Celletti, *A History of Bel Canto*, trans. Frederick Fuller (OUP, 1991), 115.

He adds in a footnote:

To a sound virtuoso he certainly must not yield, for he would then spoil his tempo rubato. What this 'stolen tempo' is, is more easily shown than described [he includes no musical examples].[3]

One of the earliest violinists to be associated with the performance of tempo rubato was Franz Benda, who probably learned the technique while studying with Johann Georg Pisendel and Johann Gottlieb Graun. Graun had been a pupil of Tartini and Pisendel; the latter had studied with Torelli in violin and Pistocchi in singing. Since Tosi considered Pistocchi the greatest singer of all, it is possible that rubato was transmitted to the violinists via Pisendel, Graun, Tartini, and Benda.[4] Benda performed until the 1760s and from 1733 until his death in 1786 was in the service of Frederick the Great. Reichardt lists in 1774 the qualities of his performance in an Adagio, which include 'the special vigour with which a note is sometimes stressed' and 'some extraordinarily meaningful deviations in the time values of the notes'.[5]

In an article in the *Allgemeine musikalische Zeitung* in 1808, Koch, himself a violinist as well as a theorist, looks back at previous uses of the term *rubato*:

Formerly, and especially in the former Berlin school they associated with the expression *tempo rubato* ... that manner of performance of this or that cantabile passage of a solo part, in which the player intentionally digressed from the assumed movement of the tempo and from the usual distribution of note values, and executed the melodic line as if without any fixed division of time, while the accompaniment played on absolutely in tempo. Among others, Franz Benda often made use of this manner of performance as a special means of expression in the Adagio movements of his concertos and sonatas.[6]

---

[3] *Gründliche Violinschule*, 3rd edn. (Augsburg, 1787; repr. Br. & H, 1983), 267; trans. of 1st and 3rd edns. (1756 and 1787) by Editha Knocker in *A Treatise on the Fundamental Principles of Violin Playing*, 2nd edn. (OUP, 1951 and 1959), 224. According to Knocker these passages appear also in the 1st edn. of 1756.

[4] Werner Freytag states parenthetically in *MGG* v, 709, concerning Graun: 'sein Rubato, das im Prinzip dem Chopins glich, war berühmt.' Although I have been unable to verify such a direct connection between Graun and rubato, an article in Hiller's *Wöchentliche Nachrichten und Anmerkungen die Musik betreffend*, i (1766/7; repr. Olms, 1970), 191 (for 16 Dec. 1766) relates that Benda admired Graun's adagio playing so much that he studied a few examples with him; see also 75 (2 Sept. 1766) concerning Graun. See Paul Nettl, *Forgotten Musicians* (New York: Philosophical Library, 1951), 223; and Carl Mennicke, *Hasse und die Brüder Graun als Symphoniker* (Br. & H, 1906), 349.

[5] Johann Friedrich Reichardt, *Briefe eines aufmerksamen Reisenden die Musik betreffend*, i (Frankfurt and Leipzig, 1774), copy on microcard from SML, 162.

[6] Heinrich Christoph Koch, 'Über den technischen Ausdruck: Tempo rubato', *AMZ* 10/33 (11 May 1808), 518; trans. mostly from Sandra P. Rosenblum, *Performance Practices in Classic Piano Music* (IUP, 1988), 376.

Koch goes on to say that there are still some performers who play in a similar manner, but with so much less deviation from the tempo that one can really consider the device obsolete. He feels that its disappearance is a good thing, partly because composers are now writing out the complete melody of the Adagio movements themselves, and partly because the device too easily sounds ridiculous.

Thirty-three of Benda's sonatas for violin and continuo, probably from the 1760s, exist in a manuscript collection copied presumably by his students.[7] For most of the movements, Benda's original score appears on the outer two staves (see Plates VI–VIII), with an embellished version added on the middle staff. The words *tempo rubato* occur twelve times below the embellished versions, the expression *tempo robato* twice, and the single word *rubato* once. Twelve different movements are involved, ranging in tempo from adagio to presto. The places marked *rubato* employ various techniques, which serve to emphasize specific musical situations. In the process, various delays and anticipations in the melody cause different types of displacement with the accompaniment. It will be worthwhile to investigate Benda's rubatos in some detail, for they involve some features we have already noted in vocal music as well as some other techniques that are different.

In two cases, tempo rubato is indicated for an embellished version that is identical to the original music or even simpler (see Plate VII, measure 5 in the fourth line). This is perhaps an invitation for the performer to improvise rubato. In a few other places rubato occurs in the more or less conventional manner we have seen in vocal music. In measure 13 of line 4 on Plate VII the B is delayed, in measures 14 and 16 the A$^\sharp$ and G are anticipated, and in measure 15 the A is delayed by grace-notes. A very slight adjustment of note values occurs in Ex. 4.1, with the rhythm of the triplet notes clearly articulated by staccato bow strokes. Neither Benda nor the copyist of the manuscript indicates triplets or other unusual metrical groups by number, as we do in modern notation.

The triplet on the middle staff in the second measure of Ex. 4.1, however, also creates a counter-rhythm with the bass's two eighth notes and causes an alternation between the notes of the melody and the bass. We have already seen three beats against two as García

---

[7] Berlin, Staatsbibliothek Preussischer Kulturbesitz, Musikabteilung, Mus. MS 1315/15. There are 34 numbered sonatas, but No. XXI is missing. Exx. 4.1 and 4.2 come from pp. 116 and 8. A fac. of Sonata I is in Hans-Peter Schmitz, *Die Kunst der Verzierung im 18. Jahrhundert* (Bärenreiter, 1955), rubatos on 136 and 139; see the modern edn. in *Anthology of Music (Das Musikwerk)*, xii: *Improvisation in Nine Centuries of Western Music*, ed. Ernest T. Ferand (Cologne: Arno Volk Verlag Hans Gerig KG, 1961), 137 and 142.

VI*a*

VI*b*

PLATE VI. Franz Benda, Violin Sonata IV, 2nd movt., lines 1–4 and 13, Berlin, Staatsbibliothek Preussischer Kulturbesitz, Musikabteilung, Mus. MS 1315/15, from pp. 22–3

included it later in his tempo rubato in Ex. 3.10, and even four against three as in Ex. 3.14 by García (at the beginning of the second half of the first measure) and Ex. 3.18*a* by Cinti-Damoreau. Nine of the places marked *rubato* in the Benda sonatas involve groups with an unusual number of notes, and six of them produce alternation between bass and melody notes.

PLATE VII. Franz Benda, Violin Sonata VIII, 3rd movt., Berlin, Staatsbibliothek Preussischer Kulturbesitz, Musikabteilung, Mus. Ms 1315/15, p. 48

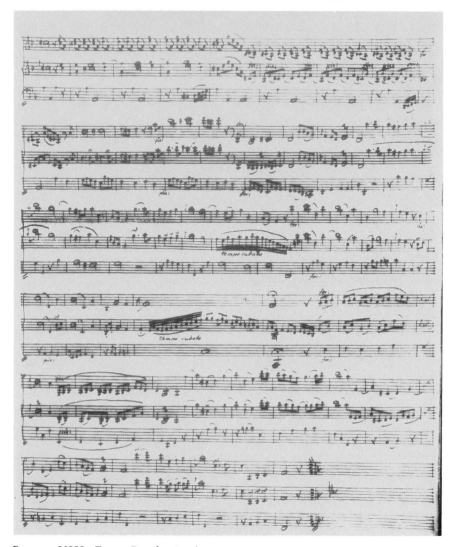

PLATE VIII. Franz Benda, Violin Sonata XXXI, 1st movt. (Allegro), concluding page, Berlin, Staatsbibliothek Preussischer Kulturbesitz, Musikabteilung, Mus. MS 1315/15, p. 177

In Ex. 4.2 four notes substitute for three eighth notes in 6/8 metre,
causing delays in some melody notes, but no alternation with bass
notes. The cadenza-like passage marked *tempo rubato* on the fourth
line of Plate VIII contains a group of eleven thirty-second notes beamed
together. Since the bass simply sustains a single note for the entire

Ex. 4.1 Benda, Violin Sonata XX, 3rd movt., c.1760

Ex. 4.2 Benda, Violin Sonata I, last movt., c.1760

measure, the irregularity of the group is again not emphasized by displacement with bass notes. On the second line of Plate VII, however, eight notes in two bars of 3/4 (the first tied over from the preceding measure) occur against six quarter notes in the accompaniment. The second, third, and fourth notes of each four-note group therefore alternate with the second and third notes in the bass. In the process the original F♯ is slightly anticipated, the following F♮ delayed.

A more complex group occurs in Plate VIb, where the rubato is graphically portrayed by beaming and slurring fifteen notes together over two barlines. Although the notes are distributed over the three measures as 4 + 7 + 4, probably fairly equal value was intended for each note, but resulting in an impression of metrical freedom. Taken literally, the seven notes of the second measure of the rubato would coincide with four in the accompaniment. Perhaps it would be more accurate to speak of fifteen notes in the melody against eight in the bass, for in the corresponding place on the previous page (Plate VIa), which is also marked *tempo rubato*, the same notes are distributed as 4 + 6 + 5. Particularly visible on Plate VI is the substantial melodic retardation, with both the descending passage and the following ascent shifted in each case into the next measure.

Another complicated group constitutes the first *tempo rubato* on Plate VIII. In the third line, sixteen notes are beamed and slurred together, and to further depict their continuity the barline has either been omitted or coincides with a note stem. This group is to fit against the equivalent of thirteen eighth notes of the regular metre (during three half notes in the bass plus another eighth note). The original descent from G to B$^\flat$ has been filled in, in the manner of Galliard's *drag* (Ex. 3.7, but without the alteration of note values), and the descending tetrachord from B$^\flat$ to F considerably delayed.

The examples of rubato in these sonatas by Benda act to intensify a number of different musical situations. They emphasize a final cadence, an important dominant chord, the return of a theme, or a chromatic passage. Two cases, first of all, involve the final cadence of a sonata. In one, the rubato is not written out,[8] and in Ex. 4.2 the sudden change from three- to four-note groups gives a sense of conclusion in a driving *presto scherzando* rhythm where a rallentando or pause on the penultimate chord (see Chapter 1) would probably not be appropriate.

Several rubatos serve to intensify the chords that precede, delay, and hence strengthen the appearance of an important dominant chord. Benda's movements are constructed in the early types of sonata form, in simple binary or ternary structures, or in simple rondo forms. The dominant chord plays an important and conspicuous role in these structures, and it is often emphasized by prolonging not only this chord but also the subdominant approach to it. In one case the measure containing tempo rubato leads to the V chord that precedes the return of the home key of A major in the second section of this binary sonata form.[9] In another, the rubato on a IV chord immediately precedes the secondary dominant that leads to the V chord near the end of a modulation to a new key.[10] The rubato in Plate VI emphasizes the secondary dominant of the V chord, first (in the third line) as part of the confirmation of the second tonal level of B flat major in this sonata movement, and later (in Plate VIb) in the corresponding place in E flat in the recapitulation.

Other examples use rubato to mark the return of a melody. The first rubato on Plate VIII marks the return of the home key of F major

---

[8] MS 1315/15, p. 99 in Sonata XVII, 3rd movt.: Presto.
[9] Ibid. 5, Sonata I, 1st movt.: Adagio un poco andante.
[10] Ibid. 62, Sonata XI, 2nd movt.: Adagio.

and the recapitulation immediately thereafter of material from the opening theme. In other sonata movements the recapitulated first theme is emphasized by a rubato at a half cadence[11] or at the climax of a phrase.[12] In one case, the second section commences with two measures parallel to the first two measures of the movement, but with rubato added.[13]

Occasionally rubato also varies the immediate repetition of a brief passage. Ex. 4.1 is the conclusion of the repetition of the opening eight bars of a brief binary section in the middle of a movement marked *Tempo di Minuetto*. The embellished version of the initial statement differs from that in Ex. 4.1 only very slightly: the tie is missing, staccato dots are only on the first two notes of the second measure, and each of the two triplets is slurred. The second rubato on Plate VIII marks the end of a repeated and recapitulated four-bar phrase and contributes to a dramatic harmonic pause that leads up to a fermata. In the opening statement of the theme, the four measures are simply repeated without interruption, since they lead directly into a modulation to the second tonal level. Still other rubato examples in the sonatas seem mainly to emphasize chromaticism. On Plate VII the three rubatos intensify such passages in the ABACA rondo which forms the opening section of a larger ternary structure. The first rubato occurs in the middle of the B section in C major and acts polymetrically to emphasize a secondary dominant of the V chord. The second rubato, although not written out, begins the C section with a modulation to E minor; the third underlines the high point of this section, which again involves a secondary dominant to V.

The diversity of the rubato passages in these sonatas enables one to gain a fairly comprehensive sense of the meaning of rubato to Benda and his students. The musical situations, as with García, include cadences and repetitions. The most striking rubato technique is the inclusion of unusual numbers of notes, especially those that cause melody notes to alternate with notes of the accompaniment. In Ex. 4.2, of course, one can speak of the embellished version displacing or shifting the last five notes in the first measure of the original melody in relation to the bass and thus producing an exhilarating broadening at the cadence. On the other hand, the four-note groups in the second line of Plate VII create quite a different effect, for the displacement gains a new dimension when the bass provides enough notes for alternation with the melody. This tends, I think, to sound more like the descriptions of metrical digression and vague rhythm given by

---

[11] Ibid. 89, Sonata XVI, 1st movt.: Allegro non molto.
[12] Ibid. 113, Sonata XX, 1st movt.: Allegretto.
[13] Ibid. 146, Sonata XXVI, 1st movt.: Allegro.

the listeners and theorists. This was probably the effect that Quantz attempted to depict in his simplified schemes in Ex. 3.8. The use of unusual groups probably comes one step closer to reality, however, and even this notation should no doubt be rendered with a certain amount of freedom. Quantz and Benda were both together at the court of Frederick the Great, along with C. P. E. Bach, who included similar unusual groups of notes in his concept of tempo rubato for the keyboard.

Turning briefly now to Italy, we find one performer and one treatise involved with the robbing of time. In 1785 a reporter describes a performance by the virtuoso violinist Regina Strinasacchi (1764–1839), who included 'some tones which were so very much emphasized by a type of *Tempo rubato*'. Her expressive playing was praised by both Leopold and W. A. Mozart. In 1784 the latter performed with her in Vienna the B flat major Sonata, K. 454, which he had composed for the occasion.[14]

Francesco Galeazzi was a theorist, composer, violinist, and violin teacher. The first volume of his *Elementi teorico-pratici di musica* (Rome, 1791) contains 'An Essay on the Art of Playing the Violin'. His second rule for making diminutions, which he illustrates with the brief melodies in Ex. 4.3, reads as follows:

Ex. 4.3 *Sincopare* (Galeazzi, 1791)

It is permissible when making diminutions to steal [*rubbare*] a little of the value of a note and to transfer it to another note; it is a laudable artifice (if it is not abused) even to have a certain flexibility of tempo, provided that at the end all is adjusted and equalized according to the values. This is called by performers *sincopare* or to play *a contrattempo*, and it is one of the finest resources of *espressione* (if always used with moderation).[15]

[14] *Magazin der Musik*, ed. Carl Friedrich Cramer (repr. Olms, 1971), ii, 353, in a report from Ludwigslust (near Hamburg) dated 4 Jan. 1785: 'einige Töne ... , welche bey einer Art von *Tempo rubato* so sehr hervorgehoben werden'. This same year she married the cellist Johann Conrad Schlick and moved to Gotha; see *MGG* xi, 1822, and *New Grove*, xviii, 274. Around 1815 she predicted (*MGG* x, 933) that Pasta would become the foremost female singer in Europe.

[15] Galeazzi, *Elementi*, i, 200–1; trans. partly from Angelo Frascarelli, '*Elementi* ...: An Annotated English Translation and Study of Volume I', D.M.A. diss. (Eastman School of Music, Univ. of Rochester, 1968), 364. See Joan E. Smiles, 'Directions for Improvised Ornamentation in Italian Method Books of the Late Eighteenth Century', *JAMS* 31 (1978), 508.

He does not mention the accompaniment at all and his examples significantly show only the violin melody. He describes the opening note as being first prolonged and then shortened, whereas the German writers, as in Ex. 3.8, would have referred instead to the second note being delayed or anticipated. According to Galeazzi, repeated passages should be varied, each time with more complex diminutions. For the violinist, such repetition might take place in a movement from a sonata or concerto—in a binary form with repeat signs, a ternary da capo form, the refrain of a rondo, or the recapitulation in a sonata form.

The Romantic Period

A number of articles in Leipzig's *Allgemeine musikalische Zeitung* during the first quarter of the nineteenth century, as we have seen, condemned tempo rubato. Haeser in 1822 and two anonymous authors in 1802 and 1813 lamented the unfortunate increase in the use and abuse of the device in vocal music. Koch condemned it with equal vehemence in violin music, but felt, perhaps hopefully, that it was by 1808 becoming obsolete. In spite of excesses and the adverse reactions they provoked, tempo rubato continued to live on in violin as well as vocal music, and in Germany as well as France and Italy. The great Italian virtuoso violinist Nicolò Paganini (1782–1840) excelled in the use of the tempo rubato by phrase (see Plate V). He performed throughout Europe from the early years of the century until 1834. Farrenc mentioned in 1861 that Paganini employed tempo rubato with much charm, and Kleczyński wrote in 1879 that Paganini, when playing with the orchestra, 'recommended that the instrumentalists should observe the time, whilst he himself departed from it, and then again returned to it'.[16] Also during the 1830s two treatises appeared which are especially significant for tempo rubato: one published by Pierre Baillot in Paris, the other by Louis Spohr in Vienna.

The Italian violinist and composer Giovanni Battista Viotti (1755–1824) was the inspiration for the nineteenth-century French school of violin playing. He influenced his student Pierre Rode, as well as disciples such as Baillot and Rodolphe Kreutzer (to whom Beethoven dedicated his Violin Sonata Op. 47). In 1803 Baillot, Rode, and

[16] Aristide Farrenc, *Le Trésor des pianistes*, i (Paris, 1861; repr. Da Capo, 1977), 4; and Jan Kleczyński, *O wykonywaniu dziel Szopena* (Warsaw, 1879), trans. Alfred Whittingham in *How to Play Chopin: The Works of Frederic Chopin, their Proper Interpretation*, 6th edn. (London: William Reeves, [1913]), 57.

Kreutzer jointly authored the *Méthode de violon*, which was adopted by the Paris Conservatoire. In this work they speak of rhythmic flexibility:

> Expression sometimes permits a slight alteration in the time, but either this alteration is gradual and almost imperceptible, or the time is only disguised, that is to say that in pretending to lose a moment, one finds oneself soon afterwards as exact in following it as before.[17]

In 1804 Baillot also co-authored the *Méthode de violoncelle*, which contains this advice for playing the bass part in chamber music:

> He who plays the principal part sometimes may, upon inspiration, abandon himself to his inspiration and assume great freedom in his playing, but the bass accompaniment, whose function is to conduct the progression of the harmony, must permit no ambiguity, irregularity or peculiar expression which would forsake the musical character and render the music diffuse by taking away the fundamental part. The bass in its grave and simple progression always ought to be clearly articulated and not only ought to be steady but, also, ought to make itself felt. The accompanist should be imperturbable during these slight alterations of time..., and, in the apparent disorder called *tempo disturbato* by the Italians, the bass must be the regulator in the ensemble piece in the same way that the left hand ought to maintain the steadiness in the playing of a piano sonata. Therefore, the first concern of the accompanist ought to be to preserve and mark steadiness.[18]

Baillot's main contribution to tempo rubato, however, occurs in his extensive work *L'Art du violon: nouvelle méthode* (1834). He describes a species of syncopation called *temps dérobé* or *troublé*, *tempo rubato* or *disturbato* (Plate IX):

> It tends to express trouble and agitation and few composers have notated or indicated it... The performer... must only make use of it in spite of himself, as it were, when, carried away by the expression, it apparently forces him to lose all sense of pulse and to be delivered by this means from the trouble that besets him.... He must preserve a sort of steadiness that will keep him within the limits of the harmony of the passage and make him return at the right moment to the exact pulse of the beat...
>
> This disorder... will become an *artistic effect* if it results from effort and inspiration and if the artist can use it without being forced to think of the means he is employing.
>
> Up to a certain point this device can be notated, but like all impassioned

---

[17] Repr. Minkoff, 1974, in *Méthodes instrumentales les plus anciennes de Conservatoire de Paris*, iii, 162; see also the trans. in Rosenblum, *Performance Practices in Classic Piano Music*, 383.

[18] Pierre Baillot, with Levasseur, Catel, and Baudiot, *Méthode de violoncelle* (Paris, [1804]; repr. Minkoff, 1974), 140; trans. mostly from Blake, 'Tempo Rubato in the Eighteenth Century', 100.

AUTRE ESPÈCE DE SYNCOPE
appelée temps dérobé.

Il est une manière d'altérer ou de rompre la mesure qui tient de la syncope et que l'on appelle *tempo rubato* ou *disturbato, temps dérobé* ou *troublé*. Ce *temps dérobé* est d'un grand effet, mais il deviendrait par sa nature, fatigant et insupportable s'il était souvent employé. Il tend à exprimer le trouble et l'agitation et peu de compositeurs l'ont noté ou indiqué; le caractère du passage suffit en général pour pousser l'exécutant à l'improviser d'après l'inspiration du moment. Il ne doit, pour ainsi dire, en faire usage que malgré lui, lorsqu'entraîné par l'expression, elle l'oblige à perdre, en apparence, toute mesure et à se délivrer ainsi du trouble qui l'obsède. Nous disons qu'il ne perd la mesure qu'en apparence, c'est-à-dire, qu'il doit conserver une sorte d'aplomb qui le retienne dans les limites de l'harmonie du passage et qui le fasse rentrer à propos dans la mesure exacte des temps. C'est ici le cas d'appliquer cette observation:

Souvent un beau désordre est un effet de l'art.

Ce désordre sera donc de nature à plaire, même à être trouvé beau; il deviendra *un effet de l'art* s'il est le résultat du travail et de l'inspiration, et si l'artiste l'emploie sans être obligé de penser aux moyens dont il se sert.

On peut noter, jusqu'à un certain point cet artifice, mais, comme tous les accens passionnés, il perdra beaucoup de son effet à être exécuté de sang froid.

Nous ne donnons ici des exemples de ce genre d'accent que pour éclairer sur son usage et pour empêcher ainsi l'abus que l'on pourrait en faire.

TEMPS DÉROBÉ.
Passage tel que l'auteur l'a noté.
(19.º Concerto de Viotti. — Ed. Pleyel.)

Apperçu de la manière de rendre ce passage.

PLATE IX. Pierre Baillot, *L'Art du violon* (Paris, 1834), from pp. 136–7

accents, it will lose much of its effect if it is performed according to the book [cold-bloodedly].

We give examples . . . here simply in order to shed light on its use and to prevent any misuse that might be made of it.[19]

In the course of this discussion Baillot makes the observation, centred conspicuously on the page, that 'often a beautiful disorder is an artistic effect'.

He illustrates this with two rather enigmatic excerpts from violin concertos by Viotti. Both consist of a long series of syncopations, which are sustained in the first as suspensions and repeated in the second as appoggiaturas. It is not clear whether he means to imply that Viotti is thereby notating tempo rubato, as he says, 'up to a certain point'. Ex. 4.4a shows the opening three measures or so of his approximation of the manner of rendering the first passage on Plate IX (but omitting his dynamic signs). In Ex. 4.4b the top staff represents the passage as the composer notated it; on the lower staff I have added the orchestral accompaniment played by strings. In Ex. 4.4c I have written an unsyncopated version of Viotti's notated melody. Since this Concerto No. 19 was also published as a piano concerto, I have included the corresponding music in (d) for comparison. In both versions this melody appears first in B flat minor, then later in G minor. In the piano concerto the passage is played both times by the piano alone.[20]

During the opening two measures of this excerpt, Baillot suggests no change in note values (compare the first two lines in Ex. 4.4). He does, however, indicate dynamic effects (see Plate IX), which include the first of his three methods for performing a syncopated note: 'swelling the note and accelerating the movement of the bow until the end of the note, but lightly'.[21] The third measure, finally, does include alterations in the note values. In comparison to the top line of (b), (a) contains a delay of the E♭, the D, and the A; in comparison to (c), an even greater delay occurs in all four notes. Thus (b) and (a) represent successive stages in the robbing of (c). The fourth measure of (a) illustrates another type of change which he frequently employs: the substitution of a rest during the syncopated portion of a note in

[19] Trans. from Robin Stowell, *Violin Technique and Performance Practice in the Late Eighteenth and Early Nineteenth Centuries* (CUP, 1985), 274. See also Pierre Marie François de Sales Baillot, *The Art of the Violin*, ed. and trans. Louise Goldberg (Evanston, Ill.: Northwestern University Press, 1991), 237–9.

[20] Viotti, *19° Concerto in sol min. per violino e orchestra*, rev. Remo Giazotto (Ricordi, 1964), 39; and *Concerto in sol min. per pianoforte e orchestra*, rev. Remo Giazotto (Ricordi, 1960), 39.

[21] *L'Art du violon*, 135: 'en enflant la note et accélérant le mouvement de l'archet jusqu'à la fin de cette note, mais légèrement'.

Ex. 4.4 Rubato in Viotti, Concerto No. 19 (Baillot, 1834)

the manner of Couperin's *suspension* in Ex. 2.9 or Marpurg's *Tonverbeissung* in Ex. 5.2*d*. The rest acts as a silence of articulation, throwing accentuation onto the following note. It should also be noted that the orchestral bass part in (*b*), which occurs in an inner voice in the piano version in (*d*), provides a clear and steady beat during the course of the syncopation. Baillot did mention the role of the accompaniment during tempo rubato in his cello method, as we have seen, but in *L'Art du violon* gives only the melodies involved.

Viotti's Violin Concerto No. 18, from which Baillot's second example comes, was first performed by Rode in 1792.[22] Shortly after the turn of the century Spohr heard Rode play in Germany and was enormously influenced by his performance.[23] This influence can be

[22] *RR in Classical Music*, v: *Giovanni Battista Viotti, Four Violin Concertos*, ii, ed. Chappell White (1976), 48.
[23] *Louis Spohr's Autobiography, translated from the German* (London: Reeves & Turner, 1878), i, 61–2, 109–10, and 161. Spohr seems not to have been influenced by Strinasacchi (see i, 88), who was both a soloist and a member of the orchestra at the ducal court in Gotha when

seen in his *Violinschule*, which was published in 1832 in Vienna and appeared a year later in English translation. In the third part, 'On Delivery or Style in General', he distinguishes between a *correct* style, which includes 'strict observance of time', and a *fine* style, which might involve 'the increasing of time in furious, impetuous, and passionate passages, as well as the retarding of such as have a tender, doleful, or melancholy character'.[24] To illustrate the elements of the *fine* style, he prints the complete solo part from Rode's Seventh Violin Concerto and below it the orchestral part reduced to a single line for the teacher to play on a second violin. He indicates the style of performance 'as near as could be done by notes, signs, and words'. Judging from the modern piano/violin reductions available for comparison, he seems to add mainly dynamic markings, slurs, and signs for articulation or accentuation, as well as writing out some of the ornaments.

The places where he suggests rubato, however, are usually described in prose at the bottom of the page rather than writing out one specific solution in notation. The first mention of rubato occurs on the second page of the Concerto (see Plate X), where he indicates, in an explanation below the score, that the ascending scale in the second half of the twenty-eighth and thirtieth measures should be played so that the first notes are lengthened and the following correspondingly shortened—a manner of delivery which, he says, is called *rubato*. In Ex. 4.5b I have shown the violin part for measures 27 and 28 and also the piano reduction of the orchestral part from a modern edition, for comparison with Spohr's violin accompaniment in Plate X.[25] Passages in even sixteenth notes often contain an inner rhythmic structure in which some notes revolve around a central note and others create motion between two notes. In Ex. 4.5a I have suggested a melodic framework which in measure 28 drives downward toward a sustained E. Rubato at this point may be an attempt to sustain the E as in Ex. 4.5a while playing the ascending scale as in (b). Spohr emphasizes the E with a crescendo towards and a decrescendo away from it—marks

he was Konzertmeister there from 1805 to 1812; by Paganini, whom he met in 1816 (i, 279–81) and heard perform in 1830 (ii, 168); or by Baillot, whose style of playing he observed in 1820 (ii, 129–30).

[24] *Louis Spohr's Grand Violin School*, trans. C. Rudolphus (London, preface dated 1833), copy at BL, 179.

[25] Pierre Rode, *VII Koncert skrzypcowy a-moll*, ed. Stanisław Mikuszewski (Polskie Wydawnictwo Muzyczne, 1966), 5 (mm. 72–3: numbered from the beginning of the piece, whereas Spohr numbered from the beginning of the solo violin section). See also the piano/violin score rev. and fingered by Ferdinand David and ed. W. F. Ambrosio (New York: Carl Fischer, 1905), and another, rev. after the edn. of Ferdinand David by Henry Schradieck (New York: G. Schirmer, 1899 and 1927).

In order to produce the shakes full and brilliant, the half of the value of the preceding note has been taken and added to the shake note. The last four notes of the 19th bar obtain again the whole bow. The shake in bar 23, begins slowly and increases gradually. The division of the bow in bar 25, is similar to the one in bar 14. The second half of the 28th and 30th bar is to be so played, that the first notes obtain a little longer duration than their value warrants, and the loss of time may be regained by a quicker playing of the following note. (This manner of deli_ very is termed rubato.) This increasing of time must be gradual, and harmonize with the decreasing of the power. For the first notes make use of much bow, in order that the latter may be the more delicate.

(W & Co. N? 819.)

PLATE X. Louis Spohr, *Grand Violin School* (London, 1833), 183

Ex. 4.5 Rubato in Rode, 7th Concerto, 1st movt. (Spohr, 1833)

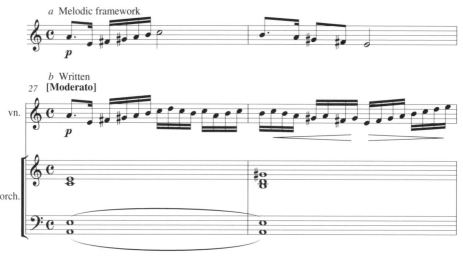

which are not in the modern editions. He states that the change of duration from note to note must be gradual and must synchronize with the decrescendo. He also mentions (see Plate X) that he has increased the duration of the trills in measures 17, 18, and 19 by stealing half of the value from the preceding note. The opening beat of each of these bars was originally written as an eighth note and two sixteenths. García, as we have seen, considered this a type of tempo rubato (see Ex. 3.14).

Two pages later Spohr refers to the excerpt in Ex. 4.6a: 'prolong the 9th note (G) a trifle, and make the loss of time good again by the increased rapidity of the following notes.' I have included the orchestral reduction in Ex. 4.6b, which shows that on the second beat a diminished chord is being sustained over a B pedal, acting to delay the arrival of a B major triad.[26] As one of the two powerful chromatic notes, the G at the beginning of the third group of sixteenth notes actually serves, in a texture suggesting two different voices, like a dotted quarter note that resolves to F♯ six notes later. To strengthen the emotional impact of the G, Spohr has added a trill (at least it is not present in the modern edition), made a crescendo toward the note, and requested tempo rubato. Unlike Ex. 4.5, where the rubato was performed while the accompaniment was stationary, Spohr has created a second violin part which is far more active than the orchestral part in Ex. 4.6b. In this case, displacement between the solo violin and that of the teacher would probably be clearly audible.

---

[26] Rode, *VII Koncert*, ed. Mikuszewski, 6 (m. 103).

Ex. 4.6 Rubato in Rode, 7th Concerto, 1st movt. (Spohr, 1833)

Ex. 4.7 Rubato in Rode, 7th Concerto, 1st movt. (Spohr, 1833)

In the excerpt in Ex. 4.7 Spohr requests the solo violinist to 'prolong the two last quavers [eighth notes] . . . a little, without disturbing the time'.[27] This is the only example, however, in which he does not indicate where to steal the time. Presumably one could commence the trill somewhat earlier or reach the E at the beginning of the next measure a bit late. In any event, the displacement with the second violin's chords would be apparent. As in the previous excerpt, modern editions do not include trills on these two notes.

He also presents several examples of rubato in the last movement of Rode's Concerto. Referring to Ex. 4.8, he writes: 'On the F sharp . . . retard a little and regain it on the five following notes.'[28] At this point in the rondo form a modulation is taking place to lead from the main

[27] Spohr, *Grand Violin School*, 186–7.
[28] Ibid. 195.

Ex. 4.8 Rubato in Rode, 7th Concerto, 1st movt. (Spohr, 1833)

key of A minor to a second idea in C major. In the bar before this excerpt the bass has first introduced F♯, then in bar 39 the violin has it for the first time and later repeats it often as part of a secondary-dominant strengthening of the G major chord which is being prolonged as the dominant of the new key. Spohr thus suggests that this sensitive appearance of F♯ in the solo part be emphasized by accentuation and rubato.

All of the examples that Spohr refers to as *tempo rubato* involve the prolonging or retarding of one or several adjacent tones. The prolonged notes invariably play a sensitive role in the sense of the music. This is similar to García's *temps d'arrêt* in Ex. 3.9, which he applies in Ex. 3.13. 'This prolongation is usually conceded', García writes (Plate V), 'to appoggiaturas, to notes placed on long syllables, and those which are naturally salient in the harmony', and, he adds in the French edition, 'to those which one wants to stand out'.[29] Like Galeazzi (Ex. 4.3), Spohr thinks of rubato for the solo performer in terms of pro-longation rather than anticipation or delay. Although he provides a second violin part that is fairly active rhythmically, the possibility for audible displacement between the two parts actually exists only in Exx. 4.6 and 4.7 and perhaps 4.8. Perhaps this is due to the fact that the orchestral parts in this music are not so often confined to the repetition of the predictable rhythmic figures so typical of Italian and French opera of the period (see Exx. 3.16 and 3.18, for example). Occasionally, as in Ex. 4.6, he writes a second violin part that is more active than the original orchestral accompaniment. In any event, he never speaks of this aspect of rubato, and one should keep in mind that he does not provide the second violin part for the purpose of illustrating tempo rubato, but rather so that the teacher can guide the student during the course of the entire piece.

Later in the book Spohr discusses tempo rubato from the point of

[29] *Traité complet de l'art du chant*, ii (Paris, 1847; repr. with i, Minkoff, 1985), 24: 'à celles que l'on veut faire ressortir'.

view of the violinist in the orchestra: 'The division of each portion of a bar, according to the value of its time, must in orchestra playing be strictly observed. The *tempo rubato* (a slight delay on single or more notes), [which] in the solo [is] of great effect, cannot here be tolerated.'[30] In this case the word *delay* refers to a delay in leaving a note, thus its prolongation, rather than a delay in its initial appearance. Referring then to the orchestra in relation to the soloist, he states that 'the solo performer must neither be hurried nor retarded by the accompaniment; he should be instantly followed wherever he deviates a little from the time. This latter deviation, however, does not apply to the *Tempo rubato* of the solo performer, the accompaniment continuing its quiet, regular movement.'[31] Since the orchestra is to follow the soloist generally in his tempo fluctuations, it would thus be necessary to inform the players, in advance, of those places where the soloist wished to make rubato.

It is Spohr the conductor who sees tempo rubato from the point of view of both the soloist and the orchestra—one deviating, the other strictly in time. Thus, although he does not seem concerned with displacement between the two, he does conceive of rubato, basically, as the simultaneous occurrence of two contrary rhythmic effects. As early as the middle of the eighteenth century it had been the practice for the principal violinist to direct in performances of orchestral music. Early in the nineteenth century many conductors, like Spohr, were violinists, and there are accounts of Spohr conducting with his bow, while keeping his violin handy in case of need. He was also one of the first to use a baton.[32]

A violinist therefore had unusually varied opportunities to observe tempo rubato: as a solo performer, as a member of an orchestra, and as a conductor. Baillot even mentions rubato, as we have seen, in the principal part of a chamber work, and Spohr describes a new type of solo quartet in which the violin should perform as in a concerto. On the other hand, he feels that in the conventional string quartet (referring, no doubt, to composers such as Haydn, Mozart, and Beethoven) no single instrument should dominate and devices peculiar to solo playing, such as rubato, are not appropriate.[33]

The history of the earlier type of rubato for the violin was heavily influenced by the mainstream development in vocal music. In neither medium was the expression *tempo rubato* used to refer to tempo

[30] Spohr, *Grand Violin School*, 231.
[31] Ibid. 232.
[32] Jack Westrup, 'Conducting', *New Grove*, iv, 643.
[33] Spohr, *Grand Violin School*, 230–1.

modification. In neither medium did the words ordinarily appear in a score. In both media rubato seems to appear first in intensely emotional pieces in a slow tempo, as in the pathetic arias of Tosi or the Adagio movements of Benda, but later occurred just as often in works of faster tempo, as in many other examples by Benda, García, Cinti-Damoreau, Baillot, and Spohr. Both solo singer and solo violinist probably perceived rubato from an almost purely melodic point of view; hence the performers and the teachers tend to include only the melody in their examples—a melody usually showing in standard notation the alterations of note value involved. An accompaniment is ordinarily present only in complete pieces, and then not for the purpose of depicting tempo rubato, but to demonstrate the broader art of embellishment (as in the Benda manuscript or the concluding pieces in García's treatise) or to provide a means for a teacher to guide a student (as with Spohr). Spohr is the only one who indicates rubato by prose instructions rather than altered note values. In this way he demonstrates, to a certain extent, the latitude within which the rubato operates and, at the same time, implies that the same sort of freedom was intended in the notated examples of García and others. It would seem to me that a student could attain a sense of spontaneity more easily with Spohr's method of presentation.

Vocal and violin rubato often involves the intensification of an important note. Emphasis, however, can be given to a note in various ways. At first, performers spoke of anticipation and delay, meaning that a note was initially sounded before or after its notated position. This method continued sometimes into the nineteenth century, as seen in the anticipation of trills with Spohr and García (Ex. 3.14) and in the anticipations of an arpeggio by Cinti-Damoreau (Ex. 3.18). In Ex. 3.15 García also includes on the third beat a delay of two notes, the first retaining its original value while the second is reduced by half. The act of delaying can thus emphasize a note even though it diminishes its duration. The baroque listeners and theorists depicted this procedure in the way in which they actually heard it—by placing bass notes where the written notes occurred, so that the delays and anticipations in the melody would be as conspicuous to the eye as to the ear. Several sources also mention that delayed or anticipated notes should receive additional emphasis by the way they are produced: with Quantz 'weak and crescendo', with Hiller 'somewhat stronger' and 'held firmly', and with Baillot 'swelling' and with accelerating but light movement of the bow.

Another method of emphasizing important notes, however, is pro-longation, which North and Alençon mentioned, which Galeazzi depicted (Ex. 4.3), and which García considered the first element in rubato. In this procedure, the note is initially sounded exactly where written, but then continues longer than its notated value. Most of the nineteenth-century examples for voice and violin are of this type: Exx. 3.12 and 3.13, for example, and all those which Spohr calls *tempo rubato*. When an important note is prolonged, later notes are correspondingly de-emphasized by both reducing their value and delaying their appearance. Thus delay may sometimes de-emphasize a note, whereas at other times, depending on the configuration of note values surrounding it, the process of delay may, as described above, do just the opposite.

The solo singer and solo violinist generally appear to take the accompanying instrumental music for granted and to let it follow its own independent rhythmic pace. Presumably, however, the great artists were able to perceive simultaneously the strict beat and the expressive alterations of it. In late baroque music, as well as in early nineteenth-century Italian and French opera, the orchestra usually gave the soloist a continuous and audible reminder of the beat, and frequently enough for most tempo rubato to cause displacement with instrumental notes. In some of Spohr's examples, on the other hand, the rubato must be performed without the aid of an audible orchestral beat. In the second measure of Ex. 4.5, for example, the orchestra sustains a chord for an entire measure, and the soloist must therefore conceive not only of the note values to alter, but must, at the same time, accurately sense the strict rhythm against which rubato is being made. One can get some idea of this dual rhythmic perception by singing the melody in Ex. 1*a* in the Introduction while clapping or snapping the fingers to the strict beat.

I suspect, however, that a great artist is not usually conscious of all the details of this complex rhythmic counterpoint, but simply imagines he is singing or playing the notes as written, but with highly intensified passion. We should keep in mind that tempo rubato was a powerful device, to be seldom used, and only for very sensitive and expressive moments in the music. The solo performer no doubt did occasionally wait for a bass note or a chord in the accompaniment to push his rubato against. But it could well be that an artist might feel that the syncopated examples of North and Quantz actually falsified or only partially portrayed the effect from the performer's point of view. Quantz himself was a flautist and a teacher of the flute. I have found no sources, however, that link the flute with tempo rubato. Quantz actually got his conception of rubato, as he reports, from listening to a singer (and perhaps also from hearing Benda play the violin), and it was as a theoretician that he formulated the tiny phrases (Ex. 3.8) that depicted what he heard—and what he heard most clearly was the alternation of notes between the solo performer and the accompaniment. I presume that those who copied the Benda manuscript were also listeners—violinists who listened to Benda play and wrote down what they heard in notation.

It seems to me that one might acquire the most vivid impression of the nature of vocal and violin rubato by combining Baillot's verbal description of the performer ('when carried away by the expression, it apparently forces him to lose all sense of pulse and to be delivered by this means from the trouble that besets him') with the graphic portrayal of metrical dissolution and the intoxicating distortion and shifting of entire passages in the Benda manuscript. I have emphasized the fact that there are quite different points of view about rubato, depending upon one's role in the musical experience. The solo performer, the member of an orchestra, the conductor, the listener, and the theorist each has his own unique viewpoint and each his own way of describing and explaining the device. The development in vocal and violin music is relatively simple and clear, however, for there were seldom any other meanings for the word *rubato*, and, in spite of diverse ways of depicting it, the basic idea of this earlier type of rubato persists fairly consistently from the end of the seventeenth century until at least the middle of the nineteenth. Keeping this development carefully in mind will be helpful as we now approach keyboard music. Here the situation is more confusing and more complex.

5 BOTH TYPES OF RUBATO IN KEYBOARD MUSIC

QUANTZ mentions tempo rubato in relation to the keyboard accompanist in 1752, and it is first presented in books on keyboard playing by Marpurg in 1755 and 1756. In keyboard music of the Classic period a homophonic style develops in which the functions of melody and accompaniment are ordinarily assumed by the right and left hands, respectively. Wolfgang Amadeus Mozart, in a letter of 1777 to his father, is the first, as far as I know, to refer specifically to the two hands:

What ... people cannot grasp is that in *tempo rubato* in an Adagio, the left hand should go on playing in strict time. With them the left hand always follows suit.[1]

This points out the special difficulty the keyboard player faced when performing tempo rubato, for a single person had to play the strict accompaniment in one hand simultaneously with the rhythmically free melody in the other. It is not surprising that many, if not most, keyboard players could not manage this dual feat. When their left hand followed their right, they produced, instead of the earlier type of tempo rubato, a general modification in the tempo of the entire musical texture. This may have been the reason why the word *rubato* began to be used in keyboard sources around the turn of the century to refer sometimes to general tempo changes. The two types of rubato then co-existed in keyboard music until at least the middle of the nineteenth century.

Keyboard sources describe the earlier type of rubato almost exclusively in terms of displacement between a melody and its accompaniment, and thus between the right and left hands. Preoccupation with displacement also produced, from around 1774 until the early years of the next century, a series of other meanings for the word *rubato*.

---

[1] *Mozart: Briefe und Aufzeichnungen*, ed. Wilhelm Bauer and Otto Erich Deutsch (Bärenreiter, 1962), 83 (letter to his father from Augsburg, 23 Oct. 1777); trans. by Carol MacClintock in *Readings in the History of Music in Performance* (IUP, 1979), 381.

The Earlier Rubato

The basso continuo existed in most vocal, chamber, and orchestral music from 1600 to about 1750 and in some cases for another fifty years or so. There were a wide variety of continuo instruments, but often the organ or harpsichord, or later the fortepiano, participated by realizing the figured bass. The keyboard player was thus often involved in maintaining the strict accompaniment while a solo singer or violinist performed the tempo rubato described in the last two chapters. He therefore heard tempo rubato executed by someone else and related it to the insistent strict beat of his own part.

In a section of his flute book entitled 'Of the Keyboard Player in Particular', Quantz emphasizes in 1752 the strict rhythm of the accompaniment:

If the accompanist is not secure in the tempo, if he allows himself to be beguiled into dragging in the tempo rubato, or when the player of the principal part retards several notes in order to give some grace to the execution, or if he allows himself to rush the tempo when the note following a rest is anticipated, then he not only startles the soloist, but arouses his mistrust and makes him afraid to undertake anything else with boldness or freedom.[2]

In his own treatise published the following year, C. P. E. Bach describes the dominating rhythmic role the keyboard played in late baroque music and in rubato:

the keyboard is and must always remain the guardian of the beat ... The keyboard, entrusted by our fathers with full command, is in the best position to assist not only the other bass instruments but the entire ensemble in maintaining a uniform pace ... Especially those parts that employ the tempo rubato will find herein a welcome, emphatic beat ... and, in addition, those performers located in front of or beside the keyboard will find in the simultaneous motion of both hands an inescapable visual portrayal of the beat.[3]

The main sources of information on the application of rubato in keyboard music during the second half of the eighteenth century are the books of Marpurg on harpsichord and clavichord playing, and the later writings of E. W. Wolf and Türk on playing the clavichord.

---

[2] *Versuch einer Anweisung die Flöte traversiere zu spielen* (Berlin, 1752); modern edn. Arnold Schering (Leipzig: C. F. Kahnt, 1926), 171; trans. from Edward R. Reilly, *Johann Joachim Quantz, On Playing the Flute*, 2nd edn. (New York: Schirmer Books, 1985), 252–3 (see also n. 1 on 253).

[3] *Versuch über die wahre Art das Clavier zu spielen*, i (Berlin, 1753; repr. Br. & H, 1969), 5–8 (footnote); trans. by William J. Mitchell in *Essay on the True Art of Playing Keyboard Instruments* (Norton, 1949), 33–5.

In three books published between 1755 and 1763 Friedrich Wilhelm Marpurg attempts to combine two different concepts of syncopated displacement: one coming directly from Quantz (see Ex. 3.8), the other from the *suspension* of F. Couperin and Rameau, which was discussed in Chapter 2 in the section on baroque arpeggiation (see Exx. 2.9 and 2.10). The French *suspension* had been preceded, as we have seen, by the *séparation*, *séparez*, or *harpègement* for lute and clavecin (Ex. 2.8) and followed around mid-century by entire pieces filled with such displacement (the practice was notated by Forqueray in 1747, as we have seen, and recommended by Foucquet in 1751 for tender pieces). Marpurg was reportedly in Paris for a long time around 1746.[4] He was also familiar, as we have seen, with the ideas of Quantz, Agricola, and other members of the Berlin school. Therefore, it is perhaps not surprising that he would relate the French practice to the Italian tempo rubato that the German writers had attempted to depict.

In addition to his 1763 book on singing mentioned in the last chapter, he published two keyboard books that refer to rubato: *Anleitung zum Clavierspielen der schönen Ausübung der heutigen Zeit gemäss* (Berlin, 1755) and its French counterpart, *Principes du clavecin* (Berlin, 1756). His ideas evolve, not without some inconsistency and confusion, during the course of these three works. I have shown the partial development of his concepts in Ex. 5.1 from the two keyboard books of 1755 and 1756, and their final resolution in Ex. 5.2 from his book of 1763.

Marpurg builds up a theory which relates all the lines in Ex. 5.1, which come from the identical tables at the ends of the 1755 and 1756 books. Both books commence from a series of main notes (*a*), each of which is divided into two repeated notes (*b*). Each pair of notes in (*b*) can then be altered by substituting a rest for the second note (which does not concern us here) or the first (resulting in Ex. 5.1*c*). He then states that (*c*) can also be notated as in (*d*), which uses the sign for the French *suspension* in Ex. 2.9. Later he speaks of syncopation (involving either suspensions or repeated notes) and its two types, anticipation and retardation.[5]

---

[4] See Ernst Ludwig Gerber, *Historisch-biographisches Lexikon der Tonkünstler*, i (Leipzig: J. G. I. Breitkopf, 1790; repr., Graz: Akademische Druck- und Verlagsanstalt, 1977), 882; and Elizabeth Loretta Hays, 'F. W. Marpurg's *Anleitung zum Clavierspielen* (Berlin, 1755) and *Principes du clavecin* (Berlin, 1756): Translation and Commentary', Ph.D. diss. (Stanford Univ., 1976; UM 77–12,641), ii, 1 and 6.

[5] *Anleitung zum Clavierspielen*, 2nd edn. (Berlin, 1765; repr. Olms, 1970), 39–40 (according to Sandra P. Rosenblum, in *Performance Practices in Classic Piano Music* [IUP, 1988], 493, this is identical to the 1st edn. of 1755 except for some minor changes unrelated to rubato); and *Principes du clavecin* (Berlin, 1756; repr. Minkoff, 1974), 52–3. Ex. 5.1 comes from table II in both volumes; lines (*b*) to (*g*) are nos. 37, 42, 49, 48, 46, and 50, respectively. See also Hays' 'Translation and Commentary', i, pp. IX-11–IX-17.

Ex. 5. 1 *Setzmanieren* or *Figures de composition* (Marpurg, 1755, 1756)

Ex. 5. 2 *Tempo rubato* (Marpurg, 1763)

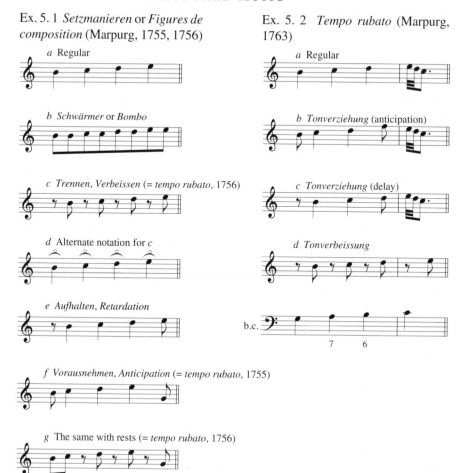

In the 1755 work he mentions rubato only in the index: 'Tempo rubato, thus the Italians designate different types of anticipation of the notes. See *Vorausnehmen*.'[6] Since he illustrates *Vorausnehmen* with Ex. 5.1*f*, he presumably limits the meaning of rubato to anticipating syncopation. In the French book of 1756, on the other hand, tempo rubato 'consists of sounding only the last half of a note divided in two in the same degree [Ex. 5.1*c*]'.[7] Relating this, then, to the French *suspension*, he adds:

---

[6] *Anleitung zum Clavierspielen*, 2nd edn., 5th unnumbered page of the index: 'Tempo rubato, so nennen die Italiener verschiedne Arten der Anticipirung der Noten. S. Vorausnehmen.'

[7] *Principes du clavecin*, 52: 'Tempo rubato ... consiste à ne faire sonner que la dernière moitié d'une note partagée en deux en même dégré.'

Some composers are in the habit of employing instead of the rest the sign which one sees [in Ex. 5.1d] . . . in order to mark the *tempo rubato* (*le tems dérobé*) of a note. But the first manner [Ex. 5.1c] is preferable to the other, in order to avoid unnecessary signs. There are already enough without that.[8]

Later, when he discusses anticipation (Ex. 5.1f), he states that 'in performing one sometimes omits the first of the two notes [on the second, third, and fourth beats of Ex. 5.1g] by means of Tempo rubato.'[9] Rubato in 1756 thus included the retardation and anticipation articulated with rests in (c) and (g), but not the corresponding sustained versions in (e) and (f). It is only in his later book of 1763 that he finally includes in tempo rubato both the *Tonverbeissung* and the two types of *Tonverziehung* shown in Ex. 5.2. In the later book he also includes a bass-line for each example.

Marpurg presents rubato in his keyboard books as a 'figure of composition', hence not as a method of delivery determined by a performer. Moreover, he seems to treat it as a purely intellectual device. This is in marked contrast to the expressive manner in which the French described the *suspension*. Couperin, as we have seen, compared it to the increasing of sound on a stringed instrument, an effect suitable for slow and tender pieces. It was invariably depicted as a delay and never as an anticipation, and occasionally the sign appeared over more than one note in succession. Couperin makes clear also that the exact duration of the delay is left to the taste of the performer and his comprehension of the sense of the music.

When several notes are thus delayed, the performer presumably controls the actual amount of silence between the notes in such a way that the listener hears the top voice as a musical phrase. This would almost certainly involve a varying amount of silence as one moved from note to note. The rests therefore actually represent a particular manner of articulation. In his *Tonotechnie* of 1775 on mechanical instruments, Engramelle discusses articulation:

All the notes in execution . . . are partly in *hold* and partly in *silence*, which means that they all have a certain length of *sound* and a certain length of *silence*, which united make the whole value of the note.

These *silences* at the end of each note fix its articulation and are as necessary as the holds themselves . . . : even the notes of the most rapid shakes are separated by very small intervals. Those intervals, more or less

[8] Ibid. 52: 'Quelques compositeurs sont dans l'habitude d'employer à la place de la pause le signe qu'on voit . . . pour marquer le *tempo rubato* (*le tems dérobé*) d'une note. Mais la première manière est préférable à l'autre, d'autant plus qu'il ne faut pas multiplier les signes sans nécessité. Il y en a dêjà assez sans cela.'
[9] Ibid. 53: 'En exécutant on passe quelquefois la première des deux notes au moyen du Tempo rubato.'

long, I call *silences d'articulation* in music, from which no note is exempt, like the articulated pronunciation of consonants in speech . . .[10]

The special interest by the French clavecinists in the expressive effect caused by the delay between two vertically aligned notes is extremely significant, I think, for the history of tempo rubato in keyboard music. It was indicated by a sign, like an agrément, and hence was not, theoretically, optional. Since it involved only a delay, the left hand always preceded the right. Conceived as an articulate melody, then, the passages in Ex. 5.1c and g come to life, for now the rests can vary, not only from performance to performance, but from note to note—in fact, the rests could be so brief that the listener would perceive no pauses at all, but only a singing melody. In addition, even the sustained notes in Ex. 5.1a, e, and f become viable musical entities, according to Engramelle, only when each concludes with a slight silence of articulation. The way Marpurg builds up his theory in Ex. 5.1 seems to suggest that he saw something like Ex. 2.8a in a French source, took it at face value, and constructed Ex. 5.2d and the bass-line below it accordingly. In any event, we can be grateful to him for relating together the Italian tempo rubato and the French *suspension*, for regardless of the extent of his own comprehension, both were described by the performers and composers who knew them best, as very similar and intensely expressive musical devices.

Ernst Wilhelm Wolf describes tempo rubato in the preface to a collection of clavichord pieces published in 1785. He was influenced by the works and ideas of C. P. E. Bach and in 1770 married the daughter of Franz Benda, whose violin rubato was considered in the last chapter. He was also for a time the brother-in-law of J. F. Reichardt, who was married to another of Benda's daughters from 1776 to 1783 and who, as we have seen, praised Benda's expressive alteration of note values. According to Wolf, tempo rubato is an ornament used in both Adagio and Allegro movements:

it consists of anticipating as well as retarding . . . , much the same as syncopation . . . The bass keeps strictly to the rhythm, while the melody in the right hand is retarded. The values of some notes are lengthened, the values of others are shortened. . . . The rhythm of the melody may wander back and forth, though it must always meet up with the bass at bar-lines, at downbeats.

He gives no musical examples in this source, but, like Kirnberger and Marpurg, considers rubato primarily as a compositional device, often

---

[10] Marie Dominique Joseph Engramelle, *La Tonotechnie* (Paris, 1775), trans. from Arnold Dolmetsch, *The Interpretation of the Music of the XVII and XVIII Centuries*, 2nd edn. (Novello and OUP, 1946), 282–3.

written out in full by the composer. He also mentions, however, that 'as an ornament of performance or as an effect in playing, it makes its best impact in adagio.'[11] In his *Musikalisches Unterricht* of 1788 he adds that skilful singers and players can utilize '*Tonverziehung* (*Tempo rubato*)' when the chords in the accompaniment are strictly equal in value to the written notes of the soloist (as in his syncopated examples, which show retardation and anticipation in the manner used earlier by Quantz). By linking rubato to this situation, Wolf ensures, of course, the presence of conspicuous and audible displacement between the melody and its accompaniment.[12]

Although Marpurg and Wolf present rubato as a mechanical device in keyboard music, it is Daniel Gottlob Türk who links it with the Italian idea of musical expression. Türk gives credit to Tosi and Quantz as well as to Hiller, with whom he studied. He joins together the theoretical syncopations coming from Quantz and the musical method of performance descended from Tosi. He published the *Klavierschule* in 1789 and a considerably revised and enlarged second edition in 1802. Around 1804 an abridged English translation appeared in London.[13] Plate XI shows most of the section on rubato from 1789, Plates XII and XIII, the entire section from 1802, and Plate XIV, the prose portion from the English version.

Türk identifies tempo rubato as one of the 'extraordinary means' of expressing the passions (see Paragraph II on Plate XIV). It should be 'left to the judgement of the performer' (Paragraph VIII on Plate XIV and page 374 on Plate XI), but it 'can be of great effect' only 'when used sparingly and at the right time'.[14] In 1802, however, he qualifies this description of rubato as a performance practice: 'Only seldom in the new compositions for keyboard may an appropriate opportunity

[11] *Eine Sonatine. Vier affectvolle Sonaten und ein dreyzehnmal variirtes Thema, welches sich mit einer kurzen und freien Fantasie anfängt und endiget* (Leipzig, 1785), copy at LC, pp. vi and ix of the *Vorbericht (als eine Anleitung zum guten Vortrag beym Clavierspielen)*; trans. by Christopher Hogwood in 'A Supplement to C. P. E. Bach's Versuch: E. W. Wolf's Anleitung of 1785', *C. P. E. Bach Studies*, ed. Stephen L. Clark (OUP, 1988), 151–3, also 141 n. 18 and 145.

[12] *Musikalischer Unterricht* (Dresden, 1788), copy at SML, i, 34–5 and Exx. Rr–Zz. See Carl LeRoy Blake, 'Tempo Rubato in the Eighteenth Century', D.M.A. thesis (Cornell Univ., 1988; UM 88–4,554), 117–19.

[13] *Klavierschule, oder Anweisung zum Klavierspielen für Lehrer und Lernende* (Leipzig and Halle, 1789), repr. in *DM* xxiii (Bärenreiter, 1962), trans. by Raymond H. Haggh as *School of Clavier Playing or Instructions in Playing the Clavier for Teachers and Students* (UNP, 1982); *Neue vermehrte und verbesserte Ausgabe* (Leipzig and Halle, 1802); and *Treatise on the Art of Teaching and Practising the Piano Forte by D. G. Turk, Professor and Director of Music at the Royal Prussian University of Hall, with Explanatory Examples Translated from the German and Abridged by C. G. Naumburger* (London, date on preface: 14 Feb. 1804), copy at BL.

[14] Türk (1789), 370; Haggh trans., 359 (see 363–5 for the main section on rubato); Türk (1802), 414.

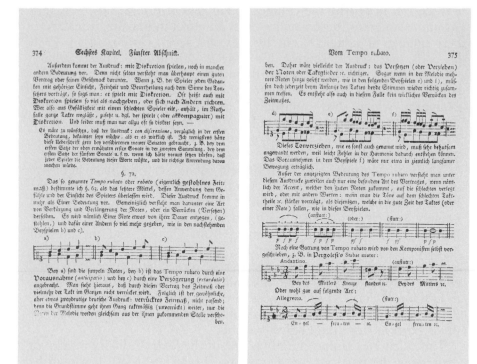

PLATE XI. Daniel Gottlob Türk, *Klavierschule* (Leipzig and Halle, 1789), 374–5

occur, since the composer indicates the tempo rubato, for the most part, himself.'

Tempo rubato, according to Türk in 1789, most often refers to 'a kind of shortening or lengthening of notes, or the displacement of these'. In 1802, however, he reverses these two aspects by describing rubato as 'the displacing of notes by means of their anticipation and retardation'. This deliberate change may reflect an increasing emphasis on displacement by the keyboard player. In both editions Türk goes on to say, in the tradition of Tosi and Galliard, that 'there is something taken away (stolen) from the duration of a note and for this, another note is given that much more'. He follows this statement immediately with examples of anticipation and retardation, which in 1789 (see Exx. *a*, *b*, and *c* on page 374 of Plate XI, almost identical to Exx. 15–17 mentioned on Plate XIV) look very much like those of Quantz in Ex. 3.8. These are replaced in 1802 (page 419 on Plate XII) by

PLATE XII. Daniel Gottlob Türk, *Klavierschule*, enlarged and improved edn. (Leipzig and Halle, 1802), 418–19

chromatic versions using the treble clef. All of these examples, of course, make the displacement of melodic notes conspicuous by providing an undisplaced bass. Türk may have intended the chromatic bass in the 1802 examples to indicate that rubato was most suitable at intensely emotional places in music. Referring to these examples, then, Türk observes that 'the tempo, or even more, the meter as a whole is not displaced', since, as he adds in 1802, 'this type of tempo rubato is based on . . . syncopation'.

In 1802 he includes also a second group of examples (*d* and *e* on page 419 in Plate XII), drawn this time from a real composition, the aria 'Ihr weichgeschaffnen Seelen' from the Passion cantata *Der Tod Jesu* by Carl Heinrich Graun. This work was popular in Germany until late in the nineteenth century and constitutes the principal product of the *empfindsamer Stil*, an intimate and subjective style cultivated in North Germany around the middle of the eighteenth

PLATE XIII. Daniel Gottlob Türk, *Klavierschule*, enlarged and improved edn. (Leipzig and Halle, 1802), 420–1

century by C. P. E. Bach and others.[15] In Ex. 5.3 I have expanded Türk's excerpts by including the complete passages as well as a reduction of the full string accompaniment.[16] The text concerns weeping and pain (*Schmerz*), and the music responds to these intensely emotional words through chromaticism and rhythmic displacement.

In Ex. 5.3*e* the tenor syncopates the melody by retardation and is accompanied during the second half of the measure by the first violins playing the same melody in unrobbed form. Later in the same aria, a longer melisma on the word *Schmerz* commences a measure before Ex. 5.3*d*, and the tenor, beginning in the second half of the first

---

[15] See Daniel Heartz's article on 'Empfindsamkeit' in *New Grove*, vi, 157–9.

[16] Carl Heinrich Graun, *Der Tod Jesu*, ed. Howard Serwer in *Collegium musicum, Yale University*, 2nd ser., v (Madison, Wis.: A–R Editions, 1975), 60 and 57. I have omitted the bass figures in Ex. 5.3.

CHAPTER XX.

OF THE NECESSITY OF A NATURAL, AND JUST FEELING, FOR ALL THE PASSIONS, WHICH ARE TO BE EXPRESSED IN MUSIC.

I. THE hints given about Expreſſion are of very little uſe to a Performer, who is without this Natural Feeling, or who poſſeſſes it only in a ſmall degree. Verbal Inſtruction may be more uſeful to him, tho' the moſt able and conſcientious Teacher will not be very ſucceſsful.

II. If the Compoſer has marked the required Expreſſion of the whole Piece, and its ſingle Paſſages and Notes, as correctly as poſſible, and the Performer obſerve ſtrictly what has been ſaid, there remain ſtill ſome caſes, in which the Expreſſion may be improved by extraordinary means. Amongſt theſe may be reckoned: 1ſt, Playing *ad libitum*; 2dly, Diminiſhing, or encreaſing the Time; and 3dly, What the Italians call *Tempo Rubato.*

III. More according to Feeling, than in Time, muſt be played Free Fantaſies, Cadences, Embelliſhed Pauſes, &c. likewiſe Paſſages marked with the word *Recitativo.* Theſe Paſſages would have a bad effect, if the Performer were to play them in ſtrict Time. The ſuperior Notes muſt be played ſlower and ſtronger, and the inferior ones faſter and ſofter, juſt as a Singer of taſte and feeling would ſing the Notes, and a good Orator would declaim by them.

IV. Tho' it is almoſt impoſſible to deſcribe every Paſſage in which retarding or haſtening the Time may take place, neverthelſs it ſhould be obſerved, that in Pieces of a fiery, violent, and furious Character, the ſtrongeſt Paſſages muſt be haſtened a little, or played *accelerando*; likewiſe thoſe Paſſages which are repeated ſtronger. In very tender, languiſhing, and plaintive Paſſages the effect may be improved by retarding the Time. Likewiſe, in Notes, before certain Embelliſhed Pauſes, the Movement is taken a little ſlower by degrees, juſt as if the ſtrength was almoſt exhauſted. The Paſſages which are marked towards the end of a Piece, or a part of it, with *diminuendo, diluendo, ſmorzando*, may be played likewiſe a little ſlower.

V. A ſoft and affecting Paſſage, between two lively ones, may be played a little ſlower; but the Movement muſt be made ſlower at once, not by degrees.

VI. To thoſe Paſſages which are not played in ſtrict Time, but a little ſlower, belong thoſe Embelliſhments, and Tranſitions to the ſubject, moſtly marked with ſmaller Notes, with the addition of the words *ſenza tempo*, or *ad libitum.* But there exiſts another ſort of Tranſition from one Period to another, which may be likewiſe played ſlower, though not marked with ſmaller Notes, nor *ad libitum*. (pl. 23, example 14.) Likewiſe a weak Paſſage may be played ſlower at the Repetition.

VII. If the Compoſer does not wiſh to have his Piece played throughout in ſtrict Time, he commonly ſignifies it by the words *con diſcrezione.* In this caſe, it is left to the taſte and judgment of the Performer, to delay the Time in ſome Paſſages, and to haſten it in others. In this caſe muſt be obſerved, what has been ſaid in the four preceding paragraphs.

VIII. *Tempo rubato*, or *robbato*, ſignifies a ſtolen, or robbed Time, the application of which is likewiſe left to the judgment of the Performer. Theſe words have ſeveral ſignifications. Commonly they ſignify a manner of ſhortening, and lengthening Notes; that is to ſay, a part is taken from the length of one Note and given to the other. In plate 23, example 15, are the plain Notes; in plate 23, example 16, *Tempo rubato* is uſed by an anticipation; and in plate 23, example 17, by a retardation. Beſide this ſignification, it is underſtood by *Tempo rubato*, that the Accent is put upon the inferior Notes inſtead of the ſuperior ones. (pl. 23, ex. 18.) The above-mentioned delaying, or haſtening the Time deſignedly, is likewiſe ſignified by *Tempo rubato.*

APPENDIX.—Abbreviations often met with in Muſic are ſhewn in plate 23, example 19.

PLATE XIV. Daniel Gottlob Türk, *Treatise on the Art of Teaching and Practising the Piano Forte* (London, preface dated 1804), 40

Ex. 5.3 Rubato in C. H. Graun's *Der Tod Jesu* (Türk, 1802)

measure in (*d*), anticipates the melody in the first violins. In both examples, the string orchestra accompanies the voice with chords repeated insistently and percussively. In the footnote at the bottom of pages 419 and 420 on Plates XII and XIII, Türk mentions the clash between the voice and the violins (marked with a cross on Plate XII), which causes, he feels, a fairly noticeable harshness. This very rare concurrence of the unrobbed melody with the robbed, however, was no doubt an intentional response, in the spirit of *Empfindsamkeit*, to the extraordinary passion of the text. In the second half of the measure in Ex. 5.3*e* the effect on each eighth-note beat is similar, except for octave displacement, to that in J. S. Bach's 'Erbarme dich' (mentioned in Chapter 2), where notes are heard simultaneously with and without appoggiaturas. Graun, like Quantz, was in the service of Frederick the Great, and *Der Tod Jesu* was first performed in 1755, three years after Quantz's *Versuch* appeared. Perhaps Graun intended through his syncopated notation to represent rubato as Quantz did in Ex. 3.8.

This may imply a somewhat freer rhythm than the precise note values indicate. In any event, he provided Türk with remarkable examples within the same piece of both anticipation and retardation.

Türk presents, finally, one more set of examples (labelled *d*, *e*, and *f* on Plate XI; *f*, *g*, and *h* on Plate XII) to show that the duration of the measure is not altered even when more notes are added to the melody. The bass has been changed in the 1802 versions and some notes dotted. Unlike the other two sets, this one presents retardation before anticipation, and the notes are re-struck like appoggiaturas. Like others before him, he warns that the device must be applied carefully to avoid errors in the harmony. All three of Türk's groups of examples, however, clearly emphasize, through syncopation, the displacement between the melody and the bass.

He also mentions rubato elsewhere in his book. In the chapter on 'Extemporaneous Ornamentation' he states that 'it is also possible to vary by displacing the notes, as when some are lengthened and others shortened.' A footnote refers the reader to the section on tempo rubato. When adding improvised embellishment, 'the counting must be maintained in the strictest manner, even for the most extensive ornaments. If some tones are played a little too soon or too late for the sake of the affect, the tempo must not be changed in the slightest degree as a result.'[17] In a section entitled 'Concerning Clarity of Execution', he discusses the prolonging of important notes:

> Because it is recognized by everyone, I do not have to provide evidence for the possibility of lingering somewhat longer on a very important note than on one less important. . . . As far as how long a note should be held is concerned, I would like to establish the rule that it should at the most not be lengthened more than half of its value. Often the holding of a note should be only scarcely perceptible—when a note has already been made important enough by an accidental, for example, by the height of its pitch, or by an unexpected change of harmony, etc. That the following note loses as much of its value as has been given to the accentuated note goes without saying.[18]

Earlier he had identified some of the important notes that could be prolonged: those on the strong beats, the first note of every phrase (if on a strong beat), appoggiaturas and other dissonant intervals, syncopated notes, notes involved in modulation, notes distinguished by length, notes at a high or low point, notes important to the

---

[17] Türk (1789), 323–4 and 325; Haggh trans., 311 and 313; and Türk (1802), 338 and 340.

[18] Türk (1789), 338–9; trans. from Haggh, 328, except for the penultimate sentence: 'Oft darf das Verweilen kaum merklich werden, wenn der Ton schon an und für sich z. B. durch ein zufälliges Versetzungszeichen, durch auszeichnende Höhe, durch eine unerwartete Harmonie u dgl. wichtig genug wird.'

harmony.[19] He does not use the word *rubato* in this discussion, however, and does not depict any of these prolongations, as he did tempo rubato, as a displacement between melody and bass. In his description of tempo rubato, on the other hand, he does not mention a variable amount of prolongation nor its often very subtle application. A singer or violinist of the time would have recognized tempo rubato in his discussion of prolongation better than in the examples of displacement from his article on rubato. The keyboard player, concerned with a relationship between the left and right hands, apparently found the displacement produced by syncopation a more manageable approach to the subject. Türk, however, also recognized the special melodic control required of a singer, for he says that 'in general . . . that instrumentalist plays best who comes closest to the singing voice or who knows how to bring out a beautiful singing tone.'[20]

There are a few reports linking tempo rubato with contemporary performers and composers. We have already noted W. A. Mozart's comments. In 1783 Christian Gottlob Neefe, who at about the same time was teaching the young Beethoven piano, organ, and composition, describes the performance of Countess von Hatzfeld: 'She plays the fortepiano brilliantly and in playing yields herself up completely to her emotions, wherefore one never hears any restlessness or unevenness of time in her *tempo rubato*.'[21] The following year a reporter writes concerning Muzio Clementi's style of playing: 'each note is separated from the others in the clearest manner, and with inimitable enthusiasm, with *crescendo* and *diminuendo*, with imperceptible *lentando, rubando*, etc., all of which it would be impossible to express on paper.'[22]

It is possible that Clementi, as well as the other keyboard composers of his time, sometimes meant to indicate *rubando* (presumably the same as tempo rubato) when they notated a syncopated melody.[23] Such a melody appears with the words *tempo rubato* in a keyboard sonata probably from the 1790s and later attributed to Hummel. The

[19] See Türk (1789), 335–8; Haggh trans., 325–8; and Türk (1802), 374–8.

[20] Türk (1789), 331; Haggh trans., 318.

[21] *Magazin der Musik*, ed. Carl Friedrich Cramer (repr. Olms, 1971), i, 388 (issue for 30 Mar. 1783); trans. in Alexander Wheelock Thayer, *The Life of Beethoven*, ed. Elliot Forbes (PUP, 1964), i, 37. Gerber quotes the passage in *Historisch-biographisches Lexikon der Tonkünstler*, i (1790), 604, and attributes it to Neefe. The latter had studied with J. A. Hiller, whose tempo rubato we considered in Ch. 3. Maria Anna Hortensia Gräfin von Hatzfeld, an accomplished singer as well as keyboard performer, was active in the thriving musical life of Bonn. Beethoven dedicated to her his 24 Variations, composed in 1790–1, on Righini's Arietta 'Venni amore'.

[22] *Magazin der Musik*, ii, 369 (issue for 11 Dec. 1784); trans. from *MMR* 24 (1894), 172.

[23] Leon Plantinga, *Clementi: His Life and Music* (OUP, 1977), 90–1, also his *Romantic Music* (Norton, 1984), 29 n. 6. Blake, in 'Tempo Rubato in the Eighteenth Century', 52–4, includes examples of such 'syncopation rubato' by Haydn and Mozart.

Ex. 5.4 Hummel (?), Sonata in C major, 1st movt., 1790s?

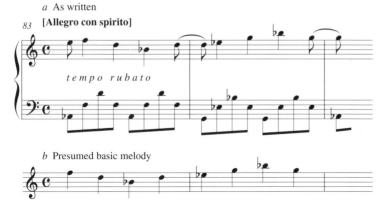

a As written

b Presumed basic melody

seven-measure passage begins as in Ex. 5.4*a*. Hummel studied as a child with Mozart and therefore was no doubt familiar with his manner of applying the earlier type of rubato.[24] A reviewer for *Le Pianiste* reports in 1834 that Dussek (1760–1812) loved rubato, although he never wrote the word in his music; he did try to make it visible by means of syncopation (presumably as in Ex. 5.4*a*), but a faithful execution of this notation did not achieve the suave and charming manner that he intended, so he contented himself thereafter with writing simply *espressivo*.[25] These comments may shed some light on the actual performance of the rubato in Ex. 5.4*a*: performing the syncopated melody in (*a*) was perhaps more like playing the presumed basic melody that I have constructed in (*b*), but with great passion and with increased intensity in the singing quality of the line. In this sense, then, the syncopations notated in (*a*) represent only an approximation of the performer's actual amount of delay and, like

[24] The sonata is repr. in *Johann Nepomuk Hummel: The Complete Works for Piano*, ed. with introductions by Joel Sachs, iv/2 (New York and London: Garland, 1990), 188 (in a section entitled 'Works Probably Not by Hummel'). See the comments in vol. i, p. xi, and in Joel Sachs, 'A Checklist of the Works of Johann Nepomuk Hummel', *Music Library Association, Notes*, 30 (1973/4), 750. See also Gerald Abraham, *Chopin's Musical Style* (OUP, 1939; repr. 1941), 7. The sonata comes from a collection ed. by Louis Winkler and published by Litolff, perhaps the one listed in Adolph Hofmeister's *Verzeichnis sämmtlicher im Jahre 1869 . . . erschienenen Musikalien*, xviii, 73. A modern edn. is in *J. N. Hummel, Complete Piano Sonatas*, ed. Harold Truscott (London: Musica Rara, 1975), ii, 84. The passage appears 5 bars into the development section of the 1st movt., and the melody does not recur elsewhere in the movement. I have changed the tied melody note at the beginning of m. 84 from E to D.

[25] *Le Pianiste* (repr. Minkoff, 1972), 1/5 (Mar. 1834), 78: 'Dussek, qui aimait beaucoup le *Rubato*, quoiqu'il n'ait jamais écrit ce mot dans sa musique; Dussek avait essayé de le rendre *visible* au moyen des syncopes; mais, lorsqu'on exécutait fidèlement ces syncopes, on était bien loin de rendre sa manière suave et délicieuse. Il y renonça lui-même, et se contenta d'écrire: *espressivo*.'

Couperin's *suspension*, imitates the swelling of a stringed instrument and approaches closer to the 'beautiful singing tone' that Türk admired.

Henri Herz, a German pianist and composer who studied at the Paris Conservatoire and later became a professor there, provides a detailed description of this technique in his *Méthode complète de piano* from around 1837:

Sometimes... the double character of the accompaniment and the melody requires from each hand a different rhythmic effect. Thus, whereas the right hand seems to lose its way in fantastic variations, the left, supporting the bass notes in a counter-rhythm, follows it reluctantly and with syncopated notes. This case, as in all those where the expression is complex, requires not only that the hands be perfectly independent from one another, but, if I can say it, [that] a different soul [be] in each of them. It is thus that Dussek poured out a hazy and melancholy tint on certain passages by letting the right hand sing in a vague and nonchalant manner, whereas the left executed the arpeggiated chords rigorously in time. I don't know why this manner of playing, so extolled a short time ago, has now been forgotten.[26]

The same independence of hands is attributed to the pianist Sigismond Thalberg (1812–71), who had studied voice 'for five years under the guidance of one of the most eminent professors of the Italian school'.[27] He was famous for his singing tone, and 'when he accelerated, retarded, or embellished the melody, the accompaniment proceeded with steady, unwavering precision, unaffected by the emotion displayed in the solo parts'.[28]

The keyboard player's concern for displacement, as depicted in the syncopated examples of Marpurg and Türk, thus naturally involved a special interest in the different functions of the two hands. In the homophonic texture of most classic and romantic keyboard music the right hand has a melody and the left, a chord activated rhythmically by some sort of arpeggiation. The Alberti bass, as well as other arpeggiated figures such as the one in Ex. 5.4, constitutes an accompaniment which, like that in the vocal and violin music discussed in the

---

[26] Op. 100 (Paris, the year 1837 appears below his portrait), copy at BL, 20: 'quelquefois... le double caractère de l'accompagnement et de la mélodie exige de chaque main un effet rythmique différent. Ainsi, tandis que la droite semble s'égarer en de folles variations, la gauche, appuyant à contre temps sur les basses, la suit à pas pesans et par notes syncopées. Ce cas, comme tous ceux où l'expression est complexe, exige non seulement des mains parfaitement indépendantes l'une de l'autre, mais, si je puis le dire, une âme différente dans chacune d'elles. C'est ainsi que Dussek répandait une teinte vaporeuse et mélancolique sur certaines périodes en laissant chanter la main droite d'une manière vague et nonchalante, tandis que la gauche exécutait des batteries rigoureusement en mesure. J'ignore pourquoi cette manière de phraser, tant prônée naguère, est tombée maintenant dans l'oubli.'

[27] *L'Art du chant appliqué au piano*, 1st–2nd ser. (London: Cramer Beale & Co. [1853]), copy in BL, p. ii of preface.

[28] Adolph Friedrich Christiani, *The Principles of Expression in Pianoforte Playing* (New York: Harper & Bros., 1885), 301.

previous chapters, provides a continuous and insistent beat. Although the gradually increasing use of the pedal in the nineteenth century softened the effect somewhat, the staccato style of playing in the Classic period made the beat even more conspicuous.[29] A strict left hand thus gave the performer a constant, audible, and tactile reminder of the beat, and since the notes were so short (eighth notes in Ex. 5.4), almost any alteration in the melody notes produced displacement with one of the bass notes. As already noted, Baillot states in his cello method of 1804 that 'the bass must be the regulator in the ensemble piece in the same way that the left hand ought to maintain steadiness in the playing of a piano sonata'.

A syncopated melody occurs also in Louis Adam's presentation of the type of articulation now known as *portato*. It is notated, as in Ex. 5.5a, with both dots and a slur above a series of notes. It lies somewhere between legato and staccato, with a special emphasis on each note.[30] In his *Méthode de piano* written for his students at the Paris Conservatoire, Adam gives a four-measure example that begins as in Ex. 5.5:

Ex. 5.5 Portato with *retard de la note* (Adam, 1804/5)

One must not jab at the key, just lift the finger. This manner of detaching adds much to the expression of the melody and is sometimes made with a little retard on the note which is being thus expressed.[31]

[29] Concerning the different styles of keyboard playing, see Carl Czerny, *Complete Theoretical and Practical Piano Forte School* (London, dedication dated 1839), copy in BL, iii, 100.

[30] See Rosenblum, *Performance Practices in Classic Piano Music*, 184–5, 438 n. 6, and 445 n. 93. The notation occurred earlier in the *Tragen der Töne* performed on the clavichord: see C. P. E. Bach, *Versuch*, i (1753), 8–9 and 126; Mitchell's trans., 36 and 156; and Robert Donington, *The Interpretation of Early Music*, new version (Faber, 1975), 577. See also Türk (1789), 354, and Haggh's trans., 343.

[31] Louis Adam, *Méthode de piano du Conservatoire* (Paris, 13th year of the French revolutionary calendar, which runs from 23 Sept. 1804 to 22 Sept. 1805), copy at UCLA ML, 156; trans. mostly from Blake, 'Tempo Rubato in the Eighteenth Century', 57.

He illustrates this *retard de la note* in Ex. 5.5*b* by showing each tone delayed by a rest equal to a quarter of its value. This reminds us once again of Couperin's *suspension* (Ex. 2.9) and Marpurg's *Verbeissen* (Ex. 5.1*c*). Francesco Pollini, influenced by Adam, includes a similar example in his *Metodo per clavicembalo* (Milan, 1810).[32] Adam also seems to be describing tempo rubato more generally (but without naming it) when he writes, after complaining about those who perform every piece in free rhythm:

Without doubt expression requires that one slow down or hurry certain notes of the melody, but these retards [and accelerations] should not be continued during the entire piece, but only in those places where the expression of a languorous melody or the passion of an agitated melody requires a retard or a more animated movement. In this case it is the melody that it is necessary to alter, and the bass ought to strictly mark the beat.[33]

Wilhelm Volckmar, finally, describes a late and unusual application of the earlier type of rubato in his *Orgelschule* of 1858. One of his methods for emphasizing two simultaneous melodic lines is 'by letting certain parts of one melody linger slightly and then hurry, while the other continues in the strictest tempo (Tempo rubato)'. He gives an example of twenty measures in which two voices are played by the right hand and two others by the left on a softer second manual. He explains that 'in the performance of the two melodies played by the right hand, one is to be set apart from the other through the *tempo rubato*'. This is the only source, as far as I know, to mention the use of rubato by organists. Although Volckmar describes the technique in terms of retardation and acceleration, an organist would surely be vividly conscious of the displacement produced between notes written in vertical alignment.[34]

Most of the keyboard sources, however, present tempo rubato primarily in terms of displacement—between a melody and its bass, between the left and right hands. They were therefore attracted to the manner in which the baroque theorists had depicted it. Thus the

---

[32] See ibid. 57–8.

[33] *Méthode de piano*, 160: 'Sans doute l'expression exige qu'on ralentisse ou qu'on presse certaines notes de chant, mais ces retards ne doivent pas être continuels pendant tout un morceau, mais seulement dans quelques endroits où l'expression d'un chant langoureux ou la passion d'un chant agité exigent un retard ou un mouvement plus animé. Dans ce cas c'est le chant qu'il faut altérer, et la basse doit marquer strictement la mesure.' See Blake, 'Tempo Rubato in the Eighteenth Century', 99.

[34] Br. & H, *Vorwort* dated 1858, copy in BL, 101: 'indem man einzelne Theile der einen Melodie um ein Geringer zögern und dann wieder eilen lässt, während die andere im strengsten Takt ihren Fortgang nimmt. (*Tempo rubato*)'; and 104: 'Bei Ausführung der beiden von der rechten Hand vorzutragenden Melodien soll die eine der andern durch das *tempo rubato* entgegen gesetzt werden.'

theorists (who were listeners), the continuo players (who played only the strict accompaniment), and the solo keyboard player (who performed with two independent hands) all tended to approach tempo rubato in a similar way. Never is keyboard rubato represented, as it is by García and Baillot in vocal and violin music, by alterations in a separate melodic line. It is usually portrayed as syncopation, and the later examples (see Exx. 5.4 and 5.5), influenced perhaps by the French emphasis on two-note arpeggiation and the *suspension*, seem to favour retardation rather than anticipation of the melody notes.

Other Kinds of Rubato

During the period from about 1770 to the first decade or so of the nineteenth century, the term *tempo rubato* was used for at least five other related musical devices, all deriving from the idea of displacement embodied in the examples of Quantz and his followers. They were not necessarily confined to keyboard music, but some are described by C. P. E. Bach and Türk in their keyboard books. They involve displacement due to an unusual number of notes, as well as displacement of a dynamic accent, a metre, a contrapuntal voice, or a syllable of text.

We have already seen unusual numbers of notes in the embellished versions of the Benda violin sonatas. We have also seen that the unusual groups sometimes cause an alternating displacement with bass notes that occur simultaneously with the group (second line of Plate VII). There seems to be a tendency at times to regard this polymetric sound by itself as tempo rubato, regardless of the extent of the rhythmic robbery and regardless of whether an unrobbed version exists for comparison. Quantz's schemes (Ex. 3.8) depicted this alternation of melody and bass notes; the unusual groups, when accompanied during their course by bass notes, produce the same general effect, but with greater rhythmic complexity. This alternation seems to have been the most conspicuous feature for those who listened to tempo rubato. It was also appealing to the keyboard soloist, who faced the challenge of producing rubato with the right and left hands.

E. W. Wolf mentions in 1785, in connection with his description of the earlier type of rubato, that 'at the suitable places, one or more notes may be added, producing an irregular number of notes to the beat [referring especially to quintuplets and septuplets]'.[35] In his book of 1788 he adds: 'One can thus fill one or several beats, as demon-

---

[35] *Eine Sonatine*, p. ix of the *Vorbericht*; trans. by Hogwood in *C. P. E. Bach Studies*, 151.

strated by keyboard sonatas by myself and Bach as well as some violin solos by Franz Benda'.[36] It is indeed C. P. E. Bach who combines the idea of the unusual group with alternating displacement as the sole definition of tempo rubato. Since Bach worked with Quantz and Benda for many years at the court of Frederick the Great, it is difficult to determine which of them may have influenced the others. It is tempting to speculate, however, that Benda, as the solo performer, was the first to be involved with rubato, and that Quantz, Bach, and the copyist of the Benda manuscript were all attempting to notate what they heard Benda play. In any event, Bach makes use of the device occasionally at sensitive moments in the formal structure of a movement in sonata form. In some cases there is a simpler version of the passage for comparison.

Ex. 5.6 C. P. E. Bach, Sonata IV, 1st movt. (1760)

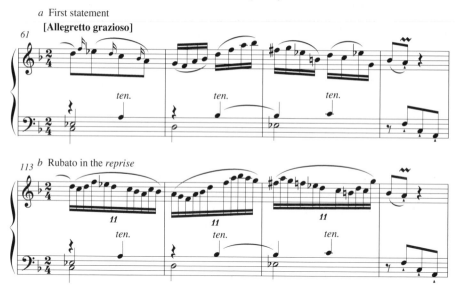

Ex. 5.6 comes from the first movement in the fourth of Bach's *Six Sonatas with Varied Reprises for Keyboard*, published in 1760 at Berlin.[37] In this collection Bach has written out the repetition of each section of the movements in sonata form for the sake of those not

---

[36] *Musikalischer Unterricht*, i, 35.

[37] *VI. Sonates pour le clavecin avec des reprises variées* (Berlin, 1760), repr. with intro. by Darrell Berg in *The Collected Works for Solo Keyboard by Carl Philipp Emanuel Bach 1714–1788* (New York and London: Garland, 1985), ii, 17–18 of original (62–3 of repr.); modern edn. in *Carl Philipp Emanuel Bach, Sechs Sonaten mit veränderten Reprisen für Clavier (1760)*,

skilled enough to improvise. When an unusual group appears in one of these varied reprises, we can therefore locate a simpler version in the original statement of the same section. Ex. 5.6a appears near the beginning of the original second section; Ex. 5.6b shows the corresponding measures in the varied reprise. This passage represents a deliberate approach to an F major dominant chord, which is later prolonged before resolving in a conspicuous cadence to its tonic of B flat major. The first time the expression is enhanced by appoggiaturas, continuous sixteenth-note movement, and chromaticism; the variant intensifies the passage even more by the eleven-note groups that momentarily threaten the regularity of the metre. Bach makes the location of the quarter notes in the left hand of (b) very clear by leaving extra space for them between the melody notes.

A comparison of Ex. 5.6b with (a) reveals all sorts of subtle displacement in the notes of the melody, but these project to the listener an effect different from the sound of the sixth and seventh notes of the eleven-note groups alternating with the left-hand notes. Bach includes unusual groups also when there is no written-out reprise or no corresponding spot elsewhere in the movement for comparison.[38] Rubato must refer in this case either to note values stolen from some imaginary simpler version, or—more probably—to metrical regularity stolen by an unusual group and made perceptible by alternation with notes in the left hand. For Bach, tempo rubato apparently did not require a simpler written or previously heard version, but involved only the sort of metrical counterpoint that set irregularity conspicuously against regularity.

Bach makes this clear in the 1787 edition of his *Versuch*. In the first edition in 1753, Bach discusses tempo rubato only in relation to the continuo player. In a second edition of 1787, however, he includes an extensive description of rubato played by a solo keyboardist. It should be noted that he never supplies musical examples and never refers to the schematic type of passages given by Quantz. Separate sentences from Bach's 1787 description of rubato seem by themselves to fit the earlier type of rubato and have often been quoted in this context. It is with considerable reluctance that I have finally come to agree with scholars who have recently pointed out that Bach was not speaking of

ed. Étienne Darbellay (Winterthur: Amadeus Verlag, 1976), 37–8. Following Darbellay, I have added 2 obviously missing notes in the middle voice of m. 63 and a tie in m. 62. See also the 7-note groups in mm. 147–8, a variant of mm. 95–6, and the 5-note groups in mm. 126–7, a variant of mm. 74–5.

[38] For examples, see *The Collected Works for Solo Keyboard*, ii, 113 of the repr. for an 11-note group, 280 for a group of 9 notes, 281 for 5 notes, and 282 for 13.

the usual tempo rubato, but rather of the use of unusual groups—specifically groups like those in Ex. 5.6 which cause displacement with notes in the accompaniment.[39]

The indication of tempo rubato, according to Bach, 'is simply the presence of more or fewer notes than are contained in the normal division of the bar'. He gives this description:

A whole bar, part of one, or several bars may be, so to speak, distorted in this manner. The most difficult but most important task is to give all notes of the same value exactly the same duration. When the execution is such that one hand seems to play against the bar and the other strictly with it, it may be said that the performer is doing everything that can be required of him. It is only rarely that all parts are struck simultaneously.... Slow notes and caressing or sad melodies are the best, and dissonant chords are better than consonant ones.... Without a fitting sensitivity, no amount of pains will succeed in contriving a correct rubato. As soon as the upper part begins slavishly to follow the bar, the essence of the rubato is lost... Other instrumentalists and singers, when they are accompanied, can introduce the tempo [rubato] much more easily than the solo keyboardist.

According to Bach, there are many examples of this type of rubato in his own keyboard works.[40] He does not speak of one note stealing time from another. He only requires the keyboard player to render the passage precisely as he has notated it, with every note of an unusual group receiving 'exactly the same duration'. Presumably the performer could also include such a group for expressive reasons when improvising a varied reprise himself. It appears to me that Bach, admitting that the real tempo rubato was far more difficult if not impossible for the solo keyboard player, tried to devise a method of performance that would approximate the effect he heard when listening to the singers and violinists. The unusual number of notes, although difficult, was at least manageable for skilled keyboard performers. Since 'it is only rarely that all parts are struck simultaneously', this method produced the desired sound of a displacement between melody and accompaniment.

Such unusual groups probably occur every time improvised embellishment becomes popular. We saw them in Chapter 2 with Ganassi

---

[39] As far as I know, Darbellay was the first to discover this; see his edn. of *Sechs Sonaten*, pp. iv, xiv, xxxv, and xxxvi in the intro., and 34, 38, 54, 61, and 63 in the music. Blake, in 'Tempo Rubato in the Eighteenth Century', 75–84, illustrates every sentence from Bach's 1787 description with examples drawn from his music.

[40] *Versuch über die wahre Art das Clavier zu spielen*, rev. edn. (Berlin, 1787), copy at the Univ. of Iowa, i, 99–100; modern printing of the section on rubato in the fac. of the 1st edns. of 1753 and 1762 (Br. & H, 1969), 13–14 at end of volume. Trans. from Mitchell's *Essay on the True Art of Playing Keyboard Instruments*, 161–2. On 162, however, he incorrectly translates 'In meinen Clavier-Sachen findet man viele Proben von diesem Tempo' as 'Most keyboard pieces contain rubato passages.'

in 1535 (Ex. 2.2c and e). They are finally included in keyboard scores by composers beginning in the second half of the eighteenth century and gradually increase in number and complexity during the nineteenth century. We are most familiar with them, of course, in the music of Chopin. Hummel describes groups of up to fifty notes per measure heard against four triple groups in the left hand. He emphasizes the strict accompaniment and the independence of the hands, but, unlike Bach, states that one 'must play the first notes [in the right hand] . . . rather slower'. With the left hand maintaining strict time, and the right varying its rate of speed, we actually have here the real tempo rubato in which the performer steals time from the written note values.[41] Although Czerny also seems to favour a generally strict rhythm for the accompaniment, some of his examples include a rallentando in both hands for the final few notes. He feels that one should avoid 'too careful a distribution' of the right-hand notes, for 'it should appear as if these embellishments had first presented themselves to the player during the performance, and that they flowed unpremeditatedly from his own fancy'.[42]

Neither Hummel nor Czerny refers to these embellishment groups as tempo rubato. A number of features, however, bear a close resemblance to rubato: the independence of the hands, the improvisatory sound, the effect of heightening the expression, almost guaranteed displacement between melody and bass, and, above all, the general sound of a simple, strict accompaniment heard simultaneously with a rhythmically free melody. As Czerny suggests, this is a means of imitating the effect obtained more easily by singers and violinists. E. W. Wolf, C. P. E. Bach, and the copyists of the Benda manuscript are the only ones, as far as I know, who actually link this practice with rubato. Bach's book, however, was enormously influential, and he himself was a sensitive observer, a thorough scholar, and an imaginative composer. His involvement with the *empfindsamer Stil* and with the clavichord reveals his special interest in expressive nuances. His approach to tempo rubato is therefore of significance to us, for it shows us the way he heard the rubato of the singers and the violinists.

Bach also employs syncopation as an expressive device, but does not include it as part of rubato. There are passages in his varied reprises that are syncopated versions of the original measures.[43] This is

---

[41] *A Complete Theoretical and Practical Course of Instructions on the Art of Playing the Piano Forte* (London, preface dated 1827), copy in NYPL, iii, 53; French version: *Méthode complète théorique et pratique pour le piano-forte* (Paris, 1838; repr. Minkoff, 1981), 452.

[42] *Piano Forte School*, iii, 43.

[43] Compare, for example, mm. 36–7 with 5–6 in the 3rd movt. of the 1st sonata from *VI. Sonates* (1760), fac. edn. in *The Collected Works for Solo Keyboard*, ii, 2–3 of original (47–8 of repr.); modern edn. in Darbellay's *Sechs Sonaten*, 6.

precisely the sort of regular alternation between melody and bass that
we have seen in the rubato of Marpurg (Exx. 5.1 and 5.2) and Türk
(Ex. 5.3 by Graun) and in the example attributed to Hummel (Ex.
5.4). Bach's unusual groups probably sounded more like the rubato of
the singers and the violinists. In spite of this, however, the traditional
syncopated melody had become so well entrenched that it continued
to be the principal method of representing tempo rubato. Furthermore,
it gave birth to still other meanings for the word *rubato*.

Occasionally rubato referred to a dynamic accent on a weak beat.
This may be related to the crescendo and swelling which Quantz,
Hiller, and Spohr associated with syncopated notes. Türk gives this in
1789 as the second meaning of rubato and in 1802 as the third (see
his examples on the second staff of page 375 in Plate XI and repeated
again in Plate XIII). In 1774 Reichardt describes the special charm in
the singing of 'Madem. Benda of Potsdam', specifically her performance of the recitative from Graun's *Der Tod Jesu* that represents
Peter weeping bitterly (see Ex. 5.7).[44] According to Reichardt, 'she

Ex. 5.7 H. C. Graun, *Der Tod Jesu*, recitative

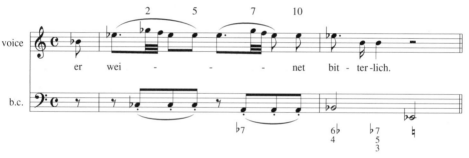

used the so-called tempo rubato, that is, she gave a strong accent to
a note which ordinarily should have none, thereby effecting true
sobbing'. He further urges his readers to try it themselves: 'Give to the
second, fifth, seventh, and tenth notes of this measure [which I have
marked in Ex. 5.7] a special emphasis while, in turn, letting the others
diminish. Then you will see what a special effect it is.'[45]

[44] Carl Heinrich Graun, *Der Tod Jesu*, ed. Howard Serwer, *Collegium musicum, Yale Uni-
versity*, 2nd ser., v, 54 (from the recitative 'Nun klingen Waffen').
[45] Johann Friedrich Reichardt, *Briefe eines aufmerksamen Reisenden die Musik betreffend*, i
(Frankfurt and Leipzig, 1774), microcard from copy at SML, 41–3; trans. mostly from Blake,
'Tempo Rubato in the Eighteenth Century', 62–3. See also Boris Bruck, *Wandlungen des
Begriffes Tempo rubato* (Berlin: Paul Funk, [1928]), 25–34. Madem. Benda (incorrectly cited as
Madame by both Bruck and Blake) may be Juliane, the daughter of Franz; she was a singer and
composer and married Reichardt in 1776.

In 1782 Petri presents such an effect as the only meaning for rubato.[46] Koch in 1802 and Schilling in 1838 include it as the second of three meanings, Koch in 1808 as the first of several.[47] Koch also includes an example in which all voices have rests on the strong beats. Composers increasingly employ weak-beat accentuation, but without considering it a type of rubato. C. P. E. Bach uses it as a method of variation.[48] By the time of Beethoven, of course, this sort of sforzando was a characteristic part of the vigorous and athletic flow of classic rhythm and was, at the same time, part of the broader concept of suddenly violating the listener's expectancy for humorous or dramatic purposes. We hear it in the music of Beethoven on single notes (first movement of the 'Eroica' Symphony, measures 40 and 42, for example) or in longer passages (measures 28–34 in the same movement). In the latter case a sforzando occurs seven times in the entire orchestra on alternate quarter notes in 3/4 metre, thus threatening a change to 2/4.[49]

An accent can be placed on a weak beat and a new metre suggested also by the succession of rhythmic figures. In 1808 Koch gives the melodies in Ex. 5.8a and b, which show that when a three-note figure consisting of two eighth notes and a quarter note recurs in 3/4 metre as in (b), it temporarily produces the actual rhythm shown in (a).[50] I have added in (c) the hemiola rhythm that seems to me to result naturally from (b). The actual effect depends to some extent, of course, upon which rhythm is emphasized by the other voices. Koch says that Joseph Haydn and Dittersdorf were the first to use this type of rubato, although sparingly, in their minuets. In his *Musikalisches Lexicon* of 1802 Koch defines tempo rubato first of all as 'a rhythmic

[46] Johann Samuel Petri, *Anleitung zur praktischen Musik*, 2nd edn. (Leipzig, 1782; repr., Giebing über Prien am Chiemsee: Musikverlag Emil Katzbichler, 1969), mainly p. 164, but also pp. 161, 166, 167, and 207. He advocates rubato for expressing the particular affect that combines lamenting and sobbing with obstinacy and hostility.

[47] Heinrich Christoph Koch, *Musikalisches Lexicon* (Offenbach am Main, 1802), 1502–3; Gustav Schilling, *Encyclopädie der gesammten musikalischen Wissenschaften oder Universal-Lexicon der Tonkunst* (Stuttgart, 1838), vi, 602; and Koch's article 'Über den technischen Ausdruck: Tempo rubato', *AMZ* 10/33 (11 May 1808), 515.

[48] For a spectacular example, compare mm. 16 ff. and 124 ff. from the last sonata in *VI. Sonates* (1760), fac. edn. in *The Collected Works for Solo Keyboard*, ii, 30 and 32 of original (75 and 77 of repr.); modern edn. in Darbellay, *Sechs Sonaten*, 65 and 69.

[49] Rosen adds that the same sort of accent on a weak beat occurs sometimes as the result of unusual slurring. He cites Chopin's Nocturne Op. 27 No. 2, m. 28 (where the 4th and 5th notes of a group of 6 8th notes in the 2nd half of a measure of 6/8 are slurred together and the last one is slurred to the 1st note in the next measure) and the 2nd movt. of Mozart's Piano Sonata in A minor, K. 310/300^d, mm. 22 and 25 (where both slurs and dynamics indicate accentuation on the 2nd of 4 16th beats).

[50] *AMZ* 10/33 (11 May 1808), 517. See Blake, 'Tempo Rubato in the Eighteenth Century', for the German (108–9), a trans. (111–12), and the examples (68).

Ex. 5.8 *Imbroglio* (Koch, 1808)

movement which has been stolen from another metre'. He refers the reader to the synonym *imbroglio*, where he prints the opening five measures from the Trio of the Minuet in Haydn's String Quartet Op. 9 No. 3.[51] This meaning of rubato recurs in the dictionaries of Danneley (London, 1825), Lichtenthal (Milan, 1826), Schilling (Stuttgart, 1838), and Gathy (Hamburg, 1840), all of whom quote either Koch's musical examples or his words.[52] A related meaning of rubato, unique to Koch, involves the imitation on a weak beat of a melodic motive that originally appeared on a strong beat.[53]

Still another meaning for tempo rubato concerns the placing of short syllables on long notes. Türk depicts it in 1789 in examples from Pergolesi's *Stabat Mater* (the last two staves on Plate XI).[54] He repeats the same examples in 1802 (Plate XIII), but mentions in a footnote that this type of rubato is controversial. In 1771 Kirnberger gave examples of such syncopation used for expressive purposes in operas by C. H. Graun and Handel.[55] Reichardt, however, complains in 1774

[51] *Musikalisches Lexicon*, 776 and 1502.

[52] John F. Danneley, *An Encyclopedia, or Dictionary of Music*, copy at Mills Music Library, Univ. of Wisconsin, Madison, repr. in *MD*, article on 'Tempo rubato'; Pietro Lichtenthal, *Dizionario e bibliografia della musica*, ii, 245; Schilling, *Encyclopädie*, iii, 691–2, and vi, 602; and Auguste Gathy, *Musikalisches Conversations-Lexicon*, 2nd edn., 461. Lichtenthal, after describing displacement of accent and metre, comments that 'the plagiarists, stealing the melodies of others, sometimes try to mask the theft with Tempo rubato' ('I plagiarj, involando le melodie altrui, cercano talvolta di mascherare il furto col Tempo rubato').

[53] *AMZ* 10/33 (11 May 1808), 516; and Blake, 'Tempo Rubato in the Eighteenth Century', 67–8, 108, and 111.

[54] Türk translates the texts into German. The first comes from the aria 'Cujus animam gementem', the other from the aria 'Quae moerebat et dolebat'. See the piano/vocal score in *Opera omnia di Giovanni Battista Pergolesi*, ed. F. Caffarelli, xxvi (Rome, 1942), 4 and 9; or the full score ed. Helmut Hucke (Br. & H, 1987), 13 and 20. Pergolesi composed the work in 1736, the year of his death.

[55] Johann Philipp Kirnberger, *Die Kunst des reinen Satzes in der Musik*, i (Berlin and Königsberg, 1771; repr. Olms, 1968), 197–8; trans. by David Beach and Jurgen Thym in *The Art of Strict Musical Composition* (YUP, 1982), 212–13.

about short syllables that are made long and awkward,[56] and Schulz states that the displacement in the Pergolesi aria (Türk's first example) is so improper that 'any linguist would shudder upon hearing it'.[57] In 1782 Reichardt, in a discussion of this sort of tempo rubato in Reinhard Keiser's opera *Die grossmüthige Tomyris*, comments that the capricious lengthening of a short syllable was probably less offensive in 1717 when the opera was written.[58] In 1798 Dittersdorf replied to Schulz's article with a defence of Pergolesi in the *Allgemeine musikalische Zeitung*. A year later, in two issues of the same journal, Schulz made a lengthy reply to Dittersdorf.[59]

All of these less usual meanings for tempo rubato largely disappear early in the nineteenth century, except for a few dictionaries that continue to quote Koch. The *imbroglio*, the long syllable on a short note, and the imitative entry on a weak beat are all compositional devices. The dynamic accent on a weak beat and the unusual groups could be added by the performer as well as notated by the composer. The word *rubato* is applied to these special types by those who heard the rubato of the singers and violinists and recognized some similarity. Although these less usual types were called *rubato* for only a very brief period of time, they do convey to us important information concerning the real rubato, for they all involve some sort of metrical disturbance in the melody which often clashes with the regularity of the accompaniment. Most of them also involve in some way an alternating displacement between notes of the melody and bass, and this was a point of view which the keyboard sources shared, as we have seen, with the listener and theorist. This may explain why some of the special types are included in the keyboard books of Bach and Türk.

The Later Rubato

During this period of many rubatos one other type was born—one which became so important in keyboard music that it lived concurrently for awhile and became confused with the earlier rubato, which

---

[56] *Briefe eines aufmerksamen Reisenden die Musik betreffend*, i, 42.

[57] Johann Georg Sulzer, *Allgemeine Theorie der schönen Künste*, ii (Leipzig, 1774), copy in UCLA URL, 1219 (in the article 'Verrükung').

[58] Johann Friedrich Reichardt, *Musikalisches Kunstmagazin*, i–ii (Berlin, 1782; repr. Olms, 1969), 35–8 (especially 36). In a review of this article in *Magazin der Musik*, i, 41–2 (issue for 15 Jan. 1783), Carl Friedrich Cramer suggests that this device could better be called 'syncopation with long syllables' (*Syncopation bey langen Sylben*), since it is quite a different thing from Bach's tempo rubato.

[59] *AMZ* 1/13 (26 Dec. 1798), 201–5, and 2/15 and 16 (8 and 15 Jan. 1800), 257–65 and 273–80.

it finally superseded in all types of music. We have previously noted W. A. Mozart's observation in 1777 that when some keyboard performers thought they were playing rubato, their left hand was actually following the right. This would have resulted in general tempo changes in the music as a whole and may explain why such changes came to be considered a type of rubato in keyboard music at a time when only the earlier type was recognized, for the most part, in vocal and violin music.

The earliest source to link tempo flexibility with tempo rubato seems to be Christian Kalkbrenner's *Theorie der Tonkunst* published at Berlin in 1789. Although this work concerns music in general, Kalkbrenner seems to have been considerably involved with keyboard instruments, for in addition to operas and symphonies he wrote keyboard pieces, keyboard trios, a keyboard concerto, and violin sonatas. Moreover, his son Friedrich became an internationally famous pianist during the 1820s and 30s. In the *Theorie der Tonkunst* Kalkbrenner names some words that composers write in the score to indicate a slower tempo or a gradual slowing. The next section, entitled '*Tempo rubato*', commences as follows:

> Sometimes *Tempo rubato* is also set above or below a place in the score, in order to indicate thereby that the performer of the music can at this spot prolong the movement a very little, and let the notes gradually last somewhat longer.

Since he does not mention any compensating acceleration nor any strictness in the tempo of the accompaniment, he is apparently not describing the earlier type of rubato. He does, however, warn against too much tempo rubato and advises, since it is such an intensely expressive device, that it be carefully restricted:

> In places where the accompanying voices are full of passage work, one may not apply it without revealing a lack of musical perception. Only in places where the accompanying voices have few or sustained notes is it to be applied with profit . . .
>
> Also the *Tempo rubato* may not be continuous, but must cease as soon as it is perceived and return to the preceding fixed movement. In order to make its effect perceptible, it requires only a few notes and a half or at most a full measure is sufficient.[60]

---

[60] *Theorie der Tonkunst*, i, copy at NYPL, 12. The entire section reads: 'Zuweilen wird auch *Tempo rubato* über oder unter eine Stelle gesetzt, um damit anzuzeigen, dass der Ausüber der Musik bey dieser Stelle die Bewegung ein klein wenig verlängern kann, und die Noten allmählig etwas länger kann dauren lassen.

'Das *Tempo rubato* ist oft eine grosse Zierde eines Tonstüks, wenn es recht gut gemacht wird. Es gehört aber sehr viele Kenntniss dazu, um es jedesmahl recht schicklich anzubringen. Bey Stellen, in welchen die begleitenden Stimmen bearbeitet und sehr figurirt sind, darf man es nicht

I have located only a single score from this period which includes such a tempo rubato: the manuscript, probably from the late 1790s, of a sonata by Hummel. Here it apparently indicates rhythmic flexibility in a tiny phrase leading to a fermata—a phrase which acts like an expressive echo of the preceding music (see Ex. 5.9*a*). Three bars later

Ex. 5.9 Hummel, Sonata in F minor, 1st movt.

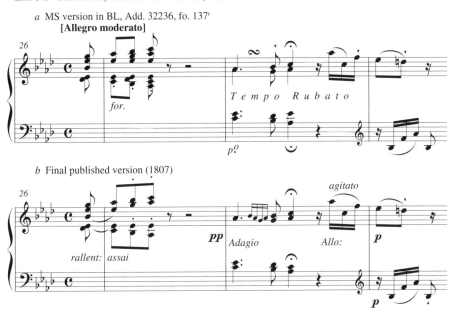

he writes *a Tempo I*. When the sonata was finally published in 1807, the corresponding passage was modified as in Ex. 5.9*b*: now, *Tempo rubato* has been replaced by *Adagio* (meaning, presumably, both slow and free), and tempo markings appear on both the following and preceding music. He apparently decided that *Tempo rubato* by itself

anwenden, ohne den Mangel musikalischer Erkenntniss zu verrathen. Nur blos in Stellen, in welchen die begleitenden Stimmen wenige oder liegende Noten haben, ist es mit Nutzen anzuwenden, und die schönste Wirkung gewiss damit zu erreichen.

'Auch darf das *Tempo rubato* nicht anhaltend seyn, sondern muss, sobald es empfunden ist, aufhören, und in die vorhergehende bestimmte Bewegung übertreten. Um seine Wirkung fühlbar zu machen, dazu braucht es weniger Noten, und ist ein halber, höchstens ein ganzer Takt hinreichend genug.

'Doch so sehr das gut angebrachte *Tempo rubato*, die leidenschaftlichen Empfindungen auf das vollkommenste ausdrückt, so behutsam muss man doch immer mit seiner Anwendung seyn. Personen, die sich allzusehr dem *Tempo rubato* überlassen, kommen sehr leicht in Gefahr, das richtige Gefühl vom Takte zu verlieren, und alsdann ist es nicht möglich, dass sie mit einem stark besetzten Orchester ordentlich begleitet werden können.'

left too much freedom to the performer. It is possible that it was meant to include not only the Adagio rhythm of the measure in which it was written, but also the following *agitato* and *allegro*, which were perhaps to be applied freely until *a Tempo I* in measure 31.[61]

Although Kalkbrenner mentions slowing and not hastening, Thomas Busby around 1801 defines tempo rubato in *A Complete Dictionary of Music* as 'an expression applied to a time alternately accelerated and retarded for the purpose of enforcing the expression'.[62] In 1802 Türk concludes his discussion of rubato by stating (at the top of page 421 on Plate XIII) that 'also the previously mentioned intentional hastening and hesitating is by some designated by the expression *Tempo rubato*, although one usually intends to indicate thereby only the hesitation or pausing (not the hastening).' (See also the last sentence on Plate XIV.)

Türk includes tempo rubato, as we have seen, as one of the extraordinary means of emphasizing the expression. He also names two other means: intentionally playing without steady time, and purposeful quickening and hesitating. The first (see Paragraph III on Plate XIV) refers to fantasies, cadenzas, and passages marked *recitativo*, which should be performed as a singer or actor declaiming a text, with the more important notes slower and louder. The other extraordinary means, the one he says is called *tempo rubato* by some, differs from this apparently only to the extent that it involves more subtle tempo changes within portions of a composition that are not generally characterized by a metrically free nature. He describes a number of situations where this means can be employed (Paragraphs IV, V, and VI on Plate XIV): one can hasten or play *accelerando* in order to express anger, to intensify a repeated idea, or to emphasize a lively passage that interrupts a gentle piece; one can retard in tender or melancholy pieces, preceding certain fermatas (as Hummel did in Ex. 5.9*a*), at the end of a composition, in passages introductory to important parts of a composition, and for the repeat of a languid phrase.[63]

---

[61] The works from which Ex. 5.9*a* and *b* come are both repr. by Sachs in Hummel's *Complete Works for Piano*, i (1989), 165 and 31. Concerning the MS sonata, which he labels S.23, see his comments on p. xi, as well as in *New Grove*, viii, 786, and in his 'Checklist' in *Notes*, 30 (1973/4), 738. In the MS, the word *Rubato* and the *a Tempo* 3 bars later are written in larger and darker letters than most of the other markings in the piece, including the *Tempo* preceding *Rubato*, as though they had been added at a later date.

[62] Repr. in *MD*, but with the incorrect date of 1786. We noted in Ch. 3 that Nathan refers in 1836 to the rhythmic alterations of singers as acceleration and retardation. Since he uses the phrase 'for the purpose of enforcing any particular passion', he may be referring to Busby's definition. It seems to me, however, that Busby, who is not confining his definition to singers or any particular medium, must be referring to the acceleration and retardation of tempo rather than notes, and thus to the later rather than the earlier type of rubato.

[63] The Ex. 14 which Türk mentions near the end of Paragraph VI on my Pl. XIV contains 3 measures marked *Larghetto*. A cadence to D minor, which begins in the 1st measure and

In the 1802 edition Türk, who was himself also a composer, warns in a footnote against the distressing misuse of this device. He complains that almost every so-called virtuoso and solo singer hurries or hesitates too often and at inappropriate places.[64] Such comments give evidence that the classic restraint of tempo was gradually being modified, for expressive reasons, to include more and more of the later type of rubato. In the process, performers and composers seem to compete with each other: the composers attempt to exert control by marking more and more tempo indications in their scores, whereas the performers increasingly include tempo changes in places not marked. The composers complain about the excessive use of later rubato by the performers; the latter complain that composers are usurping their independence in the area of execution. Together, nineteenth-century performers and composers gradually increase the flexibility of the tempo so that by the second half of the century it is probably the chief element in the service of musical expression.

The development, however, happened very slowly, so that the excesses Türk complained about in 1802 might seem very slight or perhaps even unnoticeable to us today. We must remember that he considered all types of rubato to be *extraordinary* means of expression. Eighteenth-century scores include very few words to indicate tempo changing, as we will see in greater detail in the next chapter on Haydn, Mozart, and Beethoven. By 1802, however, a few such words had begun to appear, so that Türk, in the second edition of his book, could add that slowing down is often prescribed by words such as *tardando*, *ritardando*, *rallentando*, and *lentando*. Concerning hastening, he suggests in 1802 that it would be better if, instead of leaving it up to the taste of the performer, the composer designated the appropriate spots himself, as some do by writing *accelerando* or *precipitando*.[65]

Türk felt the need to be more precise, however, and invented his own system of notation. In his own compositions, three parallel horizontal lines above the top staff, each beginning at a different point, signify a gradual slowing (later indicated by *ritardando* and *rallentando*), with its conclusion marked by a downward turn of the lines. Two parallel lines that turn downward at both ends mean a

---

concludes on the 1st beat of the 2nd, is followed by an unaccompanied ascending chromatic run in faster note values in the right hand; this leads to a new idea in F major in the 3rd measure. The ascending passage should be retarded, since it is introductory to an important part of the composition. The example also appears in Türk (1789), 372; Haggh's trans., 361; and Türk (1802), 416.

[64] Türk (1802), 415.
[65] Ibid. 415–16.

suddenly slower tempo (which we indicate by *ritenuto*). Türk cautions that in both cases the slowing should not be great, but rather almost imperceptible. His signs for acceleration are the same, but concluding with an upward turn. In addition, he marks notes which are to get a special expressive stress with a caret, and explains that such notes are to be played not with a violent jab, but rather with a gentle pressure and with a seeming prolongation of the tone.[66] In addition to these signs, he also includes terms such as *con espressione, ad libitum*, and *senza tempo*.

Cramer considered the lines unnecessary as well as offensive to look at and suggested that it would be simpler to use the customary expressions *tenuto, rallentando, smorzando*, and *accelerando*.[67] Türk answers Cramer's objections in a footnote in the 1802 edition of his book by pointing out that with these words one does not know how many notes are included, and one does not know whether the tempo change should be gradual or sudden.[68] As we now know, the words won out over Türk's signs and the termination of their effect was eventually marked by *a tempo*. Türk's other concern, however, continued to be troublesome, for it was only late in the nineteenth century that a distinction between *ritardando* and *ritenuto* was clarified.

Türk's writings and musical compositions thus reflect with unusual clarity this period in the history of keyboard music when a more subjective attitude was just beginning to demand greater flexibility of tempo within previously strict musical forms. We can see the conflict between composer and performer: how much flexibility should be marked by the former and how much left to the latter? Both points of view are presented also by Philip Antony Corri in *L'Anima di musica, An Original Treatise Upon Piano Forte Playing* (London, 1810). Corri was a composer, singer, and teacher; his sister was the wife of Dussek, who was in the publishing business with Corri's father Domenico. We have already noted that Dussek was skilled in the performance of the earlier type of rubato and that he struggled in vain to find a method of notating it. We have also seen (in Plate III) the clear and concise manner in which Domenico in 1810 describes this type of rubato, separating it completely from the 'Quickening or Retarding of Time'. It is therefore surprising to find Philip Antony, in his book published the same year, omitting any mention of the earlier rubato and applying the expression *tempo rubato* exclusively to the later type.

---

[66] The signs are explained and utilized occasionally in *Sechs leichte Klaviersonaten* (Leipzig and Halle, 1783) and the *Klavierschule* (1789), 373 (Haggh's trans., 362) and (1802), 417.
[67] *Magazin der Musik*, i, 1278–85 (issue for 7 Dec. 1783).
[68] Türk (1802), 417.

The third part of *L'Anima di musica* is entitled 'Of Expression and Style'. 'Expression', writes Philip Antony, 'consists of touch, emphasis, and modulation of sound—that is: the crescendo, diminuendo, and degrees of forte and piano—besides which, expression is produced by the protraction of time, by the Italians termed TEMPO RUBATO, but more properly should be styled TEMPO PERDUTO [lost time].'[69] He then pursues the latter more extensively in the section 'Of ad libitum'. He represents the composers' point of view when he states that the slackening of time is more often marked by words such as *slentando*, *ritardando*, *calando*, *smorzando*, *morendo*, and *perdendosi*, and the resumption of the former time by *tempo primo* or *a tempo*. 'When ad libitum is mark'd', however, 'it is left to the performer's discretion to introduce ornaments, and is generally intended to protract or slacken the time.'[70] His main description of *tempo perduto* or *rubato* is on the following page (Plate XV) and is concerned with tempo changing added by the pianist as part of performance practice. He lists a number of musical situations where slackening of time is appropriate: preceding a held note (as in Ex. 5.9a), concluding a cadence, and in certain places within 'pathetic airs'.

His examples of 'pathetic airs', an expression reminiscent of Tosi's terminology, occur in Andante and Allegro movements from compositions by Dussek. With the letter A, the word *calando*, and a decrescendo sign (on Plate XV), he points out the half cadence in the first excerpt, where the melody, after reaching its structural goal on the pitch C, quietly expires on the broken chromatic octaves before attacking with renewed strength the parallel beginning of the next phrase. Presumably the first lower E on the upper staff should also have a natural sign. Both of his examples show, at the letter B, an unaccompanied melodic passage leading to a contrasting and gentler theme. The first such passage he marks with a decrescendo and *perdendosi*; the second constitutes the conclusion of a longer *ritardando*.

Kalkbrenner and Türk, as we have seen, both emphasized hesitating rather than hastening. Through his description and the word *perduto*, Corri also emphasizes almost exclusively the act of retarding. He does, however, cautiously mention the possibility (at the bottom of Plate XV) of enhancing the retard by accelerating the previous few notes. His illustration shows an acceleration preceding the ritardando that leads to a gentler theme—actually an embellished recapitulation in the

---

[69] *L'Anima di musica*, copy at BL, 68. See J. Bunker Clark, *The Dawning of American Keyboard Music* in *Contributions to the Study of Music and Dance*, xii (Greenwood, 1988), 276–80. Around 1813 to 1817 Corri moved to Baltimore and changed his name to Arthur Clifton.

[70] *L'Anima di musica*, 78.

The TEMPO PERDUTO, or RUBATO is the protracting or slackening of Time and may be used with effect (tho' not mark'd) in pathetic airs at particular places, where the melody seems to be expiring, (as at A in the following Example) or leading to a delicate Piano subject (as at B)

After the slackening of time, or tempo perduto — the former time must be resumed, usually marked "tempo primo"—or "a tempo"

The Tempo Perduto seldom lasts longer then a Bar or two or a few notes, and if done to excess is caricature . —

Examples from DUSSEK.

To give more effect to the Ritardando, just described, Accellerando may be used, that is; accelerating or hurrying a few notes before, so that there may be more compass for a gradual diminution of time. But this is so nice a point of Expression that I only submit it to the notice of those already proficients, Example;

PLATE XV. Philip Antony Corri (Arthur Clifton), *L'Anima di musica* (London, 1810), from pp. 79–80

home key (E flat major) of the theme which he showed in the previous example in the dominant of B flat. All of Corri's examples occur when the left hand is resting, an illustration of Kalkbrenner's admonition to use rubato only when the accompaniment has 'few or sustained notes'.

Corri reminds us once again of the restraint with which *tempo perduto* should be applied. If Türk's slowing, like Kalkbrenner's, was to be almost imperceptible, Corri's seldom lasts longer than one or two measures or a few notes, and 'if done to excess is caricature'. It was the excess that concerned a number of writers. Both Hummel and Czerny finally use the expression *tempo rubato* in a derogatory sense. Hummel, who, as we have seen, indicated the later type of rubato in a score (Ex. 5.9*a*) and perhaps also the earlier type (Ex. 5.4*a*), sadly states in 1827 that 'in the present day, many performers endeavour to supply the absence of natural inward feeling by an appearance of it'; he gives as examples, distortions of the body, constant use of pedals, excess embellishment, and 'a capricious dragging or slackening of the time (*tempo rubato*), introduced at every instant and to satiety'.[71] Although he recommends 'some little relaxation as to time' for singing passages in an Allegro, for example, he continually stresses moderation. In 1840, Chorley describes the 'admirable self-control' of Hummel's style as a performer—a control 'displayed in a measurement and management of *tempo* unequalled by any contemporary or successor'.[72]

Czerny echoes Hummel's reactions to excessive tempo changing:

We have almost entirely forgotten the strict keeping of time, as the *tempo rubato* (that is, the arbitrary retardation or quickening of the degree of movement) is now often employed even to caricature. . . .

In this way a solid composition is often so disfigured as not to be recognizable, and, although in the present day, a higher degree of expression is certainly required and can be introduced, we must nevertheless make a distinction between a fantasia and a regularly constructed work of art . . .[73]

On the other hand, he devotes a considerable portion of the third volume of his *Complete Theoretical and Practical Piano Forte School* to the acceptable types of rhythmic flexibility, which, significantly, he describes as 'perhaps the most important means of expression'. His presentation is so extensive and systematic that it is a most important source for understanding the application of the later type of rubato during this early part of the Romantic period. He gives a list of eleven situations for retarding—a list often quoted by later nineteenth-century

[71] *Art of Playing the Piano Forte*, iii, 40; *Méthode complète*, 438.
[72] Henry F. Chorley, *Modern German Music* (London, 1854; repr. Da Capo, 1973), 8.
[73] *Piano Forte School*, 2nd supplement, 29.

theorists as well as twentieth-century scholars. The situations are mostly structural in nature, such as the return of an important subject or a transition to a different rhythmic movement. One usually retards also, he says, when the composer writes *espressivo*; terms such as *rallentando, ritenuto, smorzando,* and *calando* 'are only distinguished from each other by the more or less degree of *ritardando*'. In addition, he urges the performer to respond to the music by holding back the time when feeling gentle persuasion, doubt, tender complaining, or grief, and by hurrying when sensing impatience, sudden cheerfulness, resolve, or pride. He even suggests particular situations that could evoke such emotions, such as 'whispering a secret', 'taking leave', 'hasty or curious interrogation', or 'timid flight'. This is the beginning of a more subjective approach by the romantic performer.[74]

Hummel and Czerny both use the expression *tempo rubato* when complaining of excessive rhythmic flexibility. Türk includes expressive tempo changing as the latest in a series of different meanings for the term. Both Kalkbrenner and Busby confine the definition only to this meaning, but without restricting it to keyboard music. The work of the younger Corri, however, is the only specifically keyboard source of this period, as far as I know, that refers to tempo rubato exclusively as the expressive employment of rhythmic flexibility.

Thus in keyboard sources spanning from around 1752 to 1840, diverse meanings arose for the expression *tempo rubato*. In the beginning, the keyboard performer, either as a continuo player looking at the score or merely as a listener, heard singers and violinists create rubato. The attempt to transfer the device to the keyboard involved a complex rhythmic relationship between the left and right hands. Only the best performers could successfully achieve the required independence of the hands.[75] Some keyboardists, admitting that they could not reproduce the exact effect, attempted to imitate the most conspicuous element they heard in vocal and violin rubato—the metrical conflict between melody and accompaniment, manifested most obviously by the sound of alternating displacement between them. Thus they were attracted to the syncopations of the theorists, as well as introducing unusual groups that would guarantee frequent displacement. Others, on the other

---

[74] Ibid. iii, 31–8.

[75] Instead of 'only the best performers', I originally wrote (considering Mozart's own comments) 'probably only a genius like Mozart'. To this Rosen replied: 'The independence of the hands needed is not all that difficult, in fact. It is the taste with which it is accomplished that is more of a problem.' I certainly agree with his emphasis on taste. Concerning independence of hands, however, we need to keep in mind that Rosen speaks as a pianist who has himself conquered far more formidable challenges of co-ordination in music by composers such as Elliott Carter (see Ch. 10).

hand, apparently continued to attempt the earlier type of rubato on the keyboard, but without success, since the left hand usually followed the right. Listeners to such a performance would hear retardations and occasionally accelerations in the general tempo of the music, and the performer would no doubt claim he was using tempo rubato. In addition, other meanings of the word, as we have seen, emerged briefly, all related in one way or another to displacement. Moreover, all of this took place during a period when the original type of rubato continued to be practised in violin and vocal music.

The variety of meanings for tempo rubato at the end of the eighteenth century and the beginning of the nineteenth, makes it difficult in some cases to determine which type is being described. Schubart, for example, gives a definition of *amoroso*, a tender, yearning method of execution which involves many subtle nuances added by the performer. The nuances include portamento, trills, sudden interruptions, and the tempo rubato, 'where the execution does not want to go on, yet does go on'. However, 'this tender hesitation of a man who is about to leave his lover . . . [is] only effective under the hand of a master'.[76] This could be, it seems to me, an eloquent description of either the earlier or later type of rubato. Schubart was a poet, journalist, and composer, as well as an accomplished performer on the organ, harpsichord, and clavichord.

Another ambiguous definition occurs in Herz's piano method of around 1837. His colourful description of earlier rubato in relation to Dussek, as we have seen, does not include the word *rubato*. Elsewhere, in the *Vocabulaire* from the same book, however, he defines *rubato* as 'robbed. Execution in which one robs, that is neglects the value of the notes and the metre' and *tempo rubato* as 'execution independent of the metre'. Taken by themselves, these definitions, again, could apply equally to either type. Perhaps in this case, however, since he omitted the word in his description of Dussek's playing, he is using the term *rubato* for the later type.

Around the same time, definitions began to appear in dictionaries of musical terms. Jousse gives a somewhat inaccurate impression by considering only two notes:

Rubato, Tempo Rubato. These words are applied to a passage in which, for the sake of effect, a note is made longer than its real value, and the succeeding note shorter, or vice versa, so that the measure is still in complete time.[77]

[76] Christian Friedrich Daniel Schubart (1739–91), *Ideen zu einer Ästhetik der Tonkunst*, ed. Ludwig Schubart (Vienna, 1806), 366–7: 'das *Temporubato* [sic], wo der Vortrag nicht fort will, und doch fort geht,—diess zärtliche Zögern eines Liebhabers, der eben von seinem Mädchen gehen will; . . . [ist] nur unter der Faust des Meisters wirksam.'

[77] John Jousse, *Compendious Dictionary of Italian and Other Terms Used in Music* (London, 1829), repr. *MD*, 103.

Hamilton includes the idea of borrowing as well as robbery, but does not restrict the note to be shortened to the succeeding one:

RUBATO, or ROBATO (*Italian*) *Robbed, borrowed.* The terms *tempo rubato* are applied to a style of performance in which some notes are held longer than their legitimate time, while others are curtailed of their proportionate durations, in order that, on the whole, the aggregate value of the bar may not be disturbed.[78]

Neither Jousse nor Hamilton, however, mentions the strict rhythm of the accompaniment; hence they may be describing an example of the later rubato in which a retard in one portion of a measure is exactly balanced by an acceleration in another.

Thus the circumstances in keyboard music that led to new meanings for *tempo rubato*, together with the theorists' application of the term for various syncopated devices, often produced great uncertainty in the meaning of the expression. Moreover, many of the types of rubato could occur either as a compositional device indicated in a score by a composer, or as an expressive nuance added at the initiative of a performer. Further confusion was caused by certain similarities between the two rubatos that finally emerged as the dominating types—those identified here as the earlier and later: they were both utilized for intensely expressive purposes in the same types of composition, and, in order to be effective, they both needed to be surrounded by music in strict time. Furthermore, retardation is generally emphasized over acceleration for the later type, and prolongation over diminution for the earlier. This no doubt reflects the natural inclination to make significant, interesting, sensitive, or striking passages last longer.

When considering the background of the later rubato in Chapter 1, we noted the declamatory, expressive, and structural purposes served by tempo flexibility in Gregorian chant, in the madrigal, monody, and recitative, in cadences and cadenzas, and in the preludial forms. Most of these forms continued, even at the end of the eighteenth century, to be types of music in which considerable rhythmic freedom could be exercised. The addition at this time of more rallentando or accelerando to *these* forms would scarcely have been noticed. It was precisely in the *other* classic forms, that is the sonata forms—those in which strict rhythm ordinarily prevailed—that the later rubato could make a vivid contrast and hence be perceived as expressive. At first, this sort of rubato was employed to enhance sensitive moments in the musical architecture: thus, at cadences, fermatas, repeated phrases, contrasting passages, and introductory sections. Increasingly, however, later rubato

---

[78] James Alexander Hamilton, *A Dictionary of Two Thousand Italian, French, German, English, and Other Musical Terms* (New York, 1842), repr. *MD*, 75 and 87.

was also employed as an expression of emotions felt by the performer in response to the music; Türk mentions anger, as well as tender, melancholy, and languid passages, and Czerny gives long lists, as we have seen, of subtle emotional nuances, some based on specific programmatic situations. Some of the earlier sources complained about applying to all music the rhythmic freedom appropriate in the preludial forms. Adam writes around 1804:

Some have wanted to make playing out of time fashionable and to play every type of music as a fantasy, prelude, or capriccio. They think they give more expression to a piece thereby, and they alter it in such a way that it is unrecognizable.[79]

Theoretically, however, later rubato was meant to be employed at this time only briefly and only for carefully selected places in the music.

In tracing the background of the earlier type of rubato in Chapter 2, we found evidence of the technique in melodic variation employed by composers and in improvised embellishment added by performers. We saw it also used for increasingly expressive purposes in the ornaments, arpeggiation, and unequal notes of the Baroque period. This type of rubato, however, was not the sort of device that could generally characterize entire genres, as could the technique used in the later rubato. The earlier type can exist only momentarily and only in a strictly rhythmic environment against which it can be measured. Thus it, too, cannot exist in the rhythmically free forms, but only in forms which guarantee the preservation of a strict metrical rhythm. Thus vocal rubato can occur in the aria, but not the recitative. Furthermore, it can appear in the *strict* sections of an aria, but not in its cadenzas. Violin rubato, in a similar way, can occur in sonatas, as well as in any part of a concerto except its cadenzas. In the operatic aria, of course, text and plot generate their own emotions to which rubato may respond. García, however, names a number of purely musical situations suitable for rubato. In general violinists and the orchestras in which they played, either for purely instrumental music or as an accompaniment for singers, seem to have been less concerned with tempo changing than keyboard performers. García, for example, describes rallentando and accelerando very briefly (Plate IV), but favours tempo rubato for places marked *ad libitum*. In addition, he advocates earlier rubato (first paragraph on Plate V) for final cadences rather than ritardando.

---

[79] *Méthode de piano*, 160: 'Quelques personnes ont voulu mettre en vogue de ne plus jouer en mesure, et d'exécuter toute espèce de musique comme une fantaisie, prélude ou caprice. On croit par là donner plus d'expression à un morceau et on l'altère de manière à le rendre méconnaissable.'

Since the later rubato at first shared with the earlier type the same expressive purposes and, like it, occurred only briefly at sensitive moments within otherwise rhythmically strict forms, there may well have been many musicians and listeners, especially in connection with keyboard music, who could not distinguish between them—or who were not even aware that there were two types. However, a few sensitive observers such as Türk did comment on both types. Czerny mentions 'the very frequent application of each kind of *tempo rubato*' in the playing of Liszt.[80] Farrenc, who felt that the later rubato used by keyboard players in 1861 was ridiculous and tiring, refers also to another type of tempo rubato which Mozart mentioned and which was employed with much charm by the celebrated violinist Paganini and the pianist Field:

[It] consisted in a type of undulation, a slight retard or a slight acceleration of certain parts of the singing phrase, during which the rhythm of the accompaniment should not undergo any alteration: the skill of the performer made up for this apparent irregularity by always finding the means of falling back into agreement with the metre; . . .[81]

During this and the preceding two chapters, we have noted considerable variety in the ways in which various persons viewed tempo rubato. For both types, there were performers who employed it freely and listeners who complained of excess; there were composers who wanted control and performers who wanted freedom of execution. Because of its essential nature, however, the earlier rubato, with its melody rhythmically different from the accompaniment, involved a greater variety of viewpoints. The last two chapters presented the point of view of the vocal and violin soloist, as well as the orchestral violinist, the conductor, the theorist, and the listener. To these can now be added the keyboard performer, either as continuo player or soloist. The metrical conflict resulting from an expressive distortion of melody heard simultaneously with a strict accompaniment can be, as we have seen, perceived in many different ways. The solo singers or melody instrument players notice the time stolen from one note and

---

[80] *Piano Forte School*, 2nd supplement, 28.

[81] *Le trésor des pianistes*, i, 3–4: 'Je sais que quelques grands artistes ont fait usage d'un *tempo rubato* qui consistait dans une espèce d'ondulation, un léger retard ou une légère accélération de certaines parties de la phrase chantante, pendant laquelle le rhythme de l'accompagnement ne devait subir aucune altération: l'habileté de l'exécutant suppléait à cette apparente irrégularité en trouvant le moyen de tomber toujours d'accord avec la mesure; mais on concevra qu'il faut laisser cette subtilité de l'exécution à des virtuoses hors ligne qui savent tirer un parti avantageux de ce qui, chez d'autres, serait un défaut. A ma connaissance, le célèbre violoniste Paganini et le pianiste Field employaient avec beaucoup de charme ce *tempo rubato*. D'ápres un passage des lettres de Mozart, il paraît que ce grand homme en faisait usage dans l'exécution de ses œuvres.'

given to another; the listeners hear the alternation between notes in the melody and the accompaniment, which *sound* as though they should, under ordinary circumstances, have been simultaneous. The latter effect was then imitated by theorists who wanted to explain the device, by composers who wanted to notate it, and by performers who wanted to reproduce a similar expressive effect on the keyboard. It is rare to find a description in a keyboard source such as Farrenc gives in the quotation above. This can be explained, however, by the fact that he himself was a flautist rather than a pianist—a fact which may explain also why he quoted extensively from the writings of García rather than from keyboard sources.

As we approach the second quarter of the nineteenth century, then, we have two main types of tempo rubato. By this time the notated types of displacement—the unusual groups of C. P. E. Bach, the syncopated dynamics, syllables, metres, and imitative entries—were generally no longer labelled *rubato*. The singer and violinist were mainly concerned with the earlier type. It was in keyboard music that the two principal meanings of the word caused confusion, and this was the situation which existed when Frédéric Chopin first wrote the word *rubato* in a score. Before turning to the music of this important pianist and composer, however, we shall briefly survey the rhythmic robbery in the music of the three best known composers of the late eighteenth and early nineteenth centuries—Haydn, Mozart, and Beethoven.

6 HAYDN, MOZART, AND BEETHOVEN

NEITHER Haydn, Mozart, nor Beethoven ever wrote the word *rubato* in a musical score. There is some evidence, however, both in notation and in accounts of performance practice, of the existence in their works of the earlier as well as the later type of tempo rubato. The later type manifests itself in modifications of tempo—partly notated by the composer and partly added by the performer. Mozart described how he employed earlier rubato as a performer (see Chapter 5). This type may perhaps also be indicated occasionally by the notation. Both types of rubato occur in classic music to reinforce formal structure. Both tend in the late works of Beethoven and in romantic music generally to appear more and more for expressive purposes.

Notated Tempo Changes

As we saw in the last chapter, Türk suggested innovative notation for tempo modification in his compositions of 1783. In the same year, Cramer stated that he preferred verbal terminology. This experimental state of affairs reveals how new and recent was the concept of notating this aspect of music. Words to designate a general tempo had first emerged during the seventeenth century and were therefore well known by the classic composers. Terms relating to the changing of tempo, however, appear only in the 1780s—and even then rarely—in works of Haydn and Mozart. With Beethoven, finally, such terms become more frequent.[1]

General retards could be marked *adagio, sempre più adagio*, or eventually *ritardando* or *rallentando*. Beethoven also uses the word *ritenente* followed by *a tempo*, referring apparently to the more expressive and intense sort of broadening involved in *tenuto* notes and

---

[1] See Sandra P. Rosenblum, *Performance Practices in Classic Piano Music* (IUP, 1988), 362–85.

portato touch.[2] Other terms such as *espressivo* sometimes imply a slowing of the tempo as part of the effect, as do *calando, mancando, smorzando, perdendosi,* and *morendo.* As one would expect, markings to indicate an acceleration—*accelerando, stringendo, immer geschwinder*—are even less frequent in classic music and begin at a later date.

Adagio, ad libitum, a piacere, or *senza tempo* may occur in company with small notes and fermatas to indicate a more dramatic type of tempo modification for a brief passage of several measures. In this way the sort of fluctuation which generally characterized the rhythmically free forms discussed in Chapter 1—the recitative, the cadenza, and the preludial forms—could be introduced briefly into the strict sonata forms in a carefully controlled manner and for specific structural purposes. Such a passage might act, for example, to delay an important and expected formal event, which, when it does arrive, receives thereby an enormously enhanced presence. These free sections often precede a recapitulation, an important theme or cadence, or the return of a rondo melody.

Tempo changing is indicated also by *col canto, avec la voix,* or *suivant le chant.* When a singer or violinist alters the rhythm for expressive purposes, it is extremely important, of course, for those who play the accompanying instruments to know whether earlier rubato or tempo changing is involved. In the case of earlier rubato, as we have seen, the accompaniment maintains its strict rhythm. When *col canto* or its equivalent is marked, on the other hand, the accompaniment must follow exactly every tempo change of the soloist.

The absence of tempo flexibility is occasionally indicated by *tempo giusto,* although it seldom appears in scores. For the theorists and lexicographers it may also refer to a proper or moderate tempo. It was still a confusing term at the end of the nineteenth century.[3] We shall encounter it later, however, when it is sometimes used in scores to indicate the absence or the conclusion of a passage in the later type of rubato.

The evolution of terminology can thus give some idea of the nature of tempo changing in classic and early romantic music. In addition to using terms, however, composers occasionally suggest a retard or acceleration by an increase or decrease in note values. This is the sort

---

[2] Rosen writes (in the typescript of his Norton lectures): 'detached notes [in portato touch and occasionally elsewhere] carry the expressive weight: playing them this way enforces a kind of *rubato,* a very slight holding-back of the tempo for a fraction of a second.'

[3] See Frederick Niecks, 'Notes on Doubtful and Often Misunderstood Musical Terms', *MMR* 18/216 (1 Dec. 1888), 265–6; he also discusses *calando* and *ritenuto.*

of notation which was used earlier to depict accelerating trills (see Chapter 2).

Except for the specifically notated adagio passages, indications for tempo changing are very rare in the music of Haydn, Mozart, and the early works of Beethoven. Later editions often add such indications, and it is important for the performer to know which are authentic. Expressive playing in the Classic period depended more, I believe, upon subtleties of touch and dynamics—the sort of subtleties, for example, that seem to occur naturally on the fortepiano. Gradually, but probably very slowly, tempo changing joined these other techniques in achieving an ever-increasing degree of expression. Tempo flexibility becomes particularly significant, finally, when applied for expressive rather than purely structural purposes.

Tempo Changes by Performers

In the same letter from 1777 in which he comments on tempo rubato (see Chapter 5), Mozart criticizes the playing of the 8-year-old daughter of the piano builder Johann Andreas Stein: 'She rolls her eyes and smirks. When a passage is repeated, she plays it more slowly the second time. If it has to be played a third time, she plays it even more slowly.' Later in the letter he describes his own playing:

Herr Stein . . . now . . . sees and hears . . . that I do not make grimaces, and yet play with such expression that, as he himself confesses, no one up to the present has been able to get such good results out of his pianofortes. Everyone is amazed that I always keep strict time.[4]

Nannette Stein was only a child, of course, but her playing must have reflected in some measure the way in which she had been taught—to be expressive visibly and by means of tempo changing. Mozart, on the other hand, performing within the constraint of a strict tempo, achieved even more effective expression through other technical means. This, then, seems to be the ideal for the classic music of Mozart and Haydn: by generally maintaining a strict beat, subtler nuances of articulation and tone colour can be produced and perceived and hence employed in the service of musical expression. And, of course, it is only in this strictly controlled rhythmic environment that the earlier rubato can exist. At a later date, García states that compositions of Mozart and

---

[4] *Mozart: Briefe und Aufzeichnungen*, ed. Wilhelm Bauer and Otto Erich Deutsch (Bärenreiter, 1962). 83; trans. from Carol MacClintock, *Readings in the History of Music in Performance* (IUP, 1979), 381, except for the omission of *can* from the last phrase: 'ich immer accurat im Tact bleybe'.

others of the period 'demand great exactitude in their rhythmic movements', and the only permissible change in note values occurs in rubato, where the tempo itself is not altered (see the top of Plate IV). Only in a few cases of florid ornamentation is a slight flexibility of tempo probably required.[5]

Regarding Beethoven, there is more information on the way he himself played and conducted, and the way in which his students thought his compositions should be performed. Descriptions of his style of performance at the piano stress classic restraint during the earlier years, but an increasing amount of flexibility later. Ferdinand Ries, a pupil of Beethoven during the early 1800s, speaks of Beethoven's playing as follows:

In general, he played his own compositions in a very capricious [*launig*] manner; he nevertheless kept strictly accurate time, occasionally, but very seldom, accelerating the *tempi*. On the other hand, in the performance of a *crescendo* passage, he would make the time *ritardando*, which produced a beautiful and highly striking effect.[6]

Anton Schindler, who was a piano student of Beethoven as well as a companion and general helper for most of the years from 1820 on, describes, on the other hand, a considerable increase of flexibility during the late period. Although he exaggerates this aspect in his biography of 1840, he is more moderate in the rewritten version of 1860. He writes in 1840:

words directing the quickening or retarding of the time, such as *accelerando*, *ritardando*, &c., do not, in their ordinary acceptation, convey an adequate idea of the wonderfully delicate shading which characterized Beethoven's performance; . . .

All of the pieces which I have heard Beethoven himself play were, with few exceptions, given without any constraint as to the rate of the time. He

---

[5] For some possible examples, see Frederick Neumann, *Ornamentation and Improvisation in Mozart* (PUP, 1986), 133 (concerning 'the soloistic freedom that would usually permit a slight rubato delay of the downbeat') and 166–7 (concerning small-note arpeggiated figures). Robert D. Levin challenges Neumann's ideas in his review in *JAMS* 41 (1988), 359–60; and Neumann replies in *New Essays on Performance Practice*, SM cviii (1989), 163–4. Levin calls the suggested flexibility 'rubato (i.e. delaying the left hand)' and claims it violates Mozart's statement about the left hand's strict time; Neumann describes it as a *caesura* or *Luftpause*—a slight tempo fluctuation distinct from Mozart's rubato.

[6] Ferdinand Ries and Franz Gerhard Wegeler, *Biographische Notizen über Ludwig van Beethoven* (Koblenz, 1838), 106; *Neudruck*, ed. A. C. Kalischer (Berlin and Leipzig: Schuster & Loeffler, 1906), 127. Quotation from Anton Schindler, *Biographie von Ludwig van Beethoven* (Münster, 1840), 227; trans. by Ignaz Moscheles in *The Life of Beethoven* (London, 1841; repr., Mattapan, Mass.: Gamut Music Co., 1966), ii, 128; (Boston: Ditson, 1841?), 156. Moscheles' trans. was published simultaneously in London and Boston, the former paged in 2 vols., the latter in 1. MacClintock, in *Readings in the History of Music in Performance*, quotes the passage on 388, but gives page numbers on 384 only for the Boston edn.

adopted a *tempo rubato* in the proper sense of the term, according as subject and situation might demand, without the slightest approach to caricature. Beethoven's playing was the most distinct and intelligible declamation . . . His old friends, who attentively watched the development of his genius in every direction, declare that he adopted this mode of playing in the first years of the Third Period of his life, and that it was quite a departure from his earlier method, which was less marked by shading and coloring; . . .[7]

In his revised biography of 1860, however, Schindler modifies his concept of rubato:

we must note that the term 'free performance' has falsely been equated with the *tempo rubato* of the Italian singer. The fact alone that the Italian term generally occurs only in *opera buffa* and hardly ever in *opera seria* [where, presumably, 'free performance' for expressive purposes would be more likely] is an indication that the two terms are not identical. Beethoven protested against the use of the Italian term in regard to his music, albeit in vain, for Italian terminology had come to dominate everything in his epoch, including his own music.[8]

Beethoven probably did know about the earlier type of rubato used by the Italian singers, for his teacher Neefe, as we saw in the last chapter, had used the expression in 1783 in reference to the strict time maintained in the keyboard rubato of a certain countess. Schindler noted correctly, however, that others did use the word *rubato* in reference to the free performance of Beethoven, and he gives as an example this quotation from the composer and conductor Ignaz von Seyfried, who recalls Beethoven's relationship with the orchestra during the period from 1800 to 1805:

he demanded great exactitude in the matter of expression, minute nuances, the balance between light and shade, as well as an effective *tempo rubato*, and would gladly speak to each member of the orchestra individually about these points without showing the least impatience.[9]

---

[7] *Biographie*, 227–8; trans. Moscheles, London edn., ii, 127 and 129–30; Boston edn., 156–7, and MacClintock, *Readings*, 387–8. Adolf Bernhard Marx, who studied with Türk, quotes Schindler's reference concerning rubato in his *Anleitung zum Vortrag Beethovenscher Klavierwerke* (Berlin, 1863). In a footnote he adds: 'The tempo rubato was a fashion of the eighteenth century, dating from the last half, and came from the singers of Italy and France. It was intended to replace the free, deep feeling which was wanting in the compositions themselves; this tempo rubato was therefore an untruth, and was soon compelled to yield to the reaction of reason. With this fashion (which amid other things became visible in Pergolesi's Stabat Mater) Beethoven had nothing in common: he followed entirely the inner impetus—the demand of the thing—when he resorted to free movement'; trans. by Fannie Louise Gwinner from the 2nd edn. of 1875 in *Introduction to the Interpretation of the Beethoven Piano Works* (Chicago: Clayton F. Summy Co., 1895), 73. Concerning Pergolesi, see Türk's examples in my Plates XI and XIII and the comments in Ch. 5.

[8] Anton Felix Schindler, *Beethoven as I Knew Him*, ed. Donald W. MacArdle, trans. Constance S. Jolly (Faber, 1966), 409.

[9] Ibid. 409, quoted from Seyfried's *Ludwig van Beethoven's Studien im Generalbasse, Contrapuncte und in der Compositions-Lehre* (Vienna, 1832). See *Louis van Beethoven's Studies*

Schindler adds a footnote that reads: '*Tempo rubato* even in orchestral music!'

In addition to these general indications of Beethoven's style of playing, both Czerny and Schindler have provided instructions for performing some of the piano sonatas, showing precisely where the performer should add rhythmic flexibility.[10] Czerny, who was a pupil of Beethoven and performed his works during the early 1800s, included in his *Complete Theoretical and Practical Piano Forte School* two chapters devoted to the proper performance of all Beethoven's works for piano. In most cases, his suggestions for tempo changing enhance the markings already in the score and seem to serve both structural and expressive purposes. He often urges restraint, however, and comments that melodies can be played 'with greater tranquillity, and yet not perceptibly slower' and that sometimes intense expression 'is to be produced more by the touch than by the employment of *rallentando*'.[11]

In his biography Schindler displays a far less subtle concept of tempo flexibility in his description of Beethoven's playing of the two Sonatas Op. 14. He bases his analysis of sonata form on the opposition of two principles: the entreating and the resisting, or the masculine versus the feminine. In addition to some momentary tempo changes, however, he advocates others of rather long duration applying to entire melodies. He also mentions Beethoven's prolonging of notes for accentuation and his application of rhetorical pauses (the lengthening of dramatic rests) and caesuras (the insertion of rests for the purpose of setting off new ideas).[12] According to an entry forged by Schindler in a Conversation Book from 1823, Beethoven felt that if he marked in the score all the tempo changes required by the two contrasting principles, 'confusion might well result from the many indications of *tempo rubato*'.[13]

Some of Schindler's ideas may well be his own or those of a later generation. He probably heard Beethoven play mainly in the 1820s, however, when he might well have performed works from his early years with the more expressive rhythmic freedom characteristic of his

---

in *Thorough-Bass* . . . , trans. and ed. Henry Hugh Pierson (Edgar Mannsfeldt) (Leipzig, Hamburg, and New York: Schuberth and Comp., 1853), p. 13 of the appendix.

[10] Rosenblum, *Performance Practices in Classic Piano Music*, 387–92.

[11] *Carl Czerny: On the Proper Performance of All Beethoven's Works for the Piano*, ed. and with commentary by Paul Badura-Skoda (Universal, 1970), 36 and 93 of the original.

[12] *Biographie* (1840), 195–6, 223–42; *The Life of Beethoven*, London edn., ii, 81, 122–51; Boston edn., 133–4, 153–69; *Beethoven as I Knew Him*, 417–20; and MacClintock, *Readings*, 385–99.

[13] *Ludwig van Beethovens Konversationshefte*, iii, ed. Karl-Heinz Köhler, Dagmar Beck, and Günter Brosche (Leipzig: Deutscher Verlag für Musik, 1983), 350 (see n. 835 on p. 487); trans. from Rosenblum, *Performance Practices in Classic Piano Music*, 388.

later period.[14] In general, notated indications for tempo fluctuation, as well as descriptions of flexibility added by performers, all seem to point toward a relatively rigorous strictness of rhythm during the Classic period, with increasing amounts of freedom slowly and gradually permitted for expressive purposes as one moves into the new century.

Alteration of Note Values in Classic Music

Mozart is the only one of the three composers who actually mentioned tempo rubato, and he stated, as we have seen, that 'in *tempo rubato* in an Adagio, the left hand should go on playing in strict time'. This clearly refers to the earlier rubato and the alteration of note values in a melody, rather than with the changing of tempo. Mozart, however, never wrote the word *rubato* in a musical score, hence we can only speculate about how he actually incorporated the device in his own music. There are two possibilities, I think: he may have notated or at least partially notated it by means of syncopation; or it might have been added by a performer even when not indicated in the score.

We considered the early history of the technique of melodic variation in Chapter 2 and noted that alteration of note values often occurs (Exx. 2.1–2.3). Mozart, Haydn, and Beethoven inherited this technique and applied it widely, not only in sets of variations, but also to vary melodies that recur within formal structures: a rondo theme, a recapitulated theme in sonata form, a passage repeated at the end of a ternary or rounded binary form. Syncopation of the note values sometimes occurs in such a way that one feels that the composer is attempting to notate the displacement involved in the earlier rubato. This can happen in the first statement of a passage, as well as in the variant of a melody previously heard without syncopation. In the last chapter we noted the widespread tendency to explain rubato through syncopation (see the notation in Ex. 5.4). We also saw that Dussek was reported to have attempted to notate rubato by syncopations, but when he realized that a strict observance of the note values did not achieve the desired effect he contented himself with writing simply *espressivo*.

Some modern writers feel that the classic composers did occasionally notate at least a rubato-like effect by syncopating the notes. Not every

---

[14] See the following works by William S. Newman: *Beethoven on Beethoven: Playing His Piano Music His Way* (Norton, 1988), 110–20 and 253–4; *Performance Practices in Beethoven's Piano Sonatas: An Introduction* (Norton, 1971), 25–6, 53–5; and 'Tempo in Beethoven's Instrumental Music: Its Choice and Its Flexibility', ii, *PQ* 30/117 (1982), 26–31.

syncopation, of course, conveys intense passion. When it occurs within a chromatic or adagio setting, however, the accompanying dissonances make a greater impact and thus intensify the expression. If this very expressive sort of syncopation is meant to represent rubato, then presumably the actual note values are only approximate and the performer should exercise some freedom in order to overcome Dussek's objection to a literal rendition.

Leon Plantinga has identified such rubato-like syncopation in the works of Clementi (see Chapter 5, above) and notes that it tends to be associated particularly with the minor mode. He cites also the minor variation in Beethoven's Sonata Op. 26 and the first variation of Haydn's F minor *Piccolo divertimento*.[15] Eva and Paul Badura-Skoda mention examples by Mozart: among them a passage in parallel thirds in the Adagio movement of the C minor Sonata, K. 457, which occurs plain in measure 3 and syncopated in measure 19; and another, with thirds in both hands, in the slow movement of the F major Sonata, K. 332 (compare measures 30–1 with 34–5).[16] Carl Blake and Sandra Rosenblum compare the syncopation in measures 85–7 of his A minor Rondo, K. 511, with the unsyncopated version in measures 5–7.[17] Both Rosenblum and William Newman cite excerpts from the third movement of Beethoven's Piano Sonata Op. 106, and Newman one from the Adagio of Op. 110.[18] Charles Rosen, finally, speaks of the written-out rubato in the second movements of Haydn's String Quartets Op. 54 Nos. 2 and 3. In Op. 54 No. 2 the second violin plays a slow melody in quarter notes while the first violin spins rhapsodic figuration in sixteenth and thirty-second notes, providing sometimes a counterpoint to the melody, sometimes a heterophonic variant, but often with syncopation and dissonance.[19]

I would like to add the brief excerpt in Ex. 6.1*b* to this list of passages that might represent a notated form of rubato. It comes from the second movement, marked 'Andante con espressione', of Mozart's

---

[15] *Clementi: His Life and Music* (OUP, 1977), 91, the ex. on 90, and 110 n. 56.

[16] Eva and Paul Badura-Skoda, *Interpreting Mozart on the Keyboard*, trans. by Leo Black of *Mozart-Interpretation* (St Martin's, 1962), 44–5. See also Blake, 'Tempo Rubato in the Eighteenth Century', 52–3, and Rosenblum, *Performance Practices in Classic Piano Music*, 379.

[17] Blake, 'Tempo Rubato in the Eighteenth Century', 54, and Rosenblum, *Performance Practices in Classic Piano Music*, 380.

[18] Newman, *Beethoven on Beethoven*, 116, 138, and 253; Rosenblum, *Performance Practices in Classic Piano Music*, 381.

[19] *The Classical Style*, 139–40. Rosen writes: 'The rubato of the classical period (as we can see from those passages where Mozart, Haydn, and Beethoven wrote it out) was used to create the most affecting dissonances: unlike the romantic *rubato* (and the one most in use today), it was not just a delaying of the melody, but a forced overlapping of the harmony as well. I should imagine that as a kind of suspension it was originally related to the appoggiatura, the most expressive of ornaments and almost always a dissonant note.'

Ex. 6.1 Mozart, Sonata in D Major, 2nd movt., 1777

a Written
[Andante con espressione]

b Variant written by Mozart

c Different notation for *b*, with small notes on the beat

Sonata in D major, K. 311/284^c.[20] The movement is in the so-called slow-movement sonata form, with an unrepeated exposition and recapitulation, but no development section. Following the recapitulation, however, the opening theme appears one final time. Ex. 6.1*a* comes from the beginning of the recapitulation, although an almost identical two-bar phrase begins the piece and also recurs many times in both tonic and dominant keys (see measures 5, 25, 27, 43, 61, and 63). Following the end of the formal recapitulation, the theme reappears, but commencing as in Ex. 6.1*b*. Now several of its notes are delayed in the manner of the earlier type of rubato.

If this notation were indeed meant to represent rubato, then the performer should play it with a certain rhythmic freedom. I would like

---

[20] *Neue Ausgabe sämtlicher Werke*, IX: 25/i, 111 and 112.

to suggest the different notation in Ex. 6.1c, in which the small notes are to be played, according to the custom of the time, on the beat. The regular notes in the left hand are to be played as strictly in time as those in (a) or (b). If this is done, then a performance of (c) could result in the music of (b), or it could, as the amount of delay between the two hands varies, also result in an endless number of variants— most too subtle to be easily notated. The amount of delay could vary not only from performance to performance, but also from note to note within a single performance. The notation in Ex. 6.1c thus allows a range of variation in the amount of delay; it depicts rubato as a displacement between right and left hands, which, as we saw in the last chapter, is the predominant way the keyboard player conceived of rubato; and, in addition, the melody looks like the melody in (a), so that the performer can feel that he is in fact playing the melody in (a) but with increased passion. The small-note notation in (c) provides the same sort of rhythmic freedom already inherent in the small-note appoggiaturas and turns notated by Mozart in (a) and (b). In (c) I have replaced the small-note turn with a sign, in order to separate it more clearly from the appoggiatura.[21]

Simply performing Ex. 6.1c intellectually, however, will not produce rubato. Playing with intensified expression requires a sensitive and complete comprehension of the sense of the music at this spot. This, in turn, depends upon an understanding of the total design of the movement and the way this one event fits in. Furthermore, I think producing rubato depends not only upon understanding the sense of the music intellectually or upon one occasion, but upon understanding it musically and vividly at *each moment of actual performance*. In the case of Ex. 6.1, a tiny two-bar melodic phrase is heard, sometimes in a simply varied form, ten times within a structural form of exposition and recapitulation. The latter concludes with an exceedingly dramatic cadence, involving an ascending chromatic scale which leads with a crescendo to the highest pitch of the entire movement and finally to an emphatic and powerful trill. At this point one might normally expect some closing material leaning to the subdominant. One certainly would not necessarily expect one more statement of the opening theme, but following the trill a measure of thirty-second notes announces exactly that event. The playing of Ex. 6.1b or c then depends completely upon the sensitivity with which one plays the concluding cadence of the

---

[21] In *W. A. Mozarts Werke* (Leipzig, 1877–83; repr., Ann Arbor, Mich.: J. W. Edwards, 1956), Serie XX, 8(98), most of the ornaments in Ex. 6.1b are written in ordinary note values: in m. 75 the turn consists of 4 64th notes; in m. 76 the appoggiatura and the following note are both 32nd notes.

recapitulation, as well as the measure of thirty-second notes. By the time one reaches Ex. 6.1*b*, the musical sense has already been determined, and if at this point one plays Ex. 6.1*a* with the very specific and intense expression demanded by this special moment in the form, then something like Ex. 6.1*b* or one of the variants suggested by (*c*), can result.

In addition to these notated examples, however, rubato may also occur in classic music when added at the initiative of a performer. Ex. 6.2 shows two measures from Mozart's Sonata in F major which might

Ex. 6.2 Mozart, Sonata in F major, 2nd movt., 1781–3

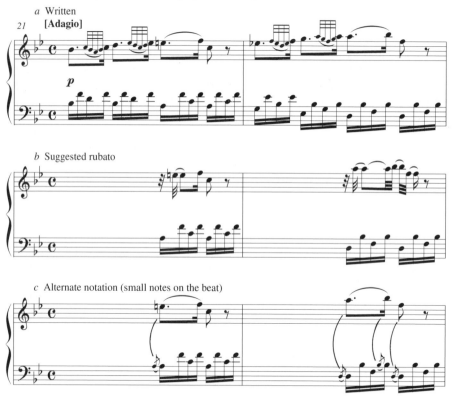

be suitable for such rubato. This Adagio is also in slow-movement sonata form. The opening two bars of the recapitulation in Ex. 6.2 are almost identical in the first edition, except for the first quarter beat, to the opening of the piece.[22] In measures 5–6 the opening theme turns

---

[22] K. 332/300^k; *Neue Ausgabe*, IX: 25/ii, 37.

to the minor mode and participates in a passionate and dramatic modulation to the dominant key. Here, however, the notes themselves seem sufficient without the addition of rubato. The rest of the exposition contains widely contrasting rhythmic motives, some in triplets, some with thirty-second and even sixty-fourth notes. The return of the main theme at the beginning of the recapitulation (Ex. 6.2a), then, is a welcome and poignant resumption of the serenely flowing melody. This is the moment where I feel that some subtle and carefully limited rubato could enhance the expression. I have shown some possibilities in regular notation in (b). Simply feeling warmer or more tender about the second turn tends to delay the E in the first phrase, thus adding further accentuation on this important dissonance. To achieve an increase of intensity in the second phrase, I have shown each of the last three notes delayed. In Ex. 6.2c I have again used my more flexible notation to indicate similar results. As with Ex. 6.1, the motivation for the rubato comes from the vivid comprehension by the performer of the musical sense of this moment, a sense that is prepared by the preceding music—especially, in this example, by the trill, then the pedal-point, and finally a series of thirty-second notes in the previous three measures.

In addition, one must keep in mind the nature of the fortepiano and the general style of playing cultivated by Mozart. Sounds decayed more rapidly on the instrument of Mozart's time and encouraged a highly articulate touch and a highly detailed sense of phrasing. Czerny describes Mozart's style as follows:

A distinct and considerably brilliant manner of playing, calculated rather on the staccato than on the legato touch [in contrast to Beethoven's 'smooth and connected cantabile']; an intelligent and animated execution. The pedal seldom used, and never obligato.[23]

This suggests that figures such as the Alberti bass in Ex. 6.2 should be played without pedal and, although probably not usually staccato, at least without holding any of the notes, even the first of each group of four, after the following note has been struck.[24] If one will play such a passage in this manner, even on a modern piano, the truly musical performer will eventually realize, I believe, that subtleties of touch and dynamics will more than make up for the lack of sonority provided by the sustaining pedal. The resulting style of playing will have more clarity and more rhythmic control. Furthermore, this is precisely the kind of left hand accompaniment I believe Mozart had in

---

[23] *Piano Forte School*, iii, 100.
[24] Rosen comments, on the other hand, that 'the practice of holding down the first note of the beat is an old one still practised'.

mind when he spoke of rubato. For when this Alberti bass 'goes on playing in strict time', without pedal and without overlapping notes, one can hear with unusual clarity the slightest amount of tempo rubato in the melody. In this restrained musical environment, such alterations in the melody notes thus have an extremely expressive effect.

I have spent hours at the piano seeking in the music a motivation for rubato and learning how to play classic music without pedal. I have attempted above to describe these important aspects in relation to Exx. 6.1 and 6.2, but it is difficult to express such things in words. If the reader is sufficiently skilled, I would invite him to make his own exploration at the keyboard—an experience more meaningful than any words could be. The excerpts in Exx. 6.1 and 6.2 illustrate the earlier rubato as we have seen it described in previous chapters: the accompaniment is in strict time; the alterations in the melody occur over a brief period of time; the purpose is to intensify the expression. When I spoke of motivation from within the music itself, and when I suggested that Ex. 6.1*b* or 6.2*b* was like playing Ex. 6.1*a* or 6.2*a* with special passion, I was thinking of Baillot's assertion (see Chapter 4) that 'this disorder . . . will become an artistic effect . . . if the artist can use it without being forced to think of the means he is employing'. If a pianist were to practise displacing melody notes in relation to bass notes in the same way he acquires a skill in touch and dynamics, then, when he is genuinely making music, I presume he would be as little aware of the precise anticipations and delays in his tempo rubato as he is of the exact degree of loudness and articulation on each note.

The earlier type of rubato may, however, have been less applicable in the more legato style of Beethoven, with its presumably less precisely articulated accompanimental figures. Perhaps this explains why the expressive power of Beethoven's declamation is described by his contemporaries in terms of the later type of rubato and its tempo flexibility.[25]

In the music of the classic composers one can also find examples of written-out accelerations in the melody while the left hand continues its regular rhythm. In the Adagio of Haydn's E flat major Sonata, a single note is repeated on the first beat of the measure with four sixteenth notes, on the second, with three triplet sixteenths followed

---

[25] See George Robert Barth, 'The Fortepianist as Orator: Beethoven and the Transformation of the Declamatory Style', D.M.A. diss. (Cornell Univ., 1988; UM 88–21, 213). He states on 81–2: 'Beethoven's declamation, because it was based on *legato*, needed highly varied agogics for the clarification of its syntax. Only in this way could the articulation of structure remain as clear and as free from other aspects of expression as it had been when silence was used [as in Mozart's staccato style] as the primary marker.'

by four thirty-second notes.[26] Such written-out accelerations occasionally occur also in the late works of Beethoven. The decreasing note values of the trill in measures 129–30 of the third movement of the Ninth Symphony are played by the first violins as a counterpoint to the strict beat of the accompanied melody in the rest of the orchestra. A passage in the first Bagatelle of Op. 126 also gives the impression of acceleration when a three-note motive (beginning in measure 21) is subjected to a metre change from 3/4 to 2/4, then syncopated, and finally reduced in note value. Finally, as an example of note values altered by Beethoven as a performer, Schindler states that in the seventh and eighth measures of the first movement of the Piano Sonata Op. 14 No. 1, Beethoven prolonged the eighth-note A on the third quarter beat and correspondingly diminished the following note.[27]

Alteration of Note Values in the Late Works of Beethoven

We have already noted the rubato-like syncopation in two of Beethoven's late piano sonatas, as well as some written-out accelerations in other late works. There are some other methods of altering note values, however, which are particularly characteristic of the late period. These involve certain ways of anticipating or delaying notes or chords so that they do not fall in their expected positions. This creates a feeling of yearning or striving which becomes a part of Beethoven's personal style of romanticism. He also becomes increasingly involved with the rhythmic transformation of entire melodies.

An anticipated note occurs in the 'Ode to Joy' theme from the choral finale of the Ninth Symphony. When the opening phrase returns in this rounded binary form, its first note commences one beat sooner than it did at the beginning of the melody. Similar anticipations occur elsewhere in the same symphony and tend to give an anxious sense of urgency and forward drive to the flow of music. They become even more powerful when, as in this case, the phrase is first heard without an anticipation. Beethoven also delays notes, in which case the effect, as in Couperin's *suspension* in Ex. 2.9, is to increase the singing quality and sense of presence of the notes delayed and thus intensify the expression.

Beethoven also delays or anticipates entire chords as well as indi-

---

[26] *Joseph Haydn: Werke*, ed. Joseph Haydn-Institut, Cologne (1962–), xviii/iii, 92 (m. 14) and 93 (m. 46).

[27] *Biographie*, 232–3; *The Life of Beethoven*, London edn., ii, 136; Boston edn., 160; and MacClintock, *Readings*, 391.

Ex. 6.3 Beethoven, 9th Symphony, 3rd movt., 1822–4

vidual notes. Ex. 6.3 occurs near the end of the opening theme of the third movement of the Ninth Symphony. A listener accustomed to the harmonic practices of classic music would normally expect the I6/4 chord to appear promptly on the first beat of the second measure. For expressive purposes, however, Beethoven prolongs the preceding secondary dominant chord and thus delays for an eighth beat the expected arrival of the cadential I6/4 chord.

An excerpt from the second movement of the same symphony in Ex. 6.4 illustrates the appearance of chords sooner than expected. Here

Ex. 6.4 Beethoven, 9th Symphony, 2nd movt.

one expects the chords to continue changing on the first beat of each measure. The chord change expected on the first beat of measure 69, however, appears earlier on the third beat of the previous measure. In addition, the change of both chord and melody on the second beat of measure 72 sounds as though it comes two beats too soon. The first beats of each measure are marked to be heavily accented dynamically, so that the syncopations created by the unexpected chord changes become, in a sense, even more powerful although not marked for dynamic emphasis themselves. In the midst of the restless energy of this movement, such syncopated chord rhythm causes a rough and boisterous effect. When applied momentarily as in Exx. 6.3 and 6.4, such chordal syncopation is similar to the unexpected sforzandi which, as mentioned in the last chapter, were for a brief period referred to occasionally as *tempo rubato*.

One of the most characteristic types of chordal shifting in Beethoven's late works occurs with the 16/4 chord, especially at cadences. Ex. 6.5

Ex. 6.5 Elements delayed or anticipated in a V⁷–I cadence

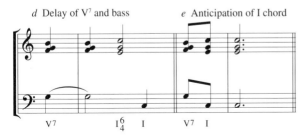

shows five different distributions of the two chords in a V^7–I cadence. The usual position for the final tonic chord is on the first beat of the last measure. In (*a*) the bass note moves to the root of I at the expected spot, but the upper voices are delayed and thus contribute to the sense of finality. This is a common cadence in classic music, and sometimes the delayed upper voices are written as small-note ap-poggiaturas. In (*a*), however, the movement of the bass note gives the ear the impression that the tonic chord has arrived at the proper time. In (*b*) the entire V^7 chord is prolonged, thus delaying the final tonic; this so-called *weak* cadence occurs in some of the baroque dance forms such as the faster type of saraband.[28] When Beethoven uses this sort of delay he often makes the second V^7 chord briefer and resolves to the tonic on some unusual subdivision of the beat. As a sort of cadential version of the anticipated chords in Ex. 6.4, Beethoven occasionally resolves the final tonic chord early as in Ex. 6.5e.

Far more characteristic, however, are the delays in Ex. 6.5c and d, which cause the momentary creation of I6/4 chords. Ordinarily in classic harmony, a 6/4 chord must be handled carefully, with the notes that provide the intervals of the sixth and fourth usually resolving downward by step in the manner of appoggiaturas. Hence the cadential I6/4 chord is in reality an ornamental form of the V chord. In the case of Ex. 6.5c, however, this expected resolution is replaced by the descent of a fifth in the bass, which simply moves the harmony from the second inversion of a triad to the root position of the same chord. The I6/4 chord had often generated unusual suspense in earlier classic music (immediately preceding a cadenza, for example): it promised certain expectancies, which could occupy some time to unfold. Beethoven now provides new possibilities for its resolution. In some cases, the ear does not know which type of resolution to expect; at other times, the location of the tonic chord is so clearly predicted, that when a I6/4 appears in its place, one tends to expect the new resolution of Ex. 6.5c.

Ex. 6.6 illustrates both types of resolution. The opening phrase of the brief sixth movement of the C sharp minor String Quartet concludes as shown in (a), whereas its repetition ends as in (b). In both cases the second chord of the first measure is a regular I6/4 chord, which resolves as expected to the V^7 chord on the third beat. In (b), however, a second I6/4 chord appears when the bass delays its downward fifth movement. Thus we see in (b), within close juxtaposition, both the usual resolution of a 6/4 chord and Beethoven's new manner. The delay of the bass in (b) also has the effect of increasing the sense of flow toward the following music. Occasionally the idea of delaying the bass (Ex. 6.5c) and delaying the V chord (Ex. 6.5b) are combined as in Ex. 6.5d.[29]

[28] See *New Grove*, xvi, 491; also my volumes entitled *The Folia, the Saraband, the Passacaglia, and the Chaconne* in AIM's series *Musicological Studies and Documents*, xxxv (Neuhausen–Stuttgart: Hänssler Verlag, 1982), ii: *The Saraband*, pp. xxii and xxiii, as well as numerous musical examples later in the volume.

[29] This occurs at the end of the opening phrase (m. 8) of the 3rd movt. (*Alla danza tedesca*) of the String Quartet Op. 130. For examples of the other cadences in Ex. 6.5, see the last measure of the Piano Sonata Op. 109 for (a), the conclusion (m. 16) of the theme of the Arietta in the Piano Sonata Op. 111 for (b), mm. 106–8 of the 'Heiliger Dankgesang' from the String Quartet Op. 132 for (c), and the end of m. 6 of the adagio intro. to the 1st movt. of the String Quartet Op. 130 for (e).

Ex.6.6 Beethoven, String Quartet Op. 131, 6th movt., 1826

Examples of the delayed bass, as in Ex. 6.5c and d, appear often in the late works of Beethoven. They tend to make the music sound as though the expression is so intense that it can continue only with difficulty and can pull only part of the chord along at full tempo. Along with the other anticipations and delays described above, these reluctant cadences infuse the music with a new tension of yearning. Sometimes restless, sometimes profoundly contemplative, such alterations in the expected or previously heard notes or chords become an important part, I think, of the romantic language in the late works of Beethoven.

In addition to these alterations in single notes and chords, Beethoven also shows an increasing interest in rhythmic transformation. In Chapter 2 we saw the background of this practice in the motet tenors of the Middle Ages, in the dance pairs and polyphonic chant settings of the Renaissance, in the canzonas of the Baroque period, and in the late works of J. S. Bach (Exx. 2.15–2.19). In general the composers of the Classic period showed little interest in the technique.[30] Romantic

---

[30] Concerning the rhythmic transformation of brief motives on a subtler level, see Rosen, *The Classical Style*, 199–204, in regard to a Mozart concerto. In his *Sonata Forms* (Norton, 1980), 171–4, he demonstrates that each motivic or thematic transformation in a sonata movement by C. P. E. Bach from 1765 helps to articulate a different moment in the form and hence 'demands a different style of playing'; one of them, for example, 'must be inflected by a slight *rubato* to bring out its vocal character'. See also Christopher Reynolds, 'The Representational Impulse in Late Beethoven, I: An die ferne Geliebte' and 'II: String Quartet in F major, Op. 135', *Acta musicologica*, 60 (1988), 43–61 and 180–94.

composers, on the other hand, were attracted by the expressive nature of such alterations. Beethoven provides examples even in works from his middle period. One of the most extraordinary and expressive examples in all of music history occurs at the end of the second movement of the 'Eroica' Symphony. Furthermore, in the last movement, the opening portion of the main theme (shown in Ex. 6.7a) is

Ex. 6.7 Beethoven, 3rd Symphony, 4th movt., 1803

rhythmically transformed during the fugal section on the inverted bass-line. Ex. 6.7b shows the transformed melody played by the flutes; on the bottom staff is the inverted fugue subject, which is played by the cellos and double basses and which provides an accentuation on the first beat of each measure that emphasizes the syncopation in the distorted melody. The transformed melody in (b) delays the melodic tones in (a) during measures 294 and 295, catches up by hastening the notes in measure 296, and finally concludes by anticipating the notes in measures 297 and 298. In the process, the original six-bar passage has been reduced to five measures. Actually, the listener knows the melody in (a) so well by the time (b) arrives, that the transformation seems shifted in a sort of polymetric counterpoint—that is, one hears the E♭ in the bass of measure 294 as a first beat while simultaneously perceiving the sforzando G in the melody also as a first beat on another rhythmic plane.

In the late works of Beethoven rhythmic transformation appears in various degrees of complexity and for various purposes. A simple technique occurs in the concluding presto section in the last movement of the Archduke Trio, where each measure of the main theme is transformed from 2/4 into 6/8 metre in order to quicken the pace. In the last movement of the Ninth Symphony, on the other hand, transformation not only alters the character of the main theme for expres-

Ex. 6.8 Beethoven, 9th Symphony, last movt.

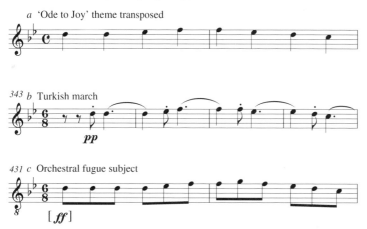

a 'Ode to Joy' theme transposed

343 b Turkish march

431 c Orchestral fugue subject

sive reasons, but also provides a more subtle means of achieving unity within a large form. We have already noted the conspicuous anticipation of one of the notes in the main theme of this movement. The first two measures are given in Ex. 6.8*a*, but transposed for the purpose of comparing the two transformations in (*b*) and (*c*). In the section marked *Alla marcia*, the entire melody is transformed in the manner shown in (*b*). Half of each original measure in 4/4 becomes a full measure in 6/8, with the first of each pair of notes diminished in value and the stolen amount added, after the opening measure, to the preceding note. The orchestral fugue subject in (*c*) derives from only the first two bars of the theme. Disregarding the extra repetitions of the D in the opening bar of (*c*) and the addition of a G in the second, the rhythmic alteration involves stealing time from the last two notes in the first bar and from the last three in the second.

The alteration of note values that we see in Exx. 6.3–6.8 represents the technique of the earlier type of rubato applied as a compositional device. Additional rhythmic flexibility by the performers is not required in these cases, for the composer has exercised control over every detail of the process through the notation. Syncopation such as Mozart briefly employs in Ex. 6.1, on the other hand, may involve subtle flexibility in the melody while the left hand 'goes on playing in strict time'.

Even this possibly more flexible notation is far more precise, however, than the methods for indicating tempo changes. Words such as *ritardando* and *accelerando* do not fix the *amount* of slowing or hastening. Even *poco ritardando* is only relative to *ritardando*, and the performer must sometimes decide whether certain words such as *calando* or *espressivo* involve any tempo changing at all. The dramatic and free adagio passages that lead to a fermata are likewise notated only in a very general way. In consideration of the lack of precision in this notation, as well as in contemporary descriptions of the rhythmic flexibility of performers, we must keep in mind the restrained rhythmic control which generally prevailed at this time. Against this strict background, very slight alterations of tempo or note value—alterations that might be barely noticeable to later generations—would be conspicuous, and hence expressive, to the listener of this period.

During the last decades of the eighteenth century and the early years of the nineteenth, an increasing interest in using the homophonic musical language of Classicism for more intense expressive purposes led to an increasing concern with the role of rhythm. The continuous flow of rhythm was the one element of music that had been most restrained in the past. Perhaps for this very reason, the manipulation of this flow eventually became the central means of expression. Both tempo flexibility and the practice of altering note values within a strict tempo extended, as we have seen, far back in history. But at this time composers and performers became more aware of their expressive value and began to expand the frequency and degree with which they were used.

We can see in the music of Haydn, Mozart, and Beethoven the beginnings of this new concern for rhythmic manipulation. It was during this period that the expression *tempo rubato* was extended in keyboard sources to include not only the original earlier type, but also the later type involving tempo flexibility. Therefore the notated and performed tempo flexibility and alteration of note values in the music of these composers seem closely related to the history we are tracing. They are not confined, as we have seen, to keyboard music. But music for the keyboard, since its performance is under the complete control of a single artist, was at the forefront of the new rhythmic developments.

As we return now to rubato in keyboard music and turn to the composer and performer who is probably the central figure in our history, it is important to keep in mind the rhythmic practices of Haydn, Mozart, and Beethoven, as well as the earlier rubato as it occurred in vocal and violin music, and both types of rubato as they finally emerged in keyboard music.

7 CHOPIN

FRÉDÉRIC Chopin (1810–49) marked the word *rubato* in fourteen different compositions which he wrote between 1828 and 1835 and which were published, except for one manuscript and two posthumous works, between 1832 and 1836. He usually wrote simply the single word *rubato*. Twice he marked *poco rubato*, once *sempre rubato*, and on another occasion *languido e rubato*. He never included the word *tempo*.

In previous chapters we have traced the history of the earlier type of rubato from its beginnings up to the time of García's work in 1847 for voice and the books of Spohr and Baillot from 1833 and 1834 for violin. We have also followed the more complex development in keyboard music, tracing the earlier rubato from its first appearance to the article about Dussek in *Le Pianiste* of 1834, and also tracing the later type of rubato from the time of Christian Kalkbrenner's book of 1789 up to the comments by Hummel and Czerny around 1827 and 1839. During the late 1820s and the 1830s, then—at the very time when Chopin included the word *rubato* in his scores—there were two different meanings for the word currently in use. In our attempt to discover which type Chopin had in mind, we will first turn our attention to tempo flexibility in his music, and then to an investigation of his possible involvement with other types of rubato.

Tempo Flexibility in Chopin's Performing Style

Chopin moved to Paris in 1831 and became well known and respected in the thriving pianistic community. He did not play often in public, and when he did it was usually for small groups in intimate surroundings. Almost everyone who heard him felt that his manner of playing was unique. Compared to the dazzling virtuosity of other pianists, his playing was elegant and refined, dreamy and introspective, and so carefully restrained in many ways that very subtle expressive nuances could be easily perceived. According to contemporary accounts, he generally played with less volume of sound than other pianists, and Berlioz mentions in 1833 his 'utmost degree of softness . . ., the hammers merely brushing the strings'.[1] He was similarly restrained in touch, avoiding the brilliant tone characteristic of most of the virtuosos of the day. He accentuated gently, and one critic was even disturbed by his 'inobservance of the accent marking the beginnings of new musical lines'.[2] He likewise displayed 'marvellous discretion' in the use of the pedals, which were playing an increasingly important role in creating the textures and tone colours of piano music.[3]

Noting the considerable subtlety that Chopin exercised in dynamics, touch, accentuation, and pedalling, we might expect a similar sort of restraint in matters relating to rhythm and time. Berlioz mentions that 'his interpretation is shot through with a thousand nuances of movement of which he alone holds the secret, and which are impossible to convey by instructions'.[4] Yet there is a considerable difference of opinion concerning the nature of Chopin's rhythm. Mendelssohn, who, like Berlioz, advocated the keeping of relatively strict time, complains in 1834 that Chopin indulged in the 'Parisian spasmodic and impassioned style, too often losing sight of time and sobriety'.[5] According to Berlioz, 'Chopin chafed under the restraints of time' and 'pushed rhythmic freedom much too far'—indeed, 'simply *could* not play in strict time'.[6] In 1861 Farrenc, who had known Chopin from the 1830s, considers the later type of rubato the central feature of his art:

The tempo rubato, of which one makes today a usage so ridiculous and tiring, and which was the basis of the method of the pianist Chopin, . . . ought

[1] *Le Rénovateur*, ii/345 (15 Dec. 1833), quoted from Jean-Jacques Eigeldinger, *Chopin: Pianist and Teacher—As Seen by His Pupils*, trans. by Naomi Shohet with Krysia Osostowicz and Roy Howat, ed. Roy Howat (CUP, 1986), 71 and 272. The original French version is *Chopin vu par ses élèves* (Neuchâtel, Switzerland: Éditions de la Baconnière, 1970), followed by the 2nd edn. in 1979 and the 3rd in 1988; the English trans. is also labelled the 3rd edn.

[2] *Wiener Theaterzeitung* for 20 Aug. 1829; trans. from Eigeldinger, *Chopin: Pianist and Teacher*, 288.

[3] Antoine Marmontel, *Histoire du piano et de ses origines* (Paris: Heugel, 1885), 256; trans. from Eigeldinger, *Chopin: Pianist and Teacher*, 58.

[4] *Le Rénovateur*, ii/345 (15 Dec. 1833); trans. from Eigeldinger, *Chopin: Pianist and Teacher*, 71 and 272.

[5] *Briefe aus den Jahren 1830 bis 1847*, ed. Paul Mendelssohn Bartholdy (Leipzig: Hermann Mendelssohn, 1863–4), ii, 41; quotation from trans. by Lady Wallace (London: Longman, Green, Longman, Roberts, & Green, 1863), 38. See also *Allgemeine Theaterzeitung* (Vienna) for 18 June 1831: 'he seems to be a little too free with his tempi'; trans. in William G. Atwood, *Fryderyk Chopin: Pianist from Warsaw* (New York: Columbia University Press, 1987), 217.

[6] *Mémoires de Hector Berlioz* (Paris: Michel Lévy Frères, 1870; repr. Gregg, 1969), 420; trans. David Cairns (Knopf, 1969), 436. I have italicized *could* to match *pouvait* in the original French. See also *Chronique de Paris* for 18 June 1837, 404; trans. in Kerry Murphy, *Hector Berlioz and the Development of French Music Criticism*, SM xcvii (1988), 150 and 263 n. 93.

to be severely excluded from the interpretation of the works of Emanuel Bach, Haydn, Mozart, Clementi, Hummel, and even Weber and Mendelssohn.[7]

Farrenc then goes on to explain the other type of rubato used by Mozart, Paganini, and Field (see Chapter 4).

More moderate descriptions of Chopin's rhythmic flexibility come from Moscheles, Chorley, and Hiller. The pianist Ignaz Moscheles writes that 'the *ad libitum* playing, which in the hands of other interpreters of his music degenerates into a constant uncertainty of rhythm, is with him an element of exquisite originality'.[8] Henry F. Chorley, in a review of a recital played by Chopin in London, writes in 1848:

After the 'hammer and tongs' work on the pianoforte to which we have of late years been accustomed, the delicacy of M. Chopin's tone and the elasticity of his passages are delicious to the ear. He makes a free use of *tempo rubato*; leaning about within his bars more than any player we recollect, but still subject to a presiding sentiment of measure such as presently habituates the ear to the liberties taken. In music not his own we happen to know that he can be as staid as a metronome.[9]

Hiller states that with Chopin 'rhythmic firmness was combined with freedom in the declamation of his melodies, so that they would seem to have occurred to him at that very moment.'[10]

In contrast to these descriptions of Chopin's rhythmic freedom, a number of observers comment on the strictness of his tempo. Friederike Streicher, one of Chopin's pupils, notes in her diary for the years 1839–41 that 'he . . . required adherence to the strictest rhythm, hated all lingering and dragging, misplaced *rubatos*, as well as exaggerated *ritardandos*.'[11] Charles Salaman, who heard Chopin play in London

---

[7] Aristide Farrenc, *Le Trésor des pianistes*, i (Paris, 1861; repr. Da Capo, 1977), p. 3 of 'Observations générales sur l'exécution': 'Le *tempo rubato*, dont on fait aujourd'hui un usage si ridicule et si fatigant, et qui était la base de la méthode du pianiste Chopin, . . . doit être sévèrement exclu de l'interprétation des œuvres d'Emmanuel Bach . . .' In 1832 Farrenc made a contract with Chopin to publish his works in France. He never actually published any of his pieces, however, because Chopin, he claimed, was 'very lazy and eccentric' and his works too difficult to play. See Jeffrey Kallberg, 'Chopin in the Marketplace: Aspects of the International Music Publishing Industry in the First Half of the 19th Century', i, *Music Library Association, Notes*, 39 (1982/3), 541–3.

[8] *Aus Moscheles' Leben nach Briefen und Tagebüchern*, ed. by his wife (Leipzig: Verlag von Duncker & Humblot, 1872), ii, 39; trans. from *Life of Moscheles*, adapted by A. D. Coleridge (London: Hurst & Blackett, 1873), ii, 52.

[9] *The Athenaeum*, no. 1079 (1 July 1848), 660.

[10] Ferdinand Hiller, *Briefe an eine Ungenannte* (Cologne: DuMont & Schauberg, 1877), 150–2; trans. from Eigeldinger, *Chopin: Pianist and Teacher*, 270.

[11] Quoted in Frederick Niecks, *Frederick Chopin as a Man and Musician*, 3rd edn. (Novello, preface dated 1902), ii, 341.

in 1848, writes: 'In spite of all I had heard of Chopin's *tempo rubato*, I still recollect noting how precise he was in the matter of time, accent, and rhythm, even when playing most passionately, fancifully, and rhapsodically.'[12]

In the preface to his edition of Chopin's piano works, Mikuli tries to correct the false impression of Chopin's playing that had become current by 1879—an impression of a dreamer whose playing was barely audible, highly uncertain and unclear, and 'distorted into something totally arhythmic by a constant rubato'. Mikuli adds that 'in keeping tempo Chopin was inflexible, and it will surprise many to learn that the metronome never left his piano.'[13] On another occasion, however, Mikuli gives a different description of Chopin's rubato (the later type):

> Chopin was far from being a partisan to metric rigour and frequently used rubato in his playing, accelerating or slowing down this or that theme. But Chopin's rubato possessed an unshakeable emotional logic. It always justified itself by a strengthening or weakening of the melodic line, by harmonic details, by the figurative structure. It was fluid, natural; it never degenerated into exaggeration or affectation.[14]

Jan Kleczyński, who studied with three of Chopin's pupils, also emphasizes the musical purpose of Chopin's tempo changing and describes even more specifically how sparingly and carefully Chopin employed flexibility in a long, singing melodic line:

> Chopin arrived at the following conclusion, to which he attached much importance: *do not play by too short phrases*; that is to say, do not keep continually suspending the movement and lowering the tone on too short members of the thought . . . Do not spread the thought out too much, by slackenings of the movement—this fatigues the attention of the listener who is following its development.[15]

In addition, Mikuli and many others, as we will see, also mention the strict rhythm in the left hand which was necessary in order to produce the earlier type of tempo rubato.

Thus the rhythmic flexibility in Chopin's playing was perceived by

---

[12] 'Pianists of the Past: Personal Recollections of the Late Charles Salaman', *Blackwood's Edinburgh Magazine*, 170/1031 (Sept. 1901), 327.

[13] Trans. from pp. vi and vii in the Dover repr. (1987) of the Mazurkas from *Fr. Chopin's Pianoforte-Werke*, ed. Carl Mikuli (Leipzig: Fr. Kistner, foreword dated 1879).

[14] Aleksander Michałowski, 'Jak grał Fryderyk Szopen?', *Muzyka*, 9/7–9 (1932), 74–5; trans. from Eigeldinger, *Chopin: Pianist and Teacher*, 50.

[15] *How to Play Chopin: The Works of Frederic Chopin, Their Proper Interpretation*, trans. Alfred Whittingham, 6th edn. (London: William Reeves, [1913]), 52; quoted also in Eigeldinger, *Chopin: Pianist and Teacher*, 43–4.

different listeners in quite diverse ways. Some thought his rhythm was strict. Others considered his freedom excessive and offensive. Still others thought the flexibility noticeable but justified for musical reasons. Most, however, seemed to agree that his rhythmic manner was a conspicuous element within his unique and personal style of performing.

As previously noted, Türk, Hummel, and Czerny all described the rhythmic freedom added by the keyboard player, and the amount of such flexibility gradually increased as it was used more and more for specifically expressive purposes. At the same time, this sort of rhythmic freedom was also applied to some extent by violinists and singers, in addition to the earlier type of rubato. García mentions (in the second footnote on page 50 in Plate IV) that both rallentando and accelerando, although not marked, are frequently required in the music of Donizetti and Bellini, and Weber states, in his letter of 1824 concerning the opera *Euryanthe*, that 'the beat, the tempo, must not be a controlling tyrant'.[16] Chopin lived at a time when classic restraint still existed side by side with the newly emerging freedoms of romanticism.

Tempo Flexibility in Chopin's Notation

In his earlier works Chopin frequently employs terms to describe mood or rhythmic flow. After the Op. 24 Mazurkas, written in 1835 and published in 1836, there is a steady decline in the use of such terms, as well as in the number of metronome markings.[17] He continues classic practice by using *rallentando*, and occasionally *ritardando*, for slowing down. In addition, he includes *ritenuto*, which seems to be a new term around this time. Charles Chaulieu, in a glossary of musical terms published in *Le Pianiste* in 1834, defines *ritardando* and *rallentando* as delaying or slowing the movement, whereas *ritenuto* means 'to hold back the [note] values while lengthening a little their duration'.[18] Thus, like Beethoven's *ritenente*, Chopin's *ritenuto* seems related to the sort of expressive broadening indicated by dots and a slur (portato), dashes (tenuto), or accents (like small decrescendo

[16] Printed in the *Berliner allgemeine musikalische Zeitung*, 4/28 (11 July 1827), 218–19, and in *AMZ* 50/8 (Feb. 1848), 127; trans. Constance S. Jolly in Anton Felix Schindler, *Beethoven as I Knew Him*, ed. Donald W. MacArdle (Faber, 1966), 410–11.

[17] See Eigeldinger, *Chopin: Pianist and Teacher*, 121 n. 99.

[18] *Le Pianiste* (repr. Minkoff, 1972), 1/7 (May 1834), 103: 'retenir les valeurs, en allonger un peu la durée'.

signs).[19] A similar touch may be intended when he marks the same finger for several notes in succession.[20]

Chopin seems to use the word *ritenuto* when the music must be held back for some intensely expressive reason, with the slowing down usually accompanied by a richness and warmth of tone colour. In contrast, he seems to use *rallentando* for a more relaxed dying out of movement after the main expressive thrust has run its course. Ritenuto, I think, might result either in a gradual slowing or in a suddenly slower tempo, depending upon the expressive sense of a particular passage. In the Mazurka published posthumously by Fontana as Op. 68 No. 2, Chopin even indicates *poco a poco riten.*[21] Especially instructive are works in which Chopin uses both *ritenuto* and *rallentando*. In the Mazurka on Plate XVI, for example, the *ritenuto* in the third line has an expressive purpose associated with the dynamic change, the accentuations, the repetition of a brief phrase, and the F♮ as it changes finally to F♯. The *rallent.* on the next line, however, serves mainly the more relaxed structural purpose of dissipating the remaining energy from the preceding section and pointing toward the return of the main theme in the following measure.[22]

There thus seems to be a rather consistent distinction at the time of Chopin between *rallentando* and *ritenuto*, a distinction somewhat different from the one applied today. When one sees *ritenuto* in the music of Chopin, one should therefore look for something especially expressive—something either coinciding with the term, or something immediately preceding or following. The expressive event will then motivate the special warmth and weight of the touch, which in turn act to create reluctance or hesitation, and hence a slowing down of the

[19] Rosen comments: 'The distinction between rallent. and ritenuto in Chopin is precisely that in Beethoven; that is, ritenuto or ritenente is immediate and expressive (see Op. 111).' See Chopin's B minor Sonata, Op. 58, mm. 69 and 177 in the 1st movt. for tenuto dashes; Mazurka Op. 6 No. 2, mm. 29–30 for accent signs, and 69–70 for portato. Concerning some disagreement on the meaning of Chopin's accent signs, see Eigeldinger, *Chopin: Pianist and Teacher*, 148 n. 172, and Jan Ekier, Wiener Urtext edn. of the Impromptus (Vienna, 1977), p. ix of preface.

[20] I am grateful to Rosen for pointing this out to me. The 5th finger of the right hand, for example, plays the 3 notes E♭, D, and C in the singing passage in m. 26 of the Nocturne Op. 9 No. 2 (Ex. 7.21). See the Nocturnes ed. by Ewald Zimmermann (Henle, 1980), 14, or *Stirling*, 24. The Stirling Collection consists mostly of editions published in Paris during Chopin's lifetime; some works lack pages, however, and some are missing altogether (see Eigeldinger, *Chopin: Pianist and Teacher*, 179–81 and 200–11). For modern editions, I cite where possible those from Henle Verlag. See Thomas Higgins, 'Whose Chopin?', *19th Century Music*, 5 (1981), 67–75; also Ewald Zimmermann, 'L'Édition moderne des œuvres de Chopin chez Henle-Verlag' in *Sur les traces de Frédéric Chopin*, ed. Danièle Pistone (Paris: Librairie Honoré Champion, 1984), 105–12.

[21] *Œuvres posthumes pour le piano* (Berlin: Ad. Mt. Schlesinger, 1855), copy at BL. See the Henle edn. of the Mazurkas, 141.

[22] For both terms, see also the Rondo Op. 16, mm. 57–60; Étude Op. 10 No. 3 (*Stirling*, 33), mm. 8, 16, and 20; and Nocturne Op. 27 No. 1 (*Stirling*, 140), mm. 93 and 96.

PLATE XVI. Frédéric Chopin, 1st page of the Mazurka Op. 6 No. 1 from
Quatre mazurkas pour le pianoforte (Leipzig: Kistner, originally 1832)

movement. Considering the difference between these various terms, it is important that one use an edition which faithfully indicates them. In the complete works of Chopin edited by Paderewski (Warsaw and Kraków, 1949–61), for example, *ritenuto* has been abbreviated to *rit.*, which a modern performer will surely misinterpret as *ritardando*.[23] Chopin himself either writes the word complete or abbreviates it, usually as *riten.*[24]

Words indicating an increase of the tempo also occur more frequently at the time of Chopin than they did in classic music. He seems to use *accelerando*, as he does *rallentando*, for less intense structural situations, whereas *stretto*, or less often *stringendo*, acts like *ritenuto* as a more expressive effect.[25] Chaulieu defines *accelerando* with the verb *presser* (to hurry), *stretto* and *stringendo* with *serrer*, which indicates a more intense tightening, compacting, or bringing things closer together.[26] With Chopin, *stretto* and *accelerando* can occur successively or simultaneously, and either can be intensified.[27] *Stretto* often seems to occur close to *ritenuto* in the same expressive passage.[28]

In addition to these words for indicating a decrease or increase of tempo, Chopin sometimes writes simply *più* or *meno mosso*, *poco* or *molto più lento*. Sometimes he simply marks *adagio* for a cadence. *Calando*, *mancando*, and *morendo*, according to Chaulieu, all involve slowing as well as softening, whereas *perdendosi* and *smorzando* do not. He states that *agitato* 'permet d'altérer le mouvement'; *precipitato* is 'hors de mesure', *abbandono* 'sans mesure précise'; and *appassionato* 'admet un peu d'exagération'. Herz adds that *abbandonatamente* subordinates the time to the expression, *liberamente* means freely, and *negligente* indicates a lack of constraint.[29] Chopin does not use all of these terms himself, but he was living at a time when they were increasingly in demand. Like the classic composers, Chopin includes adagio sections, sometimes marked *senza tempo* (see the free

---

[23] Compare, for example, the ritenutos in the Mazurka Op. 6 No. 1 or Op. 7 No. 3, or the Nocturne Op. 15 No. 3 in the Stirling Collection or the Henle editions with the markings of *rit.* in the Paderewski *Complete Works*.

[24] In his glossary Chaulieu gives the abbreviation *ritard.* for *ritardando*, and simply *rit.* for *ritenuto*.

[25] See the dramatic *stretto* in the Prélude Op. 28 No. 4, for example, or the Mazurka Op. 7 No. 1, where the *stretto* struggles against the left hand's temporary shift to 2/4.

[26] *Le Pianiste*, 1/7 (May 1834), 98 and 103; see also *affrettando* on p. 98 and *animato* on p. 99.

[27] See the Étude Op. 10 No. 9 for both words in succession; the end of the Polonaise Op. 26 No. 2 for *accel. e stretto*; the Nocturne Op. 27 No. 1 and Étude Op. 10 No. 9 for *sempre più stretto*; and the end of the Valse Op. 64 No. 3 for *poco a poco accelerando al fine*.

[28] See the Étude Op. 10 No. 3, mm. 7 and 8; and the Nocturne Op. 15 No. 2, m. 24.

[29] Henri Herz, *Méthode complète de piano*, Op. 100 (Paris: J. Meissonnier, *c.*1837), copy at BL, 137–8.

cadenza in small notes at the end of the Nocturne Op. 9 No. 2), sometimes without barlines (end of the Nocturne Op. 32 No. 1), and often with fermatas.

Following the free introductory section in the Polonaise-Fantaisie Op. 61, which contains many small notes between fermata-held chords and ends with a *rallentando*, rhythmic figures in octaves in the left hand announce the arrival of an organized theme. These fanfare-like figures are marked *a tempo giusto*, where the word *giusto* no doubt follows Chaulieu's definition: 'going rigorously in time, precise'. On the other hand, when Chopin writes *Tempo giusto* as the main tempo marking for the Waltz Op. 64 No. 2, he may refer in this case to a 'moderate tempo', as the expression is defined by Herz. This expression apparently continues to have two meanings, as it did also, as we noted in the last chapter, in the preceding century.

Chopin reflects in a general way, then, the steadily increasing frequency of terms denoting tempo flexibility, as well as the inclusion of new words that prescribe increasingly subtle shades of meaning. He seems to be especially concerned with terms, such as *ritenuto* and *stretto*, which involve, at the same time, subtle nuances of touch, articulation, and tone colour—all employed along with tempo changing to achieve a special intensity of expression. From a study of those works available in more than one source, it is evident that Chopin gave considerable thought to every small detail marked in his scores.[30] It is equally apparent that every marking has significant musical meaning. Perhaps even more than for other composers, the careful following of Chopin's marks is essential, I believe, for the proper understanding and performance of the music. This extends also to other types of marks—those for dynamics, pedalling, articulation, etc.—for sometimes a faithful observance of these other elements determines the manner in which one performs a rallentando or ritenuto, an accelerando or stretto. All the elements of Chopin's music, when properly restrained and balanced, fit together to produce natural and unified musical ideas. His compositions seemed, even to his contemporaries, to be unique in the sincerity of their expression and in the power and special nature of the emotions evoked.

The tempo modifications considered so far have been of a relatively general nature. In the Mazurkas, however, Chopin seems to have used a special type of tempo alteration. Its effect was apparently so remarkable that we will consider it here as a separate type of rhythmic robbery.

---

[30] The Henle editions often include several different versions of the same piece; see, for example, the Valse Op. 69 No. 1 or the Mazurka Op. 6 No. 1.

Alteration of Note Values in the Mazurkas

Music based on dance forms sometimes reflects animated choreographic movements by means of special accentuations. These are sometimes exaggerated by lengthening the value of certain notes in the written notation. We saw in the section on inequality in Chapter 2 how the dotted second beat was probably extended beyond its notated value in the main French type of baroque sarabande, as well as the folies d'Espagne, passacaille, and chaconne. From a later period, we are probably most familiar with the rhythmic alterations in the Viennese waltz. Polish dances also had their characteristic rhythmic qualities. The mazurka, in particular, seems to have certain accentuations and distortions of the beat that give to its music a striking rhythmic animation.

This sort of rhythmic alteration does not involve a free melody against a strict accompaniment, for vertical parts are sounded together. On the other hand, it is not the same as the usual later rubato, since the adjustments of the note values are absorbed within each single measure. The first beat of each measure is therefore theoretically in the proper time. Some other beat, however, may be extended, with still another losing the corresponding amount.

Mazurka was the name for a group of dances that originated in the area near Warsaw. They had triple metre and a strong accent on the second or third beat to coincide with a tap of the heel.[31] Eventually the folk version of the dance was transformed into a ballroom dance, which became popular elsewhere in Europe during the eighteenth and early nineteenth centuries. Chopin, with intimate knowledge of the folk tradition, incorporated the flavour and some of the musical characteristics of the dance into his piano pieces. He clearly had a special enthusiasm for composing, teaching, and playing the mazurka. His fifty-seven works in this genre amazed his contemporaries as

---

[31] A number of scholars have studied Polish folk dances in relation to rubato. See especially Marian Sobieski and Jadwiga Sobieska, 'Das Tempo Rubato bei Chopin und in der polnischen Volksmusik', *The Book of the First International Musicological Congress Devoted to the Works of Frederick Chopin, Warsaw 1960* (Warsaw: Polish Scientific Publishers, 1963), 247–54. Roderyk Lange writes out the choreography in dance notation in the following articles: 'Der Volkstanz in Polen', *Deutsches Jahrbuch für Volkskunde*, 12 (1966), 342–57; 'The Traditional Dances of Poland', *Viltis*, 29/1 (May 1970), 4–14; and 'On Differences between the Rural and the Urban: Traditional Polish Peasant Dancing', *Yearbook of the International Folk Music Council*, 6 (1974), 44–51. See also Maciej Golab, 'Research on Changes in Traditional Wedding Songs in the Opoczno Region' and Tomasz Szachowski, 'An Attempt to Apply a Method of Measurement in Research on Tempo Rubato', both in Polish in *Muzyka*, 19/3 (1974), 49–62 and 63–75; and Kay Bedenbaugh, 'Rubato in the Chopin Mazurkas', D.M.A. final project (Stanford University, 1986; UM 86–19,849). For a brief description of the folk mazurka, see *New Grove*, xi, 865–6, and Eigeldinger, *Chopin: Pianist and Teacher*, 145–6 n. 169.

much as they amaze us by their modality, modulations, drones, and especially their rhythmic peculiarities. Many contemporary listeners to Chopin's performances simply felt a strange and wonderful rhythmic exuberance. The reviewer in the *Athenaeum* of 1848, quoted above, also stated that the Mazurkas 'lose half their characteristic wildness if played without a certain freak and licence—impossible to imitate, but irresistible if the player at all feel the music'.[32]

More specific information on the rhythmic features comes from Moscheles, Lenz, and Hallé. Moscheles reports that his daughter (Emily Roche), who took lessons from Chopin in 1848, 'had played . . . a new Chopin Mazurka with such a rubato that the entire piece gave the impression of being in 2/4 instead of 3/4'.[33] Lenz tells a story about a disagreement between Meyerbeer and Chopin. Lenz was playing a Mazurka for Chopin at a lesson when Meyerbeer appeared and announced that he heard the music in 2/4 time. Chopin tapped out the 3/4 rhythm with his pencil, then played the music himself and stamped the time with his foot. Neither would give in and they parted in anger. Lenz adds, however, that 'Chopin was right. Though the third beat in the composition . . . is slurred over, it no less exists.'[34]

Fortunately for us, Lenz identifies the piece he was playing as the Mazurka Op. 33 No. 3 in C major. The piece is marked *semplice* and, following a cadence in C major at the end of the opening section, turns abruptly to A flat major for the middle section. Ex. 7.1*a* shows the first phrase of the middle section as it appears in the Paris edition in the Stirling collection. Other editions also indicate a dynamic change—from *piano* in the opening section to *forte* for the middle. It seems to me that the considerable change of texture in the middle section, combined with its change of dynamics and key, might evoke the sort of exuberance that a performer might express by means of the

[32] *The Athenaeum*, no. 1079 (1 July 1848), 660.

[33] Moscheles told this to Hans von Bülow; see the latter's *Briefe und Schriften*, ed. Marie von Bülow, 2nd edn., i (Br. & H, 1899), 133 (in letter to his mother dated 19 Nov. 1848). Trans. from Eigeldinger, *Chopin: Pianist and Teacher*, 73.

[34] Wilhelm von Lenz, 'Die grossen Pianoforte-Virtuosen unserer Zeit aus persönlicher Bekanntschaft: Liszt, Chopin, Tausig', *Neue Berliner Musikzeitung*, 22/38 (1868), 302; and a book with the same title but adding the name of Henselt (Berlin, 1872), 46. Quotation from *The Great Piano Virtuosos of Our Time from Personal Acquaintance: Liszt, Chopin, Tausig, Henselt*, trans. Madeleine R. Baker (New York: G. Schirmer, 1899), 67–8; for another trans., see Eigeldinger, *Chopin: Pianist and Teacher*, 73. In connection with this dispute between Chopin and Meyerbeer, Wanda Landowska describes a scene from her childhood in which she saw 'a farm girl milking cows in 2/4 time while soothing them by singing a *mazur* in 3/4 time'. She comments that 'this, which seems incredible to a stranger, is natural for a Pole because of the peculiar accentuation of the mazurka . . . What a pity that Meyerbeer did not see the Polish milkmaid at work and hear her singing!'; *Landowska on Music*, collected, ed., and trans. by Denise Restout assisted by Robert Hawkins, 3rd impression (New York: Stein & Day, 1964), 288–9.

Ex. 7.1 Chopin, Mazurka Op. 33 No. 3 (Paris: M. Schlesinger, 1838; *Stirling* 192)

rhythmic alterations shown in Ex. 7.1*b*. In this case the third beats of measures 17, 18, and 19 are somewhat shortened, as Lenz describes, and the second beat lengthened, resulting theoretically in measures of duple metre. I presume, however, that the amount of prolongation is variable, both from measure to measure and from performance to performance. Therefore, one might occasionally, as I show in measure 20, reaffirm the triple metre in some measures played precisely in time. The same effect that I show in Ex. 7.1*b* may have been applied by Lenz and Chopin to the opening section, for here the second beats of many of the bars are marked with an accent sign.[35]

[35] Gastone Belotti, in 'Die rhythmische Asymmetrie in den Mazurken von Chopin', *Chopin-Jahrbuch* (1970), 112–15, writes out the right-hand part of the opening section in various ways: in 2/4, then 4/4, and finally in 13/16 to illustrate Paderewski's assertion that 'thanks to this Tempo rubato, we may lengthen the second beat in a Chopin Mazurka by about a sixteenth note in order to produce a stronger accent on the last beat.' Although this particular Mazurka seems at first glance to evoke an innocent, childlike happiness, Lenz felt it was full of grief and sorrow—an indication, perhaps, of a very slow tempo and a response to the bittersweet dissonance of the 3rd degree of the scale on accented 2nd beats.

I have indicated the source of each example by Chopin as in Ex. 7.1, with the publisher's location and name as well as the date of his 1st edn. of the work, and, where applicable, the page number in the Stirling Collection. I have printed main tempo markings, according to usual modern practice, with only the 1st letter capitalized and transitory indications, such as *ritardando* or *rubato*, in italics. Kistner and Wessel (Chopin's principal publishers in Leipzig and London) capitalized all the letters in the main markings, and Wessel sometimes capitalized the 1st letter of transitory terms (such as *Rubato*). M. Schlesinger (in Paris) seems to use larger and smaller versions of the same type style for the two categories, whereas Kistner (see Pl. XVI) comes closer to our italics for the transitory words. All 3 publishers place a period after every word, whether an abbreviation or not.

The left hand in Ex. 7.1a plays what we recognize as a waltz bass. In addition, the alterations in Ex. 7.1b are similar to those we associate with the Viennese waltz. In regard to gypsy music, Liszt writes: 'One does not find there the trepidations, the swayings, full of hesitation and of agitation, similar to those of the waltz or the mazurka.'[36] W. H. Sherwood, one of Liszt's students around 1876, describes the performance of waltz rhythm as follows:

Liszt had a habit frequently of dashing the wrist abruptly from the chord at the second beat of the measure, with more or less accent, sometimes almost prematurely, the movement being correspondingly retarded, before playing the chord on the third beat of the measure, with another less conspicuous *up* stroke. Such treatment certainly lent a piquancy and sparkle to the performance. As it was never twice alike there was no objectionable mannerism therein.[37]

The Mazurkas, however, show far more variety than the waltz in their accompanimental patterns and only occasionally employ the bass of Ex. 7.1a. Furthermore, the second beat is not the only place where rhythmic exaggeration can be applied. Charles Hallé, who was a friend of Chopin and heard him perform often between 1836 and 1848, describes a different manner of accentuation:

A remarkable feature of his playing was the entire freedom with which he treated the rhythm, but which appeared so natural that for years it had never struck me. It must have been in 1845 or 1846 that I once ventured to observe to him that most of his mazurkas..., when played by himself, appeared to be written, not in 3/4, but in 4/4 time, the result of his dwelling so much longer on the first note in the bar. He denied it strenuously, until I made him play one of them and counted audibly four in the bar, which fitted perfectly. Then he laughed and explained that it was the national character of the dance which created the oddity. The more remarkable fact was that you received the impression of a 3/4 rhythm whilst listening to common time. Of course this was not the case with every mazurka, but with many.[38]

Since Hallé does not name a specific composition, I have chosen the Mazurka Op. 7 No. 1 to illustrate his remarks. Chopin wrote the

[36] Franz Liszt, *Des Bohémiens et de leur musique en Hongrie* (first published 1859), *Nouvelle édition* (Br. & H, 1881; repr., Bologna: Forni, 1972), 399: 'On n'y retrouve pas les trépidations, les balancemens pleins d'hésitation et de trouble, propres à ceux de la valse ou de la mazoure.'

[37] 'Student Days in Weimar with Liszt', *The Etude*, 26 (1908), 285. According to William J. Maloof, in *Standard Score Format: Layout and Notation* (Boston: the author, 1981), 106, 'the *rubato* employed in ... the "Viennese Waltz"' involves performing ♩ ♩ ♩ approximately as ♩ ♩ ♩. Fritz Spiegl describes the 'Viennese lift' in *Music through the Looking Glass* (London: Routledge & Kegan Paul, 1984), 311: the 2nd of 3 quarter notes in 3/4 'is played a little early, the third a little late ... Thus the rubato is kept within each bar and does not affect the overall pulse of the music.'

[38] *The Autobiography of Charles Hallé, with Correspondence and Diaries* (originally published at London in 1896), ed. Michael Kennedy (New York: Harper & Row, 1973), 54.

Ex. 7.2   Chopin, Mazurka Op. 7 No. 1 (Paris: M. Schlesinger, 1833; *Stirling* 12)

opening two bars as in Ex. 7.2*a*, whereas in (*b*) I have doubled the first beat of each bar in relation to the other two beats, resulting in 4/4 time. This passage seems particularly suited for this alteration, for it is *vivace, forte,* and exuberant.[39]

The prolonging of the second or first beats that we see in Exx. 7.1*b* and 7.2*b* is a form of agogic accentuation. It projects a special vigour or exuberance in imitation of some powerful choreographic thrust of energy. The performer, like the appreciative listener, perceives the music, as written, in 3/4; the rhythmic alterations occur as an expressive rendition of the original notation. The music appears as in Ex. 7.1*b* or 7.2*b* only to a trained musician, like Moscheles, Meyerbeer, or Hallé, when he is giving particular analytical attention to rhythmic details. Lenz felt that Chopin was right when he insisted that the Mazurka was in 3/4; Hallé admitted that it was years before he noted this peculiarity, and even after analysing it, felt that the music was still perceived in 3/4 time.[40]

---

[39] I am indebted to Rosen for the idea that this alteration is particularly well suited to the rhythm in Ex. 7.2. He suggests that Chopin might want the 1st note lengthened when he follows it by a 16th rest, and not when he writes the rhythm as a dotted 8th note and 16th without a rest.

[40] I have shown the effect of note-lengthening in Exx. 7.1*b* and 7.2*b* by quadruplets within 3/4 metre rather than by a change of metre. In Ex. 7.1*b* especially, it seems to me that this more accurately depicts what happens when a measure of 3 beats (m. 20) follows a quadruplet. The

Charles Rosen has informed me that his teacher Moriz Rosenthal, who studied with Chopin's pupil Mikuli as well as with Liszt, always played the rhythm of two eighth notes followed by two quarters as two eighth notes, a half, and a quarter (as I show in Ex. 7.1). Yet when faced with the possibility of doubling the length of a note value, he reacts as Chopin did:

> [Hallé] states that Chopin [in a Mazurka] lengthened the first quarter-note beat of each measure to double its value. . . . With all due respect to Hallé, this tale seems to me simply incredible. The results of such Procrustes-like distortions are simply pitiful. In my opinion, all these offsprings of a misunderstood *tempo rubato* should be minimized. There is not a single mazurka which would gain by an exaggerated *tempo rubato* treatment.[41]

He simply confirms, as Chopin, Lenz, and Hallé did, that the performer conceives and projects the music in triple metre—but in a triple metre so forceful that it can actually be strengthened by prolonging certain beats. If upon analysis the prolongation seems to cause a mathematical relationship between the notes similar to duple metre, then this is simply a curious and astonishing feature.[42]

Alteration of Note Values While the Accompaniment Keeps Strict Time

Turning now to that other method of rhythmic robbery, we will investigate a number of ways in which Chopin introduces flexibility into a melody while the accompaniment maintains a strict beat. In issues of *Le Pianiste* for 1834, there are descriptions of Chopin's rubato. Chaulieu gives the following definition, explaining in a footnote that the name of a composer in parentheses indicates that he is the first to use the word or is one who uses it most frequently:

prolongation of the 1st beat, however, seems to present a more complex situation. It might be possible that in some cases (such as the opening measure of a piece, for example) the prolongation adds to the total time of a measure rather than causing internal adjustments. One might notate this more faithfully by changing the metre to 4/4 for the measures involved and setting them next to a regular measure in 3/4, where the quarter note retains the same value. It is difficult in such circumstances to separate the metrical effect of the prolongation from a possible simultaneous effect produced by the later type of rubato. Actually, the amount of lengthening involved can vary so much that any method of notation is merely approximate and theoretical.

[41] 'Mazurka Op. 24 No. 4 in B flat minor by Frédéric Chopin: A Master Lesson by Moriz Rosenthal', *Etude Music Magazine*, 58 (1940), 130.
[42] See David Fuller's comments in *Performance Practice: Music after 1600*, ed. Howard Mayer Brown and Stanley Sadie (Norton, 1989), 120.

RUBATO or ROBATO,—unspecified note values borrowing from one [note] to another; *lusingando* [which he defines as coaxing, caressing, gracious] and *espressivo* combined (Chopin).[43]

Another issue of the same periodical contains a review of the three Nocturnes in Chopin's Op. 15. Preceding the comments on Dussek's futile attempt to notate rubato by syncopation (see Chapter 5), the author complains about the way Chopin spoils his fresh and pleasing ideas by great difficulties of execution, as well as by

a manner of affectation to write the music almost as it should be played (we say *almost*, for *completely* is impossible)—to write this swaying, languid, groping style, this style which no known arrangement of note values can well express; the *Rubato*, finally, this *Rubato*—the terror of young women, the bogeyman of beginners.[44]

The author must be referring here to the second Nocturne of Op. 15, which includes the notation shown in Ex. 7.3. The beginning of the piece appears in (*a*), the parallel beginning of the second phrase in (*b*). By means of a combination of large and small notes, Chopin makes clear that the A$^\sharp$ in the melody at the beginning of measure 9 must occur later than the first C$^\sharp$ in the bass. In addition, the Paris edition prints the second bass note in the measure vertically below the G$^\sharp$ rather than the E$^\sharp$. The two notes of anacrusis are similarly delayed by the arpeggiated chord and the small notes. If the bass notes maintain a steady beat, then this notation represents a remarkable attempt to portray the sort of alteration of melodic note values involved in the earlier type of rubato.[45] In addition, the same Nocturne includes unusual groups of notes: thirty in the melody against four in the bass, groups of seven or eight notes in a chromatically descending *portamento* to be freely inserted between two notes, and, in the entire middle section, quintuplets in an inner voice against two eighth notes in the bass. The unusual groups, as we saw in Chapter 5, were called *tempo rubato* by C. P. E. Bach and the method of execution explained

---

[43] *Le Pianiste*, 1/7 (May 1834), 103: 'valeurs indécises empruntant l'une à l'autre; *lusigando* [sic], *espressivo*, réunis, (*Chopin.*)', and 101: 'LUSINGANDO,—en cajolant, en flattant; gracieux'.

[44] *Le Pianiste*, 1/5 (Mar. 1834), 78: 'une sorte d'affectation à écrire la musique presque comme il faut l'exécuter,—(nous disons *presque*, car *tout-à-fait* est impossible.)—à écrire ce genre balancé, languissant, tâtonné, ce genre qu'aucun arrangement de valeurs connues ne peut bien exprimer; le *Rubato* enfin, ce *Rubato* l'effroi des jeunes filles, le *Croque-Mitaine* des mazettes!' The word *mazette* seems to refer at that time to a novice, who is inexperienced and unskilled and hence clumsy or awkward.

[45] Donald N. Ferguson, in *Piano Music of Six Great Composers* (Prentice–Hall, 1947), 197, describes this notation as 'a graphic picture of rubato'. He is quoted by George A. Kiorpes in 'Arpeggiation in Chopin: Interpreting the Ornament Notations', *PQ* 29/113 (1981), 56; Kiorpes advocates the addition of some 'free time' for the small notes.

Ex. 7.3 Chopin, Nocturne Op. 15 No. 2 (Paris: M. Schlesinger, 1834; *Stirling* 70)

by Hummel and Czerny. All in all, this particular composition illustrates most vividly the 'swaying, languid, groping style' which the reviewer for *Le Pianiste* calls *rubato*.

In addition to these contemporary accounts, there are descriptions of Chopin's rubato recorded later by his students and by those who had heard him perform. These tend to emphasize the strict left hand which contrasted with the free melody. Lenz, who became Chopin's student in 1842, writes in 1868:

That which particularly characterized Chopin's playing was his *rubato*, whereby the rhythm and time through the whole remained accurate. 'The left hand', I often heard him say, 'is the conductor, it must not waver, or lose ground; do with the right hand what you will and can.'[46]

---

[46] *Die grossen Pianoforte-Virtuosen* (1872), 47; trans. from *The Great Piano Virtuosos*, 68. See Eigeldinger, *Chopin: Pianist and Teacher*, 50, for a trans. of the corresponding passage from Lenz's original article in *Neue Berliner Musikzeitung*, 22/38 (1868), 302. Similar ideas occur in a letter included by Arthur Hedley in *Selected Correspondence of Fryderyk Chopin* (New York: William Heinemann, 1963; repr. Da Capo, 1979), 216, and supposedly written by Joseph Filtsch (brother of Karl, who was Chopin's student) on 8 Mar. 1842. Eigeldinger, in *Chopin: Pianist and Teacher*, 142 n. 157, questions the authenticity of the letter, which states: 'To his pupils he [Chopin] says: "Let your left hand be your conductor and keep strict time." And so his right hand, now hesitant, now impatient, is nevertheless constrained to follow this great rule and never weakens the rhythm of the left hand.' The letter does not include the word *rubato*, but, if it is authentic, would be the earliest source (and the only one while Chopin was still alive) to mention the strict left hand as a conductor.

He quotes a more poetic definition, presumably describing the same effect, which Liszt gave to a student at Weimar in 1871: 'Do you see those trees? . . . The wind plays in the leaves, stirs up life among them, but the *tree remains the same*—that is the Chopin *rubato!*'[47]

In another source from 1872 Lenz describes the Valse Op. 42 in A flat major:

This waltz . . . should evoke a musical clock, according to Chopin himself [referring, presumably, to the two even quarter notes in each measure of the opening section which are played against the three beats of the waltz bass]. In his own performance it embodied his rubato style to the fullest; he would play it as a continued *stretto prestissimo* with the bass maintaining a steady beat.[48]

By 'rubato style' he may be referring to the polymetric section, to delays and anticipations added by the performer to notes in the warm melody in measures 121 to 165, or to written-out note alterations when a melody is repeated.[49]

The analogy between the left hand and the orchestra or its conductor, and between the right hand and a singer, continues in later writings. Moscheles compares Chopin as a performer to 'some singer who troubles himself very little about the accompaniment, and follows his own impulses'.[50] Maurycy Karasowski, a Polish music critic and author of the first biography of Chopin based on authentic documents, writes in 1877:

The *tempo rubato* was a special characteristic of Chopin's playing. He would keep the bass quiet and steady, while the right hand moved in free *tempo*, sometimes with the left hand, and sometimes quite independently, as, for example, when it plays quavers, trills or those magic, rhythmical runs and *fioritures* peculiar to Chopin. 'The left hand', he used to say, 'should be like a bandmaster, and never for a moment become unsteady or falter.'

By this means his playing was free from the trammels of measure and acquired its peculiar charm.[51]

---

[47] *Die grossen Pianoforte-Virtuosen*, 47; most of the trans. is from *The Great Piano Virtuosos*, 68–9, but I have replaced the phrase 'life unfolds and develops beneath them' with 'stirs up life among them' from Niecks, *Frederick Chopin as a Man and Musician*, ii, 101, and quoted in Eigeldinger, *Chopin: Pianist and Teacher*, 51. Kleczyński, in *How to Play Chopin*, 57, gives what seems to be a variant of this quotation: 'Suppose a tree bent by the wind; between the leaves pass the rays of the sun, a trembling light is the result, and this is the *rubato*.'

[48] Eigeldinger, *Chopin: Pianist and Teacher*, 87, trans. from 'Übersichtliche Beurtheilung der Pianoforte-Kompositionen von Chopin', *Neue Berliner Musikzeitung*, 26/38 (1872), 298.

[49] Cf. mm. 141–2 and 125–6, for example.

[50] *Life of Moscheles*, trans. A. D. Coleridge, ii, 53; *Aus Moscheles' Leben*, ii, 39.

[51] *Friedrich Chopin: Sein Leben, seine Werke und Briefe* (Dresden, 1877), trans. Emily Hill in *Frederic Chopin: His Life and Letters*, 3rd edn. (London: William Reeves, 1938; repr. Greenwood, 1970), 320–1.

Kleczyński, who studied with three of Chopin's best students, reports in 1879: 'Some of Chopin's pupils have assured me that in the *rubato* the left hand ought to keep perfect time, whilst the right indulges its fancy; and that in such a case Chopin would say, "The left hand is the conductor of the orchestra." '[52] In his preface to the keyboard works of Chopin, Mikuli, who studied with Chopin from 1844 to 1848, states:

Even in his much-slandered rubato, one hand, the accompanying hand, always played in strict tempo, while the other—singing, either indecisively hesitating or entering ahead of the beat and moving more quickly with a certain impatient vehemence, as in passionate speech—freed the truth of the musical expression from all rhythmic bonds.[53]

Hugh Reginald Haweis, who apparently heard Chopin perform, writes later that 'from Chopin, Liszt and all the world after him got that *tempo rubato*, that playing with the duration of notes without breaking the time'.[54] Camille Dubois, Chopin's student during the period 1843 to 1848, quotes him as saying 'Let your left hand be your conductor and always keep time', and Elise Peruzzi, who studied with him around 1836–8, reports that 'he would bring out certain effects by great elasticity' and 'got very angry at being accused of not keeping time, calling his left hand his *maître de chapelle* and allowing his right to wander about *ad libitum*'.[55] Valentin Alkan, a fellow pianist and friend of Chopin, repeatedly quoted Chopin in connection with tempo rubato by saying 'the left hand must act as conductor, regulating and tempering any involuntary inflexions of the right hand.'[56]

Georges Mathias studied with Chopin for five years or more beginning around 1838. In an essay written in 1897 he describes first the later type of rubato, then states:

There was another aspect: Chopin...often required simultaneously that the left hand, playing the accompaniment, should maintain strict time, while the melodic line should enjoy freedom of expression with fluctuations of

[52] *How to Play Chopin*, trans. Alfred Whittingham, 57.

[53] *Fr. Chopin's Pianoforte-Werke*, i: *Mazurkas*; trans. from Dover repr., p. vii. Following the description of Julius Stockhausen's rubato quoted in ch. 3, Bernhard Scholz states: 'That is exactly what Chopin, according to the testimony of his student Mikuli, required in the performance of his music'; see his *Verklungene Weisen* (Mainz, 1911), 127, and Edward F. Kravitt, 'Tempo as an Expressive Element in the Late Romantic Lied', *MQ* 59 (1973), 498.

[54] *My Musical Memories* (New York: Funk & Wagnalls, 1887), 268.

[55] Both quoted in Niecks, *Chopin as a Man and Musician*, ii, 101–2 and 339.

[56] A. de Bertha, 'Ch. Valentin Alkan aîné: Étude psycho-musicale', *Bulletin français de la Société Internationale de Musique*, 5/2 (Feb. 1909), 146; trans. from Ronald Smith, *Alkan, the Reluctant Virtuoso*, i (New York: Crescendo Publishing, 1977), 98. See Hedley, *Selected Correspondence of Fryderyk Chopin*, 195, 279, and 375.

speed. This is quite feasible: you can be early, you can be late, the two hands are not in phase [*en valeur*]; then you make a compensation which re-establishes the ensemble.[57]

He reports further that Chopin recommended this manner of playing in the works of Weber, specifically in the solo piano passage in A flat major that commences in measure 57 (counting from the marking *Allegro passionato*) of the *Konzertstück*. Here the left hand provides a bass note and seven eighth-note chords in each measure while the florid melody in the right hand moves first in even sixteenth-notes, then descends chromatically in thirty-second-note figures.

In 1910 *Le Courier musical* devoted special attention to Chopin on the centennial of his birth. Two articles include reminiscences by Chopin students on rubato. Francis Planté, the French pianist, transmits ideas from Chopin's cellist friend Auguste Franchomme and from two of Chopin's pupils, Countess Delfina Potocka (to whom he dedicated the Concerto Op. 21 and the Valse Op. 64 No. 1) and Princess Marcelina Czartoryska (whose playing was regarded by contemporaries as much like Chopin's):

That which has been called 'Tempo rubato' . . . was, according to his faithful listeners, only a great liberty and fantasy in the melodic design with its fiorituras . . .
    The accompaniment, on the contrary (that is to say, the bass parts) remained rhythmic *in spite of all*, so much so that when he played with [orchestral] accompaniment, his left hand always fell, it appeared, exactly with the baton of the conductor.[58]

In another issue of *Le Courier musical*, Camille Saint-Saëns describes the important ideas on Chopin's method of playing related to him by Marie Félicie Clémence de Reiset, Vicomtesse de Grandval (1830–1907) and by Pauline García Viardot (1821–1910). The Vicomtesse had studied with Chopin sometime during or before 1848, when Saint-Saëns recalls hearing her play. She later became a composer and a pupil of Saint-Saëns. Pauline Viardot was a friend of both Chopin and George Sand and studied piano informally with him during the summers from 1841 to 1845. She was the sister of the

---

[57] Preface, dated 1897, to Isidore Philipp's *Exercises quotidiens tirés des œuvres de Chopin* (Paris: J. Hamelle), p. v; trans. from Eigeldinger, *Chopin: Pianist and Teacher*, 49–50.
    [58] *Le Courier musical*, 13/1 (1910), 36: 'Ce qu'on a tant appelé . . . le "Tempo rubato" . . . n'était d'après ses fidèles auditeurs, qu'une grande liberté et fantaisie dans le dessin mélodique avec ses fioritures . . .
    'L'accompagnement, au contraire (c'est-à-dire, *les basses*) restait *quand même* rythmique, à ce point qu'alors qu'il jouait avec accompagnement, sa main gauche tombait, paraît-il, toujours impeccablement avec le bâton du chef d'orchestre.'

famous singer Maria Malibran and of Manuel García the younger, who wrote the treatise discussed in Chapter 3. Her father was the Manuel García who 'excelled in the use of the *tempo rubato*' (see the middle of Plate V). Pauline eventually became herself a famous singer, appeared in concerts with Chopin, and even published music that includes indications for rubato. She was thus in a unique position to understand the meaning of tempo rubato—both from the singer's point of view and from Chopin's. The comments transmitted by Saint-Saëns therefore carry, I believe, a special significance. He describes Chopin's rubato as follows:

I have learned the true secret of *Tempo Rubato*, already advocated by Mozart, necessary even with Sebastian Bach, indispensable in the music of Chopin.

Ah! This *Tempo rubato*—what errors are committed in its name! for there is the true and the false, as in jewels.

In the true, the accompaniment remains undisturbed while the melody floats capriciously, rushes or retards, sooner or later to find again the support of the accompaniment. This manner of playing is very difficult, requiring a complete independence of the two hands; and when some cannot achieve this, they give the illusion to themselves and to others by playing the melody in time and dislocating the accompaniment in order to make it fall at the wrong time; or else—and this is the worst of all—they are content to play the two hands one after the other. It would be a hundred times better to play everything evenly in time and the two hands together, but then they would not have the 'artistic air' . . .

The high point of difficulty in this style is found in the Etude in C sharp minor [Op. 25 No. 7]. In this Etude, which one may say was conceived for the cello, the melody is in the left hand; it is that [hand] which ought to sing freely and the right hand accompanies it, keeping a regular rhythm. It is generally the contrary effect which is produced.[59]

---

[59] 'Quelques mots sur l'exécution des œuvres de Chopin', *Le Courrier musical*, 13/10 (1910), 386–7. Since the passage is attributed solely to Viardot in Eigeldinger, *Chopin: Pianist and Teacher*, 49, I include here also the paragraphs that lead up to the description of rubato: 'J'avais douze ans, lorsque j'entendis pour la première fois la vicomtesse de Grandval qui en avait dix-huit. C'était dans une matinée musicale, chez le violoniste de Cuvillon. Elle chanta une délicieuse chose de sa façon, *la Source*, en s'accompagnant elle-même; et je fus frappé et charmé par la tranquillité, la fluidité de son jeu pur, sans nuances inutiles, qui s'accordait si bien avec ma manière de voir.

'Ce style uni et tranquille, elle le tenait de Chopin dont elle avait été l'élève.

'Par elle, par Mme Viardot,—qui aurait pu être, si elle l'avait voulu, aussi célèbre comme pianiste qu'elle l'a été comme cantatrice,—j'ai eu sur Chopin, sur son exécution, les renseigne-ments les plus précieux; j'ai su que sa manière était beaucoup plus simple qu'on ne se le figure généralement.

'J'ai su que lorsque d'immenses liaisons s'étendent sur des périodes entières, c'est une indication de ce jeu *spianato*, sans brisures de rythme et sans nuances, impossible à ceux dont la main n'est pas douée d'une souplesse parfaite.

These comments thus confirm again what many other accounts also state: that in Chopin's rubato, the accompaniment was strict, the melody free. Significantly, Saint-Saëns mentions also some of the bad effects resulting when a pianist imitates the sound of someone else's rubato without understanding its inner meaning. Since a listener notices most easily the displacement between notes in the accompaniment and the corresponding note in the melody, that is the effect imitated by those who cannot achieve a perfect independence of the hands or those who do not understand the essential nature of rubato. At worst, this may lead, as he states, to simply alternating the hands. In 1879 Kleczyński lists as one of the ways bad pianists inject false 'feeling' into the playing of Chopin's music: 'striking the chords with the left hand just before the corresponding notes of the melody'.[60] As early as 1853, however—just four years, that is, after the death of Chopin—Thalberg describes the same effect, contrasting his own subtle approach with the manner in which it was employed by insensitive performers:

It will be indispensable in playing to avoid that manner, which is ridiculous and in bad taste, of delaying with exaggeration the striking of the melody notes long after those of the bass, and producing thereby, from one end of a piece to the other, the effect of continuous syncopation. In a slow melody written in long notes, it is effective, especially on the first beat of every measure or at the beginning of each phrase, to attack the melody after the bass, but only with an almost imperceptible delay.

He also mentions arpeggiation: 'Chords that bear the melody in the upper notes should be performed in very close arpeggio, . . . and the

---

'J'ai connu le vrai secret du *Tempo Rubato*, déjà préconisé par Mozart, nécessaire même chez Sébastien Bach, indispensable dans la musique de Chopin.

'Ah! ce *Tempo rubato*, que d'erreurs on commet en son nom! car il y a le vrai et le faux, comme dans les bijoux.

'Dans le vrai, l'accompagnement reste imperturbable, alors que la mélodie flotte capricieusement, avance ou retarde, pour retrouver tôt ou tard son support. Ce genre d'exécution est fort difficile, demandant une indépendance complète des deux mains; et quand on ne peut y parvenir, on en donne à soi-même et aux autres l'illusion, en jouant la mélodie en mesure et en disloquant l'accompagnement pour le faire tomber à faux; ou bien encore,—c'est le dernier degré,—on se contente de faire arriver les deux mains l'une après l'autre. Mieux vaudrait cent fois jouer tout uniment en mesure et les deux mains ensemble; mais alors on n'aurait pas "l'air artiste". . . .

'Le comble de la difficulté dans ce genre se trouve dans l'*Etude* en *ut dièze mineur*. Dans cette étude, qu'on dirait conçue pour le violoncelle, la mélodie est à la main gauche; c'est celle-ci qui doit chanter librement et la main droite l'accompagne en gardant un rythme régulier. C'est généralement l'effet contraire qui se produit.'

[60] *How to Play Chopin*, trans. Alfred Whittingham, 19.

melodic note should be dwelt upon more than the other notes of the chord.'[61]

By the end of the century the delaying, and sometimes anticipating, of a melody note and the similar effect caused by arpeggiating the accompanying chord had become a mannerism.[62] This 'breaking of the hands' was employed finally so frequently that if often ceased to have any special expressive value. The pianists themselves no longer connected it with rubato, for this word had by that time acquired its later meaning. It is significant, however, that Sainst-Saëns, perhaps reflecting the ideas of Viardot or the Vicomtesse, recognized that this practice derived from a bad imitation of the earlier type of rubato. Even if such performers are not themselves producing a genuine rubato, their faulty imitation reveals to us what they heard when they listened to someone like Chopin.

Ornamental Notes

The quotations above from both Karasowski and Planté mention the *fiorituras* or ornamental passages which occur while the accompaniment keeps strict time. Such *fiorituras* in the music of Chopin include unusual groups of notes, written in either large or small notes, as well as shorter ornamental figures notated by small notes or signs. In Chapter 2 we observed the rhythmic robbery involved with baroque ornaments, and Ex. 7.3 illustrated some of the effects created by small ornamental notes in the music of Chopin. We shall consider now, in greater detail, various aspects of the rhythmic robbery caused by these

[61] *L'Art du chant appliqué au piano*, Op. 70, Ser. 1–2 (Paris: Heugel, n.d.), *Préface*, trans. of the 5th of his *règles générales de l'art de bien chanter* (except for a portion of the opening sentence) from E. Douglas Bomberger, 'The Thalberg Effect: Playing the Violin on the Piano', *MQ* 75 (1991), 205; the quotation on arpeggiation from the English edn. (London: Cramer Beale & Co., n.d.), *Preface*, p. ii; copies of both edns. from BL. Although the work is undated, the German edns. of the first two series by Br. & H are cited in Hofmeister's *Kurzes* or *Jahresverzeichnis* (repr., New York: Johnson Reprint Corp., 1968), [2] (1853), 60 (under two-hand piano music). The English edn. is reviewed in *The Athenaeum*, no. 1344 (30 July 1853), 922, and the presence of a preface confirmed (the author 'having prefixed a few intelligent remarks'). *DJM* 4/2 (15 Oct. 1853), 12–13, and 4/26 (1 Apr. 1854), 203, gives 1853 as the date of publication and reprints part of *The Athenaeum*'s review. The work was published initially in Paris, London, Leipzig, and Milan, and somewhat later in Boston.

[62] Rosen brought to my attention the relation of this practice to rubato. In the typescript of his Norton lectures he mentions 'the older form of rubato so important to Mozart' and adds: 'In this form, the melody note in the right hand is delayed until after the note in the bass.... We associate this manner of playing with the early twentieth century... An allied form of this rubato is the arpeggiation of the chords, thereby delaying the melody note.'

brief ornamental figures in the music of the early nineteenth century and in the works of Chopin in particular.

Small notes increase in number as the century progresses, and most ornaments formerly indicated by sign are now written out in small notes. Every small note or every note still indicated by a sign must steal its rhythmic value from either the preceding or the following note. It is not my purpose to determine how every ornament in the music of Chopin should be executed; a number of books have already attempted that task.[63] I would like to identify here, however, three related aspects of ornamental notes in his music: (1) their role in projecting the long, singing line; (2) the coexistence of those that rob the previous note and those that rob the following; and (3) an increasing intensity in the gesture of reaching to a higher note.

The evolution of the singing line is part of an increasing preoccupation during this period with melody in general. Melodies become longer and more conspicuous; they include more and more nonharmonic notes and are accompanied by increasingly complex accompanimental figures. Sonata form is eventually conceived primarily in terms of melody and melodic events.[64] Brief character pieces for piano allow melody to dominate almost entirely. The detailed phrasings of classic keyboard music are replaced by longer legato lines. Chopin, who often acknowledged influence from vocal music, said: 'It is necessary to sing with the fingers.'[65] The long, singing lines required an instrument that would not damp the tone immediately and completely as the classic fortepiano did when a key was released, but would have, as Czerny desired, 'a sustained and singing quality of tone, so that even in slow melodies it may be played upon in a connected and interesting manner'.[66] *Le Pianiste* states in 1834: 'One would thus give an Érard piano to Liszt..., but one would give a Pleyel to Kalkbrenner, Chopin, and Hiller; one needs a Pleyel in order

[63] See John Petrie Dunn, *Ornamentation in the Works of Frederick Chopin* (Novello, [1921]; repr. Da Capo, 1971); Maria Ottich, 'Chopins Klavierornamentik', *Annales Chopin*, 3 (1958), 7–62; and Paul Badura-Skoda, 'À propos de l'interprétation des œuvres de Frédéric Chopin' in *Sur les traces de Frédéric Chopin*, ed. Danièle Pistone (Paris: Librairie Honoré Champion, 1984), 113–22. Most comprehensive, as far as I know, is George A. Kiorpes, 'The Performance of Ornaments in the Works of Chopin', D.M.A. diss. (Boston University, 1975; UM 75–20,936)— an excellent study, which seems to me to combine thorough scholarship with a sensitive approach to performance. See also David Fuller, 'Ornamentation', *NHD* 597.

[64] See, for example, Jane R. Stevens, 'Theme, Harmony, and Texture in Classic-Romantic Descriptions of Concerto First-Movement Form', *JAMS* 27 (1974), 25–60, especially 44 ff.

[65] Quoted by his student Emilie von Gretsch in Maria von Grewingk, *Eine Tochter Alt-Rigas, Schülerin Chopins* (Riga: Löffler, 1928), 20; trans. Eigeldinger, *Chopin: Pianist and Teacher*, 45.

[66] *Piano Forte School*, iii, 126.

to sing a romance of Field, caress a mazurka of Chopin, sigh a nocturne of Kessler.'[67]

The long, singing line was thus enhanced by changes in the instrument, as well as by the touch and approach of the fingers of the performer and the careful use of the sustaining pedal. In addition, two ornaments, it seems to me, contribute in a special way to the flowing quality of a melody: the trill and the turn. Chopin's trills do not ordinarily occur at cadence, as they did in baroque and classic music, but during the course of a melody for the purpose of intensifying the sense of flow. Thus, in the well-known E flat Nocturne, Op. 9 No. 2, the melody of the third measure is varied, when it recurs in measure 7, by adding a trill on the long note. Although trills by mid-century usually commenced on the main note, Chopin seems to have preferred beginning either on the upper note or on the note below as shown in Ex. 7.4.[68] In addition, Ex. 7.4c suggests that the preparatory small notes in Ex. 7.4a occur on and not before the beat, thus robbing the following note. Also stealing time from the same note is the termination of the trill, consisting in Ex. 7.4a of the two small notes (slurred or not) that are familiar from the baroque trill. When Chopin expands the termination, as in Ex. 7.4c, the theft of time from the previous note is more substantial and the effect on the singing quality of the melody more intense.

The melodic power of the trill in Ex. 7.4 may at least partly be explained by the fact that it incorporates the inverted turn in its opening four notes (see Ex. 7.4b) as well as the standard turn at the end. A turn contributes to the singing quality of a melody by generating a powerful momentum for forward movement to the next note. Chopin's turns usually occur between two full-sized notes, thus robbing time from the first one. They are most often written in small notes and take the basic five-note form shown by the bracket at the end of Ex. 7.4b. Furthermore, Chopin often incorporates this configuration in ordinary notes as a characteristic feature of melodic

[67] *Le Pianiste*, 1/9 (July 1834), 130: 'Vous donnerez donc un piano d'*Érard* à Listz [*sic*]...; mais vous donnerez un piano de *Pleyel* à Kalkbrenner, à Chopin, à Hiller; il faut un piano de *Pleyel* pour chanter une romance de Field, caresser une mazure de Chopin, soupirer une nocturne de Kessler...' Concerning the construction of the Pleyel piano, which was Chopin's favourite, see Claude Montal, *L'Art d'accorder soi-même son piano* (Paris, 1836; repr. Minkoff, 1976), 222–4, 230–1, and 240–3, some of which is quoted in Eigeldinger, *Chopin: Pianist and Teacher*, 91–2 n. 7. See also Robert Winter's description in Brown and Sadie, *Performance Practice: Music after 1600*, 358–62.

[68] Concerning Chopin's trills, see Eigeldinger, *Chopin: Pianist and Teacher*, 58–9 and 131–3 n. 126; the same author's 'Chopin et l'héritage baroque', *Schweizer Beiträge zur Musikwissenschaft*, 2 (1974), 67–8; and Kiorpes, 'The Performance of Ornaments in the Works of Chopin', 1–88.

Ex. 7.4 Trill beginning on note below

a Usually written

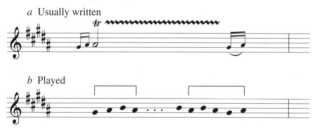

b Played

c Chopin, Nocturne Op. 32 No. 1 (Paris: M. Schlesinger, 1837; *Stirling* 186)

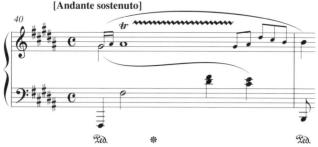

Ex. 7.5 Chopin, Impromptu Op. 29 (Paris: M. Schlesinger, 1837; *Stirling* 174)

construction (see the first five pitches in the melody on Plate XVI).[69]
The pattern is sometimes expanded by repeating the first small note
once or twice or by adding other notes at the end (Ex. 7.5). The
inverted turn can likewise be expanded in length and in like manner
enhances the singing quality of the melody by taking time from the
preceding note.

[69] For other examples, see Kiorpes, 'The Performance of Ornaments in the Works of Chopin',
315 (Ex. 214) as well as 297 n. 2; and Eigeldinger, *Chopin: Pianist and Teacher*, 133 n. 127.

Small notes in the music of Chopin may thus steal time from either the preceding or the following main note. In general, the tendency during the nineteenth century was for an increase and eventual domination of small notes that rob the previous note. This occurred partly because the long dissonant appoggiaturas were increasingly written out in ordinary notation. Chopin, however, was often influenced, as in the case of his trills, by earlier practice. We have already seen pre-beat robbery in his turns (Ex. 7.5) and trill terminations (Ex. 7.4), and it is sometimes indicated by a small note before a barline (Ex. 7.6). Sometimes a downward chromatic sweep in imitation of the vocal *portamento* (Ex. 7.7) also takes its time from the

Ex. 7.6 Chopin, Nocturne Op. 72 No. 1, 1827 (Berlin: Ad. Mt. Schlesinger, 1855)

Ex. 7.7 Chopin, Nocturne Op. 15 No. 2 (Paris: M. Schlesinger, 1834; *Stirling* 70)

preceding note (compare Caccini's *cascata* in Ex. 2.11 as well as Galliard's *strascino* or *drag* in Exx. 3.6 and 3.7).[70]

On the other hand, ornamental notes that rob the following main note still play a significant role in Chopin's music. The trill prefixes

[70] See Ottich, 'Chopins Klavierornamentik', *Annales Chopin*, 3 (1958), 24–9; Kiorpes, 'The Performance of Ornaments in the Works of Chopin', 148 and 348; Eigeldinger, 'Chopin et l'héritage baroque', *Schweizer Beiträge zur Musikwissenschaft*, 2 (1974), 61–5; and Eigeldinger, *Chopin: Pianist and Teacher*, 113–14 n. 82.

that lean on the beat (Ex. 7.4c) have already been mentioned. Similarly, there is impressive evidence, in the scores Chopin marked for his pupils, that many appoggiaturas—whether single, in groups (Ex. 7.8a and b), or occurring simultaneously in pairs or chords—should occur

Ex. 7.8 Chopin, Ballade Op. 47 (Paris: M. Schlesinger, 1841; *Stirling* 258, 262; lines from Dubois score)

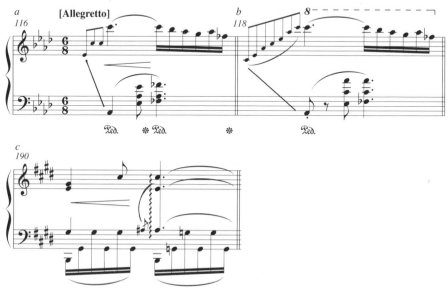

on the beat.[71] The notation in Ex. 7.3 seems to confirm the practice, and in Ex. 7.6 the placement of the small notes implies that the second should be played on the beat. The groups of small notes often involve arpeggiated chord notes (as in Ex. 7.3 or 7.8a and b), suggesting that when Chopin wrote a wavy line or an arc to the left of a chord (as in Ex. 7.8c), he may have intended that the arpeggio commence here also on the beat.[72] Chopin's markings in his students' scores sometimes also show long appoggiaturas on the beat.[73] Whereas the pre-beat ornaments act to increase the general sense of flow between notes, those that occur on the beat create an expressive accentuation by

[71] In the scores owned by Camille O'Meara Dubois, the composer frequently indicates on-beat performance by writing a line connecting a small note to the bass note with which it should sound. For a list of examples see Eigeldinger, *Chopin: Pianist and Teacher*, 133–4 n. 128, the musical examples on 245–66 *passim*, and the description of the scores on 212–19.

[72] See George A. Kiorpes, 'Arpeggiation in Chopin: Interpreting the Ornament Notations', *PQ* 29/113 (1981), 53–62; and 'The Performance of Ornaments in the Works of Chopin', 179–273.

[73] Eigeldinger, *Chopin: Pianist and Teacher*, 134, last paragraph of n. 128.

delaying the appearance of the main note—a process that reminds us somewhat of the earlier type of rubato.[74]

Small-note robbery may also involve the act of reaching. Early in the Romantic period writers recognized a special mood of infinite longing and yearning, of striving for the unattainable.[75] This feeling is embodied most vividly, it seems to me, in the musical gesture of reaching from a lower to a higher note. Later in the century ascending scalar movement would often lead to an upward leap to a dissonance within a new and sometimes exotic chord, followed by a descending stepwise resolution of the dissonance.[76]

Earlier in the century, however, the main note of the reaching gesture was usually approached directly as a consonance rather than delayed by an appoggiatura, and the approach was less dramatic. Ex. 7.9 shows some of these early nineteenth-century reaching motives.

Ex. 7.9 Basic reaching figures

The main note could be approached by a direct leap from a small note (Ex. 7.9*a* and *b*), by an arpeggio (Ex. 7.9*c*, *d*, and *e*),[77] or by a circling of the lower note through a turn (*f*) or a trill (*g*) in order to generate a more energetic thrust to the higher note.

These types of reaching were increasingly used by composers such as Dussek, Hummel, and Field. Chopin continued the process by expanding the preparation for the leap, using two repeated notes (Ex. 7.10), two small notes in an octave leap (Ex. 7.11), or the three- and seven-note configurations in Ex. 7.8*a* and *b*. In Ex. 7.8*c* he combines a small note with an arpeggiated chord that commences with the same pitch. Such repeated notes seem to impart a particularly urgent quality to the sense of yearning created by the small notes. This is especially

---

[74] See Kiorpes, 'Arpeggiation in Chopin', 55: 'The procedure often recommended, not only in Chopin, that arpeggios for the right hand should begin on the beat has considerable merit. Among other advantages, the heightened expressive effect of the resulting *rubato* delay of the principal melody note is universally acknowledged.'

[75] Described, for example, in the essay on Beethoven written in 1813 by E. T. A. Hoffmann, trans. in *Source Readings in Music History*, ed. Oliver Strunk (Norton, 1950), 775–81.

[76] See, for example, the Prelude to Wagner's *Tristan und Isolde*, mm. 16 and 93; or Mahler's 4th Symphony, 3rd movt., mm. 326 ff.

[77] A chord marked by a wavy line or arc may, of course, be arpeggiated very quickly and produce primarily a rhythmic effect, or more slowly and cause an expressive delay in the melody note. It is the latter type I refer to here.

Ex. 7.10 Chopin, Scherzo Op. 31 (1837)

Ex. 7.11 Chopin, Étude Op. 25 No. 7 (Paris: Henry Lemoine; *Stirling* 122)

intense when the expected appearance of the main note is delayed by small notes *on* the beat.

The device of reaching is used for those melody notes that require, for expressive reasons, a special sense of colour, warmth, or accentuation. In Exx. 7.10, 7.11, and 7.8*a* and *b*, it points out the beginning or recurrence of a melodic idea, whereas in Ex. 7.8*c* it emphasizes a mysterious change of chord. In addition, the use of this procedure between two notes of the left hand, as in Ex. 7.12, can have a similar

Ex. 7.12 Chopin, Mazurka Op. 56 No. 1 (Paris: M. Schlesinger, 1844; *Stirling* 309)

effect on the melody in the right hand; whether the lowest note actually precedes or coincides with the right hand, it is perceived in some sense, I believe, as a reaching from the lowest note in the left hand toward the highest pitch in the right.

The similar reaching motives in French baroque music described in Chapter 2 were also used for expressive purposes: to enhance the flowing quality of a series of notes (Ex. 2.8) or to accentuate a single note (Ex. 2.10*a*). Couperin further intensifies the expression, when the phrase recurs, with the left-hand arpeggio in Ex. 2.10*b*, an effect not unlike that by Chopin in Ex. 7.11. We have seen that Marpurg linked the *suspension* in 1755 with the earlier type of rubato (see Ex. 5.1*d*) and that around the same time French keyboard composers such as Foucquet and Forqueray were notating almost continual displacement between the hands in tender and expressive pieces.

Considering the special effect of the reaching motive, whether notated in small notes or as an arpeggiated chord, it is not surprising that Chopin, according to Mikuli, permitted 'breaking the chord... only where the composer himself specified it'; otherwise, 'in double notes and chords, he demanded precisely simultaneous attacks'.[78] This attitude seems to contrast with general practice, however, for both P. A. Corri in 1810 and Czerny around 1839 identify situations in which the performer may arpeggiate unmarked chords. Their examples show that arpeggiation can provide an expressive accentuation on long chords, especially if dissonant, or the music is marked *con espressione*, *con anima*, or *dolce*. Both writers, however, warn against abuse of the device, and Czerny, as though predicting one of the mannerisms of the late nineteenth and early twentieth centuries, states that 'many players accustom themselves so much to arpeggio chords, that they at last become quite unable to strike full chords or even double notes firmly and at once'.[79]

The reaching motives and the other small notes that occur on the beat cause the following main note to be delayed in a manner similar to the earlier tempo rubato and, to a certain extent, with a similar freedom in determining the amount of delay. Unlike the earlier rubato, they cannot force a note to appear too soon. Both techniques, however, tend to appear at similar spots in the music, spots that demand a particular expressive warmth.

The Word Rubato in the Scores

There are thus many rubatos in the music of Chopin. There is the rubato of tempo changing, as added by the performer or notated by the composer. There is the dance rubato, in which certain beats of a measure are accentuated through prolongation. There are the different rubatos that can occur over a strict accompaniment: the earlier type as practised by the singers and violinists, the polymetric type caused by

[78] *Fr. Chopin's Pianoforte-Werke*, i: *Mazurkas*, trans. from Dover repr., pp. vii–viii.

[79] Czerny, *Piano Forte School*, iii, 55–6; and Philip Antony Corri, *L'Anima di musica* (London, 1810), copy at BL, 74–8. Czerny, in accord with the pre-beat tendency of the time, specifies that chords must be 'sprinkled' in such a way that 'the upper or melodial note shall never come in out of its time'.

unusual groups of notes, and the ornamental type produced by small notes or arpeggios on the beat. Chopin surely employed all these rubatos when he performed, no doubt with two or more types sometimes occurring simultaneously.

Each of these expressive effects has been referred to, at one time or another, by the word *rubato*. Several writers even use the word for both the earlier and later types. Lenz, for example, after describing the left hand as the conductor, states:

> He [Chopin] taught: 'Supposing that a piece lasts a given number of minutes; it may take just so long to perform the whole, but in the details deviations may occur! . . .
> In the fluctuation of the tempo, . . . in the *rubato* of his conception, Chopin was ravishing.[80]

He makes clear that there are two different types of rubato, however, when he describes the playing of the pianist Adolf Henselt:

> his *rubato* is not the Chopin *rubato*, it is a shifting of the tempo [the later type], not a general dislocation of the visual angle for the phenomenon as a whole, like the effect of a scene viewed through the small end of an opera-glass.[81]

These unusual comments make two important points, I think. First of all, Lenz states that Chopin's rubato is not the later type involving tempo changing. On the other hand, it is like the visual effect produced by perspective: consider a group of actors standing on a stage in a line parallel to the rows of seats. They would appear in one way when viewed from the middle of the theatre; when seen from the extreme left side of the room, however, the relative distance between them would seem different and those at the left would seem relatively larger than those on the right. This changed perception is then like musical notes which are made to sound longer or more briefly, or sooner or later than expected. In addition, the fact that the persons on the stage do not alter their position during the process provides a fixed control like the strict accompaniment in rubato. Looking through the wrong end of an opera-glass may be like the shortening of a group of notes that have been delayed by rubato. If my interpretation of his remarks is correct, then this is indeed a curious and imaginative analogy.

---

[80] *Die grossen Pianoforte-Virtuosen* (1872), 47; trans. from *The Great Piano Virtuosos*, 68. In the original article from the *Neue Berliner Musikzeitung*, 22/38 (1868), 302, the passage reads: 'Er lehrte: ein Stück dauert, angenommen 5 Minuten, wenn das Ganze nur genau so lange gedauert; im Einzelnen kann's anders sein. Das ist *rubato*.

'In dem Schwanken der Bewegung, in diesem Hangen und Bangen, im *rubato*, war Chopin hinreissend.'

[81] *Die grossen Pianoforte-Virtuosen*, 102; trans. from *The Great Piano Virtuosos*, 144–5.

In a similar way, Kleczyński, although dutifully reporting that the left hand should keep perfect time, traces the history of rubato back to Gregorian chant, recitative, and the fantasia—the sort of background traced in Chapter 1 for the later type of rubato. In an analysis of Chopin's Nocturne Op. 55, he marks the unaccompanied sixteenth-note descent in measures 69–70 *con disperazione e molto rubato*, an effect which surely, without a strict left-hand part to control it, produces tempo flexibility and hence the later tempo rubato.[82] In addition, Moscheles, who, as we have seen, describes the mazurka accentuations as rubato, also seems to use the word for a broader sort of rhythmic flexibility when he advises Bülow to play a certain Chopin Nocturne 'still more lively and with more rubato, quite fantastically, almost never holding to the beat'.[83]

Liszt, who became well acquainted with Chopin and his style of playing when they both lived in Paris during the 1830s, describes Chopin's rubato in the following poetic passage from his book on Chopin (first published in 1852, revised in 1879):

He always made the melody undulate like a skiff borne on the bosom of a powerful wave; or he made it move vaguely like an aerial apparition suddenly sprung up in this tangible and palpable world. In his writings he at first indicated this manner which gave so individual an impress to his virtuosity by the term *tempo rubato*: stolen, broken time—a measure at once supple, abrupt, and languid, vacillating like the flame under the breath which agitates it, like the corn in a field swayed by the soft pressure of a warm air, like the top of trees beat hither and thither by a keen breeze.

But as the term taught nothing to him who knew, said nothing to him who did not know, understand, and feel, Chopin afterwards ceased to add this explanation to his music, being persuaded that if one understood it, it was impossible not to divine this rule of irregularity. Accordingly, all his compositions ought to be played with that kind of accented, rhythmical *balancement* [wavering], that *morbidezza* [softness], the secret of which it was difficult to seize if one had not often heard him play.[84]

---

[82] *How to Play Chopin*, trans. Alfred Whittingham, 56–7 and 66. See also Kleczyński's own 'reflections on the *rubato* of Chopin' on 58–9, which involve, in general, his exaggeration of those moments in the flow of the music which tend toward retardation or acceleration, and more specifically the prolongation of the more important tones.

[83] Hans von Bülow, *Briefe und Schriften*, 2nd edn., i, 133 (letter of 19 Nov. 1848): 'noch lebhafter . . . und mehr rubato, ganz fantastisch, fast nie Takt haltend'.

[84] Franz Liszt, *Frédéric Chopin*, 2nd edn. (Br. & H, 1879), 115; trans. from Niecks, *Chopin as a Man and Musician*, ii, 101. The 1st edn. of Liszt's book, which was published at Paris in 1852 after being serialized the year before in *La France musicale*, included neither the opening sentence nor the references to apparitions, corn fields, or tree tops. For translations from the earlier text, see *Life of Chopin*, trans. John Broadhouse, 2nd edn. (London: William Reeves, [1913]), 83–4; *Frederic Chopin*, trans. Edward N. Waters (London: Free Press of Glencoe, 1963), 81; and, for an abridged version, *DJM* 1/5 (8 May 1852), 37. See also Waters, 'Chopin

The image of a skiff or an apparition suggests the sort of expressive lingering or hesitation involved in the later type of rubato or the accentuations in a dance rhythm. The flame, the cornfield, and the tree tops, on the other hand, are all objects which are fixed at their lower extremity (analogous, presumably, to a strict accompaniment in the left hand) and vacillate only in their upper portion (like a melody consisting of an unusual number of notes, or one displaced by ornaments or by the earlier type of rubato).

Using the word *rubato* to describe collectively all the unique rhythmic features of Chopin's music, however, or using it to refer to more than one of his rhythmic techniques, does not lead to the sort of analytical precision useful to the performer and the scholar. Determination of the meaning of the word *rubato* in the scores depends, of course, on what the word meant to Chopin himself. We have some clues to show us where some of the types of rubato were incorporated into his music. Notation for tempo flexibility of the later type occurs, of course, in the scores, and presumably additional fluctuations added by the performer would follow some of the general principles suggested by this notation. The prolonged beats to accentuate dance rhythms occur mainly in the Mazurkas, although it still remains for the performer to determine the precise location and to select the first or second beat for emphasis. In regard to location of the rubatos that require a strict accompaniment, the unusual groups are clearly visible in the notation, and the small notes and arpeggios, although they offer some freedom in their rhythm, are equally evident in the score. The only type of rubato, finally, for which we might seem to have no clue at all in the scores, is the earlier type of tempo rubato—and it is precisely this type that I believe Chopin had specifically in mind when he wrote the word *rubato*. Some modern writers feel that Chopin sometimes used the word to refer to one type and at other times to refer to another.[85] I would like to suggest, on the contrary, that Chopin had a definite technique in mind when he wrote the word *rubato* and that he learned about this technique directly from intimate acquaintance with those who performed the earlier type of rubato most effectively.

Several circumstances seem to point toward this conclusion. First of all, Chopin was deeply involved with opera. It was fashionable at that

by Liszt', *MQ* 47 (1961), 170–94. For the involvement of Princess Carolyne von Sayn-Wittgenstein, see *The Early Correspondence of Hans von Bülow*, ed. by his widow, trans. Constance Bache (New York: Vienna House, 1972), 111; and Alan Walker, *Franz Liszt*, i: *The Virtuoso Years (1811–1847)* (Knopf, 1983), 20–3, and ii: *The Weimar Years (1848–1861)* (Knopf, 1989), 379–80.

[85] See, for example, Eigeldinger, *Chopin: Pianist and Teacher*, 121–2 n. 99.

time for pianist−composers to improvise on themes from current operas or incorporate them into compositions. Chopin has pieces on themes from Hérold's *Ludovic*, Meyerbeer's *Robert le diable*, and Mozart's *Don Giovanni*, as well as contributing to the *Hexameron*, a set of variations on the march from Bellini's *I puritani* composed jointly with Liszt, Thalberg, Pixis, Herz, and Czerny. It was Chopin's friend Pixis who made the piano reduction of *Robert le diable* which I employed with Cinti-Damoreau's rubato in Ex. 3.18. In addition, Chopin sketched a piano accompaniment for the cavatina 'Casta diva' from Bellini's *Norma*.[86] In this way, the sounds of the operatic aria became incorporated into keyboard music: the long cantabile melody, as well as accompanimental figures, such as we saw in Exx. 3.13 and 3.18, which provide such an effective and continuous strict beat against which tempo rubato can be performed.

Chopin was far more intimately involved with such music, however, as an avid opera lover. He never composed an opera, and he never, as far as I know, participated in any capacity in a performance. Yet wherever he lived or visited he lost no opportunity of attending as many performances as possible. He admired the great Italian singers whom he heard at Warsaw's National Theatre and later at the Théâtre-Italien in Paris. His letters are filled with comments on the operas and singers he heard—comments of the sort that opera lovers of any age would make. He describes and compares the vocal and physical qualities of various singers. At one moment Sontag is the best, then it is Malibran (sister of Manuel García the younger), Pasta, or Cinti-Damoreau.[87] All of these singers, as noted in Chapter 3, were on García's lists of experts in the singing styles that incorporated tempo rubato. In addition, we saw in Ex. 3.16 an example of Pasta's rubato and in Ex. 3.18 an example by Cinti-Damoreau. In his letters Chopin describes meetings with Sontag and later with Jenny Lind, one of García's pupils.[88] In addition, Chopin knew Manuel García himself and played in concerts during the 1840s at which García's sister Pauline Viardot sang, sometimes accompanied by the composer.[89] He

[86] Act I, sc. iv, m. 55. See Eigeldinger, *Chopin: Pianist and Teacher*, 111, last paragraph in n. 75, and Wojciech Nowik, 'Do zwiazków Chopina z Bellinim . . .' (Connections between Chopin and Bellini: Chopin's Manuscript of Bellini's 'Casta diva'), *Pagine, Polsko-Włoskie materiały muzyczne*, 4 (1980), 241−71.

[87] See Hedley, *Selected Correspondence of Fryderyk Chopin*, 46−9 (letter of 5 June 1830 concerning Sontag) and 100−1 and 104 (letters from Dec. 1831 on Malibran, Pasta, Cinti-Damoreau, and others). See also Niecks, *Chopin as a Man and Musician*, i, 140−2 and 227−9; and Eigeldinger, *Chopin: Pianist and Teacher*, 110−11 n. 75.

[88] Hedley, *Selected Correspondence*, 314−16, 318, 321, and 334−5 (letters from England and Scotland, May−Aug. 1848).

[89] See Atwood, *Fryderyk Chopin: Pianist from Warsaw*, 140, 150, and 170 for programmes of concerts, and 239−40, 242, 243, and 248−51 for reviews.

also played in the same concerts with singers from García's lists such as Cinti-Damoreau, Giulia Grisi, Fanny Persiani, and Antonio Tamburini.[90] Furthermore, Chopin was personally acquainted with many of the composers from whose operas García and Cinti-Damoreau drew examples of rubato, including Rossini (see Exx. 3.12 and 3.15), Meyerbeer (Ex. 3.18), and Bellini (Ex. 3.16).[91]

Chopin felt that his playing style was modelled on the great singers, and he taught his students that the best way to learn how to perform the long, singing melodies was to listen to and imitate the singers. Lenz reports, in a description of Chopin's teaching of the E flat Nocturne, Op. 9 No. 2, that 'the style should be modelled upon Pasta and the great Italian school of singing'.[92] Czartoryska describes the way the fingers should work 'to bring out from even the least melodious instrument a singing quality close to that of the Italian singers whom Chopin recommended as models'.[93] In addition to incorporating the vocal cantabile style and its sense of declamation and phrasing, Chopin also seems to have borrowed some specific ornamental techniques from the vocal style. We have already seen his keyboard imitation of *portamento* in Ex. 7.7, as well as the on-beat performance of trills and several types of appoggiaturas. He also employs the vocal ornament in which a small note before the beat repeats the pitch of the preceding main note.[94] Considering the enormous extent of the operatic influence on Chopin—as a composer, a teacher, and a performer—it would not be surprising if he also knew what Pasta and the other singers meant when they spoke of tempo rubato, and he no doubt recognized the effect when he heard it performed.

He was also acquainted with some of the instrumental composers and performers who were involved with the earlier type of rubato, including a number of the violinists mentioned in Chapter 4. He probably heard Paganini play in Warsaw in 1829 (see García's description of his rubato on Plate V) and during this same year composed

---

[90] Atwood, *Fryderyk Chopin*, 70–1, 73–5, 132, 136, 150, and 163–4; reviews on 233, 236–7, 243, and 245. See the comments in ch. 3 n. 54 connecting Persiani with the Cavatina from *Lucia*, from which García drew his example of tempo rubato in Pl. V.

[91] Hedley, *Selected Correspondence*, 92, 98, and 100 (on Rossini), and Niecks, *Chopin as a Man and Musician*, i, 285–7 (Bellini), and ii, 169–70 (Meyerbeer).

[92] 'Übersichtliche Beurtheilung der Pianoforte-Kompositionen von Chopin', *Neue Berliner Musikzeitung*, 26/38 (1872), 297; trans. from Eigeldinger, *Chopin: Pianist and Teacher*, 77 (and, concerning the influence of the vocal bel canto style, see 14–15, 44–5, and 85).

[93] Adam Czartkowski and Zofia Jeżewska, *Fryderyk Chopin*, new edn. (Warsaw: PIW, 1970), 374; trans. in Eigeldinger, *Chopin: Pianist and Teacher*, 31.

[94] Eigeldinger, *Chopin: Pianist and Teacher*, 114–15 n. 82. The device is called here *cercar della nota*. Concerning the influence of operatic singing on Chopin, see Eigeldinger, 'Chopin et l'héritage baroque', *Schweizer Beiträge zur Musikwissenschaft*, 2 (1974), 60–8.

his set of variations entitled *Souvenir de Paganini*.[95] He later compares Paganini's ornamentation to that of Sontag in one letter and praises his style in others.[96] He appeared in concerts with 'Paganini's famous rival Baillot' (see Ex. 4.4), as he describes him in a letter, and often played chamber works by Spohr (Exx. 4.5–4.8).[97]

Keyboard influence on Chopin comes from several sources: from Mozart (whose comments on the strict left hand in tempo rubato have already been noted) and his pupil Hummel (the possible composer of the sonata which includes the syncopated tempo rubato in Ex. 5.4); from Clementi (whose *rubando* playing was praised) and his student Field (who, according to Farrenc, employed the type of rubato used by Mozart and Paganini). Chopin reports in a letter from 1831 that he was delighted when Frédéric Kalkbrenner, after hearing him play, asked if he were a pupil of Field, since he had the same touch.[98] Chopin often taught his pupils works of Hummel (whom he knew personally), Clementi, J. S. Bach, and Weber. He was not attracted, in general, to the style of Beethoven or his pupil Czerny, whom he met in Vienna.

As we have already seen, Hummel, who may have indicated the earlier rubato in Ex. 5.4, also used the term for the later type in Ex. 5.9 and in his *Art of Playing the Piano Forte*. Czerny also used the word *rubato* for the later type, but was not likely to have influenced Chopin. Influence might possibly have come from Joseph Elsner, Chopin's composition teacher in Warsaw. Elsner knew Christian Kalkbrenner in Paris in 1805 and hence may have been familiar with his treatise of 1789, which, as we have seen, was the earliest source that seems to use the word *rubato* for the later type.[99] On the other hand, Elsner was a violinist and, as the director of the National Theatre in Warsaw could just as easily have transmitted to Chopin the

---

[95] Based on the theme 'O mamma mamma cara', which Paganini used for his set of variations *Il carnevale di Venezia*, Op. 10, composed the same year. See Maria Rosa Moretti and Anna Sorrento, *Catalogo tematico delle musiche di Niccolò Paganini* (Genoa: Comune di Genova, 1982), M. S. 59 on pp. 186–91. Chopin's *Souvenir de Paganini* was first published as a supplement to *Echo muzyczne i teatralne*, v (Warsaw, 1881) and appears in the Paderewski *Complete Works*, xiii; see Eigeldinger, *Chopin: Pianist and Teacher*, 119 n. 95. There is also a fragment from 1826 of a set of variations for piano 4 hands by Chopin on the same theme; see Krystyna Kobylańska, *Frédéric Chopin: Thematisch-bibliographisches Werkverzeichnis*, trans. Helmut Stolze (Munich: G. Henle Verlag, 1979), 217.

[96] Hedley, *Selected Correspondence*, 48 (from Warsaw, 5 June 1830), 71 and 73 (Vienna, Dec. 1830), and 98 (Paris, 12 Dec. 1831).

[97] Ibid. 98–9 and 104 concerning Baillot; 35, 55, and 104 on Spohr. For programmes and reviews of concerts at which Chopin and Baillot both performed, see Atwood, *Fryderyk Chopin*, 58–60, 93, 219, and 224–5.

[98] Hedley, *Selected Correspondence*, 98; see also 64.

[99] Ibid. 96 (letter from Elsner, 27 Nov. 1831).

earlier meaning of rubato employed by the singers and violinists. The decisive influences on Chopin seem to have exerted themselves during his years in Warsaw, and they seem to involve ideas from the Baroque and Classic periods and composers such as J. S. Bach and Mozart.[100] We have already seen a number of details in his ornamentation that derive from earlier rather than contemporary practice. The bel canto singing style which he admired so much, and which García described in such detail, had its origins in the late seventeenth century with singers such as Tosi. This retrospective point of view may be one more factor in suggesting that the rubato he knew by name was the earlier type.

Other factors leading to this same conclusion are the contemporary comments in *Le Pianiste*: Chaulieu's definition of rubato linked with the name of Chopin, and the reviewer's complaint about Chopin's complex attempt to notate the sort of rubato that Dussek had tried in vain to write in syncopation. Enormously important also are the numerous comments by Chopin's students linking the word *rubato* with the strict left hand and its role as conductor. It is no doubt true that the pupils of Chopin tried, after his death, to dispel the feeling that all subsequent exaggeration of the later type of rubato could be blamed on their master. The comments on rubato, however, come from so many different sources and are so remarkably similar in nature, that they must surely have their origin in Chopin's teachings.

Most of Chopin's pupils did not possess the highest degree of talent. The few who could have transmitted his ideas most faithfully either died too young or did not pursue a professional life as performers or teachers. Everyone agreed that Chopin himself had a unique style of playing, a style that deeply moved all who heard him. It was sincere and honest, and every detail of execution was done for purely musical reasons and not for virtuosic display. It involved a broad sense of flowing melody moving horizontally along the keyboard and not interrupted by heavy accentuations. He believed that the effortless flight of the long, singing line was achieved by the power of the imagination rather than by long hours of finger exercises. Apparently, however, Chopin had only one student who had an imagination original and musical enough to motivate this manner of playing. This was Karl Filtsch, who died when only 15 years of age.[101] Chopin taught the other students to follow his own interpretations and con-

[100] Eigeldinger stresses the importance of baroque and classic influence on Chopin; see 'Chopin et l'héritage baroque', *Schweizer Beiträge zur Musikwissenschaft*, 2 (1974), 51–74, and *Chopin: Pianist and Teacher*, 14, 109–10 n. 70, and 118–19 n. 95.

[101] See ibid. 140–2 n. 157, and Atwood, *Fryderyk Chopin*, 150 and 243.

sequently illustrated at the keyboard frequently himself by playing excerpts or entire pieces. Thus it appears that the teaching process was mainly one of imitation. There is a danger, however, in this method of teaching, for some students are very clever at learning, like a parrot, all the nuances appropriate to a particular piece, but, when faced with a new composition, cannot by themselves comprehend the musical sense which determines all the details of execution. When one imitates, one reproduces only the exterior results of musical thought.

The comments of Saint-Saëns disclose the results of such imitation when it involves rubato. Chopin's students seem to have remembered his verbal description concerning the free melody over the strict accompaniment. When it came to producing rubato themselves, however, they could either motivate the effect, as Chopin himself did, by personal comprehension of the expressive sense of the music, or, if this were not possible, by imitating what they heard when he played. Saint-Saëns describes what happens when pianists merely imitate: they simply play the 'two hands one after the other'. Viardot comes from a remarkable family of experts on the earlier rubato: her father and sister could perform it; her brother describes it in his treatise; Pauline herself understood it as a singer, as a pianist, as a composer, and as a listener to Chopin's playing. That Saint-Saëns, in relaying her ideas, links the earlier rubato so clearly to Chopin and also links it with its false imitation, seems to me highly significant. It means that the mechanical mannerism of 'breaking hands', although not a true rendition of the earlier rubato, is, indeed, evidence of its earlier existence.

Thus even this late mannerism, as well as student comments about the left-hand conductor, joins the other factors mentioned above—influence from operatic singers, violinists, and other keyboard players, and influence from earlier baroque and classic practice—to suggest that the word *rubato* signified to Chopin the earlier type. One final evidence of this conclusion concerns the location of the word in his scores. It does not appear in cadenza-like sections or in connection with fermatas or the marking *senza tempo*. It does not occur when only the right hand is playing. It does not appear near the end of a musical idea where one might retard,[102] nor when pushing toward a high point where one might accelerate. In fact, it often occurs at the beginning of a phrase where one might expect *a tempo*. In only a few

---

[102] Lucian Kamieński, in 'Zum "Tempo rubato"', *Archiv für Musikwissenschaft*, 1 (1918), 119–22, examines the way Chopin marks accelerations and retards and notes that *a tempo* never follows the word *rubato*. Actually it does happen in one piece, the posthumous Polonaise in G sharp minor (Paris: Mayence, 1854), copy in BL. Unlike any of the other Chopin examples, the word *rubato* does appear here at cadence (mm. 12 and 27) and probably should have been *rallent.* instead. Eigeldinger expresses similar doubts in *Chopin: Pianist and Teacher*, 121 n. 99.

cases does it coincide with an expressive high pitch or dissonance that might be suitably rendered by a portato-like *ritenuto*. It does not indicate the prolongation of beats in mazurkas. It does not even occur when some of the traits of the earlier rubato, such as polymetre or the delay of notes, are present. Thus it does not appear when there are unusual groups of notes to synchronize with the bass (the twenty melody notes in measure 73 of the Nocturne Op. 9 No. 1, for example, to be played against nine notes in the left hand). It does not occur with expanded trill terminations (as in Ex. 7.4*c*), elaborated turns (Ex. 7.5), or *portamento* (Ex. 7.7). It does not accompany small notes that lean on the beat and cause expressive delays in melody tones (Exx. 7.3, 7.4*c*, and 7.8), and it does not coincide with syncopation.

Chopin does mark the word *rubato*, however, for three main purposes: to articulate the repetition of a unit of music, to intensity an expressive high point or appoggiatura, and to establish a particular mood.[103] García, as we have already seen, used the earlier rubato to emphasize a repeated passage, and in the Benda manuscript it marked the return of a theme. In addition, García, the copiers of the Benda manuscript, Spohr, and Türk all suggest the use of the earlier type of rubato for a special expressive effect on intense appoggiaturas and other important notes or chords. Chopin therefore seems, in most cases, to mark the word in locations similar to those where the singers, violinists, and keyboard players were accustomed to performing the earlier type of tempo rubato. Furthermore, these are the same sort of locations, as we have seen, where the most urgent reaching motives occur.

A number of modern writers have concluded that Chopin's marked rubatos are the earlier type. Few, however, attempt to describe more precisely how this earlier rubato is to be performed. Some have suggested that an alteration of note values can safely take place on those notes that do not coincide with bass notes, thus avoiding any displacement between right and left hands.[104] I would like to suggest, on the contrary, that it is precisely where the two hands are written in vertical alignment that earlier rubato is most effective. This is the

---

[103] See the categories of Eigeldinger, ibid. 121–2 n. 99; and of Robert Winter, 'Performing Practice (After 1750)', *The New Grove Dictionary of Musical Instruments* (1984), iii, 58.

[104] See Marion Louise Perkins, 'Changing Concepts of Rhythm in the Romantic Era: A Study of Rhythmic Structure, Theory, and Performance Practices Related to Piano Literature', Ph.D. diss. (Univ. of Southern California, 1961; UM 61–6,302), 270: 'In the designated rubato passages [of Chopin], the accompaniment is, at least in part, invariably written in longer note values than the melodic line, enabling the performer to modify the notated rhythm of the melody between the chords without influencing the simultaneity of the right and left hand parts.' See also Robert H. Goepfert, 'Ambiguity in Chopin's Rhythmic Notation', D.M.A. diss. (Boston Univ., School for the Arts, 1981; UM 81–26,665), 45–6 and 50–6.

situation in which audible displacement occurs between the hands, and this, of course, is the effect that is imitated when less talented pianists simply 'play the two hands one after the other'.

We have noted in previous chapters the different points of view regarding the earlier rubato. The singers and violinists conceived it in terms of alterations of melodic note values; those who listened to them—the music lovers, theorists, accompanists, conductors—heard mainly the expressive effect caused by melody notes which sounded as though they should have occurred simultaneously with notes in the accompaniment, but which actually happened too soon or too late. The solo keyboard player, due to the unique circumstance that he must himself perform both the free melody and its strict accompaniment, tended to view rubato not like the singers and violinists, but rather, like the listeners, as an expressive device in which the left and right hands, although sounding as though they should have been together, are in fact played one before the other. It is unusual when a keyboard source, such as Chaulieu's glossary, defines Chopin's rubato as the singer would perceive it—as the borrowing of note values from one note to the other. More typical is the reviewer of Chopin's Op. 15 Nocturnes in the same journal, who, as we have seen, emphasized the displacement between the hands in Dussek's syncopations. Chopin's knowledge of rubato must have come to him primarily in two ways: as a listener to operatic singers, and as a keyboard performer, both activities encouraging him to think of the device mainly as displacement between melody and accompaniment.

Chopin's powerful sense of melodic flow may have made it possible for him, like Mozart, to perform with such independence of hands that he could conceive of rubato, like a singer, in a completely linear fashion. However, he was certainly aware of the expressive power of the small notes and arpeggios as they delayed notes of the melody through their reaching motives (Exx. 7.3 and 7.8). In the preceding chapter I employed such small notes to show how alterations in the note values could be rewritten (Ex. 6.1c) and how a rubato added by a performer could be notated (Ex. 6.2c). This method of presentation has several advantages: it uses symbols already familiar from the same music; it depicts a variable amount of displacement between the hands; and it preserves the original melody in ordinary notes, so that eventually the performer can feel that he is simply playing this melody but with intense expression. I will therefore attempt to explore at least some aspects of Chopin's use of the earlier rubato by means of this notation. Without implying that he himself necessarily conceived it in exactly this fashion, I think such a presentation can reveal to modern musicians some possibilities that have not previously been considered.

When Chopin marks *rubato* in a score, he is usually indicating a momentary effect, the termination of which is not marked. On one occasion, however, he wrote *sempre rubato*, with dashes to show a duration of about two and a half measures. I presume, therefore, that the word *rubato* by itself indicates considerably less time. If one assumes that the rubato ends, in general, at the barline following the word, the effect would ordinarily involve a half or a full measure. All of the places marked *rubato* have a melody in the right hand and a sufficient number of notes in the left-hand accompaniment to produce significant displacement. I will therefore approach Chopin's rubato by proposing certain ways of delaying or anticipating those melody notes which are written vertically above notes in the accompaniment. With conspicuous displacement between the hands ensured, then the other notes can fall naturally into place. I think such initial concern with displacement can lead to a linear concept of the melody and eventually perhaps to a true independence of the hands.

The vertical situations that occur in Chopin's rubato passages are shown in the first measure of each line in Ex. 7.13. In (*a*) a single bass note appears with the melody note, in (*d*) a three-note chord; in (*b*) the melody and in (*c*) the bass is doubled at some interval. The second measure in each line then shows how a delay in the melody note can be notated, in the manner already employed in the Mozart excerpts in Exx. 6.1*c* and 6.2*c*, by means of small notes that occur on the beat. The accompaniment in the left hand thus appears exactly where originally notated, but the small notes cause a delay in the melody tone. The slur from the small notes to the melody conveys the idea that there is a reaching gesture from the left hand to the right. In the third measure on each line of Ex. 7.13 small notes played before the beat cause an anticipation of the melody note; again, the left-hand part appears precisely on the beat. In the following exploration of Chopin's rubato, then, I will employ a notation in which the small notes on the lower staff for the left hand are played *on* the beat, whereas those on the top staff for the right hand occur *before* the beat. In this way the accompaniment in the left hand will remain undisturbed while delays or anticipations occur in the melody notes.

Chopin marks *rubato*, then, to articulate a repetition, to intensify an expressive effect, or to establish a mood. There are various situations involving repetition. Rubato may occur, first of all, at the beginning of the second of two parallel phrases in a main theme. The Mazurka Op. 6 No. 1 begins with a measure like the second one in Ex. 7.14*a*, but with the opening melody note tied back to a quarter-note anacrusis (see Plate XVI). Ex. 7.14*a* shows the last bar of the first eight-measure unit and the first bar of the second. Measure 9 is marked *rubato* in

Ex. 7.13 Vertical situations in Chopin's rubato passages

order to emphasize the fact that a repetition is about to commence. Using the notation in Ex. 7.13, Ex. 7.14*b* shows a delay of the first melody note, (*c*) a delay also in the E#. Ex. 7.14*d*, on the other hand, depicts anticipation of the same notes. Presumably, one might anticipate only the first, for it is actually displacement on the opening beat that articulates the fact of repetition most effectively. It is considerably easier to accentuate the opening F# when it is anticipated, but it should be similarly emphasized when delayed. An anticipation on this note creates an effect similar to the large-note anacrusis at the beginning of the piece (see Plate XVI).[105]

---

[105] In the edns. of Schlesinger (*Stirling*, 10) and Kistner (Pl. XVI), there is no tie between the F# at the end of m. 8 and the beginning of m. 9, whereas there is in the Wessel edn. (BL,

Ex. 7.14 Chopin, Mazurka Op. 6 No. 1 (Paris: M. Schlesinger, 1833; *Stirling* 10)

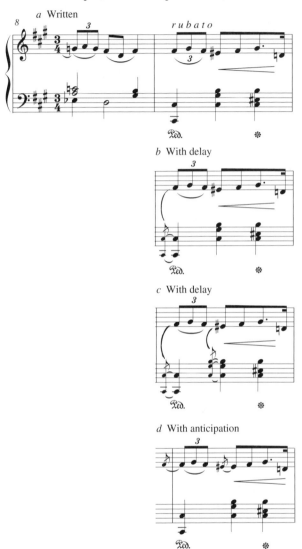

h.472.[3]). None of these edns., however, contains a tie at this point when the theme returns at the end of mm. 32 (on the last line of Pl. XVI) and 64. A tie creates a parallel with the beginning, but all other statements of the theme enter without anacrusis. If a tie were intended, I think that Chopin would not have slurred the last 3 melody notes of m. 8 together and would have begun the word *rubato* on the last quarter beat of this measure. The articulation of the repeated phrase would then have commenced with the tied F♯, in which case one might delay its entrance until after the simultaneous G♯ and B in the left hand. There is a tie in the later edns. of Liszt (Ch. 8), Mikuli, and Debussy (Ch. 9 n. 174), but not in that of Brahms (Ch. 9 n. 1). For a considerably different version without rubato, see the holograph in Ferdinand Hiller's album repr. in the Henle edn. of the Mazurkas, 4–5.

By delaying or anticipating the notes that sound as though they should have coincided exactly with the accompaniment, one thus establishes points of audible rhythmic robbery. The other notes fall easily into place in the linear flow of the melody. One needs to experiment, I think, with both delay and anticipation in order to develop the technique and in order to decide which works best in a particular passage. It is important, however, to keep in mind that rubato must not be performed mechanically, but almost unconsciously as a response to events in the music. Therefore, it is one's perception of the sense of the music that will ultimately determine the type of robbery most appropriate. The first sixteen measures of this piece are marked for immediate repetition, so the performer would no doubt perform the rubato somewhat differently the second time, certainly with different tone colour, and perhaps with less or more time between the small and large notes. Perhaps the rubato could even be omitted the second time. The same music occurs two more times during the course of the Mazurka, but without rubato (see, for example, the last line on Plate XVI).[106]

Ex. 7.15 shows a similar situation of repetition. The Mazurka Op. 7 No. 3 begins the main theme after an eight-bar introduction. Starting in measure 17 (Ex. 7.15a), the parallel second half of the theme repeats almost the entire first half, but commencing with rubato. No really apparent robbery can take place on the first beat (the F is slurred from the C below), since there is no bass note. The chords on the second and third beats are arpeggiated, probably rapidly, and more for rhythmic reasons than as reaching gestures. The rubato could involve a delay of the first A♭ and the C, as in (b), or of just the A♭. Anticipation seems less suitable in this case. A similar situation occurs upon repetition of the theme near the end of the piece (c), but this time there are no arpeggio signs on the left-hand chords. Again, one or both notes might be delayed (d). One must be aware, however, that the theme has different dynamic markings the second time, with a sudden change from *f* to *p* coinciding with the word *rubato*. It is my experience that a careful and faithful following of every small marking

---

[106] Rosen feels that 'most of the indications for rubato in the Mazurkas, although probably indicating the earlier form, also demand a slight slackening of the tempo in order to be performed convincingly.' In fact, he states that 'the two kinds of rubato are not always separate in Chopin', and that my 'emphasis on the earlier rubato for Chopin is good, but it should not be exclusive even for his use of the term'. On the other hand, the accompaniment's strict rhythm during rubato is the most frequently and consistently reported aspect of Chopin's teaching. I suspect he intended, in those cases in which the performer feels a tempo flexibility is desirable or necessary for expressive or technical reasons, that the listener at least hear the struggle of the left hand attempting (but perhaps in vain) to assert its strict rhythm.

Ex. 7.15 Chopin, Mazurka Op. 7 No. 3 (Paris: M. Schlesinger, 1833; *Stirling* 15, 17)

in the music, whether for dynamics, pedalling, phrasing, or mood, has a crucial effect on the way the rubato is performed.[107]

Another occasion for rubato involves the repetition of a theme that recurs again at the end of a piece. In the Mazurka Op. 6 No. 2, the main melody occurs first in measures 9–16 and is marked with repeat signs. When it returns in measure 57 (Ex. 7.16*a*), its repetition begins as in Ex. 7.16*b*, with altered note values and portato touch, as well as the word *rubato*. One might perform the rubato with a delay on the first note (*c*) and perhaps also the F♯. One could likewise anticipate these two notes (*d*). In addition, an exhilarating effect occurs when the first note is anticipated and the F♯ delayed (*e*). It is of crucial importance in executing the rubato to sense keenly the changes between (*a*) and (*b*).[108]

In three different pieces Chopin articulates the third of four parallel units with rubato, thus clearly marking the midpoint between the two structural halves of the theme. In the Mazurka Op. 7 No. 1 the word

[107] I have consulted copies of the 3 early edns. from BL: h.473.a.(6.), h.471.aa.(3.), and h.472.(3*.). Only in Wessel's version are the chords in mm. 93–6 arpeggiated, and there is a *dim.* commencing on the 2nd beat of m. 92 leading to the *p* on the 2nd beat of m. 93. In an autograph MS in the Stiftelsen Musikkulturens Främjande in Stockholm, the word *rubato* is missing.

[108] In the holograph version in Stockholm, printed by Ewald Zimmermann in the Henle edn. of the Mazurkas, 8–9, the repetition, at the end of the piece, of the introduction and main theme is not written out, but indicated only by *Da capo al fine*. Hence there are no changes and no rubato.

Ex. 7.16 Chopin, Mazurka Op. 6 No. 2 (Paris: M. Schlesinger, 1833; BL, h.471.y.[1.])

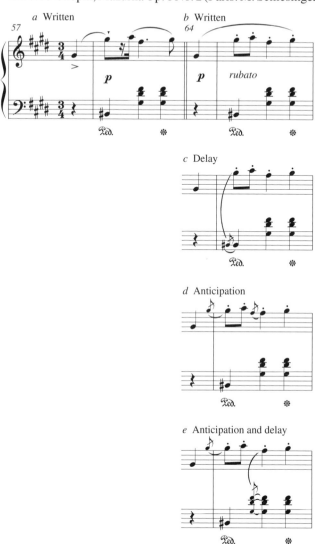

rubato commences below the second melody note of the measure (Ex. 7.17*a*). If the word, however, is meant, as I think it is, to apply to the entire measure, then one could delay, as in (*b*), the F and perhaps also the E and even the Dᵇ, but with diminishing displacement. Anticipation could occur as in (*c*). If rubato is to begin later, as actually marked, then displacement would commence with the E and tend to accentuate the agogic stress already inherent in this longer note; however, Chopin

Ex. 7.17 Chopin, Mazurka Op. 7 No. 1 (Paris: M. Schlesinger, 1833; *Stirling* 13)

did not place an accent mark on this note as he did on other notes two and three bars later. Important in the proper execution of either (*b*) or (*c*) are the dynamics (***pp*** and *sotto voce*) and the fact that the sustaining pedal is depressed four bars before the rubato and not released until two bars later. The section from measure 45 to the end of the piece, which includes also a return to the opening theme, is repeated, and the rubato should, of course, be different the second time.[109]

The Mazurka Op. 24 No. 2 presents a similar situation (Ex. 7.18*a*),

Ex. 7.18 Chopin, Mazurka Op. 24 No. 2 (Paris: M. Schlesinger, 1836; BL, h.471.y.[4.])

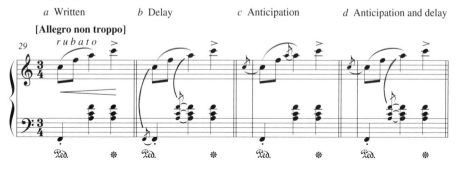

but here an accent mark appears above the melody on the third beat. Ex. 7.18 shows two delays (*b*), two anticipations (*c*), and both an anticipation and a delay (*d*). For proper motivation in playing the rubato, the performer must carefully observe the *ritenuto* marked two bars earlier to emphasize an unusual B♮. Although *a tempo* is not

---

[109] The word *rubato* begins below the 2nd note of the bar in the early edns. of Schlesinger, Wessel, and Kistner, as well as the later edn. of Liszt. It appears below the 1st note in edns. by Mikuli, Brahms (Ch. 9 n. 1), and Debussy (Ch. 9 n. 174).

written in measure 29 as it is later in measure 37, it must be understood that the rubato coincides with a decisive return to the original tempo.[110]

The third example of rubato on the third of four almost ostinato phrases occurs in the last movement of the Second Piano Concerto. Ex. 7.19 shows the piano part and an orchestral reduction. Since the

Ex. 7.19 Chopin, Concerto Op. 21, 3rd movt. (Paris: M. Schlesinger, 1836; BL, h.471.j.[13.])

piano acts here exclusively as a melody instrument, some of its notes can be displaced in relation to the strictly measured chords of the orchestra in the same way they were in the earlier and contemporary rubato for accompanied violin or voice. One can emphasize the enormous leap to the C by means of delay or anticipation, with perhaps a lesser displacement on the following A♭. In an autograph manuscript of the full score, Chopin has written the word *rubato* further to the right, commencing with the A♭;[111] in this case, one would be confined, for the most part, to a rather substantial anticipation of this note. The rubato coincides with a decrease in dynamics in the orchestra and is therefore not only a means of articulating a repeated phrase, but is also part of the preparation that leads toward the long, sustained tonic chord that commences in measure 177.

---

[110] According to Rosen, the accent mark on the 3rd beat implies a lengthening of the beat. If the performer wishes to continue the rubato this far, then a lengthening could be accomplished, it seems to me, by anticipating the C over the steady accompaniment.

[111] Warsaw, Biblioteka Narodowa, MS Mus. 215, p. 24 of the 3rd movt.

Ex. 7.20 Chopin, Mazurka Op. 24 No. 1 (Paris: M. Schlesinger, 1836; *Stirling* 102)

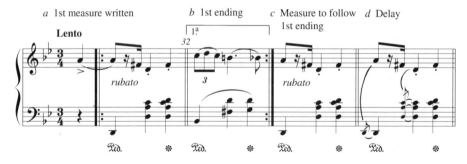

The final example of repetition involves the Mazurka Op. 24 No. 1, in which the opening thirty-one bars are followed by first and second endings. Ex. 7.20*a* shows the opening measure, Ex. 7.20*b* the first ending. When the pianist begins to play the opening measure, he soon realizes that there is very little worth robbing there. However, upon completing the first ending, it becomes apparent that the repetition of the opening measure is quite different from its first appearance: now the tie back to an anacrusis is gone and the first melody note occurs vertically above the bass note (Ex. 7.20*c*). Now effective robbery can take place for the purpose of announcing the repetition of an entire section of music. The A may be delayed, or both A and D as shown in (*d*). To make the rubato effective one must keenly sense the accented dotted quarter notes in the two preceding measures. If anyone should feel obliged to rob something from the first appearance of the passage (Ex. 7.20*a*), then I suppose the D, and perhaps the following note as well, could be delayed.[112]

The second main purpose for which Chopin uses rubato is to emphasize an expressive moment such as the high point of a melodic line or unusual non-harmonic notes. Ex. 7.21*a* shows *poco rubato* marked in the Nocturne Op. 9 No. 2. After the main melody has been repeated and varied a number of times, the music turns in measure 25 to the minor subdominant to express the idea of conclusion. Measure 26 repeats the brief melodic passage from measure 25 and extends it by reaching to a high E♭ by means of a turn, a leap, and *poco rubato*. In this case only the delaying of notes seems appropriate, as shown in Ex. 7.21*b*. In this case the displacement of the last three melody notes in the measure creates an intensely singing effect (much like that in Ex. 5.4), which is enhanced by the subtle nuance of touch produced by using the fifth finger for each note. The effect, in fact, can be so

---

[112] Rosen states that 'delay would be most convincing on [the] third beat'.

Ex. 7.21 Chopin, Nocturne Op. 9 No. 2 (Paris: M. Schlesinger, 1833; *Stirling* 24)

intensely expressive that Chopin must have realized that the player had to be warned that the highest point of the phrase did not occur here, but one measure later on the G. Restraint is therefore enforced at the end of measure 26 by the word *poco*, by the dynamics (*sempre pp*), and by the sustaining of an E♭ pedal-point in the bass up to the very moment the high G appears. The G itself is then approached not only by a large leap, but by a turn written in ordinary notes and a small-note appoggiatura, and the note itself is a dissonance within a chromatic harmony. One will note a similarity between the way Chopin notated the appoggiatura to G in Ex. 7.21*a* and the manner in which I have indicated the rubato to the high E♭ in Ex. 7.21*b*. An important element in the second half of measure 26 in Ex. 7.21*a* is also the sustaining pedal, which is depressed during the entire rubato passage.[113]

Another example of rubato used for a warm, singing effect at a high point occurs in the Second Concerto. The second theme begins with two statements of the two-measure unit in Ex. 7.22*a* and moves after two more measures to a half cadence on the dominant; its second half commences, like the first, with two statements of Ex. 7.22*a*, but is

---

[113] Rosen feels that a ritenuto is demanded by the repeated fingering (easily accomplished, I think, over a strict accompaniment) as well as by the portato signs at the end of m. 27 (this time, I presume, with both hands participating, since the rubato has probably ceased by this time).

Ex. 7.22   Chopin, Concerto Op. 21, 3rd movt. (Paris: M. Schlesinger, 1836; BL, h.471.j.[13.])

then followed by the measures in (*b*), which lead to a final cadence two bars later. In the measures involved with Ex. 7.22*a*, the violins and violas are playing *col legno* and the cellos *pizzicato*, but in the measures that follow, including those in Ex. 7.22*b*, the strings are all marked *arco*. In the *col legno* measures, the harmony alternates between I and V chords, whereas the *arco* coincides in Ex. 7.22*b* with a written-out appoggiatura in the second violins followed by a chromatic turn toward the subdominant. Thus at the very moment when the melody reaches its structural climax on A♭ and accentuates this pitch through repetition and portato touch, the orchestration and the harmonic movement suggest a particularly warm and tender expression. This is further intensified by the rubato in the piano part. Either delay or anticipation is possible on the A♭s, and the amount of displacement could increase as one approached the longest note on the last beat, or it could be greatest on the initial A♭ in order to emphasize the leap. Another possibility is to anticipate the first A♭ and delay the third. In the holograph manuscript mentioned above, the word *rubato* begins with the second of the three repeated notes, which must imply a delay of only the third A♭.[114] Chopin probably wanted to emphasize

[114] Warsaw, Biblioteka Narodowa, MS Mus. 215, p. 22. Note that the 3rd A♭ in Ex. 7.22*b* is tied over the bar in the right hand, but not in the left. This, as well as the dots above the 3 A♭s, occurs in both the Schlesinger (p. 25) and Br. & H edns. (p. 27) in BL (h.471.j.[13] and h.473.c), as well as in MS 215. In the Paderewski *Complete Works*, xiv (5th edn., 1958), 168, there is a tie in both hands, but dots only above the first 2 A♭s.

here the singing quality of the three notes in the same way, as Couperin noted in connection with the *suspension*, that a stringed instrument does when it increases the volume of sound.[115]

Rubato is used in three of Chopin's works to intensify the effect of large-note appoggiaturas. In the opening movement of the Trio Op. 8 for violin, cello, and piano, rubato occurs in the repeated exposition and again in the recapitulation of a sort of modified sonata form. It appears during the second half of an *espressivo* melody in order to enhance the singing quality of a series of appoggiaturas. Here it is marked, as in Ex. 7.23*a*, *sempre rubato*, since it spans more than a

Ex. 7.23   Chopin, Trio Op. 8, 1st movt. (Paris: M. Schlesinger, 1833; BL, h.471.s.[10.])

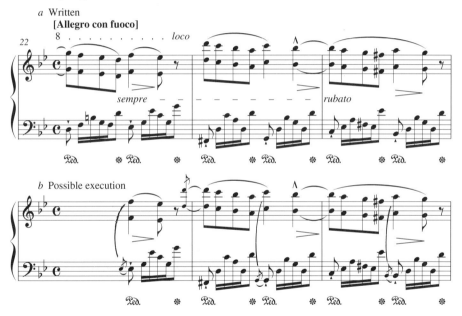

measure in duration. However, there are only selected vertical combinations that seem appropriate for displacement. For purposes of illustration in Ex. 7.23*b*, I show a delay for each of the three quarter-note appoggiaturas (F, C, and A) and an anticipation on the first beat of measure 23 to articulate the repetition of a sequential unit. The violin and cello are merely sustaining chords in half notes, so that the process of rubato is a matter mainly for the pianist. When the

[115] Rosen writes that Ex. 7.22*b* 'is impossible without a ritenente'. The broadening indicated by the portato marking, however, may be achieved in the melody, I think, without disturbing the rhythm of the orchestra.

exposition is repeated, the rubato, it seems to me, can either be omitted or performed with a different intensity.

When the same melody recurs in the recapitulation, the notes themselves are exactly the same, but now only the word *rubato* occurs, spanning from about the last quarter beat of measure 159 (corresponding to measure 22) to the first two or three eighth notes of the following measure. Thus it begins later and ends sooner than the rubato in Ex. 7.23*a*. For variety, one might delay both the E♭ at the end of measure 159 and the D at the beginning of the following measure. This is, after all, the third time the passage will be heard in the piece, so I think Chopin's markings should be carefully observed.[116] Chopin adds a similar type of singing rubato to chromatic appoggiaturas in his Rondeau Op. 16 (see Ex. 7.24) and the song 'Smutna rzeka' (Sad River), Op. 74 No. 3 (Ex. 7.25). In the latter, the placid nature of the preceding music (measures 1–3, in fact, are identical to

Ex. 7.24 Chopin, Rondeau Op. 16 (Paris: M. Schlesinger, 1834; BL, h.473.b.[2.])

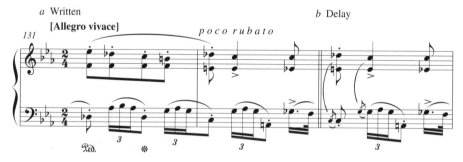

Ex. 7.25 Chopin, Song Op. 74 No. 3, 1831 (Österreichische Nationalbibliothek, Fontana MS)

---

[116] In the Paderewski *Complete Works*, xvi (2nd edn., 1959), 24, the word *rubato* appears instead of *sempre* in m. 22, followed by dotted lines that end halfway through m. 24. In m. 159 (p. 34), *rubato* is printed as in the edn. of M. Schlesinger (BL, h.471.s.[10]), but is followed by dotted lines that extend to the end of m. 161.

4–6 except for a repetition of the melody's F♯ in measure 3), seems to require the sort of surge of energy produced by the process of anticipation.

Another rubato passage containing expressive non-harmonic notes appears in a manuscript version of the Valse Op. 34 No. 1. We are so accustomed to hearing measure 85 of this well-known work performed as in Ex. 7.26*b* that it is difficult to adjust to the earlier version in (*a*).

Ex. 7.26 Chopin, Waltz Op. 34 No. 1

b Published version (Paris: M. Schlesinger, 1839)

Played without rubato, the G♮ in this measure creates in (*a*) a startling dissonance with the left-hand chord on the second beat. Rubato, however, displaces the hands, softens the effect of the dissonance, and adds a singing quality to the inner voice as it moves from A♭, G♮, G♭, and F to G♭ in the next measure. If the published version in (*b*) represents an approximation of the way Chopin conceived the rubato, then the right-hand chord containing the G♮ should anticipate the left hand. Perhaps the four chords in the right hand of measure 85 could be played approximately in the manner of a quadruplet against the three strict beats in the accompaniment. Since Chopin actually begins the word *rubato* before the barline, perhaps the opening chord could also be anticipated. Other markings in the manuscript are crucial to the performance of the rubato: *dolce* marked in measure 81; the accent sign on the second quarter note of measure 84 which, although gentle, acts to motivate the rubato; and the accent marks over the right-hand chords in measures 86, 87, and 88—all a consequence of the rubato in measure 85. All of these accents, as well as the *dolce* and signs for crescendo and decrescendo, are missing in the revised version. It is also of significance that the left hand, which begins the section in measure 81 with a single note on each quarter beat, changes suddenly in measure 85 to the typical 'oom-pah-pah' waltz bass which had animated most of the measures preceding measure 81. The rubato is thus triggered by the contrasting nature of the new melody which commences in measure 81, by the second-beat accent in measure 84, by an anticipation of resuming the waltz bass in measure 85, and, finally, by a desire to make the chromatic inner voice sing with a special intensity and to give expressive rhythmic freedom to the written-out turn in the melody that reaches up to G^b in the next measure. This is the one rubato of Chopin that seems to me to be obligatory, for he must surely have been counting on some degree of displacement between the hands on the second beat. It may not be coincidental, then, that this is also the one rubato that Chopin did not publish.[117]

Chopin thus uses rubato to emphasize repetition or expressive notes. On two occasions, however, he uses rubato for still another purpose: to set a mood at the beginning of a piece. In the Mazurka Op. 67 No. 3 (see Ex. 7.27a), the rubato marks neither a repetition nor an expressive note. Here, I believe that the intent is for a particular mood to be established: imagine a pianist sitting before a keyboard in reverie, perhaps dreaming of a sentimental dance recalled from the evening before, and only half in reality and still half in dream reaching the hands toward the keyboard to commence playing—at first languid and with hesitation, but quickly, as the sounds reach the ear, gaining control of the instrument. I have suggested in Ex. 7.27b a displacement

[117] See *Waltzes of Fryderyk Chopin: Sources*, i: *Waltzes Published during Chopin's Lifetime*, ed. Jan Bogdan Drath (Kingsville, Tex.: Texas A & I University, 1979), which contains facs. of the following: (1) holograph dated 15 Sept. 1835 from the album of Countesses A. and J. von Thun-Hohenstein, now lost, copied from photographs at the Biblioteka Uniwersytecka in Lodz, Poland (Ex. 7.26a comes from the 2nd page of the MS on p. 59); (2) the holograph from which the published versions were made, now at the Library of the Warszawskie Towarzystwo Muzyczne in Warsaw (m. 85 on p. 64, curiously, is crossed out and a new version written below); (3) the version published by Schlesinger in Paris, from the copy of Camille O'Meara Dubois now in BN (Ex. 7.26b from p. 74); and (4) versions published by Br. & H in Leipzig and by Wessel in London, with m. 85 on pp. 85 and 95 is essentially the same as in the Paris edn. In the MS of 1835 the 16-measure section from mm. 81 through 96 is marked with repeat signs, whereas the repeat is written out in the published versions; therefore, m. 85 in the MS corresponds to both mm. 85 and 101 in the others (as indicated in Ex. 7.26b). The MS is also reproduced by Jaroslav Procházka in *Chopin und Bömen* (Prague: Artia, 1968), 164, and its rubato discussed by Bertrand Jaeger in 'Quelques nouveaux noms d'élèves de Chopin', *Revue de musicologie*, 64 (1978), 85 (he places the rubato in m. 86).

Ex. 7.27 Chopin, Mazurka Op. 67 No. 3, 1835 (Berlin: Ad. Mt. Schlesinger, 1855; BL, h.472.a)

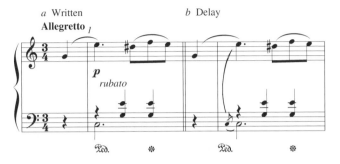

only on the first beat, but one might also delay the F on the third quarter beat. Delay in this case seems more expressive than anticipation. The theme continues for sixteen bars, presenting two almost identical halves, and then recurs exactly, but without rubato. Playing measure 17 without rubato becomes as expressive, in its own way, as the technique of rubato in measure 1, especially if one observes the sudden change from *ff* to *p*. If Chopin had indicated rubato for measure 17 rather than measure 1, the effect would, of course, have been entirely different, for the rubato would then have had the function of pointing out a musical repetition rather than setting an initial mood.

The Nocturne Op. 15 No. 3 is a similar example, I believe. *Lento* is the tempo for the piece. *Languido e rubato* is marked in the opening measure (Ex. 7.28*a*). Since the only vertical coincidence between the melody and the accompaniment occurs on the first beat, this presents the only opportunity in this measure for displacement. Ex. 7.28*b* shows a delay, (*c*) an anticipation. If rubato continues into the following measure, then its first melody note could be delayed or anticipated; (*d*) combines a delay in the opening bar with an anticipation in the second.[118] The same reviewer for *Le Pianiste* in 1834 who complained of Chopin's tendency to notate rubato, stated that this nocturne was 'in rubato from one end to the other'.[119] This is clearly incorrect, however, for the entire second half of the piece is mostly chordal and is marked *religioso*. Furthermore, almost all descriptions of rubato speak of this device as having extraordinary power, which should be used sparingly and only for brief periods of time. Even

[118] According to Rosen, 'the anticipation would mar the mazurka rhythm', which is indicated at the beginning of m. 2.
[119] *Le Pianiste*, 1/5 (Mar. 1834), 79: 'Il est en *rubato* d'un bout à l'autre.'

Ex. 7.28 Chopin, Nocturne Op. 15 No. 3 (Paris: M. Schlesinger, 1834; *Stirling* 73)

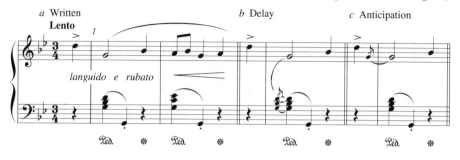

Chopin's *sempre rubato* in the Trio (Ex. 7.23*a*) spans only two and one half bars. Therefore, I see no reason for assuming that the word *rubato*, when it appears in the opening measure of a piece, indicates anything more than a momentary effect, as it does elsewhere.[120] In the case of Op. 15 No. 3, a languid manner of performing the rubato would affect not only the amount of displacement, but also the tempo, dynamics, touch, and tone colour. Such a beginning would, of course, set a mood that would influence the rest of the composition.

It is significant, I believe, how sparingly Chopin uses the word *rubato*. In only four works does the word occur more than once: in the Mazurka Op. 7 No. 3 (Ex. 7.15) and the Trio Op. 8 (Ex. 7.23) for repetition of the same music, in the Second Concerto for two different passages (Exx. 7.19 and 7.22), and in the song Op. 74 No. 3 (Ex. 7.25) as part of a ritornello that occurs three times.[121] When rubato

---

[120] The word *rubato* occurs in the opening measure of the Mazurkas Op. 24 No. 1 (see Ex. 7.20) and Op. 67 No. 3 (Ex. 7.27) and the Nocturne Op. 15 No. 3 (Ex. 7.28). I know of no other way Chopin could, in fact, have indicated his usual brief rubato in this measure. In *Chopin: Pianist and Teacher*, 121 n. 99, however, Eigeldinger concludes that when marked in the 1st bar rubato 'applies to the entire piece or at least to its first section' and 'therefore concerns agogic fluctuations'. Concerning the rubato in Op. 15 No. 3, see Jeffrey Kallberg, 'The Rhetoric of Genre: Chopin's Nocturne in G Minor', *19th Century Music*, 11 (1987), 246–7.

[121] In Julian Fontana's MS of the song at the Österreichische Nationalbibliothek, the ritornello is written only at the beginning, with indications for repetition between and after the 2 strophes.

appears within a repeated section—in the Mazurkas Op. 6 No. 1 (Ex. 7.14) and Op. 7 No. 1 (Ex. 7.17) and the Trio Op. 8 (Ex. 7.23)—it should probably, as I have indicated above, be omitted or at least executed differently the second time.

Exx. 7.14–7.28 present, then, the places where Chopin actually marked the word *rubato*. I have suggested some possible methods of displacing melody notes in relation to notes or chords in the accompaniment. In most cases, either delay or anticipation could probably be made to work. Anticipation, however, seems to me to project, in general, a feeling of urgency and renewed energy. Therefore, it seems most appropriate for articulating repeated units, as in Exx. 7.14, 7.16, 7.17, and 7.18. It is also effective for outbursts of emotion, such as the passionate impatience of the octave D's at the beginning of measure 23 in Ex. 7.23. Delay, on the other hand, seems to express particularly well the languid mood—suitable especially for the beginnings of pieces such as the Mazurka Op. 67 No. 3 in Ex. 7.27 and the Nocturne Op. 15 No. 3 in Ex. 7.28. Furthermore, a series of consecutive delays produces the singing effect we noted on the last three melody notes in measure 26 of the Nocturne Op. 9 No. 2 in Ex. 7.21.

In all the examples above, delays and anticipations have occurred in such a way that every note in the left-hand accompaniment continues in its exact time. Thus, as Saint-Saëns states: 'The accompaniment remains undisturbed while the melody floats capriciously, rushes or retards'. One special difficulty occurs in delaying a note, especially one, such as the first note of a unit, which should receive more emphatic accentuation. When one sharply attacks the opening F♯ on the first beat of Ex. 7.14*b*, for example, it is easy to hear this accentuated note as the first beat of the measure, in which case the octave C♯s in the left hand sound as though they occurred before the beat. There is a real danger that the performer will acquire the habit of actually playing the left hand too soon and leaving the melody in its notated position. This must have been the case with many who mechanically imitated Chopin's rubato, for Saint-Saëns also says, as we have seen, that when some cannot execute the true rubato, 'they give the illusion . . . by playing the melody in time and dislocating the accompaniment in order to make it fall at the wrong time'.[122]

It is thus a question whether the rubato should occur in each ritornello. If it does, it should, of course, respond to both its formal position and to the meaning of the text (in this case, a sad one of a woman weeping by a river over the loss of seven daughters).

[122] The same effect is achieved when anticipation by the left hand does not involve robbery of the previous note, but simply the addition of free time. See Kiorpes, 'The Performance of Ornaments in the Works of Chopin', 136–8. He applies the concept occasionally to notated small notes, a process related to the later type of rubato.

Another questionable method of playing involves a breaking up of the notes of the accompaniment so that only the lowest note actually occurs at the proper time. Although such a practice does disturb the accompaniment in violation of Saint-Saëns' description, it is a natural sort of thing for a pianist to do, especially considering the presence of arpeggiation elsewhere in the same music. This would occur in Ex. 7.14c, for example, if the lowest note of the C♯ octave in the left hand were played exactly on the beat and the upper C♯ delayed so that it sounded simultaneously with the delayed F♯ in the melody. On the second beat of the same example, the lowest note of the chord could similarly fall precisely on the beat, with its two upper notes being delayed along with the E♯ in the melody. One might even arpeggiate the entire chord. This method of delaying, of course, can also lead easily to the habit of anticipating the left hand and playing the melody in time. One is tempted to wonder if the later mannerism of chordal arpeggiation, like the alternation of hands, was also the result of imitating a mechanical feature from someone else's inspired rubato.

A similar method could also be applied when the melody occurs in octaves in the two rubato passages in the Second Concerto. In Exx. 7.19 and 7.22 the lower note of each octave could anticipate the upper, with the lower note either on or before the beat. This would probably be more effective in Ex. 7.19, but in either case, the accompaniment in the orchestra remains undisturbed while a portion of the octave melody is either delayed or anticipated. Furthermore, the pianist is thereby creating rubato in the manner to which he is accustomed—by displacement between the two hands. We have already noted Mikuli's report that Chopin did not want chords arpeggiated unless specifically marked. Perhaps these methods of breaking up the right or left-hand parts should therefore be considered only remote possibilities.

Within the range of possibilities suggested in Exx. 7.14–7.28 a number of variables occur. Although the amount of time involved in the rubato is approximately indicated by the composer, the actual duration may be less, perhaps confined to displacement on a single beat. An important variable is the amount of time between the small notes and the large ones in my examples, and this, as I have pointed out from time to time, might change as one moves from beat to beat in a measure, or as one repeats a rubato passage for a second time. In some cases, as in the Trio of Ex. 7.23, only selected vertical combinations seem to be involved. Although both the listener and the performer give particular attention to the points of right and left hand displacement, the ultimate goal, of course, is to perceive the melody in a flowing, linear manner. Another variable, then, is the way the other notes fit in—those notes that do not cause displacement with a note or chord of the accompaniment. One of the most significant variables, finally, is the attack, touch, volume, pedalling, or mood—in short, the total tone colour—with which the performer plays the notes involved. Theoretically, this may vary widely—from soft to loud, from rough to smooth, from subtle to bold, from calm to sparkling, from boisterous or passionate to languid. As we have seen, however, Chopin's rubato, according to Chaulieu, ordinarily combined *lusingando* and *espressivo*.

All of these variables can be properly controlled only when one is completely motivated by musical events. This is why I have so often spoken above of the musical situation before, as well as during and after the rubato passages. Although all musical elements must be vividly lived by the performer in order to be properly played, this particular device seems to depend especially upon genuine motivation and good taste. Almost every writer from Tosi to Baillot and García has warned against applying rubato mechanically and without feeling and against its excess use. The performer should have explored all the possibilities (such as those in Exx. 7.13–7.28) in advance and have them all as working techniques, so that when he executes rubato in a performance, it is an almost unconscious response to the keenly perceived sense of the music. When Saint-Saëns states that the worst type of false rubato is 'to play the two hands one after the other', he may be referring not only to the later custom of breaking the hands, but also to the fact that simply alternating the hands mechanically does not constitute rubato. Rubato is produced first of all in the imagination of a sensitive performer. It is realized during the course of a flow of melody and becomes a part of this continual linear flow. The identification of specific points of displacement must not divert the mind from the essential linear quality of the music. The brief dislocations act within the melodic continuum to articulate significant structural points, heighten expressive moments, or intensify the singing quality.

I am somewhat concerned that illustrations such as those in Exx. 7.14–7.28 may be misunderstood. Although displacement may look like a primarily vertical event, it acts only as a sort of temporary modification of the melodic flow and must not impede its flowing forward movement. It is not my intention in these examples to provide a definitive solution for each case of rubato, as one might illustrate the execution of an ornament, nor to suggest that they can now be played mechanically without inner motivation. I offer to the performer not a solution for each rubato, but a means of exploration whereby he might be able to discover the unique rendition that is appropriate for him during a particular performance. We may wish that Chopin had specified in greater detail exactly what he envisioned for each rubato—

at least whether it involved anticipation or delay. He apparently pre-ferred, however, to offer the performer a wide variety of possibilities from which *one* would ultimately be selected according to the performer's perception of the living music.

For those readers who recall with animosity the offensive habit of some later pianists of continually playing the hands separately or arpeggiating every chord, one must emphasize that Chopin's rubato was a subtle nuance within an extremely restrained style of playing. Judging from the few times he marked the word in scores before 1835, the device was seldom employed and then only for special situations. In addition, other aspects of his performing style, as we have noted, were highly restrained—especially the dynamic level, and, I believe, also the tempo fluctuations. When one is playing with restraint in tempo and dynamics, very slight nuances become apparent and hence expressive. I have the feeling that within Chopin's carefully controlled style both the earlier and later types of rubato were applied so rarely and at such sensitive places that they had, especially at that time in history, enormous expressive effect. I see no reason why, if pianists do not continually play with a loud and harsh tone, the same result could not be produced today in renditions of his music. Margaret Chanler describes how Liszt taught Chopin's rubatos in 1877:

When Chopin was being played, only the most delicate precision would satisfy him. The *rubatos* had to be done with exquisite restraint and only when Chopin had marked them, never *ad libitum*.[123]

Chopin apparently played with an unusually elegant and powerful sense of melodic flow—a flow which did not contain the conventional accentuations of other pianists of his day. His music moved horizon-tally along the keyboard. His imagination was so powerful and his body so relaxed and receptive that ideas flowed directly and perfectly from his mind onto the keyboard. The conception by his imagination of long, singing melodies over long spans of time controlled and imposed restraints on all elements of the music and produced incredibly subtle nuances—nuances that the intellect by itself could not create. At the same time, his intensely linear approach facilitated an indepen-dence of the hands, a skill developed perhaps from his study of the contrapuntal works of J. S. Bach.

Within this flowing, elegant, subtle style lived many types of rubato: the tempo flexibility of the later type, the prolonged beats of the dance

---

[123] Margaret Terry (Mrs Winthrop Chanler), *Roman Spring* (Boston: Brown, Little, & Company, 1934), 85. At the age of 15 the author attended a class lesson given by Liszt in Rome; she had previously studied piano with Giovanni Sgambati, a pupil of Liszt.

rhythms, the metric exhilaration of the unusual groups, the expressive delays of the small notes and arpeggios, and, finally, the displacement and melodic shifts of the earlier type. It is this last type which is most mysterious, however, and the type least able to be counterfeited. The singer and violinist conceived of this kind of rubato primarily as a matter of one note stealing time from another; the displacement that occurred between the soloist and the strict accompaniment happened naturally as a result. Their musical examples thus depict small amounts of time added to or subtracted from certain notes, often in company with additional notes and ornaments. Keyboard players, on the other hand, tended to perceive the same rubato from the opposite point of view, considering first of all the displacement between left and right hands, with the small adjustments in the note values simply resulting as a consequence.

Historically a delay of melody notes seems to occur, as we have seen, more often than an anticipation. This was true in the vocal and violin examples as well as those for keyboard (Ex. 5.4, for example). In addition, there is a long tradition in French lute and keyboard music of retarding notes—in the arpeggiation of two or more vertical notes (Ex. 2.8), in the *suspension* of Couperin (Exx. 2.9–2.10), and in the portato of Adam (Ex. 5.5). We have seen expressive delays by Mozart (Ex. 6.1). We have noted numerous reaching motives that cause delays in melody notes. Therefore one might expect, at first, that Chopin's passages marked *rubato* would likewise favour the delay of notes, and the typical singing melody in the Nocturne Op. 9 No. 2 (Ex. 7.21) offers impressive evidence of the power of such delays. With Chopin, however, the melody, according to Mathias, Saint-Saëns, and Mikuli, 'can be early or late', 'rushes or retards', 'hesitates or enters ahead of the beat'. Therefore anticipation is apparently also a possibility at times, both for passionate outbursts and for the repetition of structural units. Thus one must explore both anticipation and delay, I think, in determining the most effective manner of performing the rubato passages of Chopin.

The basic act in the process, in any case, is mainly a matter of vertically aligned notes sounding as though they are striving—but striving ultimately in vain—for simultaneity. Therein lies the yearning emotion that gives such great power to the device. At the same time, however, the resulting rubato of the pianist must enhance the forward movement of the melody and make it flow and sing as much as if it had been produced by a singer or violinist.

8 CONTEMPORARIES OF CHOPIN

A NUMBER of composers with whom Chopin was acquainted were also involved with tempo rubato. The manner in which they used the device sheds some additional light, in some cases, on Chopin's own practice. These composers include Kessler, whom Chopin knew in Warsaw; Filtsch, Méreaux, Pauline Viardot, Ernst, Franck, Gottschalk, Heller, and Liszt from his years in Paris; Clara Wieck and Robert Schumann, whom he visited in Leipzig in 1835.

Karl Filtsch

Karl Filtsch (1830−45), Chopin's most talented pupil, studied with Chopin from December 1841 until April 1843, except for the summer of 1842 when he worked with Liszt. Chopin is quoted in February 1843 as saying:

Never has anybody understood me like this child, the most extraordinary I have ever encountered. It's not imitation, it's an identical feeling, instinct, which makes him play without thinking . . . He plays me almost all my own compositions without having heard me, without my showing him the least thing—not completely like me (for he has his own style) but certainly no less well.[1]

In view of Filtsch's mastery of Chopin's style and his own creative ability, his example of rubato on Plate XVII has a special significance. It appears in the third of three pieces that constitute his *Premières pensées musicales*, one edition of which was published in London by Wessel & Stapleton. The latter were the publishers, as it states on the title page 'of the entire piano-forte works of Frederic Chopin'. They

---

[1] Quoted in French within an article in German entitled 'Pariser Tabletten: Karl Filtsch' in *Der Humorist* (Vienna), 7/37 (22 Feb. 1843), 155; trans. from Eigeldinger, *Chopin: Pianist and Teacher*, 142 n. 157. In the *Revue et gazette musicale de Paris*, 18 (1843), 150, Henri Blanchard writes that Filtsch's delicate and entrancing touch and profoundly artistic feeling made one feel as though Chopin himself were playing.

PLATE XVII. Karl Filtsch, Mazurka from *Premières pensées musicales* (London, between 1841 and 1845)

had reached Op. 42, as an advertisement on another page declares and announce that the series is 'to be continued'. The composer is named on the title page as 'Charles Filtsch, élève de Chopin'. The work was thus published sometime after Filtsch began his studies with Chopin and before 1845, the date Stapleton left the partnership with Wessel.[2]

Many features of this piece (see Plate XVII) remind us of Chopin: the title 'Mazurka', the Polish dance which was a special favourite of Chopin; the typical left-hand accompaniment, with a bass note on the

---

[2] See *New Grove*, xx, 368. A date of 1844 is given in the *Catalogue of Printed Music in the British Library to 1980*, xxi, 68; the call number is h.702.(23.), the plate number 5929. According to O. W. Neighbour and Alan Tyson, in *English Music Publishers' Plate Numbers in the First Half of the Nineteenth Century* (Faber, 1965), 42, the 5000s were assigned in 1843. Bernhard Lindenau, in 'Carl Filtsch', *Archiv für Musikforschung*, 5 (1940), 51, lists 7 works: the first 3 (including *Premières pensées musicales*) published by Pietro Mechetti qdm. Carlo in Vienna, the last 3 posthumously by C. A. Spina in Vienna. Lindenau mentions that the 6 published works were at that time in the possession of August Cranz Musikalien-Verlag in Leipzig. I have been informed in a letter from the present office of August Cranz GmbH in Wiesbaden, however, that their archives were unfortunately destroyed during the war. I have not been able to locate any of Filtsch's other works.

first beat and the remainder of the chord repeated on beats 2 and 3; small-note embellishments in measures 9 and 11 and again at the top of the second page, to be played *pp* when the opening phrase recurs; dynamics on the soft side, ranging from *p* to *pp*; and, finally, the word *Rubato*. The expression *un poco più animato* coincides with the anacrusis to the section which follows the repeat bars on the first page and therefore presumably refers generally to the eight measures that form a transition between B flat minor and the return to E flat minor at the top of the second page. Since it is not followed eventually by *a tempo*, it is presumably manifested more by an articulate touch than by an increase in tempo. The word *Rubato*, on the other hand, does not occur above the anacrusis, but above the opening beats of the following measure—an indication, surely, of its momentary nature, and a recognition that the earlier type of rubato can only be perceived when the strict left-hand accompaniment is present.

Like Chopin, Filtsch uses rubato here to articulate the beginning of a phrase. Unlike the Chopin examples, however, the new phrase is not a repetition of material already heard, but a modulating transition between statements of a main theme. In the spirit of *un poco più animato*, the rubato probably involves an impatient melodic anticipation made conspicuous by displacement—most likely on the first and second beats of the measure—with the strict accompaniment in the left hand. By means of the rubato, the transition is launched as a separate section which has a musical sense different from the statements of the main theme which surround it.

Thus at around 14 years of age, Filtsch composed a work that incorporated, but in unique ways, many of the elements he had learned from his teacher. It is, indeed, a sort of special tribute to Chopin. Filtsch performed in Paris, London, and Vienna during 1843 and 1844. He was gaining publicity, for there is an announcement at the bottom of the title page of *Premières pensées musicales* that his portrait was also for sale. If he had not died at the young age of 15 and had been able to transmit the Chopin style faithfully to succeeding generations of pianists, one wonders how different the history of piano playing might have been, or how different the history of tempo rubato.

Kessler, Méreaux, Viardot, and Ernst

Joseph Christoph Kessler (1800–72) was a German pianist and teacher living in Warsaw during 1829. Chopin participated in the weekly musical evenings at his home, and later dedicated the German edition of his Préludes to him.[3] We have already noted that a writer in *Le Pianiste* for 1834 felt that a Pleyel instrument was required 'in order to sing a romance of Field, caress a mazurka of Chopin, sigh a nocturne of Kessler'. In a review of the Nocturnes of Kessler's Opp. 27, 28, and 29 in another issue of *Le Pianiste* from the same year, the author complains in a footnote about some mistakes in spelling, among them 'rudato, pour *rubato*'.[4] This error must have occurred in one of the six Nocturnes of Op. 27, for among the six in Opp. 28 and 29 available to me from the Bibliothèque Nationale, the word is spelled correctly in its single appearance in Op. 28 No. 2. In measure 17 (see Ex. 8.1) a singing melody commences, with the opening bar

Ex. 8.1 Kessler, Nocturne Op. 28 No. 2 (Paris: Richault, *c.*1834)

marked *dolcissimo* and *Tempo rubato*. The latter may appear above the upper staff to indicate that it applies mainly to action taken by the right-hand melody. If the earlier type of rubato is intended, then significant melody notes could be delayed: the opening F♯, the A♯, and perhaps other notes in the opening measure, and possibly the opening B of the next. Since the texture in measure 17 contrasts markedly with the preceding music, one might, on the other hand, anticipate the first note as a means of articulating the beginning of the sort of movement that characterizes the rest of the piece. Every note of the melody coincides with a note in the accompaniment, so there is ample opportunity for displacement. In measure 19, eleven notes in the melody occur against nine in the left hand; in measure 21, fifteen against three, and in measure 23, eleven against six. Unlike Chopin, Kessler includes the word *tempo* as well as *rubato* and generally employs descriptive terms more frequently. They were both influenced obviously by Field. Kessler, however, was ten years older than Chopin, and it is not clear whether one of them may have influenced the other.

[3] Hedley, *Selected Correspondence of Fryderyk Chopin*, 35–6.
[4] *Le Pianiste* (repr. Minkoff, 1972), 1/8 (June 1834), 124.

Included in Fétis and Moscheles' piano method of 1840 is an 'Elegia' written especially for this volume by the French pianist and composer Jean-Amédée Le Froid de Méreaux (1802–74). He includes numerous expressive terms in the piece, among them *rubato* in measure 26 (see Ex. 8.2).[5] The word appears above a single G♯ in the left hand,

Ex. 8.2 Méreaux, *Elegia* (1840)

apparently for intensifying the sort of *espressivo* already marked for the descending arpeggio on the opening beat of the same measure. The accompaniment which should maintain strict time in the rubato, is in the right hand, I think: the A and the seven thirty-second notes. The melodic G♯ on the lower staff can then slightly precede the A, and the anticipation could perhaps continue, possibly even as far as the B♯ in the left hand. At the same time, the displacement acts to intensify the dissonant quality of the appoggiatura on A. In any event, the composer is not asking for a ritenuto here, as he is at the end of the measure. Chopin knew and apparently influenced Méreaux. He played one of his duos with Chopin in a private concert in 1833 and contributed a fantasy on a Chopin Mazurka to the *Album des pianistes*, a collection of previously unpublished works by Chopin, Liszt, and others.[6]

Chopin's Mazurkas were transcribed for a number of other media. Most of the transcriptions available to me merely repeat the word *rubato*, for the most part, as it occurs in the original work.[7] Far more

[5] François-Joseph Fétis and Ignaz Moscheles, *Méthode des méthodes de piano* (Paris: Maurice Schlesinger, 1840; repr. Minkoff, 1973), ii, 47.

[6] Atwood, *Fryderyk Chopin*, 69–72; and Niecks, *Frederick Chopin as a Man and Musician*, ii, 15.

[7] *Mazurkas von F. Chopin für Violoncell mit Pianofortebegleitung bearbeitet von C. Davidoff* (Br. & H, 1874–5), h.475.(1.) in BL includes Op. 24 Nos. 1 and 2 (transposed), with the word *rubato* only below the cello part. *Mazurkas de Fr. Chopin transcrites pour violon et piano par A. Schulz* (Brunswick: Henry Litolff, [1880]), g.375 in BL has Op. 6 Nos. 1 and 2, Op. 7 Nos. 1 and 3, and Op. 24 No. 1. Most are transposed to new keys, and the word *rubato* appears in both

PLATE XVIII. Pauline Viardot, first two pages of 'Plainte d'amour' from
Six mazourkes de F.ic Chopin arrangées pour la voix (Paris, 1866)

creative, however, are Pauline Viardot's transcriptions for voice and
piano, for she sometimes adds introductions, as well as marking *rubato*
in works that did not have it before or moving it to a new location in
a work that did. In three different letters from 1848 Chopin mentions
hearing Viardot sing his Mazurkas in London.[8] A set of six were
published at Paris in 1866 with French texts by Louis Pomey; the
same set, as well as some additional Mazurkas, were printed later in
Warsaw with Polish translations.[9]

the violin and piano parts, although the latter plays usually only the accompanimental figures. In
Op. 6 No. 1, rubato occurs also in m. 65, which is a repetition of the music in m. 9 (see Ex. 7.14
and Pl. XVI).

[8] Hedley, *Selected Correspondence*, 316, 318–19, and 324–5; see also 322. For a programme
and review, see Atwood, *Fryderyk Chopin*, 170 and 250.

[9] *Six mazourkes de F.ic Chopin arrangées pour la voix par M.me Pauline Viardot, paroles de
Louis Pomey* (Paris: Ancienne Maison Meissonnier, E. Gérard, [1866]), H.1774.j.(23.) in BL.
The Frederic Chopin Society in Warsaw (Towarzystwo im. Fryderyka Chopina) possesses a copy
of the set published by Gebethner & Wolff in Warsaw (p. 10 in the 2nd song gives a printer's
date of 1887). According to Eigeldinger, in *Chopin: Pianist and Teacher*, 188, the same firm
published a set of 15 in 1899. I have received from the Frederic Chopin Society a copy of No. 9,
which bears the date of 1897.

Plate XVIII shows the opening two pages of her transcription of Op. 6 No. 1 (on Plate XVI). Viardot has changed the key to F minor and added an introduction from measures 13–16 of the original. The word *rubato* appears in the same position and the F, like the F♯ in Plate XVI, is not tied back. The rubato is presumably performed in the manner her brother described in his treatise published in 1847 at Paris (Plates IV and V, Exx. 3.9–3.17).[10] Because of the sudden *forte* for the passionate exclamation on the preceding two notes, the rubato itself may favour a delay of the initial note. Six bars later she marks *rit.* in the voice, with *suivez* to indicate that the piano should follow the voice and not keep strict time as it did for the rubato.

The same sixteen-measure melody that contains the word *rubato* recurs two more times in the piece. Chopin includes no special indications at all for these repetitions. The first time Viardot repeats the text, she marks *retenez* and *suivez* at the beginning of the measure that includes 'Hélas' and indicates 'a Tempo' instead of *rubato* at 'loin'. For the second repetition she writes *retenez* and *suivez* on the second syllable of 'Hélas' and 'a Tempo' on the note E before 'de toi'. She thus presents three ways of intensifying the expressive sense of the same passage.

Two of Viardot's songs are based on Mazurkas—Op. 50 No. 2 and Op. 33 No. 2—which originally had no rubato. In the first she adds the word at the place (corresponding to measure 93 of the original piece) where the parallel second half of the repetition of the main theme begins; it follow *rit.* and *suivez* for the preceding cadence. She adds *rubato* in the other song at a point (measure 105 of the original) where the final vocal repetition of the main theme commences and where the text requires adjustment for both rhetorical and expressive purposes.

Chopin's Mazurka Op. 24 No. 2 originally contained the rubato shown in Ex. 7.18. Viardot omits this rubato, but adds one of her own (Ex. 8.3*a*) in the middle section of the piece. Ex. 8.3*b* gives the same measures from Chopin's original work for comparison. The accent signs in measure 70 of Ex. 8.3*a* might indicate a delay of the G♮ on the second quarter beat and perhaps a delay or anticipation of the D♭ on the third. The same signs are used by Faure (Ex. 3.19) to indicate *stentato*. Both the accents and the rubato act to intensify the meaning of the text. The middle section commences in measure 58 *avec une expression exagérée* and with the following words:

---

[10] In a letter of 8–17 July 1848, Chopin mentions that Pauline Viardot's brother was with him in London; see Hedley, *Selected Correspondence*, 325.

Ex. 8.3 Viardot, transcription No. 9: 'La Jeune fille' from Chopin's Mazurka Op. 24 No. 2

a Viardot's version

b Chopin's version (Paris: M. Schlesinger, 1836)

'Take pity, unfeeling beauty, on my pain!
Or end my martyrdom, or I die!'
But of this delirium, it is better to laugh.
Yes, of this delirium, it is necessary to laugh.[11]

The rubato thus not only emphasizes the repetition of the words 'de ce délire', but also provides, in company with the rapid rise and fall of the melodic line, a depiction of its meaning.

In her transcription of Op. 7 No. 1, Viardot omits the original rubato in measure 49 (see Ex. 7.17) and adds a new one in measure 29 (Ex. 8.4). In this case, however, the word *suivez* appears in the piano part at the same spot (and in both Paris and Warsaw editions). If the pianist does indeed follow the singer, then, of course, we no longer have the earlier type of rubato (as explained so carefully by her brother in the first paragraph on Plate V). One wonders if this is a printer's error (since there are many cases of *ritenuto* or *retenez*

---

[11] 'Prends pitié, belle inhumaine, De ma peine! | Ou termine mon martyre, ou j'expire! | Mais de ce délire, Le mieux est de rire. | Oui, de ce délire, Il faut rire.'

Ex. 8.4 Viardot, transcription No. 4: 'Coquette' from Chopin's Mazurka Op. 7 No. 1

occurring together with *suivez*), or if this is an intentional use of the word *rubato* for the later type—a sort of transition, as it were, from the earlier meaning of rubato to the later. She also seems to indicate the later type in one of the original songs in her settings of twelve Russian poems by Pushkin, Feth, and Turgenev (1864). Here, following six measures of introduction by the piano, the voice enters with three notes of anacrusis marked *rubato*—three notes, however, which are unaccompanied and hence have no strict beats against which to produce displacement.[12]

As we noted in Chapter 3, scores of vocal music ordinarily did not contain the word *rubato*, since the device was one of the embellishments added by the singer. In such a case, however, the conductor or the keyboard player had to be warned in advance, so that he would maintain strict rhythm during the rubato. Viardot's scores, on the other hand, make it very clear what the pianist should do, but, at the same time, she seems to remove rubato from the realm of performance practice and to place it under the control of the composer. This was no doubt due to the influence of Chopin and the rubatos in his Mazurkas.

Chopin also knew the Moravian violinist Heinrich Wilhelm Ernst (1814–65) and played several times on the same concert with him.[13] Ernst was influenced by Paganini and even performed with him

---

[12] The song was published with a German trans. in *12 Gedichte* (Br. & H, 1865), No. 6: 'Die Beschwörung' (on a text by Pushkin), repr. in *Historical Anthology of Music by Women*, ed. James R. Briscoe (IUP, 1987), 154–7. For a fac. of the opening page, see Angela F. Cofer, 'Pauline Viardot-Garcia: The Influence of the Performer on Nineteenth-Century Opera', D.M.A. thesis (Univ. of Cincinnati, 1988; UM 89–8,464), 185 (Ex. 3).

[13] Atwood, *Fryderyk Chopin*, 68–9, 90, 99, 105, 119, and 132; and reviews on 223, 226, 227, 229, 233, 236, and 237. Concerning concerts in 1834–5, see George Sand's letter to Pauline Viardot from 18 Apr. 1841, trans. in Hedley, *Selected Correspondence*, 193.

in 1837.[14] Berlioz admired Ernst and preferred his more rigorous rhythmic sense to that of Chopin. For the expression of passionate feelings 'Ernst is capable of abandoning strict time', he writes, 'but only so that the underlying pulse may be felt all the more strongly when he returns to it.' He is expressive, but does not neglect the 'disciplining art of music'.[15]

In view of Berlioz's emphasis on discipline and moderation, it is perhaps surprising to discover a superlative example of rubato in Ernst's *Airs hongrois* from 1850 for violin with piano or orchestra. Ex. 8.5*a* shows the opening two measures of the theme, which

Ex. 8.5 Ernst, *Airs hongrois* (Br. & H, 1850), beginning of *Tema II*

cadences on I at the end of measure 4. The next phrase presents a varied version of the first, commencing as in (*b*). These two phrases are then repeated very softly, the first exactly, the second beginning as in (*c*) with a two-and-a-half-measure *rubatissimo*. In his *Elégie* of

[14] *New Grove*, vi, 238.
[15] *Mémoires de Hector Berlioz* (Paris: Michel Lévy Frères, 1870; repr. Gregg, 1969), 420; trans. David Cairns (Knopf, 1969), 436–7.

1840 for violin and piano Ernst marks *suivez le violon* in the piano part when the violin has accent signs, *rit.*, or *con somma espressione* followed by *in tempo*. Therefore in Ex. 8.5c, where there is no such indication, I presume that the piano plays in strict time while the violin produces an abundant amount of rhythmic robbery of the early type. Ernst was influenced, of course, by the long tradition of rubato in violin music, which involved, as we saw in Chapter 4, Franz Benda, Strinasacchi, Paganini, Baillot, and Spohr.

Another musician active in Paris while Chopin lived there was the opera composer Ferdinand Hérold. Chopin apparently did not know him personally, and he died early in 1833, the year in which Chopin published his *Variations brillantes* on a theme from Hérold's opera *Ludovic* (an unfinished work completed by Halévy and first performed in May 1833). An article on Hérold's piano music in *Le Pianiste* in 1835 mentions his frequent use of the 'means known now under the name of *rubato*'.[16] I have not been able to find the word in his scores, however; instead, one finds numerous indications of *ritenuto*, *rallentando*, *stringendo*, *accelerando*, *serrez*, *pressez*, *precipitato*, as well as *ad libitum* and *con abbandono ed espressione*.[17] Chaulieu, in his glossary in *Le Pianiste*, attributes the word *abbandono* to Hérold and defines it as 'avec abandon, sans mesure précise'.[18] In this case, we seem to be dealing with the later type of rubato, a type different from that defined by Chaulieu and different from that apparently used by those who seem to have influenced Chopin the most.

Clara Wieck and Robert Schumann

In 1835 and again in 1836 Chopin visited Leipizig and met Clara Wieck, her father, and her future husband Robert Schumann. Clara was already known as a pianist and had composed a number of pieces. In 1836 she published a set of six *Soirées musicales*, three of which— Notturno, Mazurka, and Ballade—display the influence of Chopin not only in their titles, but also in their use of rubato.[19] The word *rubato* is coupled with *stretto* in two places in the Notturno. In measure 15 (Ex. 8.6a) the syncopated and accentuated Cs sound as though they occur too soon. The rubato may act in this case to

---

[16] *Le Pianiste*, 2/15 (5 June 1835), 116: 'le moyen, connu maintenant sous le nom de *Rubato*.'
[17] I have looked at some 31 of his works in BL. The marking *con abbandono* occurs in the *Rondo romantique*, Op. 31, published in Paris around 1825.
[18] *Le Pianiste*, 1/7 (May 1834), 98.
[19] Repr. of the original publications discussed here are contained in *Clara Wieck Schumann, Selected Piano Music* (Da Capo, 1979).

Ex. 8.6 Clara Wieck, *Soirées musicales*, Op. 6 No. 2: 'Notturno' (Leipzig: Hofmeister, 1836)

exaggerate the effect already notated, with each C occurring slightly before the corresponding left-hand note—perhaps with an increasing displacement as the melody cumulates energy for the octave leap in measure 17. A similar passage in measure 43 (Ex. 8.6*b*) is notated somewhat differently; here the indication is *stretto e rubato* and the repeated Cs are not syncopated. Presumably one could again anticipate the second, third, and fourth Cs, but probably not as much as indicated by the notation in (*a*). The expressions *rubato stretto* and *rubato e stretto* may specifically mean rubato with anticipation, or they may refer to rubato occurring simultaneously with stretto.

The first Mazurka of *Soirées musicales* begins with two statements of the phrase shown in Ex. 8.7*a* accompanied by the left hand from (*b*). The varied repeat of this melody commences as in (*b*), with a momentary rubato on the third beat. Anticipation of the beat by the A in the melody might provide a renewal of energy to mark the fact of repetition; *con dolore*, on the other hand, may suggest a delay. When the same melody returns at the end of the piece, the fourth measure is marked *rubato* (Ex. 8.7*c*), in which case the entire right-hand part can be delayed on the first beat and perhaps also on the second and third. In the same piece rubato also appears on the second beat of measure 13. Here one could anticipate or delay the quarter note in the melody,

Ex. 8.7 Clara Wieck, *Soirées musicales*, Op. 6 No. 3: 'Mazurka' (Leipzig: Hofmeister, 1836)

as well as several succeeding notes that rise to a climactic tenuto note. In the Ballade from the same set, rubato occurs twenty bars from the end. In this case, the preceding two measures are marked *stretto* and *crescendo* and reach for a high point. The rubato marks the beginning of a rapidly descending line, and anticipation may be the most suitable form of displacement.

Clara Wieck's *Scherzo* Op. 10 from 1838 or 1839 contains a longer rubato passage of five measures (Ex. 8.8). It involves a singing, cello-like melody in octaves in the inner voices. The notes of the melody appear on beats 1 and 3 of each measure, the chordal accompaniment on beat 2. Each hand plays a part of the melody and a part of the accompaniment. Displacement can occur only between the two melodic notes in an octave, and I presume that the lower should be heard first, thus producing a sort of reaching motive. The lower note could precede the beat or the upper could be delayed; it might even be difficult for a listener to determine which actually occurred. Since the rubato commences in the middle of measure 231, one could displace either the octave B♭ or the following G—or both, but with different amounts. What one does here sets a pattern for the next statement of

Ex. 8.8 Clara Wieck, *Scherzo* Op. 10 (Br. & H, 1838 or 1839)

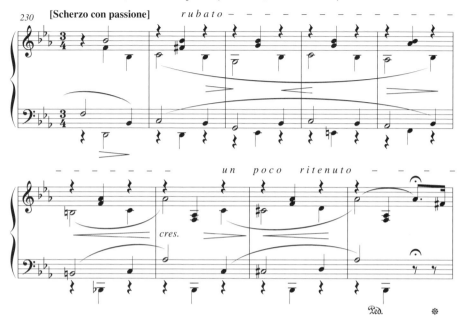

the musical motive, and I would think that an increase in the amount
of displacement and in the number of notes involved would be appropriate as one approached the A♭ in measure 236. Like Couperin's
suspension, such displacement acts like the swelling of sound on a
string instrument. One chooses the notes of special warmth, of course,
according to one's perception of the declamatory sense of the melody.
After the rubato ends, the phrase is completed finally by a tempo
change: *un poco ritenuto* without rubato.

Clara Schumann's playing was generally characterized by strict
tempo. In a review of a concert in 1856, Eduard Hanslick writes:

As compared with the common misuse of rubato [the later type], she maintains, almost without exception, a strict conformity of measure. Some may
have been surprised by her metronomical playing of the middle movement of
Chopin's D flat Impromptu [presumably the melodious D flat section in the C
sharp minor Fantaisie-impromptu Op. 66], sharply marked even in the bass.[20]

This tendency toward rhythmic strictness may be one more reason to
presume that her rubato was the earlier type. This is suggested also by

---

[20] Eduard Hanslick, *Geschichte des Concertwesens in Wien*, ii: *Aus dem Concertsaal* (Vienna,
1870; repr. Gregg, 1971), 105; and 2nd edn. (Vienna and Leipzig: Wilhelm Braumüller, 1897),
113. Quotation from *Vienna's Golden Years of Music 1850–1900*, trans. and ed. Henry
Pleasants (S & S, 1950), 42–3.

the notation in Ex. 8.6*a*, as well as by the location of the word in Ex. 8.7 near the beginning of a repeated melody. In addition, her father seems to be describing an unsuccessful application of the earlier type of rubato when he refers in his book of 1853 to an overly sentimental pianist who employs 'this *rubato* and distortion of musical phrases' and to an incompetent one who plays with 'continual *rubato* and unmusical syncopation and anticipation'.[21] It is curious to note that the last composition in which Chopin included the word *rubato* was written in 1835, the very year he visited the Wiecks and the year in which Clara probably began to write the pieces in Exx. 8.6 and 8.7 which were published the following year.

Robert Schumann also met Chopin in 1835. He had already expressed his admiration in 1831 in an essay in the *Allgemeine musikalische Zeitung* and continued to support his music as editor of the *Neue Zeitschrift für Musik* from 1834 to 1844.[22] Although he imitated Chopin's style in one piece from the *Carnaval*, which he wrote during 1834 and 1835,[23] his compositions, in general, were not influenced by Chopin in the way that Clara's were. He was more inclined toward the literary, the programmatic, and the autobiographical. He did share with Chopin, however, a special interest in singing melodies, arpeggiated accompanimental figures, and brief, simple forms.

Although Schumann never employed the word *rubato*, he continued to indicate tempo fluctuations related to the later type,[24] and notated some rhythmic robbery that resembles the earlier. Expressive delays occur as syncopated notes in his *Papillons* when the initial melody recurs in measure 45 of the Finale. An anticipation appears as a small note tied over a barline in the third full measure of *Humoreske*.[25] More complex anticipation commences in measure 299 of the same

[21] Friedrich Wieck, *Klavier und Gesang* (Leipzig, 1853), trans. Henry Pleasants in *Monographs in Musicology*, ix: *Friedrich Wieck, Piano and Song (Didactic and Polemical)* (Stuyvesant, New York: Pendragon Press, 1988), 100 and 139. For a different trans. by Mary P. Nichols, see *Piano and Song* (Ditson, 1875), 146 and 177.

[22] Robert Schumann, *Gesammelte Schriften über Musik und Musiker* (Leipzig, 1854); see the selections in *On Music and Musicians*, ed. Konrad Wolff, trans. Paul Rosenfeld (Norton, 1946), 126–46.

[23] *Robert Schumann's Werke*, Serie VII, ed. Clara Schumann (Br. & H, 1879–87; repr. Gregg, 1968), *Band* ii, 14–15 (piece entitled 'Chopin'). Note that series and page numbers are different in the Dover repr. of the *Piano Music of Robert Schumann* in 3 vols. (1972–80).

[24] According to Rosen, Schumann's use of the later type of rubato 'must have been very different from Chopin's' (he cites the 'continuous rubato' implied at the opening of the *Davidsbündler*, No. 7) and 'closer to Beethoven's'.

[25] In some of the MS sources and early edns., the small note is tied; in others it is not. See the comments in *Humoreske* ed. by Hans Joachim Köhler (Leipzig: Edition Peters, 1981) 47, and by Wolfgang Boetticher (Henle, 1989), p. iv.

piece, where the three-note groups in the right hand (marked 'Wie ausser Tempo') sound, because of unusual dissonances with the left-hand chords (marked 'Im Tempo'), as though they occur three sixteenth beats too soon.[26]

Schumann also displays a notated type of robbery in the rhythmic transformations in *Carnaval*. The work is subtitled *Scènes mignonnes sur quatre notes*—the four notes in Ex. 8.9a, which spell, in German

Ex. 8.9 Schumann, *Carnaval*, Op. 9 (1837) : Rhythmic transformation

nomenclature, the name of the city where a young lady lived. These four notes, along with two other 'sphinxes', provide the pitches which are disposed in various rhythmic configurations at the beginning of most of the separate pieces (see Ex. 8.9b, c, and d). Ordinarily composers do not present the variable material as a non-rhythmic tone-row, so that one would be more accustomed to relating each transformation to the first one. Thus, time is stolen from the A, E♭, and B in Ex. 8.9c when compared with (b), whereas the C is lengthened. At the same time, a change of accentuation and metre causes an individual note such as C to play quite a different role in (d) than it does in (c). Although in this case only a few notes at the beginning of a melody are involved, such a procedure continues the tradition that reaches back, as we have seen, to the motet tenors of the Middle

[26] *Schumann's Werke*, vɪɪ/iv, 9. Geoffrey Chew includes the passage as an example of 'melodic rubato' in *The New Grove Dictionary of Musical Instruments* (1984), iii, 266.

Ages (Ex. 2.15) and the paraphrase technique and dance pairs of the Renaissance (Exx. 2.16 and 2.17), and to the works of J. S. Bach (Ex. 2.19) and Beethoven (Ex. 6.8). The device obviously inspired Schumann's inventive imagination, and it provides a subtle sort of unity for the work as a whole. As we will see, it becomes increasingly important in the works of later romantic composers.[27]

Liszt as Performer, Conductor, and Writer

Of all the musicians acquainted with Chopin, however, it was Franz Liszt who was influenced most by the concept of rubato. Liszt lived until late in the century, from 1811 to 1886, and transmitted his ideas to countless colleagues and students. He was involved with rubato as a performer, conductor, and writer, and as an editor, composer, arranger, and teacher. He is a key figure in the history of tempo rubato, for it was his diverse and changing concept of the device which influenced later performers and composers.

The year 1832 seems to mark the beginning of Liszt's involvement with rubato. In April of that year he first heard a performance by Paganini,[28] whose rubato, as we have seen, was later described by García, Farrenc, and Kleczyński. Chopin arrived in Paris in September 1831, and Liszt attended his début concert in February 1832.[29] During this same year Chopin published in Leipzig (and later in Paris and London) his first four works that contain the word *rubato*: the Mazurkas Opp. 6 and 7, the Trio Op. 8, and the Nocturnes Op. 9. For over three years the two men were closely involved with one another. During the period from 1832 to 1835 they performed together in public concerts, either in works for piano four hands or works for two or more pianos,[30] and both performed at social gatherings in private homes.[31] They wrote joint letters together to Ferdinand Hiller.[32]

[27] *Schumann's Werke*, vii/ii, 2–29. The chant excerpts in Exx. 2.15*a* and 2.16*a*, with their succession of equal note values, look remarkably like the sphinx in Ex. 8.9*a*. In the case of chant, however, this notation indicates the method of performance, whereas Schumann's sphinxes, which appear between 2 of the pieces near the middle of the work, are there, presumably, not to be performed, but simply to reveal the method of composition.

[28] Alan Walker, *Franz Liszt*, i: *The Virtuoso Years (1811–1847)* (Knopf, 1983), 173.

[29] See his description of the concert in Atwood, *Fryderyk Chopin*, 60–1.

[30] Ibid. 68–9, 73, 79, 83, 89–90, and 113; for reviews, see 221–3. See also Niecks, *Chopin as a Man and Musician*, i, 253 (concerning concerts in Dec. 1832 and Apr. 1833), and 279–80 (a *matinée musicale* in 1834 at Pleyel's rooms), and 281 (a concert in Apr. 1835). Concerning the 1834 performance, see also *Portrait of Liszt by Himself and His Contemporaries*, ed. Adrian Williams (OUP, 1990), 64–5.

[31] See Niecks, *Chopin as a Man and Musician*, i, 244 and 255–6.

[32] Hedley, *Selected Correspondence of Fryderyk Chopin*, 112 (letter of 2 Aug. 1832) and 117–18 (letter of 20 June 1833 from Chopin, Liszt, and Auguste Franchomme, the cellist).

Chopin, who had a special talent for mimicry, included Liszt as one of the pianists whose playing and manner he imitated for the amusement of friends.[33] Even in later years, after Liszt had left Paris, the two maintained contact from time to time.[34] As we have already seen, Liszt describes Chopin's rubato in his biography of 1852, and in an edition of Weber's sonatas published in 1870 he mentions that the expression *tempo rubato* was not used (or, perhaps, not usual) before Chopin (see the footnote on Plate XIX). Apparently, then, Liszt consciously learned about rubato from Chopin and knew nothing of its previous history. Considering the fact that Liszt first heard Paganini and first became involved with Chopin during 1832, it is perhaps not surprising to discover that the first composition in which Liszt wrote the word *rubato* was the *Grande fantaisie de bravoure sur la Clochette de Paganini*, written during this same year and based on the Italian melody 'La campanella' which Paganini used in his B minor Violin Concerto.[35]

Paganini influenced Liszt mainly as a performer, however, and inspired him to cultivate to a superlative degree the theatrical and virtuosic elements already natural to him.[36] Liszt had studied with Czerny, who in turn was a student of Beethoven. This represented an approach to music quite different from that of Chopin, who was influenced more by Field, a pupil of Clementi, and by Hummel, the student of Mozart. Chopin seldom practised; Liszt repeated exercises for hours on end.[37] Chopin preferred the pianos built by Pleyel, with their light, responsive touch, Liszt, the instruments of Érard.[38] Chopin was the introspective dreamer, Liszt the flamboyant showman. Chopin enjoyed performing only for small, intimate groups, whereas Liszt preferred large, impersonal audiences. Even though Liszt was totally different from Chopin in some very conspicuous ways, however, he was apparently attracted by the rubato in Chopin's playing and adopted it as part of his own arsenal of expressive devices. In order

[33] Niecks, *Chopin as a Man and Musician*, ii, 71 and 145–6.

[34] Hedley, *Selected Correspondence*, 141 (an invitation from Chopin, dated 13 Dec. 1836, to a social gathering at his home, where Liszt will be one of those to perform) and 259 (a letter to his family in Dec. 1845 mentioning that 'Liszt has also called'). See Serge Gut, 'Frédéric Chopin et Franz Liszt: Une amitié à sens unique' in *Sur les traces de Frédéric Chopin*, ed. Danièle Pistone (Paris: Librairie Honoré Champion, 1984), 53–68.

[35] Walker, *Franz Liszt*, i, 175. Paganini includes the melody in the last movement of his 2nd Violin Concerto in B minor, Op. 7, which was written in 1826 and published in 1851. As we saw in Ch. 7, Chopin, after hearing Paganini in 1829, also composed a piece based on a theme used by that composer.

[36] See Liszt's letter of 2 May 1832, printed in Walker, *Franz Liszt*, i, 173–4, in which he expresses his enthusiasm for Paganini's playing.

[37] See Eigeldinger, *Chopin: Pianist and Teacher*, 27, also 23 and 193; and Walker, *Franz Liszt*, i, 175 and 297.

[38] Eigeldinger, *Chopin: Pianist and Teacher*, 25–6 and 92 n. 9.

PLATE XIX. Carl Maria von Weber, *Ausgewählte Sonaten & Solostücke für das Pianoforte, bearb. von Franz Liszt* (Stuttgart: Cotta, 1870), last movt. (Rondo. Moderato e molto grazioso) of Sonata No. 2 in A flat major, mm. 167 ff.

to understand what Liszt meant by the word *rubato*, we shall first explore his role as performer, conductor, and writer. In the following section, we shall investigate the way he uses the word in his scores.

Liszt performed extensively during the years 1839 to 1847. The few sources that use the word *rubato* in connection with his playing, however, give little insight into the meaning of the term. Czerny, in his piano method from around 1839, reveals that his former student employed more than one type of rubato:

The very frequent application of each kind of *tempo rubato* is so well directed in Liszt's playing, that, like an excellent declaimer, he always remains intelligible to every hearer, and therefore invariably makes the greatest impression on all classes of the public.[39]

In the same treatise, as we noted in Chapter 5, Czerny complains about '*tempo rubato* (that is, the arbitrary retardation or quickening of the degree of movement)'. His mention of 'each kind of *tempo rubato*' may therefore refer to retardation as one type and quickening as another. On the other hand, it may refer to our earlier and later types, to the other types we identified in the music of Chopin, or to still other types peculiar to Liszt. Even less information occurs in a review from 1842, which states that 'one willingly tolerates some tempo rubato and excess of speed from this virtuoso, who understands how to discover the heart of the composition.'[40] Finally, in a letter from 1854 to his pupil William Mason, Liszt himself refers facetiously to rubato in order to urge moderation in the consumption of cognac:

Let me again recommend *measure* to you, an essential quality for musicians. In truth, I am not too well *qualified* to extol the *quantity* of this *quality*, for, if I remember rightly, I employed a good deal of *Tempo rubato* in the times when I was giving my concerts...[41]

There are numerous reports, on the other hand, that refer, without using the word *rubato*, to the general rhythmic flexibility of Liszt's performing style. Madame Auguste Boissier observed her daughter's lessons with Liszt in the winter of 1831–2, during the period when he was first getting to know Chopin and just before his first contact with

[39] Carl Czerny, *Complete Theoretical and Practical Piano Forte School* (London, dedication dated 1839), copy in BL, 2nd suppl., 28.

[40] *AMZ* 44/7 (Feb. 1842), 144: 'Einiges tempo rubato und Übermaass der Schnelligkeit lässt man sich von diesem Virtuosen gern gefallen, der den Kern der Komposition aufzufinden versteht.'

[41] *Franz Liszt's Briefe*, ed. La Mara (Ida Maria Lipsius) (Br. & H, 1893), i, 186 (letter No. 129 of 14 Dec. 1854). My quotation comes partly from *Letters of Franz Liszt*, ed. La Mara and trans. Constance Bache (London: H. Grevel, 1894), i, 226, and partly from William Mason, *Memories of a Musical Life* (New York: The Century Co., 1901), 179.

Paganini. She describes the intense emotions he expressed and the freedom of his rhythm:

'I don't play according to the measure', he said. . . . Music must not be subject to a uniform balance; it must be kindled, or slowed down with judgement and according to the meaning it carries. This goes for all romantic music of the present time. The old-fashioned classics must be rendered with greater regularity.[42]

Liszt sometimes applied considerable rhythmic flexibility, however, even in his playing of the works of Beethoven. Berlioz complains of his 'excessive rhythmic fluidity and the liberties he sometimes takes in interpreting the great masters'.[43] In a review from 1837, Berlioz compares two different renditions by Liszt of the 'Moonlight' Sonata: in an earlier performance in 1829 or 1830 Liszt 'added trills and tremolo, rushed and slowed the pace', whereas in 1837 the same work 'that he had previously disfigured so curiously, stood out in its sublime simplicity'.[44] In this same year, Liszt himself published a statement repudiating such liberties:

I frequently played the works of Beethoven, Weber, and Hummel, . . . and I confess to my shame [that] in order to extract bravos from a public ever slow to perceive things of beauty, I had no scruples about changing the tempo and the composer's intentions. I even arrogantly went so far as to add a lot of brilliant passages and cadenzas . . . You will never believe how much I deplore those concessions to bad taste . . .[45]

Charles Salaman, however, described a recital Liszt gave in London three years later:

---

[42] Mme. Auguste Boissier, *Liszt pédagogue: Leçons de piano données par Liszt à Mlle. Valérie Boissier en 1832* (Paris, 1927); trans. in *The Liszt Studies*, ed. and trans. by Elyse Mach (AMP, 1973), pp. xv–xvi (lessons 11 and 12 on 31 Jan. and 7 Feb. 1832).

[43] *Memoirs of Hector Berlioz*, trans. and ed. David Cairns, 551. The quotation is included in an erased but still legible passage on the MS, written, presumably, sometime between 1848, the date of the preface, and 1870, the date of publication.

[44] *Journal des débats*, 12 Mar. 1837, quoted by Jacques-Gabriel Prod'homme in *Les sonates pour piano de Beethoven, 1782–1823* (Paris: Delagrave, 1937), 125–6; trans. by William S. Newman in *Performance Practices in Beethoven's Piano Sonatas: An Introduction* (Norton, 1971), 15, and in 'Liszt's Interpreting of Beethoven's Piano Sonatas', *MQ* 58 (1972), 194. Concerning the 'rhythmic liberties' in the last movt. during a performance in 1840, see Carl Reinecke, '*Und manche liebe Schatten steigen auf*': *Gedenkblätter an berühmte Musiker* (Leipzig, 1910), 11–13, trans. by Adrian Williams in *Portrait of Liszt*, 145.

[45] *Revue et gazette musicale de Paris*, 12 Feb. 1837; quoted by Robert Wangermée in 'Tradition et innovation dans la virtuosité romantique', *Acta musicologica*, 42 (1970), 20. My English trans. suggested partly by Charles Suttoni in Franz Liszt, *An Artist's Journey, Lettres d'un bachelier ès musique* (UChP, 1989), 17–18, and partly by William S. Newman in *MQ* 58 (1972), 198. For a German trans., see *Gesammelte Schriften von Franz Liszt*, ed. and trans. Lina Ramann, ii (Br. & H, 1881), 129 (in the 2nd 'Reisebriefe eines Baccalaureus der Tonkunst' to George Sand, Jan. 1837).

Yet, magnificent as was Liszt's playing, the works of such great masters as Beethoven, Weber, and Hummel needed no such embellishments as the pianist introduced. I suppose, however, that these excesses of virtuosity belonged to Liszt's flamboyant personality; his temperament compelled them.[46]

Even during the period when he was most active as a virtuoso pianist, Liszt began his activities as a conductor. He conducted in Budapest in 1840, in various German cities from 1844 on, and finally accepted a full-time position at Weimar in 1848. Following a performance of Beethoven's Ninth Symphony at Karlsruhe in 1853, a reviewer wrote:

The unanimous opinion was that he [Liszt] was not fit to wield the baton...It is not merely that in general he does not mark the beat..., but rather that by his baroque animation he continually, and sometimes dangerously, causes the orchestra to vacillate. He does nothing but keep changing the baton from one hand to the other—sometimes, indeed, laying it down altogether—giving signals in the air with this or that hand, or on occasion with both, having previously told the orchestra 'not to keep too strictly to the beat' (his own words at a rehearsal)...[47]

Liszt replies to such criticism in a letter in which he describes the new manner of performance required by the works of Wagner, Berlioz, and Schumann, as well as the late works of Beethoven. He feels that a strict beat often 'clashes with the sense and expression' and that conductors should not perform 'like a sort of windmill'. 'We are pilots', he writes, 'not mechanics.'[48] Ambros summarizes the rhythmic effect of these new ideas in conducting as follows: 'Anyone who has ever seen Liszt conduct an orchestral work will recall how he would push forward in one passage and hold back in another, here maintaining the tonal flow at an even pace for a time, there suddenly interrupting the movement.'[49]

Liszt thus transfers to the conducting of orchestral music the sort of expressive rhythmic freedom that he had earlier cultivated as a pianist. With the possible exception of Czerny's book, there is no source

[46] 'Pianists of the Past', *Blackwood's Edinburgh Magazine*, 170/1031 (Sept. 1901), 315.

[47] *Niederrheinische Musikzeitung* (1853), 139; see Walker, *Franz Liszt*, ii: *The Weimar Years* (Knopf, 1989), 280–1 and n. 32. The article, signed 'H', is generally presumed to have been written by Ferdinand Hiller. The trans. comes from Harold C. Schonberg, *The Great Conductors* (S & S, 1967), 162.

[48] *Franz Liszt's Briefe*, i, 144–5 (letter No. 104, dated 5 Nov. 1853, to Richard Pohl); trans. by Constance Bache, *Letters of Franz Liszt*, i, 175–6. See also Walker, *Franz Liszt*, ii, 281.

[49] August Wilhelm Ambros, *Culturhistorische Bilder aus dem Musikleben der Gegenwart*, 2nd edn. (Leipzig, 1865), 254; trans. from Marion Louise Perkins, 'Changing Concepts of Rhythm in the Romantic Era: A Study of Rhythmic Structure, Theory, and Performance Practices Related to Piano Literature', Ph.D. diss. (Univ. of Southern California, 1961; UM 61–6,302), 273–4.

known to me, however, either by Liszt himself or by those who describe his performance, in which Liszt's tempo flexibility is referred to as *rubato*. Even in his book on Chopin (both the original version in 1852 and the later edition of 1879), Liszt uses poetic images in his description of Chopin's rubato, as we saw in Chapter 7, which could refer to other musical effects just as well as to tempo changes.

It is significant, I think, that in his book on gypsy music Liszt frequently mentions tempo flexibility, but never uses the word *rubato*. At the end of 1839 and early 1840 Liszt returned briefly to his native Hungary after an absence of sixteen years. During this visit he became reacquainted with the music of the gypsies, which inspired him later to compose works such as the Hungarian Rhapsodies and to write the book entitled *Des Bohémiens et de leur musique en Hongrie* (first published in 1859, with a later edition in 1881). Referring to gypsy music in general, Liszt speaks of the 'extremely flexible [*fléchissans*] rhythms', of 'rhythms with their vacillations [*vacillations*]', and of the 'liberty [*liberté*] and richness' of rhythm 'distinguished both by a multiplicity and a flexibility [*souplesse*] nowhere else to be met with in the same degree'. He describes the rhythms as full of 'fire, flexibility [*souplesse*], dash, undulation [*ondulation*]', and assuming 'a gait which is not only free [*franche*] in itself but freely [*franchement*] treated'. One does not find here, as we noted in Chapter 7, the trepidations, swayings, hesitation, and agitation of the waltz or the mazurka. He adds in regard to the rhythms, however, that 'their diversity is infinite', 'their rule is to have no rule', and they are 'flexible [*flexibles*] as the branches of the weeping willow bending under the sway of the evening breeze'.

In the gypsy orchestra the first violin and the cimbalom are the most conspicuous; the latter 'supplies the rhythm, indicates the acceleration [*accélération*] or slackening [*ralentissement*] of time, as also the degree of movement'. In the *friska*, the fast second half of a gypsy piece, 'accelerations . . . , both sudden and gradual, lead up to rhythms too furious and excited ever to be applied to any of the dances used in civilized society. . . . There is something brusque, abrupt, irregular, and intermittent; it is interrupted by sudden starts, stops suddenly and then rushes off again with redoubled fury.'[50]

I have included in the foregoing quotations the French words that Liszt employed to indicate tempo changing in order to emphasize the

[50] Franz Liszt, *The Gipsy in Music*, Englished by Edwin Evans (London: William Reeves, [1926]), ii, 300, 304, 313, 314, and 330. These page numbers correspond with *Des Bohémiens et de leur musique en Hongrie, nouvelle édition* (Br. & H, 1881; repr., Bologna: Forni, 1972), 394, 398–400, 369, 372, and 526, respectively. Some sections of the English trans. appear in a different order in the French version.

fact that the word *rubato* does not occur. This becomes even more remarkable when we note that later authors tend to associate rubato regularly with gypsy music. Franklin Taylor, in his article on 'Tempo rubato' in the first edition of *Grove's Dictionary* in 1883, gives tempo flexibility as his second definition and states that 'perhaps the most striking instances of the employment of *tempo rubato* are found in the rendering of Hungarian national melodies by native artists.'[51] In 1885 Adolph Christiani describes Hungarian music, in which the various musical elements 'are so strangely intermingled by surprising changes in tempo, in rhythm, and accentuation, that the chief characteristic in the movements of their music... is an almost constant rubato, an alternate change of extremes in tempo, of retardation and acceleration'.[52] Modern scholars, such as Bálint Sárosi, still refer to the rubato nature of gypsy music.[53]

Liszt considered gypsy music to be the true Hungarian folk music, an opinion which caused controversy when the book appeared, and one which was corrected later by the research of Bartók and Kodály. The music that attracted the attention of Liszt, Brahms, and other composers, was the so-called *verbunkos*, which originated late in the eighteenth century and consisted of gypsy arrangements of music by contemporary Western composers.[54] In any event, Liszt and all the later writers who describe this music agree on its spectacular rhythmic freedom. Liszt, however, did not apply the expression *tempo rubato* to it. Neither did Liszt or others employ the term, apparently, for equally conspicuous tempo changing in his own playing or conducting. These circumstances suggest that for Liszt rubato referred to something different from general freedom of tempo. We shall investigate the precise meaning of his rubato, then, in the next section, which is devoted primarily to his activities as a composer.

Liszt as Composer, Editor, Arranger, and Teacher

I have found the word *rubato* forty-three times in forty different works composed or arranged by Liszt from 1832 to 1870 and pub-

[51] *A Dictionary of Music and Musicians*, ed. George Grove (London: Macmillan), iii (1883), 85. Similarly, Paderewski writes, in an article in Henry T. Finck's *Success in Music* (New York: Charles Scribner's Sons, 1927; copyright 1909), 458: 'It is Tempo Rubato which makes the Hungarian dances so fantastic, fascinating, capricious'.

[52] *The Principles of Expression in Pianoforte Playing* (New York: Harper & Bros., 1885), 97.

[53] See his article in *New Grove*, vii, 865 (Ex. 1), and especially his book *Cigányzene* (Budapest, 1971), trans. by Fred Macnicol as *Gypsy Music* (Budapest: Corvina Press, 1978), 26, 115, and 245–6.

[54] See *New Grove*, vii, 868; Sárosi, *Gypsy Music*, 114–17; and Walker, *Franz Liszt*, i, 334–42.

lished between 1834 and 1871.[55] These works display, at the same time, a characteristic romantic increase in the number of terms that indicate mood, dynamics, touch, or tempo. The style of execution is also suggested sometimes by programmatic titles, excerpts from literary works, or references to other works of art. A programmatic image could evoke in the performer a specific mood, which, in turn, would cause him to play the music in a special manner. This was also an important aspect of Liszt's teaching.

Liszt indicates tempo changes with the usual terminology: *rallentando*, *ritardando*, and *ritenuto*, or *accelerando*, *stretto*, and *stringendo*. He experimented briefly with special signs to represent tempo flexibility on an even subtler level: a single line above the staff indicated slowing, double lines closed off at both ends to form a rectangle meant accelerating, and a short open set of double lines designated a hold of briefer duration than a fermata (compare Türk's signs described in Chapter 5). Two of the *Grandes études* composed in 1837 (Nos. 5 and 9) include both the word *rubato* and, elsewhere in the scores, these special signs. When Liszt revised the pieces in 1851, however, he simply removed the signs and left their effect unmarked.[56] The works of Liszt that include rubato also contain elsewhere a wide range of terms referring to free rhythm: *a capriccio*, *a piacere*, *con abandono*, *senza tempo* or *senza tempo deciso*, *recitativo* or *recitando*, *cadenza* or *quasi cadenza*, *improvvisato* or *quasi improvvisato*. The original version of Liszt's *Harmonies poétiques et religieuses*, for example, contains *a capriccio* in measure 101, *Reciativo* in measure 102, *rubato* in measure 111, and *Senza tempo* in measure 122.[57]

Thus it is clear from the musical scores themselves, that when Liszt wrote the word *rubato* he did not mean tempo changes of the sort indicated by *rallentando* or *accelerando*, nor the subtler type which he marked for a time with special signs—the type ordinarily added by a performer without any special instructions—, nor even the free type appropriate in a cadenza, a recitative, or an improvisation. Furthermore, he does not include the word *rubato* extensively in his works

[55] Additional examples will no doubt come to light as new volumes of the *Neue Ausgabe* appear, especially in the series entitled *Freie Bearbeitungen* (Free Arrangements). At the time of writing only the 7th of 15 volumes in this series had appeared. Editio Musica Budapest publishes each volume simultaneously as part of the complete works and as part of the performing series *Klavierwerke* (Piano Works).

[56] For the signs, see the footnote on p. 4(36) of *Franz Liszts musikalische Werke* (Br. & H, 1907–36; repr. Gregg, 1966), ii/i. Compare the 1st page of Nos. 5 and 9, for example, with the corresponding pages of the *Études d'exécution transcendante* in ii/ii or in the *Neue Ausgabe sämtlicher Werke* (Budapest, 1970–), i/i. See also Walker, *Franz Liszt*, i, 310, where he states that 'these . . . symbols gave Liszt exact control over tempo rubato' (using the term, of course, in the later sense).

[57] *Neue Ausgabe*, i/ix, 147.

based on gypsy music: only three of his forty-three marked rubatos occur in Hungarian compositions, and then for passages not unlike those in his other works.[58]

If Liszt's rubato was not equivalent to any of the tempo modifications marked in his scores, nor to the rhythmic flexibility in gypsy music or in his own playing or conducting, then it must refer to some other musical effect. Although his usage of the term seems to change somewhat between 1832 and 1870, some important clues are contained, I think, in two later sources of information: Carl Lachmund's description of a lesson in 1882, and Liszt's edition of the Weber sonatas in 1870. Lachmund, who studied with Liszt from 1881 to 1884, reports in his diary as follows concerning a lesson on 12 May 1882 at which another student played Liszt's *Consolation* No. 6 in E major:

On this occasion we received an important insight into the Lisztian rubato—that is, the subtle variations of tempo and expression within a free declamation, which are entirely different from Chopin's rubato of hastening and lingering. The Liszt rubato is more like a sudden, light suspension of the rhythm on this or that significant note, so that by this means the phrasing is clearly and convincingly brought out. In his playing Liszt seemed to pay little attention to a steady beat, and yet neither the aesthetic symmetry nor the rhythm was disturbed.[59]

Liszt's edition of Weber's sonatas was published by J. G. Cotta of Stuttgart in 1870. It was part of a larger project entitled *Instructive Ausgabe klassischer Klavierwerke*, whose general editor was Siegmund Lebert. Letters between Liszt and Lebert beginning in 1868 document the progress of the work. They must have discussed rubato at some point, for Lebert states in a letter to Liszt of 29 November 1869, that 'it pleases me especially that . . . you agree with me regarding the tempo rubato.'[60] In a letter of 10 January 1870 Liszt reports to Lebert

---

[58] See the Hungarian Rhapsodies Nos. 1 and 12 (composed in 1846 and 1847) in the *Neue Ausgabe*, I/iii, 10, and I/iv, 33, and the earlier *Magyar rhapsodiák (Rapsodies hongroises)*, *Cahier* 10 No. 17 in the *Neue Ausgabe*, I/xviii, 156.

[59] Carl V. Lachmund, *Mein Leben mit Franz Liszt: Aus dem Tagebuch eines Liszt-Schülers* (Eschwege: G. E. Schroeder-Verlag, 1970), 62: 'Dabei erhielten wir einen wichtigen Einblick in das Liszt'sche Rubato, das heisst, die kleinen Abweichungen im Tempo und im Ausdruck im freien Vortrag, die ganz verschieden vom Chopin'schen Rubato des Eilens und Zögerns sind. Das Liszt'sche Rubato ähnelt mehr einem plötzlichen, leichten Anhalten der Zeit bei dieser oder jener bezeichnenden Note, so dass damit die Phrasierung erst in richtig verständlicher und überzeugender Weise herausgebracht wird. Beim Spielen schien Liszt überhaupt auf das Zeitmass wenig zu achten, und doch wurden dabei weder die ästhetische Symmetrie noch der Rhythmus gestört.' My trans. occasionally follows Eigeldinger, *Chopin: Pianist and Teacher*, 122 n. 100.

[60] *Briefe hervorragender Zeitgenossen an Franz Liszt nach den Handschriften des Weimar Liszt-Museums*, ed. La Mara, ii (Br. & H, 1895), 339: 'Dass Sie . . . bezüglich des Tempo rubato mit mir übereinstimmen, freut mich ganz besonders.'

that he has just sent back the final proofs, to which he has added many fingerings and pedalling marks in order to make the edition practical for teachers and players. He also states:

With regard to the deceptive [*verfängliche*] *Tempo rubato*, I have settled the matter provisionally in a brief note (in the finale of Weber's A flat major Sonata); other occurrences of the *rubato* may be left to the taste and momentary feeling of gifted players. A metronomical performance is certainly tiresome and nonsensical; time and rhythm must be [adapted to and] identified with the melody, the harmony, the accent and the poetry. . . . But how indicate all this? I shudder at the thought of it.[61]

Plate XIX shows the page in the Rondo of Weber's Second Sonata, Op. 39, in which the following footnote appears:

The designation *Tempo rubato*, which was not used [or usual] before Chopin, would be appropriate at this and other places in Weber. It is left to the taste and feeling of the player to perform correctly this seductive [*verführerische*] *Tempo rubato*.

The asterisk which refers to this footnote is precisely marked above the middle of the fourth measure in the third line of the score. A *grazioso* melody commences softly in the third measure of the first line, with the two dotted eighth notes in each measure being dissonant appoggiaturas. The dynamics increase rapidly to a high point in the second measure of the third line, suddenly followed by *p* and *ritenuto*, which immediately precede the tempo rubato. Since this is the only spot in the volume marked for rubato, Liszt must have felt that it was particularly appropriate here and that it would be an especially instructive example in this practical edition for teachers and players. This is the most precisely located rubato Liszt ever marked, since the terms *rubato* or *tempo rubato* in his own compositions usually span several notes of a melody. In most cases Liszt clearly indicates his additions to Weber's score: *molto legato* in smaller letters in the first line, *subito* in the third, and the variant above the staff in which the melody is doubled in octaves. In the same line, he has apparently also

---

[61] *Franz Liszt's Briefe*, ii, 156 (letter No. 98); trans. by Constance Bache in *Letters of Franz Liszt*, ii, 194. I have added brackets to show an important German word as well as 3 English words she added to the original. William Newman's partial quotation of this passage in 'Liszt's Interpreting of Beethoven's Piano Sonatas', *MQ* 58 (1972), 201, is misleading, for he places an editorial colon after the expression *Tempo rubato* and deletes everything from there to the sentence commencing 'A metronomical performance'. In any event, these remarks of Liszt did not refer to the Bülow–Lebert edition of the Beethoven sonatas. For earlier letters from Liszt to Lebert concerning the Weber edn., see *Franz Liszt's Briefe*, ii, 126 (10 Sept. 1868), 129 (19 Oct. 1868), and 133 (2 Dec. 1868).

added the dashes following *riten.* and the dashes that extend the word *crescendo.*[62]

Both Lachmund's comments and the Weber score suggest that Liszt's rubato happens suddenly and instantaneously, as a sort of intensely expressive articulation of precise and special moments in the music. Both sources seem to link the device with a slight suspension or modification of the tempo. Its effect is clearly so extraordinary that it can be described only by such powerful adjectives as 'deceptive', 'seductive', or—if we may infer it from Liszt's letter concerning Mason's drinking—'intoxicating'. With these general impressions in mind, we shall now explore the specific examples of rubato in the compositions and arrangements of Liszt. The changing manner in which Liszt uses rubato suggests four periods: from 1832 through 1836 (the time when he was most influenced by Chopin and Paganini), 1837 to 1849 (the years spent mostly as a touring virtuoso pianist), 1850 to around 1853 (years during which he revised many earlier scores and experienced a renewed influence of Chopin), and 1853 to 1870 (a period during which he incorporated rubato in orchestral as well as piano works). Examples from all the periods tend to favour soft dynamics, moderate tempo, and a warm expression suggested by terms such as *dolce, con amore,* and *con grazia.*

Since Liszt believed that Chopin was the first to use the word *rubato,* and since he had ample opportunity to become acquainted with Chopin's performance of the earlier type, he must have had in mind a similar method of execution when he himself first wrote the word in a score. Exx. 8.10–8.14 present examples from his early period. The melody in Ex. 8.10 first occurs nine bars earlier, marked *sempre dolce e grazioso.*[63] The third time it recurs (Ex. 8.10) the rubato can respond to the *poco animato* by nervously anticipating the G♯ in the melody, perhaps so that the last three melody notes of the bar even resemble a triplet. In addition, the B at the beginning of the following bar might be accentuated through a delay. As in Chopin's earlier rubato, the left hand, presumably, maintains a generally strict tempo.

---

[62] In the modern Schirmer edn. of the Weber sonata by Giuseppe Buonamici (1903), *cresc.* occurs below the 2nd and 3rd 16th notes of the group on the 2nd half of the bar, whereas in the Peters edn., prepared by Louis Köhler and Adolf Ruthardt, it appears at the beginning of the next measure, below the word *agitando.*

[63] *Franz Liszts musikalische Werke,* ii/ii, 26(124). I cite Liszt's works in the *Neue Ausgabe* except in cases, such as this, in which the piece is either not planned to be included or the volume has not yet appeared. Since contemporary editions of Liszt's works were not ordinarily available, I have simply followed modern practice in matters of type style and capitalization. The *Neue Ausgabe* includes italics very seldom, whereas the older edn. uses them, in the more usual way, for most transitory effects such as *ritardando.*

Ex. 8.10 Liszt, *Grande fantaisie de bravoure sur la Clochette de Paganini*, 1832

Ex. 8.11 Liszt, *Harmonies poétiques et religieuses*, 1st version, 1833

Ex. 8.11 presents a passage in which a transition, marked *dolce con amore*, is made between 2/4 and 6/8 metre.[64] The last phrase in 2/4 ends on the first eighth beat of measure 111. Thus the beginning of the 6/8 section (which continues for many measures thereafter) can be gently but conspicuously articulated by delaying the right hand on the second eighth beat of the first triplet (the first B♭ in the melody). A number of Liszt's examples involve more than one note in the right hand at the moment of rubato. When Chopin does this (in Exx. 7.23, 7.24, and 7.26), the doubled melody in the right hand is clearly distinguished from the accompaniment in the left. In Ex. 8.11 displacement is still effective, I believe, even though the melody and its accompaniment are not rhythmically differentiated.

In Ex. 8.12 a rubato displacement of the first three notes of the melody can give a speech-like (*parlante*) emphasis to the real metrical rhythm in contrast to the contradictory three-note ostinato figure in

---

[64] *Neue Ausgabe*, i/ix, 147.

Ex. 8.12 Liszt, *Apparitions* No. 1, 1834

Ex. 8.13 Liszt, *Fantaisie sur une Valse de Schubert*, 1834

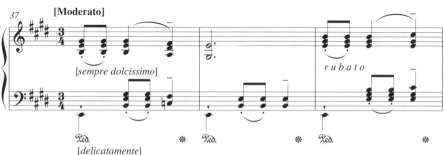

the left hand.[65] In Ex. 8.13 an anticipation of the first right-hand chord in measure 39 can articulate the recurrence of a four-bar phrase first heard eight bars earlier. Important in the execution of the rubato is the exhilarating rhythm caused by the accentuation on the third beats.[66] In Ex. 8.14 the word 'Rubato' appears above the beginning of a melody. In this case it should be treated perhaps more like Chopin's *sempre rubato*, which acts over a longer period of time. A delay of the $A^\flat$ in measure 95, and especially in measure 96, can contribute to the sad, weeping effect desired. If the rubato is to continue, then a delay of the high $A^\flat$ in measure 98—and perhaps the anticipation of the three preceding notes—can emphasize the accentuation marked on this note.

Most of these early rubatos by Liszt seem to reflect the practice of Chopin. The term *rubato* appears by itself, without the word *tempo*, and seems most often to refer to momentary displacement on one or

---

[65] Ibid. i/ix, 12.
[66] Ibid. i/ix, 23. For the original Schubert Waltz, from *36 Originaltänze*, Op. 9, see *F. Schuberts Werke* (Br. & H, 1884–97), XII/i, 12 (No. 33); repr. in *Schubert Complete Works* (Dover, 1965), v.

Ex. 8.14 Liszt, *Grande valse di bravura*, 1836

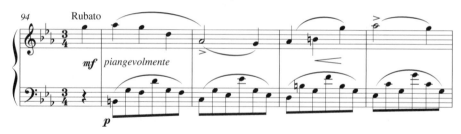

only a few beats. Like Chopin, Liszt uses the device to articulate the repetition of a unit (compare his rubato in Ex. 8.13 with Chopin's in Ex. 7.14, for example), to increase the singing quality of a melody (compare Exx. 8.12 and 7.21), to emphasize expressive dissonance (Exx. 8.14 and 7.23), and to create a special mood (Exx. 8.14 and 7.27). In addition, Liszt employs the same sort of momentary rubato for dramatic reasons (Ex. 8.10) and for accentuating the commencement of a new rhythmic movement (Ex. 8.11). In the 'Ranz de chevres', the twelfth piece in his *Album d'un voyageur* composed in 1835–6, Liszt marks *rubato* the second time a passage occurs in order to accentuate the beginning of a single three-beat phrase which follows two repeated phrases of two beats. This is all notated in 2/4 metre, and the rubato involves arpeggiated three-note chords in each hand (reminding one of Chopin's arpeggiated rubato in Ex. 7.15*a*).[67] Here, as also in Exx. 8.11, 8.12 and 8.13, the rubato concerns melody notes marked for portato touch.

During the period from 1837 to 1849 Liszt introduced a number of changes in his rubato notation. The word *rubato* may now occur in three different ways: (1) alone (as in the early period), or now sometimes combined with other words, as a momentary instruction whose termination is not indicated; (2) in company with tempo changing terms which require 'a tempo' or a new marking to terminate; and (3) either alone or in combination with other words as the main tempo marking of a piece or an internal section, in effect until a new marking appears. Liszt frequently uses the full term *tempo rubato* now, influenced perhaps by German musicians such as Seyfried and Schindler (who used the term in connection with Beethoven, as we noted in Chapter 6, in 1832 and 1840), Kessler (who wrote it in a Nocturne: see Ex. 8.1 from around 1834), and Czerny (in his piano method around 1839). Liszt applies rubato now over longer spans of time and to entire melodies or sections thereof. 'Sempre rubato' or 'Tempo

---

[67] *Franz Liszts musikalische Werke*, ii/iv, 156.

rubato' in a main tempo marking probably means the selective application of the device on 'this or that significant note', as Lachmund
explained or as we demonstrated for Chopin's Trio in Ex. 7.23.

Some of the main tempo markings from this period read: 'Tempo
rubato' (by itself); 'Quasi presto (Tempo rubato)'; 'Tempo rallentando, sempre rubato'; 'Più moderato. Tempo rubato'; 'Cantabile con
moto (sempre rubato)'; 'And^te con sentimento. Tempo rubato'; and
'Tempo rubato e molto ritenuto'.[68] The combination of *rubato* and
ritenuto sometimes occurs also in the tempo changing instructions, as
in '*ritenuto e rubato*'; '*in tempo ritenuto e rubato*'; '*un poco ritenuto
il tempo e sempre rubato*'; and '*più tosto ritenuto e rubato quasi
improvvisato*'.[69]

With Liszt the word *ritenuto* still seems to refer, at least to some
extent, to a manner of touch that is 'held back'. Hans von Bülow,
who studied piano in the early 1850s with Liszt, gives the modern
definition in a footnote to one of the Beethoven sonatas he published
in 1871 with Lebert: 'attention is called to a distinction founded
in grammar, but unfortunately seldom observed, between *ritardando*,
a gradual and successive slackening of the tempo, and *ritenuto*,
a sudden drop to a slower and continuous rate of speed'.[70] Although
Liszt preferred this edition for teaching his own students,[71] he seems
sometimes to indicate through his own markings a different con-

---

[68] These 7 markings occur, respectively, in the following works: the *Grandes études* Nos. 5, 9,
or 11 in *Franz Liszts musikalische Werke*, II/i, 34(66), 74(106), and (131)99; the *Grandes études*
No. 11, ibid. (133)101; *Rapsodies hongroises, Cahier* 10 No. 17 in the *Neue Ausgabe*, I/xviii,
156; 2. *Grande marche de François Schubert* in the *Neue Ausgabe*, II/vii, 89; *Années de
pèlerinage*, 1st year, No. 9 in *Neue Ausgabe*, I/vi, 56; *Feuille morte, Élégie pour piano d'après
Sorriano*, fac. of edn. published in Paris by E. Troupenas in *The Liszt Society Journal*, 14 (1989),
Music Section, 6; and *Années de pèlerinage*, 2nd year, No. 7 in *Neue Ausgabe*, I/vii, 50.

[69] *Harmonies poétiques et religieuses* No. 9 in the *Neue Ausgabe*, I/ix, 90; *Air du Stabat mater*
[of Rossini] in *Werke für Klavier zu 2 Händen von Franz Liszt*, ed. Emil Sauer (Leipzig: C. F.
Peters, n.d.), viii, 165; *Hungarian Rhapsody* No. 1 in *Neue Ausgabe*, I/iii, 10; and *Années de
pèlerinage*, 2nd year, No. 7 in *Neue Ausgabe*, I/vii, 41. In addition, the expression '*Ritenuto il
tempo, sempre rubato*', which occurs in the *Hungarian Rhapsody* No. 12 in *Franz Liszts
musikalische Werke*, II/xii, (127)5, is printed in the *Neue Ausgabe*, I/iv, 33, with the comma
following the 1st word. With these markings, the diverse ways of printing the word *rubato*
become more conspicuous; it may or may not be capitalized or in italics.

[70] Ludwig van Beethoven, *Sonatas for the Piano*, rev. and fingered by Hans von Bülow and
Sigmund Lebert, trans. by Theodore Baker (New York: G. Schirmer, 1894), ii, 403, with
reference to the section from the 9th to the 12th measures from the end of Op.
53. The same distinction occurs on p. 342 in connection with the *ritardando* in mm. 3–5 of the
opening movt. of Op. 31 No. 3. At the end of the century writers were still contrasting the
correct meaning of *ritenuto* with its common usage as a synonym for *ritardando*: see Frederick
Niecks in *A Concise Dictionary of Musical Terms*, 2nd edn. (London: Augener, 1884), 207, and
in *MMR* 18/216 (1 Dec. 1888), 267; also Theodore Baker, *A Dictionary of Musical Terms*, 16th
edn. (New York: G. Schirmer, n.d.; original edn. 1895), 167.

[71] See William Newman, 'Liszt's Interpreting of Beethoven's Piano Sonatas', *MQ* 58 (1972),
203–5.

cept of ritenuto. In his arrangement in 1879 of the Polonaise from Tchaikovsky's *Eugene Onegin*, for example, he twice marked, following *poco rall.*, the instruction *a tempo ma un poco ritenuto*.[72] He may make a distinction between *ritenuto* by itself (a held-back touch) and *ritenuto il tempo* (the modern type). In any event, the numerous times in which he couples *rubato* with *ritenuto* or *ritenuto il tempo* may suggest that important notes are to be delayed rather than anticipated.

Only nine of the twenty-three rubatos from this period are disposed in such a way that the right and left hands could conceivably be displaced in the manner described by Chopin. Several facts, however, suggest that Liszt may have modified his original concept of rubato during this period to include a brief addition of time, either in company with or as a substitution for a displacement between the hands. First of all, I know of no comment by Liszt or anyone connected with Liszt that refers to the strict, conductor-like left hand so often mentioned by Chopin's students. Secondly, there are six examples from this period in which right and left hands do not coincide, making displacement by the performer impossible. Finally, there are several examples in which the playing of both a melody and a more active accompanimental figure in the right hand cannot easily be accomplished without briefly modifying the tempo.

The rubato Liszt marked in Weber's sonata (Plate XIX) illustrates the last situation. In a traditional rubato rendition the E below the asterisk would no doubt be delayed until after the left-hand chord is sounded. Following the E, however, the pianist would also want to give the sixteenth notes their due time, avoiding, for example, sounding the E simultaneously with the succeeding A. A delay of the E and the rhythmic integrity of the accompanimental figure can be observed, however, if a slight amount of extra time is added in order to accommodate them both. See Ex. 8.15, where the variable amount of additional time is simply represented as a sixteenth note. The 'sudden, light suspension of the rhythm' mentioned by Lachmund can be even further emphasized by manipulating the sustaining pedal in such a way that an articulating silence occurs in the melody at the moment the chord is struck in the left hand (also shown in Ex. 8.15). These techniques might indeed produce an agogic accentuation on the E that one might describe as 'deceptive', 'seductive', or 'intoxicating'. In the Weber example, the addition of time is suggested by the

---

[72] Franz Liszt, *Piano Transcriptions from French and Italian Operas*, with an Introduction by Charles Suttoni (Dover, 1982), 240 and 242; originally published in F. Liszt, *Opernye transkriptsii dlya fortep'yano*, ed. V. S. Belov and K. S. Sorokin, iii/2 (Moscow, 1964).

Ex. 8.15 Possible performance of Liszt's rubato in the Weber sonata on Plate XIX

ritenuto that precedes the E, and the full rhythmic articulation of the following sixteenth notes seems required by the *crescendo* which begins on the A. The precise determination of all the variables of performance, however, depends, as Liszt tells us and as we noted with Chopin's rubato, upon 'the taste and momentary feeling of gifted players'. This means, of course, that the performer has been intimately experiencing the *grazioso* melody from its beginning, and vividly feeling the powerful crescendo, the sudden dynamic decrease for the chromatic passage, and the moment, finally, which is to be accentuated by rubato, when one feels certain at last that a strong dominant chord in C major is approaching.

A similar situation occurs in Liszt's First Piano Concerto, where the warm melody shown in Ex. 8.16 follows the opening cadenza-like material.[73] Although the word *rubato* is part of a longer phrase, it may be significant that it actually appears (at least in the Complete Works) above measure 33, which contains the most 'significant' notes: the E♭ certainly, as indicated by Liszt's dynamic marks, and possibly, to a much lesser extent, the opening C. Both of these melodic notes sound with only a single note in the left hand and hence could be emphasized by a rubato delay. As in Weber's sonata, however, the eighth-note figure in both hands demands its own rhythmic presence and hence tends to require a performance somewhat like Ex. 8.15. In either example, the arrival of the delayed melody note tends to sound like the beginning of a new quarter beat, with the left-hand note or chord seeming retrospectively like a rhythmic anticipation occurring before the beat. One might include other displacement in Ex. 8.16, such as an anticipation of the second C in measure 33. This could even be facilitated, I presume, by playing both eighth-note voices in

---

[73] *Franz Liszts musikalische Werke*, I/xiii, 5. For a similar example, see Liszt's arrangement of Schubert's song 'Ständchen'; Emil Sauer presents two versions in *Werke für Klavier zu 2 Händen*, ix: *Lieder Bearbeitungen* (Leipzig: Peters, n.d.), 106. The more complex version contains portions of the 8th-note accompanimental figure in each hand, whereas the simpler one has only the melody in the right hand.

Ex. 8.16 Liszt, 1st Piano Concerto, c.1849–56

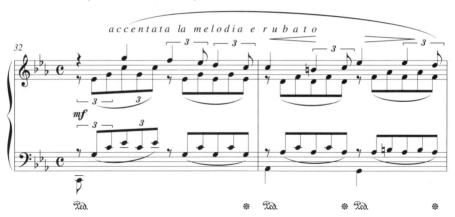

the left hand. I believe, however, that Liszt carefully distinguished the hands by the use of the two staves.

In the case of the Weber sonata, one might possibly conceive of two effects taking place simultaneously: displacement due to the rubato, and tempo modification due to the ritenuto. Regardless of the manner in which the musical effect was actually perceived, it is clear in some other pieces that displacement between the melody and its accompaniment is no longer a requirement for rubato, and that the 'sudden, light suspension of the rhythm' can be achieved, at least on occasion, by application of other means. The melody which commences as in Ex. 8.17 continues thus for eighteen measures without the accompaniment ever coinciding with the melodic notes. The chords of the accompaniment are staccato, which allows the melody to sing out conspicuously. Significant notes to accentuate by a delay and a silence of articulation might be the F♯ on the first beat of measure 24 (in order to emphasize its dissonance against the chromatically altered chord), the F♯ on beat 3 (to recognize the resolution back to the tonic chord), and the third F♯ of measure 23 (to respond to its greater length as well as to the chromatic chord which precedes).[74] A similar example is the sixth *Consolation*, the work which illustrated Lachmund's remarks quoted above. Judging from his description, the single word *rubato* at the beginning applies not momentarily, but, like *sempre rubato*, during a longer section of music. Only in measure 8 does one finally reach a place where a note in the melody is notated to sound simultaneously with a note in the left hand.[75] In both these pieces the rubato is

---

[74] *Neue Ausgabe*, ɪ/ix, 90.
[75] Ibid. ɪ/ix, 112.

Ex. 8.17 Liszt, *Harmonies poétiques et religieuses* No. 9, 1845–52

apparently achieved mainly by agogic hesitations. When the left hand does not impose a strict beat, of course, we no longer have the earlier type of rubato. The adding of brief units of extra time introduces elements of the later type and tends to cause one to consider the rubato in relationship to other tempo modifications in the melody as a whole.

In contrast to these examples, Ex. 8.18 presents an excerpt in which the melody rather than the accompaniment is syncopated. The fact that the melody in (*d*) is a rhythmic transformation of the theme in (*a*) heard earlier in the piece makes the syncopation particularly conspicuous.[76] None of the melody notes occurs with a bass note, so the rubato apparently consists of a 'sudden, light suspension of the rhythm' on the three significant notes: the C♯ and C♮ marked with a dash and a dot, and the accented F♯. There is also the possibility that *tempo rubato* refers here simply to the notated displacement in (*d*) of notes previously heard as in (*a*). Considering Ex. 8.17 and other pieces in which displacement is not possible, however, it would appear that we are indeed dealing here with a new and modified concept.[77]

The rubato of this period must come primarily out of Liszt's theatrical performing style. Reinecke recalls that 'if ... it came into his head to dazzle the ignorant throng a little, he would allow himself to be carried away by all manner of fantastic tricks'.[78] Rubato was

[76] *Franz Liszts musikalische Werke*, i/xiii, (59)1 for Ex. 8.23a, 2(60) for (*b*), 50(108) for (*c*), and (115)57 for (*d*).

[77] For other examples, see *Franz Liszts musikalische Werke*, ii/i, (133)101: the *Grandes études* No. 11, the section marked 'Quasi presto (Tempo rubato)', in which the hands rapidly alternate the melody notes at different octave levels for 11 measures; and i/xiii, (199)17: *Malédiction für Klavier solo und Streichinstrumente*, section marked 'Sempre moderato; a tempo rubato'. See also the *Neue Ausgabe*, i/vii, 41 and 50: *Années de pèlerinage*, 2nd year, No. 7: 'Après une lecture du Dante'.

[78] Carl Reinecke, '*Und manche liebe Schatten steigen auf*', *Gedenkblätter an berühmte Musiker* (Leipzig, 1910), 11–13; trans. in Williams, *Portrait of Liszt*, 145.

Ex. 8.18 Liszt, 2nd Piano Concerto, 1839, rev. 1849–61

probably one of those 'tricks' which he employed, as we have seen, 'to extract bravos from a public slow to perceive things of beauty'. It could well have been accompanied by equally seductive body gestures or facial expressions. I am reminded of a flamboyant organist of our own time who, during an agogic silence between the penultimate dominant chord of a piece and its final tonic, looked up at those seated in the balcony and flashed a mischievous smile. The audience, of course, responded with great delight. Such activity is natural to a virtuoso performer, I suppose, and Liszt seems to have been the most spectacular of them all. In this context, then, I presume that his execution of rubato, during this period of his life, was extremely deceptive and extremely seductive, that he used it frequently, and that he exaggerated the amount of time a note was delayed, the amount of silence that preceded it, or possibly the amount of displacement between it and the left-hand accompaniment. This was the period he referred to in his letter to Mason, as we noted above, in which he felt ill qualified to encourage moderation when he himself had 'employed a good deal of Tempo rubato' in his concerts.[79]

Around 1850, however, a new attitude is apparent in the rubato markings in Liszt's scores. He had ended his stage performances in 1847 and seems to have drawn back from the excesses and exaggerations of that period. He was by this time fully involved at Weimar as a conductor. Chopin's death in October 1849 seems to have exerted a profound influence on him. His biography of Chopin was first published, as we have seen, in 1851. Aspects of Chopin's musical style seem to be incorporated into some of his own compositions from this period. Now, for the first time, he employs some titles used by Chopin: *polonaise*, *berceuse*, *mazurka*, and *ballade*. In addition, he seems to return to Chopin's practice in regard to rubato, for he omits the word *tempo* and writes simply *rubato* or *sempre rubato*. It may still be linked with ritenuto, but the word *rubato* does not appear at this time in a main tempo marking. He probably considers it now a subtler, less theatrical device and perhaps even intends that it sometimes involve the strict accompaniment and displacement of Chopin's earlier type of rubato.

The *Polonaise mélancolique*, composed in 1851, commences, after a seven-bar introduction, with a singing melody marked *rubato espressivo*. It is the C at the beginning of measure 9 in Ex. 8.19 that should receive the most significant accentuation, in order to mark the unexpected chord change as well as the agogic stress created by the preceding shorter notes.[80] This example is disposed, like many of Chopin's rubato passages, with a single line of melody in the right hand, and with an accompanimental figure in the left hand in which a bass note is heard alone on the first beat of each bar, followed by repeated chords. Therefore, one could easily apply the earlier type of rubato, with the left hand keeping strict time while the C is delayed, and perhaps the opening A in measure 8 anticipated. Liszt may, of course, have conceived of more tempo flexibility, even during this period of Chopin influence. We are also faced here once again with rubato marked at the beginning of a melody. Since Liszt wrote simply the word *rubato* rather than *sempre rubato*, as he still does in other

---

[79] For other examples from this period, see the *Neue Ausgabe*, i/iv, 33 (in Hungarian Rhapsody No. 12), i/vii, 108 (*Tarantelles napolitaines*), i/xi, 44 (*Romance*), i/xviii, 156 (*Rapsodies hongroises, Cahier* 10, No. 17), and ii/vii, 106 (*3. Grande marche caractéristique de Schubert*). See also *Glanes de Woronince* No. 2: 'Mélodies polonaises', fac. of Kistner edn. (Leipzig, 1849) in *The Liszt Society Journal*, 15 (1990), Music Section, 45.

[80] *Neue Ausgabe*, i/xiii, 76. See Walker, *Franz Liszt*, ii, 145–7, where he quotes excerpts from this piece as a typical example of 'a Chopinesque turn of phrase'.

Ex. 8.19 Liszt, *Polonaise mélancolique*, 1851

scores from this same period, I presume that the effect is to be momentary rather than referring to the entire theme. When the melody returns later in the piece, as a matter of fact, it is highly embellished and marked *un poco ritenuto il tempo, sempre rubato*.[81] Liszt's Second Ballade, composed in 1853, also includes the word *rubato*. The passage, marked *appassionato*, seems to lend itself, however, to Liszt's seductive suspension of rhythm rather than Chopin's more restrained displacement against a strict accompaniment.[82]

Probably from the same period come his arrangements for solo piano of six of Chopin's Polish songs. Ex. 8.20 shows the first half of the opening ritornello to the song 'Moja pieszczotka' (My Darling), Op. 74 No. 12. In Chopin's version there is no rubato, and the melody on the top staff of Ex. 8.20*a* is accompanied by the left-hand part shown in Ex. 8.20*b*. Liszt subtitles his arrangement 'Nocturne' and arpeggiates the left-hand chords accordingly. Again he writes the single word *rubato*. Presumably this is a momentary effect that places an almost reluctant accentuation on the second melody note by delaying it in relation to or in company with the A♭ in the left hand. This effect sets a special mood for the piece, somewhat like Chopin's own rubato in Ex. 7.27.[83]

It was probably during this same period that Liszt was preparing his edition of Chopin's Mazurkas and Waltzes.[84] Although he remains

[81] *Neue Ausgabe*, ı/xiii, 78. One of Liszt's pupils writes in his diary concerning a study of this piece in 1885: 'Das 1. Thema sehr rubato'; see *Franz Liszts Klavierunterricht von 1884–1886 dargestellt an den Tagebuch-aufzeichnungen von August Göllerich*, ed. Wilhelm Jerger in *Studien zur Musikgeschichte des 19. Jahrhunderts*, xxxix (Regensburg: Gustav Bosse, 1975), 116.

[82] *Neue Ausgabe*, ı/ix, 135.

[83] *Werke für Klavier zu 2 Händen*, ed. Emil Sauer, ix, 177; see also p. 172, where Liszt has added *rubato* twice in Chopin's 14th song, 'Pierścień' (The Ring). Ex. 8.20*b* comes from the Fontana MS in the Österreichische Nationalbibliothek. Humphrey Searle, in *New Grove*, xi, 68, and in *The New Grove Early Romantic Masters*, i (Norton, 1985), 359, gives 1847–60 as the dates of composition for Liszt's arrangements, which were published in 1860.

[84] *Complete Collection of Mazurkas & Waltzes by Fr. Chopin, to which is Prefixed his Biography, and a Critical Review of his Works by F. Liszt* (Ditson, n.d.), copy in the Alderman

Ex. 8.20 Liszt, arrangement of Chopin's Polish Song No. 12, 1850s

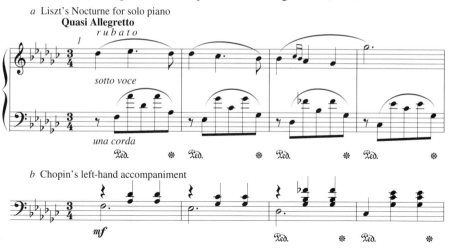

a Liszt's Nocturne for solo piano

b Chopin's left-hand accompaniment

fairly faithful, in general, to Chopin's scores, he does add a few marks affecting the rubatos. In the Mazurka Op. 6 No. 1, he adds a tie on the F♯ between measures 8 and 9, as well as in the corresponding spots in measures 32–3 and 64–5. This, as we have seen, alters the nature of the rubato marked at the beginning of measure 9 (see Ex. 7.14, Plate XVI, and footnote 105 in Chapter 7). In the Mazurka Op. 7 No. 3, Liszt adds a crescendo sign two bars before the first rubato (Ex. 7.15*a*), *Dim.* on the trilled note in the following measure, and *p* on the final note, which acts as an anacrusis to the melodic repetition marked *rubato* in the next measure. Since the approach to a rubato, as we have noted previously, affects its performance in very crucial ways, Liszt's more precise indication of the way he played the two bars which Chopin marked merely *con forza*, gives us additional informa-tion on his special concept of the device. Liszt modifies Chopin's apparently sudden change from *f* to *p* in the return of the theme by adding *Dim.* in the bar before the rubato; in measure 93 (see Ex. 7.15*c*) he arpeggiates the chords in the left hand and moves the word *rubato* to the right so that it commences on the third note, the A♭, rather than slightly to the right of the opening F. Liszt leaves the other Mazurka rubatos essentially as Chopin notated them.

It was during this same period that Liszt revised some of his earlier

Library at the Univ. of Virginia. The biography comes from the 'Preface to the English edition of Chopin's Mazurkas edited by J. W. Davison'; Liszt's *Critical Review* consists of the condensed version of his book as trans. in *DJM* for 8 May 1852 (see n. 84 in Ch. 7).

scores, making them, in general, less virtuosic and more practical for playing on the heavier action of the later pianos.[85] At the same time, a number of his rubato markings undergo some significant changes. In his *Sonetto 47 del Petrarca*, the concluding section of the earlier version of 1844–5 presents a singing melody which moves from hand to hand to accommodate sweeping arpeggios that overlap with it. In the later version, included in the second year of the *Années de pèlerinage*, the passage has been simplified by keeping the melody in the left hand and by substituting new accompanimental figures in the right. The early version is marked *sempre rubato*, the later (following *molto riten.*), 'in tempo ma sempre rubato'.[86] In this case, Liszt recognizes that rubato, far from being the contrary of 'in tempo', often takes place in company with it. This demonstrates, it seems to me, that the accelerandos and ritardandos that modify the tempo actually operate on quite a different level from the brief rhythmic accentuations involved in rubato.

During the course of revision, Liszt sometimes eliminates the rubato. In the *Grande valse di bravura*, the notes of the passage beginning in Ex. 8.14 remain the same in the revision of around 1850, but 'Rubato' and *piangevolmente* have been replaced by simply *espressivo*.[87] In the *Études d'exécution transcendante*, a revision in 1851 of his *Grandes études* of 1837, Liszt substitutes *un poco riten. (a piacere)* in No. 5 to replace 'Tempo rubato' and *dolce piacevole* for almost the same music. In Étude No. 11 he provides a new accompaniment for the melody beginning in measure 59 and changes the marking from 'Tempo rubato' and *molto espressivo il canto* to 'Più lento con intimo sentimento'. He seems to have deleted the music which was originally marked 'Quasi presto (Tempo rubato)' in measure 98 of the same piece.[88]

The ninth Étude in this set presents a particularly eventful history of

---

[85] Walker, *Franz Liszt*, ii, 147–9.

[86] The earlier version is in *Franz Liszts musikalische Werke*, II/v, (59)7; the later in the *Neue Ausgabe*, I/vii, 18. Concerning a 1st edn. of the early version with Liszt's autograph corrections, see Mária Eckhardt, *Liszt's Music Manuscripts in the National Széchényi Library, Budapest* in *Studies in Central and Eastern European Music*, ii (Budapest: Akadémiai Kiadó, 1986), 100–2; note on 102 the 'Tempo I^{mo} sempre rubato' subsequently added in bar 63 (the passage I refer to).

[87] Revised version in *Neue Ausgabe*, I/xiii, 57.

[88] For the *Grandes études*, see *Franz Liszts musikalische Werke*, II/i, 34(60) in No. 5, (131)99 and (133)101 in No. 11; for the *Études d'exécution transcendante*, see the *Neue Ausgabe*, I/i, 38 in No. 5 and 100 in No. 11. Somewhat later, in 1859, Liszt revised his *Tarantelles napolitaines* of around 1840, and included it as the 3rd piece in *Venezia e Napoli*, a supplement to the 2nd year of the *Années de pèlerinage*. The accompaniment for the 'Canzone napoletana' has been changed completely and the original markings 'Più animato', 'tempo rubato', and 'sciolto' replaced by 'Molto ritenuto il tempo' and 'cantando'. See the *Neue Ausgabe*, I/vii, 108 (m. 112) for the earlier work, 75 (m. 200) for the revision.

Ex. 8.21 Liszt, Étude No. 9

a *Étude en 12 exercises*, 1826

b *Grandes études*, 1837

c *Études d'exécution transcendante*, 'Ricordanza', 1851

revision. Ex. 8.21 shows three steps in the process. The piece first appears in the *Étude en douze exercises* from 1826, where it begins with the melody in Ex. 8.21*a*. In the revision of 1837 (Ex. 8.21*b*) the same melody follows fifteen measures of free, cadenza-like introduction; the accompaniment has been expanded, significant notes in the melody marked with accent signs and tenuto, and the expression 'Tempo rubato' added. The final revision of 1851 (Ex. 8.21*c*) presents an accompaniment which is simpler and which is confined to the left hand. The expression 'Tempo rubato' is missing, but the melody now contains portato touch for the last two notes of the anacrusis and, most significantly, delays by an eighth beat the notes previously marked for accentuation. The delayed notes are further emphasized by arpeggiating the two-note chords below them.[89] Such a change causes one to wonder whether Liszt had similar displacement in mind when he indicated 'Tempo rubato' in 1837. In any event, the syncopation notated in Ex. 8.21c continues later for many measures and recurs each time the melody is repeated.[90]

The *Valse mélancolique* of 1839 contains a melody, marked *sempre rubato* and *con molto sentimento*, which commences with a single note on the first beat of each measure against a waltz bass. When the parallel second section begins nine bars later, the same melody notes occur on the second beats of the measure following a rest on the first beat. In Liszt's revision of the piece around 1850 as the *Caprice-valse* No. 2, the melody begins immediately with the delayed notes; this time the markings are *sempre rubato ed espressivo* and *dolce*.[91] The sort of syncopation used in this piece and in the Étude in Ex. 8.21c occurs, of course, elsewhere in the works of Liszt. In *Liszt-Pädagogium*, Lina Ramann, who studied with Liszt and was his first biographer, compiled student recollections concerning Liszt's teaching of his own compositions. Concerning the passage in the *Valse-impromptu* from around 1850 in Ex. 8.22b, which follows the unsyncopated version in (a), she includes the following note: 'The syncopated notes *rubato* legato and singing, the rhythm of the accompaniment: gently rocking'. It is unclear whether she refers to syncopation as a type of rubato or whether rubato, in this particular case, should be added to the syncopated notes. In any event, she somehow links syncopation with rubato, and one is therefore reminded of the long tradition, beginning with North and Quantz, of depicting the earlier type of rubato by means of syncopated notes.[92]

---

[89] Rosen has called my attention to the original version of *Sonetto 47 del Petrarca*, where *quasi arpeggiando* is marked in m. 8 for a melody accompanied by chords in each hand. When repeated in m. 31 the passage is *più arpeggiando* and some of the chords are individually marked with a wavy line (*Franz Liszts musikalische Werke*, II/v, 2[54] and 4[56]). In the revision Liszt discards the terms, but syncopates every note in the right hand by delaying it an 8th beat against the left (*Neue Ausgabe*, I/vii, 14 and 16: mm. 12 and 36). This suggests that the arpeggiation in the early version should commence on the beat and hence cause a similar delay in the melody notes, with the delay becoming greater as one moves from *quasi* to *più* and finally to the wavy line. It is only later in the piece, however, that Liszt marks the rubato mentioned above.

[90] Ex. 8.21a and c come from the *Neue Ausgabe*, I/xviii, 24, and I/i, 74; Ex. 8.21b is from *Franz Liszts musikalische Werke*, II/i, 74(106).

[91] Ibid. II/x, (35)3 and (41)3.

[92] *Liszt-Pädagogium: Klavier-Kompositionen Franz Liszt's nebst noch unedirten Veränderungen, Zusätzen und Kadenzen nach des Meisters Lehren pädagogisch glossirt* (Br. & H, 1901), iv, 14: 'Die Synkopen *rubato* sehr gebunden und singend, die Rhythmen der Begleitung: wiegend.' The 1st half of the sentence might be interpreted as (1) the syncopated-notes *rubato* [should be played] very legato and singing; or (2) the syncopated notes [should be played] *rubato*, very legato, and singing. Ramann never capitalizes the word *rubato*, even when it is clearly used as a noun; this is the only time, however, that the word is in italics. Ex. 8.22 is from the *Neue Ausgabe*, I/xiii, 131.

Ex. 8.22 Liszt, *Valse-Impromptu*, *c*.1850

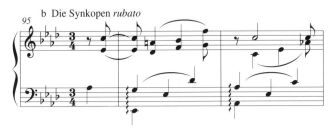

Elsewhere in the same source, Ramann again links rubato with a syncopated delay in connection with the execution of the small notes in Ex. 8.23*a*. According to Ramann, 'the Master usually plays this, as opposed to the accompaniment, with rubato and uses *Bebung* for the main note', as shown in Ex. 8.23*b*.[93] In this case, the accompaniment, as with Chopin, keeps strict time, while the melody is altered. When the turn commences on the third quarter beat of the measure, it must steal time, of course, from the following main note in the manner of a baroque on-beat ornament. The vertical alignment with the accompaniment in Ex. 8.23*b* also shows additional delaying of the accented C#. This is actually an example of the earlier type of rubato, but described by using the word *rubato* in the later sense. The later rubato, applied only to the melody, is restrained by the strict rhythm of the accompaniment, thus producing the earlier type of rubato.

The final example from this period is an arrangement Liszt made in 1852 of one of Schubert's waltzes. It includes a passage that moves downward through a circle of major thirds: from A major to F major, then to D flat major, and back to A. As the music moves to A minor in preparation for the first turn to F major, Liszt emphasizes the sweet sound of this modulation by marking *dolce appassionato sempre*

---

[93] *Liszt-Pädagogium*, i, 9–10. Ex. 8.23*a* comes from the *Harmonies poétiques et religieuses* No. 3: 'Bénédiction de Dieu dans la solitude', in *Neue Ausgabe*, i/ix, 51, and the trans. of Ramann's comments from the footnote on this page. Some inexact note values have been 'corrected' in the footnote; I have left them as they are, however, since the displacement between the 2 parts is clear enough. In any event, she states that 'the mordent [meaning, I presume, the turn] is outside mathematical precision'.

Ex. 8.23 Liszt, *Harmonies poétiques et religieuses* No. 3, 1852

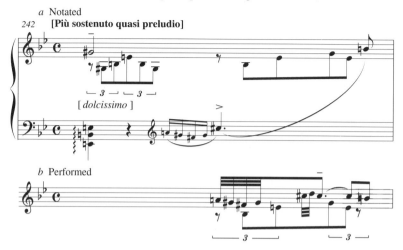

a Notated
242 [**Più sostenuto quasi preludio**]

[*dolcissimo*]

b Performed

rubato.[94] The melody in the right hand moves in parallel sixths. The left hand has the typical waltz bass, which was, as we have seen in examples of Chopin's rubato, especially effective in producing displacements between the hands. Whether Chopin's displacement or Liszt's sudden suspension of rhythm is applied, it surely is accompanied at the same time by the dance-like accentuations peculiar to the waltz. As we saw in Chapter 7, Liszt was described as 'dashing the wrist abruptly from the chord at the second beat . . . , sometimes almost prematurely'.

The years around mid-century of revision and special preoccupation with Chopin were followed by a late period from 1853 to 1870. From these years come only six examples of rubato, three of them, for the first time, in orchestral works. Returning to the practice of his virtuoso period, Liszt three times uses the full expression *tempo rubato* and sometimes employs it in a main tempo marking. In his piano arrangement of *Gaudeamus igitur* from 1870, he again links rubato with ritenuto and dolce by marking a section 'Tempo rubato e un poco ritenuto' as well as *dolcissimo.*[95] In his seventh symphonic poem *Festival Sounds*, a section marked 'Allegretto. (Tempo rubato)' contains an *espressivo* melody played first by a solo cello, then by an oboe.[96] In the middle of the first movement of the Dante Symphony

[94] The *Valse-caprice* No. 6 from *Soirées de Vienne*, printed in Franz Liszt, *Twenty Piano Transcriptions*, ed. August Spanuth (Ditson, 1903), 96.

[95] *Neue Ausgabe*, I/xvi, 116.

[96] *Franz Liszts musikalische Werke*, I/iv, 20.

a section apparently representing the past happiness of Paolo and Francesca da Rimini is marked 'Andante amoroso. Tempo rubato'; individual parts are to be played *dolce soave, teneramente*, and *dolce con intimo sentimento*. Two violins play a syncopated phrase in 7/4 metre, which is immediately repeated by all the violins with mutes. The syncopated portato notes in the melody are supported by triplet figures in the harp and violas so that notes of the melody and accompaniment never coincide during the opening half of the phrase. Perhaps this is a notated version of rubato; perhaps the conductor is supposed to indicate still further hesitations to the violinists.[97]

The nature of rubato must change, it seems to me, as it moves from the realm of the solo performer into music that involves more than one instrument. The solo pianist can control both melody and accompaniment. When a solo cello or oboe plays a brief melody in an orchestral piece, the conductor can either leave it free to play as it wishes, or guide it according to his concept of rubato. When more than one instrument play a melody simultaneously as in the Dante Symphony, however, the conductor must necessarily determine every detail of execution. Thus the performance of rubato changes from a very personal response by the solo performer to a more general fulfilment of directions given by a conductor and shared by a group of musicians.

Seyfried, as we saw in Chapter 6, recalled in 1832 the effective tempo rubato that Beethoven achieved as a conductor. He was no doubt speaking here, however, of expressive tempo flexibility rather than the sudden rhythmic suspension that Lachmund associates with Liszt. It could well be that Liszt's orchestral rubatos represent a transition to a more general concept of tempo modification—less like his pianistic rubato and more like the rhythmic flexibility which characterized his conducting in general. As a matter of fact, Lachmund may be using the word *rubato* in this manner when he describes Liszt's performance in 1882 of a long series of sixty-fourth notes in a left-hand accompaniment: the notes 'hurried here a little and then lingered elsewhere, indeed sometimes broke completely off'. In the next sentence he refers back to 'such an artistic fluctuation' and 'such a continuous declamatory rubato'. Of course, *rubato* may refer specifically to the notes breaking completely off, that is, to a suspension of the rhythm. On the other hand, it may include now also the hurrying and lingering, or expressive tempo fluctuation in general.[98]

[97] *Franz Liszts musikalische Werke*, i/vii, 47.
[98] Lachmund, *Mein Leben mit Franz Liszt*, 108–9. The work referred to is the 'Canzone' Liszt composed in 1859 as the middle movt. of the revised version of *Venezia e Napoli*, a supplement to the 2nd volume of *Années de pèlerinage*; see *Neue Ausgabe*, i/vii, 63 ff., where the

One orchestral work from this late period appears also in a version for solo piano: 'Der Tanz in der Dorfschenke' or First Mephisto Waltz from *Two Scenes from Lenau's Faust*. Liszt apparently worked on both between 1859 and 1861, and it is difficult to determine which came first. In spite of many differences, the two versions follow the same basic musical outline. Following a furious 'Allegro vivace' section, the music turns to *un poco meno mosso* and a new melody marked *espressivo amoroso*. The first appearance of this melody in both versions is like Ex. 8.24*a*, except that it starts a measure later on

Ex. 8.24 Liszt, *Mephisto Waltz* No. 1, 1859–60

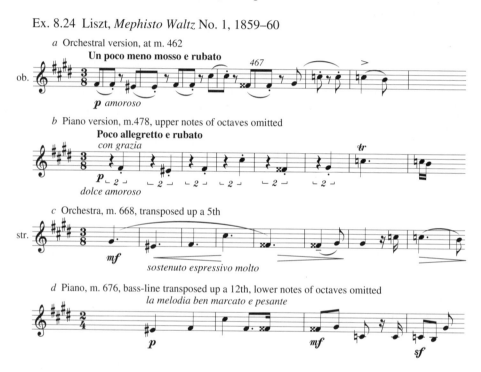

the E♯, dips down to a lower C♮, and then leaps back up for an added measure of G♯ at the end. The second appearance of the melody is different in each version (Ex. 8.24*a* and *b*), but each is marked *rubato*. In comparison to the later orchestral statement of the melody in (*c*), which corresponds to the piano version in (*d*), the notes of (*a*) seem

key is E flat minor (Lachmund gives E minor) and the accompaniment consists of 64th rather than 32nd notes. Lachmund writes: 'die Gruppe der Zweiunddreissigstel-Noten eilten hier ein wenig und zögerten wieder anderswo, ja brachen manchmal völlig ab[:] . . . ein solches kunstvolles Schwanken, ein solches wiederholtes deklamatorisches Rubato'.

anticipated, those in (*b*) delayed.[99] The anticipation and delay may constitute a sort of notated rubato, as we have seen it before in the works of Liszt. The piano melody in (*b*) is accompanied at first by quadruplet arpeggios; after two measures in which steady sixteenth notes reaffirm 3/8 metre, (*b*) recurs with quintuplets in the left hand. The introduction into 3/8 metre of figures in 2/4 and the portato touch work together to induce a provocative mood particularly well suited to Liszt's pianistic type of rubato. Displacement with the left-hand quadruplets could even take place, especially with the dissonant E$^\sharp$ and F$^{\sharp\sharp}$. When the left hand changes to quintuplets, every note of the melody, of course, is automatically displaced.

In the orchestral version, on the other hand, the full second violin section alternates measures of six sixteenth notes with the full viola section to provide the accompaniment for the oboe melody in (*a*). I presume an agogic accentuation could mark the change to the tonic chord at the beginning of measure 467 and emphasize the third C$^\natural$ and perhaps to a lesser extent also the first one. The conductor must obviously direct very clearly, either following the oboist or initiating the rubato himself, in order to synchronize properly the steady sixteenth-note accompaniment in the strings. In this situation the oboist has much less freedom, however, than the solo pianist. The latter can, on the spur of the moment and in immediate response to a particular performance, suddenly create a rubato accentuation on a certain note or in a certain manner different from any previous performance. The oboist, on the other hand, must either follow carefully the conductor or produce a type of rubato that is sufficiently familiar to the other musicians in the orchestra that they can add their own parts at the correct time.

In considering Liszt's late period, we must also keep in mind that his edition of Weber's sonatas in Plate XIX and the letter relating to it date from 1870. In addition, Lachmund's comments come from lessons in 1882, Göllerich's diary covers the years 1884 to 1886, and most of the ideas in Ramann's *Liszt-Pädagogium* seem also to come from those who studied with Liszt in his later years. The student comments in these sources use only the single word *rubato*, sometimes as a noun and sometimes as an adjective or adverb. In the original Italian, of course, *Tempo* is the noun, *rubato* a past participle used as an adjective. Lachmund, when he speaks of 'das Liszt'sche Rubato', however, uses the word *rubato*, as I have in this book, as a noun to replace the full expression *tempo rubato*. Göllerich, on the other hand,

[99] The orchestral excerpts come from *Franz Liszts musikalische Werke*, i/x, 30(66) and 44(80); those for piano from the *Neue Ausgabe*, i/xv, 129 and 134.

uses the word *rubato* as an adjective or adverb when he quotes Liszt's suggestion following a particular student performance: 'nicht so affectirt und rubato'. He also makes an enigmatic statement concerning Liszt's 'Ricordanza', with reference perhaps to the delayed notes in Ex. 8.21*c*: 'Beim 2. As-dur rubata. Thema: *und es gibt keine Tonart*'.[100] Ramann uses *rubato* as an adverb when she refers to Ex. 8.23: 'Der Meister pflegte ihn [the small-note ornament], gegenüber der Begleitung, rubato ... auszuführen', and when she describes a passage in another work which is to be played 'spitz und nüancelos, ... aber ein wenig rubato'. On the other hand, she also employs the word as a noun, and with its plural formed by Italian rules, when she states that the running eighth-note bass in Liszt's *Variationen über das Motiv von Bach* (from the Cantata *Weinen, Klagen, Sorgen, Sagen*) should proceed 'ohne rubati'. In this case she does not capitalize 'rubati' according to German practice as Lachmund did.[101]

It was Bülow, however, who was no doubt Liszt's most important pupil. As we have already seen, Liszt preferred his edition of the Beethoven sonatas (1871) for teaching. Bülow mentions rubato five times in the second volume. In a footnote to the second theme in the opening movement of the 'Waldstein' Sonata he writes: 'The second subject [in E major] ... calls for a quieter movement than the first— one which, in measure 9 [measure 43 from the beginning of the movement], with its expressive [triplet] figuration of the soprano, may broaden to the characteristic freedom of the *rubato*, of course "without any exaggeration".' A footnote to measure 14 of the Adagio movement in Op. 101 reads: 'A moderate *tempo rubato* is surely admissible here, where the effect of a dreamy improvisation should be produced.'

In the final movement of the same sonata he actually writes *Quasi Tempo rubato* in italics in the score and cancels it four bars later by marking *ritornando al Tempo giusto*. The passage occurs at the beginning of the coda and acts, according to Bülow, as if it were a 'humoristic menace to the hearer that he may expect a repetition of the *fugato*'. It also involves a threat to the tonality by turning suddenly to the key of the flatted sixth—a gesture often employed by Haydn and Beethoven near the end of a piece. The powerful momentum unleashed by the motive in measure 318 must then be suddenly brought under control during the course of four brief measures and

---

[100] *Franz Liszts Klavierunterricht von 1884–1886*, 44 and 141. I cannot explain the last letter of the word 'rubata', nor the period following. Göllerich seems to be referring to the tonally ambiguous section from mm. 59 through 71, which intervenes between the 2nd and 3rd statements of the A flat major theme in Ex. 8.30*c*.

[101] *Liszt-Pädagogium*, i, 10 and 18; iii, 10.

the movement returned to its calm and steady course in measure 325. This involves an almost narrative or speech-like declamation in measures 320–4, with forceful touch, suitable hesitations, and possibly even displacement between the hands. The rubato must absorb the energy generated by the expectancies aroused in measures 318 and 319. The great sensitivity of this moment must have impressed Bülow, for this is the only time he actually adds the word *rubato* in Beethoven's score.

On two occasions he warns the pianist in footnotes against the use of rubato: at measure 80 in the last movement of Op. 54, 'the author's direction "*espressivo*" must not tempt to a sentimental conception or to a *tempo rubato*', and, in regard to the concluding nine bars of the first movement of Op. 90, 'this closing refrain, or epilogue, may be played somewhat broader, as if accompanied by a deep inspiration, but in an even *ritenuto* rather than with *rubato*.'[102] Although Bülow had a reputation for intellectualized playing, his relative restraint in the application of rubato in the Beethoven sonatas may, however, reflect Liszt's own more moderate approach during his later years.

As we have seen, Liszt changes his approach to rubato somewhat from period to period. During the virtuoso years, as well as his later period, he almost always marks an entire melody for rubato. In only a few cases does he limit it to a portion of the melody in which the harmony conveys a special emotion. During his early period, and probably also again around mid-century, he seems to mark rubato more often as a momentary effect on one or only a few beats in order to articulate, somewhat like Chopin, a repetition or change of movement, to emphasize an important note, or to set a mood. In the Weber sonata, on the other hand, he employs rubato for a new dramatic purpose. He might have followed his past practice and marked *sempre rubato* for the entire melody (in the third measure of the top line in Plate XIX), or indicated a momentary rubato to articulate the repetition in the second bar of line 2 or to emphasize the chromaticism and sudden dynamic change in the third bar of the third line. Instead, he chose to employ rubato to intensify the cadential drama; thus, he placed it on the ii^6 chord which follows the chromatic passage and which, in effect, announces at last the imminent arrival of the structural dominant chord. This declamatory use of rubato at a sensitive spot in the musical structure seems related to Bülow's rhetorical *quasi rubato* in Beethoven's Op. 101.

Whatever the musical purpose or however long it lasted, Liszt's

---

[102] *Beethoven Sonatas for the Piano*, ed. Hans von Bülow and Sigmund Lebert, trans. Theodore Baker (New York: G. Schirmer, 1894), ii, 390, 551, 561, 435, and 531.

rubato seems to have involved, perhaps from the beginning, some sort of tempo alteration—but of a special type essentially different from the general rhythmic flexibility used in the free forms or even the expressive modifications ordinarily added by a performer. Liszt's special suspension of time was applied selectively to single significant notes; it was sudden, light, deceptive, seductive, intoxicating. It occurred almost always when the music was soft and dolce. The touch was often portato or ritenuto. The appropriate significant notes are sometimes marked with an accent sign (see Exx. 8.14, 8.21*b* and *c*, and 8.23), with tenuto (Exx. 8.18*d*, 8.21*b*), or with signs indicating the high point of a crescendo (Ex. 8.16). They are sometimes emphasized by arpeggiation in the right hand, the left hand (Ex. 8.21*c*), or both. More substantial displacement between the melody and its accompaniment often involves a juxtaposition of different metrical units such as two against three (Ex. 8.18*d*) and five against six or two.[103]

Moreover, many examples of rubato include notated syncopation in the melody (Exx. 8.18*d*, 8.21*c*, 8.24*b*) or in the accompaniment (Ex. 8.17). In almost all cases of melodic syncopation (except for a few such as Ex. 8.24*a*), the melody is delayed rather than anticipated. This notated syncopation, together with Lachmund's phrase 'suspension of the rhythm' and the frequent linkage of rubato with ritenuto and portato, suggests that Liszt's rubato, unlike Chopin's 'hastening and lingering', most often involves the delay of a melody note. Such a delay produces an unexpected agogic accentuation, often emphasizing a long dissonant appoggiatura on a strong beat—the type which would have been indicated during the Baroque period by a small note. The syncopated notation, in company with the comments of students and revisions such as that in Ex. 8.21, also suggests that, at least on occasion, Liszt's rubato involves a displacement between melody and accompaniment. Exx. 8.21*c* and 8.22*b* remind us of similar examples by North (Exx. 3.2–3.4), Quantz (Ex. 3.8), Marpurg (Ex. 5.2), Türk (Plates XI and XII), and perhaps Hummel (Ex. 5.4).

It is difficult to determine, however, whether Liszt has completely incorporated such displacements, as he has the appoggiaturas, into his regular notation, or whether the performer can displace still other vertically aligned notes as part of tempo rubato. There are, indeed, many situations, as we have seen, in which the latter would be impossible or unlikely: when none of the melody notes coincides with a note in the accompaniment, when the accompaniment occurs in

---

[103] See *Malédiction* in *Franz Liszts musikalische Werke*, I/xiii, (199)17; and the First Mephisto Waltz in *Neue Ausgabe*, I/xv, 129.

both hands, when delays are already notated by means of rests, and when the rubato appears in an orchestral score—especially when the melody is played by more than one person. When a performer cannot initiate displacement, then he must perform the deceptive and seductive accentuation by simultaneously delaying the melody note and all notes notated vertically with it. This adds a slight amount of time to the measure, producing, as Lachmund states, 'subtle variations of tempo and expression within a free declamation'. This particular type of later rubato seems to have been characteristic of Liszt as a solo pianist and is therefore, no doubt, the type he had in mind in his piano compositions. With the orchestral works, however, he may have necessarily expanded his concept of rubato to refer more broadly to the general prolongation or slackening of time described by C. Kalkbrenner, Türk, P. A. Corri, Czerny, and Hummel.[104]

Liszt also applied the various types of rhythmic robbery in the transformation of entire melodies. He is sometimes considered the inventor of the process, since he employed it so effectively in the tone poems composed during the 1840s and 1850s. As we have seen, however, Schumann used a similar technique in *Carnaval*, and this in turn was preceded by examples of Beethoven, Bach, and others, reaching back to the twelfth century. Liszt certainly popularized the device among romantic composers and demonstrated its usefulness in programmatic music. We have already noted Ex. 8.24a and b from the First Mephisto Waltz, which, when compared with an unsyncopated version in (c), have been transformed by anticipating or delaying almost every note. The version in (d) represents a duple transformation of the melody. The Second Piano Concerto provides even more elaborate examples: the original 3/4 melody in Ex. 8.18a is later adapted to what amounts to 6/4 metre in (b) and 4/4 in (c). In (d) another version in 4/4 includes syncopation and, as we have seen, tempo rubato. Curiously, the two melodies in Exx. 8.18a and 8.24c are themselves related: each commences on scale degree 3, moves in the middle to the flatted third (or, enharmonically, the raised second degree), and emphasizes at the end the flatted sixth.

Ordinarily Liszt uses rhythmic transformation in the direct manner shown in these two examples. The listener can easily relate the transformed versions to the original melody. Examples abound in such well-known works as *Les Préludes* and the Faust Symphony. Particularly effective, it seems to me, is the transformation by the tenor soloist, at the end of the latter work, of the Gretchen theme from the second movement.

[104] According to Rosen, rubato in Liszt means 'only a certain evident and expressive freedom in interpreting the rhythm'.

Franck, Gottschalk, and Heller

Franck and Gottschalk were more distantly acquainted with Chopin. Rubato influence may have come to them mainly through Liszt or through others we have already mentioned.

César Franck (1822–90) lived in Paris from 1835 to 1842 and from 1844 on. From 1837 to 1842 he studied at the Conservatoire, preparing himself as a pianist and composer. The subscription list for the publication of his Trios Op. 1 in 1843 included the signatures of both Chopin and Liszt.[105] Around this same time he met Liszt, who later became a friend and supporter. Two of his compositions from this period include rubato. In the Ballade from 1844 he borrows the title as well as the rubato technique from Chopin. The single word *rubato* appears at the beginning of a singing melody which descends stepwise through an octave and a third in the right hand. It is accompanied by an arpeggiated figure in the left hand that provides two notes for every note in the melody. It is a typical example of the sort of accompaniment that can keep a strict beat and, at the same time, offer ample opportunity for the displacement of melody tones against it. The marking *sempre pp e rubato* four bars later suggests, moreover, that the first rubato is only momentary and acts to emphasize the beginning of a cantabile melody.[106]

In the *Premier grand caprice*, Op. 5 composed in 1843, the expression *sempre rubato* marks a passage in which harmonic movement quickens and motives are abstracted from the preceding melody. Although the accompanimental figure is broken between the hands, each note in the melody sounds with only a single bass note. Later in the piece portato eighth-note chords in the right hand, some marked with a tenuto sign, are to be played *espress. e rubato* against sixteenth notes in the left. Although one could theoretically maintain the strict rhythm in the left hand of both examples, one wonders if Liszt's deceptive suspension of time in both hands might be a possibility.[107]

At a much later date, Franck includes an example in his *Prélude, choral et fugue* written in 1884. Bach and baroque music are suggested not only by the title, but also by the figuration in a long section marked *come una cadenza* (Ex. 8.25a). When six measures from the beginning of this section are repeated, but with changed dynamics and mood, this fact is articulated by rubato (Ex. 8.25b).[108] This probably

---

[105] Léon Vallas, *César Franck*, originally *La Véritable Histoire de César Franck* (Paris, 1950), trans. Herbert Foss (London: George G. Harrap, 1951; repr. Greenwood, 1973), 48–50.

[106] *Piano Compositions by César Franck*, ed. Vincent d'Indy (Ditson, 1922; repr. Dover, 1976), 63 and 71.

[107] Ibid. 26 and 36.   [108] Ibid. 97–8.

Ex. 8.25 Franck, *Prelude, choral et fugue*, 1884

involves a substantial delay of the E directly below the word *rubato*, and perhaps also a conspicuous declamatory delivery of the following two notes as well, played as though one needed to take a new breath after the first half of the measure and reluctantly exerted new energy to get the sixteenth notes gradually moving again. Such a situation is clearly related less to the earlier rubato of Chopin than to the later type of Liszt. It is, indeed, more like the *quasi rubato* Bülow applied to the Beethoven sonata.

The American pianist Louis Moreau Gottschalk (born 1829 in New Orleans) was in Paris from 1842 until 1849 or so, and then briefly around 1851. He studied first with Charles Hallé, who, as we noted in Chapter 7, often heard Chopin perform from 1836 to 1848 and who described the special rhythmic effect in the Mazurkas. He later worked with Camille Stamaty, who had been a student of Frédéric Kalkbrenner, and who had performed in concerts with Chopin in 1832 and 1835.[109] It was Stamaty who prepared Gottschalk for a private recital in April 1845 at the Salle Pleyel, at which he played Chopin's E minor Concerto as well as works by Liszt and others. After hearing the recital, Chopin told Gottschalk, according to

---

[109] Atwood, *Fryderyk Chopin*, 59–60, 95–7, and reviews on 219 and 225. See also Hedley, *Selected Correspondence of Fryderyk Chopin*, 99 (letter of 12 Dec. 1831); and Niecks, *Chopin as a Man and Musician*, i, 241.

the latter's sister, that he would become the 'king of pianists'.[110] Gottschalk made his public Parisian debut in April 1849, and his style of playing was compared to that of Chopin.[111]

Like Chopin, he incorporated in some of his compositions the style and flavour of the folk-music of his native land. A preface included in several of his pieces reads:

The author in this morceau (which is entirely original) has endeavored to convey an idea of the singular rhythm and charming character, of the music which exists among the Creoles of the Spanish Antilles. Chopin it is well known transferred the national traits of Poland, to his Mazurkas and Polonaises, and Mr. Gottschalk has endeavored to reproduce in works of an appropriate character, the characteristic traits of the Dances of the West Indias.

In a 'Note by the Author', which also appears in several different pieces, Gottschalk explains further:

The characteristics of mingled sadness and restless passion which distinguish the piece would be utterly lost were not the accuracy of each changing rhythm fully sustained. The melody should stand out in bold relief from the agitated but symmetrical back-ground of the bass with the singing sonorousness and passionate languor which are the peculiar traits of Creole music. To give entire scope to the 'Ad Libitum' and 'Tempo rubato' and at the same time not to transcend the extreme limits of the time, is the principal difficulty as well as the great charm of the music of the Antilles...[112]

In his later years, Gottschalk charged his friend R. B. Espadero 'to do for him what our friend, Jules Fontana, has done for Chopin', namely, to arrange after his death for the publication of his remaining works. In an essay printed in the posthumous editions of some of Gottschalk's

---

[110] Louis Moreau Gottschalk, *Notes of a Pianist*, originally trans. from French by Robert E. Peterson (Philadelphia: J. B. Lippincott, 1881), ed. Jeanne Behrend (Knopf, 1964), pp. xxi–xxii. Curiously, Saint-Saëns reports in *Le Courrier musical*, 13/10 (1910), 386, that he studied with Stamaty when he was 10 years old (thus in 1845, the same year as Gottschalk's recital) and that Stamaty threatened to send him away if he ever learned that he had heard Chopin play ('he feared, for good reason, the comparison'), and that Stamaty constantly required him to play *molto espressivo* when he did not want to. Bitter that he had never heard Chopin play, Saint-Saëns then goes on to describe the aspects of his playing that he learned, as we saw in Ch. 7, from Grandval and Pauline Viardot.

[111] *New Grove*, vii, 571. See also Robert Offergeld, 'The Gottschalk Legend' in *Louis Moreau Gottschalk: The Piano Works*, ed. Vera Brodsky Lawrence (New York: Arno Press and *The New York Times*, 1969), i, pp. xv–xvi: 'Chopin also expressed his pleasure with Gottschalk's early compositions... In *Mes souvenirs* [Paris, 1863], Léon Escudier, the editor and music publisher, speaks of Chopin's regard for Gottschalk, and Antoine Marmontel, of the Conservatoire, reports it as coming from Chopin that he recognized in the American a sensitivity akin to his own.'

[112] *The Piano Works*, ed. Lawrence, iv, 114–15, for example, preceding *O, ma charmante, espargnez moi!* The compositions in this set are all reprints of early editions. I have corrected the spelling of a few words in the 2 quotations.

works, Espadero writes that 'in elective affinity he [Gottschalk] was, doubtless, nearer to Chopin than [to] any other artist', but also had 'enthusiastic admiration' for Liszt.[113] In his intense and wide-ranging activity as a performer throughout the world, Gottschalk certainly followed in the footsteps of the latter.

It is tempting to see a description of the earlier type of rubato in Gottschalk's words on Creole music quoted above. One wonders if the 'symmetrical back-ground of the bass' refers to a strict left hand, against which the 'singing sonorousness' of the right-hand melody 'stands out in bold relief'. In *La Bamboula, danse de Nègres*, composed around 1844–5, Gottschalk marks a melody 'legato il canto e tempo rubato', while 'la basse' (presumably the entire left-hand accompaniment) should be 'toujours rythmée'. Moreover, many notes in the single-line melody in the right hand bear caret-like signs of accentuation—an indication, perhaps, that they are the ones to be emphasized, probably by a delay in time. In the same way that Liszt's students described their master's rendition of Ex. 8.23, Gottschalk uses the word *rubato* to refer only to the flexibility in the melody and not to the total effect.[114] He must have felt that he needed to distinguish the effect he desired here from that other type of rubato with which either he or those who might perform the piece were more familiar.

Gottschalk writes the full expression *tempo rubato* in all examples published during his lifetime. For two passages within his *Solitude*, composed presumably between 1853 and 1856 and published two years after his death, he marks 'poco rubato'.[115] In his *Souvenir de Lima*, composed in 1860, and in the Sixth, Seventh, and Eighth Ballades (dates of composition unknown) he writes simply *rubato*.[116] *Bamboula* is the only piece in which rubato refers to an entire melody. Elsewhere it applies only momentarily to a few specific notes. In most

[113] Ibid., iv, 12, for example. Julian Fontana (1810–65) and Chopin had both been students of Elsner in Warsaw. Between 1832 and 1837 Fontana was in France and England, and from 1842 to 1851 in New York and Havana. Gottschalk must have met him in 1851 when both had returned to Paris. Espadero states that he himself knew Gottschalk for 18 years, thus from 1851 until the latter's death in 1869.

[114] Ibid., i, 93 (m. 151). In the edn. by William Hall in this volume, as well as in an early edn. from Paris at the UCLA Music Library, all the words quoted are printed in type of the same size. In a modern edn. by Eberhardt Klemm, in *Louis Moreau Gottschalk: Kreolische und karibische Klavierstücke* (Leipzig: Peters, 1974), 6, however, 'Tempo rubato' is printed in larger letters, with *legato il canto* and *la basse toujours rythmée* smaller and in italics. In this case 'Tempo rubato' appears to be the name for the total effect, with the other phrases separate instructions for producing it.

[115] *The Piano Works*, v, 128 and 132.

[116] Ibid., v, 189 and 190 in *Souvenir de Lima*; i, 47 in the 6th Ballade; i, 57, 58, 59, 61, and 64 in the 7th; and i, 69 and 71 in the 8th.

cases displacement between the right and left hands could take place, but occasionally it is already notated.[117] Rubato coincides in a number of cases with a vigorously articulated touch, indicated by vertical wedges or by a horizontal dash above a dot. Gottschalk could be considered a reformer of 'piano touch', according to Espadero, for among his bold innovations were his 'new ways of attacking notes, of intensifying effects, of using the pedals'.

Gottschalk borrowed a number of titles from Chopin, such as *berceuse*, *ballade*, and *mazurka*. A work entitled *Polonia* from 1859 is marked 'Tempo moderato di Mazurka' and contains the rubatos in Exx. 8.26 and 8.27.[118] Both have left-hand accompaniments typical of

Ex. 8.26 Gottschalk, *Polonia*, 1859

Ex. 8.27 Gottschalk, *Polonia*, 1859

Chopin and therefore might involve displacement between the hands. In Ex. 8.26 the vertical wedges presumably indicate a reluctant portato-like touch for this plaintive passage suggestive of Chopin. In this case

---

[117] See the *Souvenir de la Havane*, in *The Piano Works*, v, 172, where two triplets in 2/4 occur against a *habanera* bass (dotted 8th and 16th on the 1st beat, 2 8ths on the 2nd); *O, ma charmante* (iv, 119), where a triplet in the melody coincides with an 8th rest and 2 16th notes in the left hand; and *Suis moi!* (v, 231), where two triplets sound with an 8th note, quarter, and 8th in 2/4 metre.

[118] Ibid., iv, 239 and 240.

the delay or the deceptive suspension of time could emphasize the melodic F# in measure 162. In Ex. 8.27 the repeated notes in measure 180 suddenly take on a forceful declamatory tone in order to articulate the beginning of a four-measure phrase. Such a series of reiterated notes, each requiring necessarily some sort of articulation, presents a fascinating study in the application of a wide range of piano touches. One could employ the main seductive articulation on either the first, second, or third eighth note in measure 180, with succeeding notes absorbing, through the *affrettandosi*, some of the energy generated thereby. In a number of other pieces, he also links rubato with repeated notes or with an acceleration of tempo.[119]

In his Eighth Ballade Gottschalk adds to a previously simple accompaniment some chromatic eighth notes to be played *rubato espressivo*. The rubato coincides with an 'a tempo' which follows a fermata. Here again, displacement between the hands could theoretically take place. If the hands remain together, then a slight suspension of time could emphasize any of the dissonant eighth notes.[120] In Ex. 8.28, on the other hand, Gottschalk seems to indicate more precisely which note should receive the seductive accent. Following two measures marked *capriccioso* and *forte* comes a measure with the same general figuration, but, in order to emphasize an expressive augmented sixth chord, to be played *piano* and with rubato. The rest which follows the second melody note in measure 42 is scarcely audible if one depresses the sustaining pedal as marked. This rest, however, may serve to point out that the following note—the E#—is the significant note to be delayed, either simultaneously with the B# in the left hand or slightly following it. Through very subtle manipulation of the pedal, one might also, in spite of Gottschalk's markings, insert a brief but audible rest before the E#.[121]

In his *Miserere du Trovatore: Paraphrase de concert*, Gottschalk substitutes his own cadenza (Ex. 8.29) for the vocal cadence which Verdi originally provided for Leonora near the beginning of Act IV, immediately preceding the singing of 'Miserere' by the chorus.[122] At the beginning of Ex. 8.29 one should play *ad libitum* and *con molto espress.*, but toward the end *tempo rubato* and *delicato*. Since there is no accompaniment, there is no question of displacement. We do have

---

[119] Repeated notes occur with dashes and dots in *Ricordati* (ibid. v, 46) and with wedges in *La Chute des feuilles* (ii, 46). *Stretto* appears in *O, ma charmante* (iv, 119), *pressez* in the 8th Ballade (i, 71), and *precipitoso* in *Souvenir de Lima* (v, 189).

[120] Ibid., i, 69, m. 41 (cf. m. 21).

[121] Ibid., iii, 276.

[122] Ibid., iv, 57. The vocal cadence occurs in m. 60 of Act IV, sc. i of Verdi's *Il trovatore*; see the full score (Ricordi, 1913), 333, and the piano/vocal (New York: G. Schirmer, 1926), 181.

Ex. 8.28 Gottschalk, *Manchega*, 1851–6

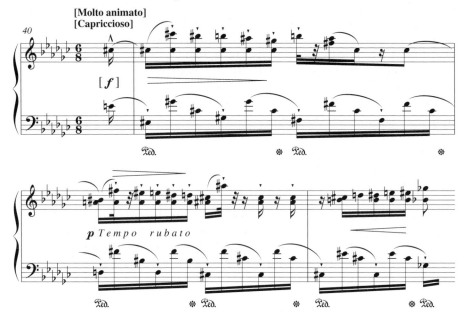

Ex. 8.29 Gottschalk, *Miserere du Trovatore*, 1856

here, however, a clear demonstration of the difference between *free rhythm*—of the sort generally found in cadenzas, preludial forms, and fantasies, and, increasingly in the nineteenth century, in the performance of music in general—and *rubato*, probably here involving the declamatory hesitations and light suspension of rhythm employed by Liszt. Gottschalk enhances the cadential drama with rubato also in other pieces.[123] Elsewhere he uses rubato to emphasize an expressive change of harmony, a chromatic line, or an unexpected rhythm,[124] and links it on occasion with *con lagrime* (with tears), *agitato*, *elegante*, or *marcato*.[125]

The pianist and composer Stephen Heller (1813–88), who knew Chopin for many years and was strongly attracted by his musical style, seems to show little influence in his use of rubato. He saw Chopin in Warsaw in 1830 when he was on a concert tour and met him frequently when he moved permanently to Paris in 1838. Although he was not a close friend, he later provided Niecks with descriptions of Chopin's playing and many anecdotes for his biography.[126] Among the compositions by Heller available to me, I have located ten between the years of 1853 and 1879 that include rubato. Like Chopin, he writes the single word *rubato*, but unlike Chopin he almost always indicates it for an unaccompanied cadenza-like melodic line in the right hand (somewhat like Ex. 8.29 by Gottschalk, but briefer). The rubato may refer to the free rhythm of the entire passage, or to just a jagged emphasis on the first few notes.

Heller notates one such section in a single unmetred bar beginning with eighteen quarter notes; other examples resemble recitative.[127] Sometimes he includes fermatas, either preceding or during the passage, sometimes marks of articulation or accentuation.[128] Often a

[123] See the 6th Ballade in *The Piano Works*, i, 47, and the 7th Ballade in i, 57 (and similar cadences on 58, 61, and the 2nd line of 64), 59 (lines 2 and 3), and 64 (line 3).

[124] Sensitive harmonic progressions occur in *Ricordati* (ibid., v, 46), *Reflets du passé* (v, 14), *Souvenir de la Havane* (v, 172), and *O, ma charmante* (iv, 119); chromatic lines in *La Chute des feuilles* (ii, 46) and *Suis moi!* (v, 231); the overcoming of inertia caused by the repetition of a rhythm in *Solitude* (v, 128 and 132).

[125] *Ricordati* (ibid., v, 46), *La Chute des feuilles* (ii, 46), *Suis moi!* (v, 231), and *Souvenir de Lima* (v, 190) or the 6th Ballade (i, 47).

[126] Niecks, *Chopin as a Man and Musician* i, 64, and especially ii, 154. See also Ronald E. Booth, Jr., 'The Life and Music of Stephen Heller', Ph.D. diss. (Univ. of Iowa, 1967, copy of typescript at UCLA ML), 43–6.

[127] *Nocturne-Sérénade* (Paris: J. Maho), copy in BL; the 4th of his *Préludes pour Mlle Lili* (Paris: J. Maho), copy at UCLA ML, modern edn. by Jean-Jacques Eigeldinger (Mâcon Cedex: Éditions Robert Martin, 1984).

[128] In 'Enfant qui pleure' from *Album dédié à la jeunesse*, modern edn. in *Stephen Heller, Ausgewählte Klavierwerke, Charakterstücke*, ed. Ursula Kersten (Henle, 1987), 65, a fermata appears above the 3rd note following the word *rubato*. In the *Sonatine* Op. 147, modern edn. by Ernst Herttrich (Henle, 1982), 21, 4 notes above the word *rubato* are marked with a caret sign; portato touch occurs in the Prélude Op. 150 No. 17 (Paris: J. Maho), copy at UCLA ML, modern edn. in *Ausgewählte Klavierwerke*, 87.

descent from a high note leads to a new theme or the return of one previously heard, or to an important cadence at the end of a sonata exposition or the end of a piece.[129] Sometimes the rubato involves a rhythmic adjustment when even eighth notes follow dotted eighths and sixteenths, or when groups of four sixteenth notes follow sixteenth-note triplets.[130] Occasionally a retard or acceleration precedes, follows, or occurs simultaneously with a rubato. In one of Heller's technical studies preparatory to playing works of Chopin, an accelerating figuration which begins the piece is marked *accel. rubato* when it returns later.[131] Although Heller wrote studies for Chopin's works and borrowed titles such as *nocturne*, *scherzo*, and *prélude*, he seems to derive his rubato from some other source.

The composers in this chapter transmitted Chopin's rubato, along with their own modifications of it, to the following generation. With the exception of Ernst and Viardot, they were all performing pianists. Most of the music in which they included rubato was written for the piano. Each, however, applied the device in his own particular manner. Touch seems often to play a prominent role. Liszt seems to prefer delaying a note, whereas Clara Wieck sometimes anticipates. Gottschalk occasionally accelerates during rubato; Liszt prefers ritenuto. Rubato may be momentary or may apply to an entire melody or section of a piece. Even when applied primarily to a single significant note, however, the accentuation caused by rubato may indeed affect succeeding notes as well.

One of the main trends was the movement away from the earlier rubato and toward a more and more general application of the later type. The displacement characteristic of the earlier type was occasionally prescribed by instructions, as in Ex. 8.23 or Gottschalk's *Bamboula*; sometimes it would be written into the notation. Gradually, however, and in different ways, such displacement was eliminated: when Viardot writes *suivez* in Ex. 8.4, when Gottschalk and Heller

---

[129] For a passage leading to a new idea, see the 4th Scherzo (London: Cramer, Beale & Wood), copy at BL; for one leading to the return of the 1st theme in a ternary form, see the Prélude Op. 150 No. 17. In the *Sonatine* Op. 147, rubato emphasizes a minor subdominant chord which is part of the dramatic cadence at the end of the exposition, and in 'Enfant qui pleure' it gives a special declamation to the opening melodic phrase when it recurs over a minor I6/4 chord to conclude the piece.

[130] *Sonatine* Op. 147; *33 Variationen über ein Thema von Beethoven*, ed. Ulrich Mahlert (Br. & H, 1985), Var. 12.

[131] *21 Technische Studien als Vorbereitung zu Werken von Fr. Chopin* (Leipzig: Kistner, 1879), copy in BL, *Heft* ii, No. 17 (preparation for Chopin's Nocturne Op. 9 No. 1), 19. See also Heller's 'Enfant qui pleure', in which the rubato and fermata in m. 29 are followed by *ritard.* and portato notes in the next.

include only a single line (as in Ex. 8.29), when Liszt uses syncopation in Ex. 8.17 that makes further displacement by a performer unlikely. Liszt himself apparently developed a personal concept of the later type of rubato, one that involved a slight addition of time for the sudden delaying of significant notes. This declamatory sort of rubato seems, however, to have become necessarily less personal and less vigorous as it was applied in orchestral works. Furthermore, it seems to me that the application of rubato to entire melodies is apt to evoke a less intense response on any specific note than a carefully marked momentary rubato. Liszt's *sempre rubato* and main tempo markings that include the word may have contributed to a weakening of the rhetorical power of the device.

As we turn to the later years of the nineteenth century, we will no longer be involved primarily with virtuoso pianist–composers, nor with a predominance of examples for the piano. These circumstances, along with the continuing increase in the amount of rhythmic flexibility desired and expected in performance, act to modify the concept of the later type of rubato.

9 THE LATER ROMANTIC RUBATO

THE second half of the nineteenth century and the early years of the twentieth represent a high point of freedom for the performer. Pianists and conductors especially—but, indeed, all performers in general—felt they had a right, if not a duty, to apply all manner of rhythmic flexibilities, and even to alter the composer's score on occasion, in order to achieve their own personal concept of expression. The word *rubato* now referred generally to the later type. The various methods of applying it were transmitted by Liszt and his numerous pupils and acquaintances to other countries and to later generations.

The musicians who suceeded Liszt, however, did not share his versatility and tended to specialize and acquire fame in a single area, as composers, performers, or writers. The relative isolation of these separate pursuits led during this period to different ways of comprehending rubato. Furthermore, rubato was conceived differently by composers in different countries and by performers on different instruments. The writers likewise had different points of view, depending upon whether they were critics, lexicographers, theorists, musicologists, or authors of *mémoires* or method books. We shall therefore consider these large categories separately, commencing with composers from the second half of the nineteenth century, moving then to conductors, writers, and performers, and dealing finally with Debussy and other composers of the early twentieth century.

Composers between Liszt and Debussy

The composers of this period apply both the momentary and the more continuous type of later rubato inherited from Liszt and the others who had known Chopin. The momentary type involves Liszt's seductive suspension of rhythm at a significant spot or a jagged unevenness on a few successive notes. The continuous type lasts longer, often throughout an entire section, and eventually seems to involve merely expressive tempo flexibility in general. This more extensive rubato was indicated, as we have seen, by the term *sempre rubato* or by the appearance of the word *rubato* in the marking for a main tempo or for a tempo change. In addition, it is no doubt intended when rubato is linked at the beginning of a section with other words, usually in small italic type, that clearly apply to the melody as a whole. When the word *rubato* or the expression *tempo rubato* occurs alone, however, it may refer to an extended section or to a momentary effect. Sometimes one cannot be certain which method is intended. In addition, it is sometimes difficult to distinguish between a momentary rubato and a brief one of one or two measures. Composers indicate the precise duration in only a few cases.

I have located almost one hundred and forty rubatos in the works of about thirty composers from this period. Less than half of the rubatos appear in piano music. Twenty-nine occur in the music of Tchaikovsky, the only one of these composers to use the word frequently over a relatively long period of time. The Russian rubatos are almost exclusively continuous; those from other countries, on the other hand, may be of either type, but after 1870 show a preference for the momentary.

From the period between 1850 and Tchaikovsky's first use of rubato in 1868 come, in addition to the examples by Liszt, Franck, and Heller already mentioned, a few others by Brahms, Bülow, Berwald, Tausig, Smetana, Saint-Saëns, and Jensen—all of which were probably directly influenced by Liszt. Brahms includes rubato in the three piano sonatas composed during 1852 and 1853. In 1853 he first met the Schumanns and might have learned from Clara about Chopin's visits in 1835 and 1836 and the subsequent rubatos in her own compositions.[1] Earlier in 1853, however, he had spent several weeks as a guest of Liszt at Weimar, and, judging from the nature of his rubatos, was influenced primarily by Liszt's concept of the device. Brahms approaches the final structural cadence at the end of the second movement of his First Sonata with the momentary rubato in Ex. 9.1. This seductive and stentato-like rubato marks a sudden change of dynamics

---

[1] Long after he had ceased marking the word in his own works, Brahms did come into contact with Chopin's rubato in his capacity as joint editor of the first *Chopin-Gesamtausgabe* (Br. & H, 1877–80). He prepared the Mazurkas, among other works, by marking corrections on what is apparently an earlier edn. by the same publishers. For a fac. of his revisions to the opening page of the Mazurka Op. 24 No. 1 (which, as we saw in Ex. 7.20, includes rubato in the 1st bar), see *Brahms-Kongress Wien 1983, Kongressbericht* (Tutzing: Hans Schneider, 1988), 356 in the article 'Johannes Brahms als Musikphilologe' by Thomas Leibnitz. In the Kalmus edn., which, according to Nicholas Temperley in *New Grove*, iv, 307, is a direct repr. of the Br. & H work of 1877–80, there is no tie between mm. 8 and 9 of Op. 6 No. 1 (see Ch. 7 n. 105, Ex. 7.14, and Pl. XVI), and in Op. 7 No. 1 *rubato* commences below the 1st note of the bar (see Ex. 7.17 and Ch. 7 n. 109).

Ex. 9.1 Brahms, Piano Sonata No. 1, 2nd movt., 1852–3

and texture, from a flowing melody in eighth and quarter notes to a vigorous portato that emphasizes each sixteenth note. It becomes part of a dramatic cadence which continues in measure 71 with similar figures over an augmented sixth chord, which leads to a high point in measure 72 with a 16/4 chord played *molto rit.* Another momentary rubato occurs with the recapitulation of the main theme in the Finale of the Second Sonata, where it is coupled with notated syncopation and *poco rit.*; the initial presentation of the theme, on the other hand, bears the main tempo marking 'Allegro non troppo e rubato'. In the Third Sonata, continuous rubato is indicated in the first movement along with a tempo change (*più vivo e rubato*) and in the Finale as part of the main tempo marking ('Allegro moderato ma rubato').[2]

Brahms met the Hungarian violinist Ede Reményi (Eduard Hoffmann) in 1850 and learned from him about gypsy music. In 1852 they gave concerts together, and together visited Liszt the following year. From 1852 to 1869 Brahms composed his twenty-one Hungarian Dances for piano four hands, but nowhere in them does the word *rubato* occur. Like Liszt, Brahms mistook gypsy music for Hungarian folk-music, and like him, does not refer to its bold rhythmic changes as rubato.[3] He simply indicates tempo changes in the usual ways with terms such as *rit., stringendo, con passione,* and *espressivo,* or by writing *sostenuto* followed by *a tempo.* His notated syncopation, like that already seen in the music of Mozart, Beethoven, and others, sometimes resembles the earlier type of rubato by intensifying a melody in comparison with an unsyncopated version heard earlier or simultaneously.[4]

[2] *Johannes Brahms sämtliche Werke* (Br. & H, 1926–7; repr., Ann Arbor, Mich.: J. W. Edwards, 1949), xiii, 13 for Ex. 9.1, 18(46) and 51(23) for the rubatos in Sonata No. 2, and 6(60) and 22(76) for Sonata No. 3.

[3] Heinz Becker writes in *New Grove,* iii, 156: 'Brahms learnt at first hand from Reményi how to play *alla zingarese* and to use rubato in ensemble playing.'

[4] Cf. mm. 1 and 11 in the Capriccio Op. 116 No. 7; see mm. 65 ff. in the 2nd movt. of the 3rd Symphony and mm. 397 ff. in the 1st movt. of the 4th.

Saint-Saëns, whom Liszt admired as a composer, pianist, and improvisor, directed performances of Liszt's music in France. He marks
'Tempo rubato' for a two-bar section of his Second Violin Concerto
composed in 1858, a cadenza-like passage containing groups of two
or ten notes for the violin against three in the accompaniment. Following the passage he writes 'Tempo deciso', presumably to cancel the
rubato. In his Piano Trio of 1863 he has an entire melody for violin
marked 'Poco più mosso quasi Allegretto, tempo rubato'.[5]
  These examples by Saint-Saëns, like Ernst's in Ex. 8.5, are some
of the few marked rubatos for violin before 1870. Adolf Jensen's
'Frühlingsnacht' of 1859 is an equally rare example for voice. In this
case, rubato involves a brief passage of about one and a half bars,
where the image of moonlight in the text is mirrored suddenly in the
music by a declamatory setting in portato repeated notes, with the
piano simply following the voice in chords. The marking 'a tempo' for
the following phrase probably indicates, like Saint-Saëns' 'Tempo
deciso', the end of the rubato.[6]
  After 1870 both the momentary and the continuous types of later
rubato spread also to other media. Tchaikovsky first marks rubato in
1868 (two years before Liszt's final rubato) and continues using the
word until 1893. Nine of his rubatos occur in piano music, and four
others are in passages for piano alone within works for piano and
orchestra. Fifteen, however, appear in works for other media: five for
orchestra and one for violin and piano, as well as five in operas and
four in songs. As a student, Tchaikovsky played piano music by Liszt
and heard his symphonic poems.[7] As a member of the Moscow
Conservatory, he was a colleague, from 1868 on, of Karl Klindworth,
one of Liszt's most talented piano pupils and editor in 1878 of the complete works of Chopin.[8] In 1855 Berlioz reported that Klindworth's
style of playing a concerto was so free that the first violinist during
a particular rehearsal shouted 'sempre tempo rubato!'[9] In 1878
Tchaikovsky dedicated his Piano Sonata Op. 37 to Klindworth. Although Tchaikovsky was not greatly influenced by Liszt's compositional style, he thus seems to have had ample opportunity to observe

  [5] Full score of concerto (Kalmus, 1900), 26–7, violin/piano version (Paris: Durand, 1950),
6–7; Trio Op. 18 (Paris: J. Hamelle, n.d.), 2nd movt.
  [6] Op. 1 No. 6. See *Ausgewählte Lieder und Gesänge* (Leipzig: F. E. C. Leuckart, 1890),
9–11; also *MMR* 17 (1887), 133, with a piano transcr. by E. Pauer on 155 (and rubato at the
same spot). In 1856 Jensen dedicated a piano trio to Liszt.
  [7] David Brown, *Tchaikovsky*, i: *The Early Years 1840–1874* (Norton, 1978, repr. 1986), 67
n. 28 and 72.
  [8] Alan Walker, *Franz Liszt*, ii: *The Weimar Years* (Knopf, 1989), 184–7.
  [9] Letter of 3 July 1855 to Theodore Ritter, in *New Letters of Berlioz 1830–1868*, trans.
Jacques Barzun, 2nd edn. (Greenwood, 1974), 140–3.

his usage of rubato—both as marked in his scores and as realized in performances by students such as Klindworth.

Almost all of the rubatos marked by Tchaikovsky and the other Russian composers extend over an entire piece or section. Most often they appear in a main tempo marking, but may occasionally be linked to a tempo or mood change. In Tchaikovsky's piano works the word *rubato* appears in the following main markings at the beginning of a piece: 'A tempo rubato non troppo mosso', 'Andante non troppo un poco rubato', 'Allegretto con moto e un poco rubato', 'Andante un poco rubato e con molto espressione', and 'Moderato mosso, molto rubato'.[10] In other pieces the return to the original tempo may be indicated by 'Tempo I, ma rubato', or a new mood described as *grazioso ed un poco rubato* or *appassionato ed un poco rubato*.[11] In the four works for piano and orchestra, cadenzas for solo piano are marked 'Cadenza a tempo rubato', 'Un poco capriccioso e a tempo rubato', 'Lo stesso tempo ma molto capriccioso e rubato', and 'Allegro non tanto, capriccioso e rubato'.[12] Only in the piano sonata from 1878, dedicated to Klindworth, does *un poco rubato* occur by itself. Here it appears at the beginning of a melody and again later for its repetition. The first time, the music changes from major to minor and from duple subdivisions to triple, so that a momentary suspension of the rhythm could be effective. The rubato could equally well apply, however, to the entire melody, which includes seven-note groups and cross-rhythms.[13]

Tchaikovsky's rubatos for other instrumental media also occur mainly in tempo changing phrases or in main tempo markings. In his *Sérénade mélancolique* he marks a section 'più mosso e agitato' in the score for violin and orchestra, but in the version for piano and violin changes the instruction, perhaps due to the more intimate chamber sound, to 'più mosso agitato e un poco rubato' and *molto espr.*[14] In his vocal duet 'Tears' he marks the lengthy piano postlude 'più mosso e un poco rubato', for it is an expressive reworking of the opening vocal melody.[15] His symphonic ballad *The Voyevoda* contains a section

[10] P. I. Tchaikovsky, *Polnoye sobraniye sochineniy* (Complete Edition of Compositions) (Moscow and Leningrad, 1940–71; repr., New York: Kalmus, 1974), *Valse caprice* in vol. li*b*, p. 3; *Six morceaux*, No. 6, li*b*, 129; *Les Saisons*, No. 4, lii, 17; *Douze morceaux* (*difficulté moyenne*), No. 12, lii, 133; and *Aveu passionné*, liii, 229.

[11] Ibid., *Dix-huit morceaux*, No. 5, liii, 126; *Six morceaux*, No. 6, li*b*, 127; and *Dix-huit morceaux*, No. 8, liii, 152.

[12] Ibid., Piano Concerto No. 1, xxviii, 82, and xlvi*a*, 67; Piano Concerto No. 2, xxviii, 198, and xlvi*a*, 154; *Concert Fantasia*, xxix, 25, and xlvi*b*, 12; and Piano Concerto No. 3, xxix, 216, and xlvi*b*, 102.

[13] Ibid., *Grande sonate*, Op. 37, 1st movt., lii, 174 and 188.

[14] Ibid., xxx*a*, 8 and lv*a*, 6.

[15] Ibid., *Six Duets*, No. 3, xliii, 114.

marked 'Moderato a tempo I' and, in smaller italics, *rubatissimo*—the latter referring probably to the entire section, which includes a *dolce cantabile* melody.[16] With the exception of these three pieces, all of Tchaikovsky's rubatos for instruments other than the piano are mentioned in a main tempo marking such as 'Adagio, ma a tempo rubato'. In other pieces Adagio is replaced by Allegro, Allegro vivo, Andante, Andantino, Moderato, or Larghetto; 'ma' is either omitted or replaced by 'e'; and 'a tempo' changed to 'un poco' or 'molto'.[17]

In applying rubato to orchestral works, Tchaikovsky was following the practice we saw in the later examples of Liszt. When he extended rubato to vocal works, however, there were few precedents beyond isolated examples such as Jensen's song or the Mazurkas of Viardot. Tempo rubato began, as we have seen, in vocal music, but there, until well into the nineteenth century, it was the earlier type and almost always a part of performance practice rather than notation. With Tchaikovsky, of course, rubato now refers to the later type and becomes part of the instructions written by the composer. In 'The Nightingale' from 1886 'Allegro molto rubato e capriccioso' may refer only to the opening piano prelude. In 'Frenzied Nights' from the same collection and in the fourth of the French Songs of 1888, however, the rubato in the main tempo marking seems to refer to the vocal as well as the piano parts.[18] His five operatic rubatos are all included in the main tempo marking for a solo section depicting an intensely emotional situation. In the first 'poco rubato' in *Mazeppa*, composed in 1881–3, Maria discloses her love for Mazeppa. Another rubato occurs in the same opera when Andrei tells Maria that he is dying and cannot help her. *The Slippers* from 1885 contains 'un poco rubato' when Oxana rejects Vakula's love. Another 'poco rubato' appears in *The Sorceress*, composed 1885–7, when Prince Yuri declares his love for Kuma. 'Larghetto, a tempo molto rubato' appears, finally, when Iolanta, in the opera from 1891 of the same name, asks Martha why he is weeping so bitterly.[19]

Tchaikovsky's vocal rubatos thus involve love or its rejection, as well as death and weeping. His instrumental rubatos sometimes appear in company with other intensifying devices such as accent signs,

[16] Ibid., xxvi, 191. See also 'Vivace, rubatissimo' at the beginning of Tchaikovsky's arrangement of Herman Laroche's Fantasy Overture *Karmozina* in lix, 60.

[17] See Manfred Symphony, ibid., xviii, 283; also *Capriccio italien*, xxv, 3, and 1*a*, 45; Suite No. 2, xix*b*, 7; and Suite No. 3, xx, 157.

[18] Ibid., *Twelve Songs*, Nos. 4 and 6, xlv, 117 and 128; *Six French Songs*, No. 4, xlv, 201.

[19] Ibid., *Mazeppa*, Act I, sc. ii and Act III, sc. xix, vi*a*, 67, vi*b*, 405, and xxxviii, 29 and 350; *Cherevichki* (The Slippers), Act I, sc. vi, vii*a*, 192, and xxxix, 93; *Charodeyka* (The Sorceress), Act III, sc. xvii, viii*b*, 189, and xl*b*, 76; and *Iolanta* (in a single Act), x, 29, and xlii, 16.

arpeggiation, cross-rhythms, or terms suggesting special warmth of expression (*espressivo, appassionato, cantabile, dolce,* or *sostenuto*). Often his rubato passages contain powerful appoggiaturas yearning to resolve into the chord below them.[20] Several times he combines *rubato* with *capriccioso.* All of these indications of intensity distinguish his rubato from the many other instructions for tempo fluctuation in his scores. In addition to the usual terms for increasing or decreasing the tempo, he also uses *a piacere, ad libitum, recit.,* or *sostenuto,* terms which are cancelled finally by *a tempo* or *tempo giusto,* and which, when applied to a solo voice or instrument, require *colla parte* for the accompaniment. Such tempo markings as 'Allegro giusto' or 'Andante giusto' presumably indicate the absence of flexibility.

Other Russian composers occasionally use rubato like Tchaikovsky, but far less often.[21] When they write *a piacere* or *ad libitum,* they grant the performer, of course, complete rhythmic freedom. Their continuous type of rubato, however, seems to be somewhat more specific: apparently it requires a further intensification of those expressive elements which the performer would have emphasized in any case, but to a lesser degree, in the absence of the word *rubato.*

In contrast to the practice in Russia, composers in other countries generally favour the momentary type of rubato after about 1870. There are a few in works by Bruckner, Mahler, Strauss, Reger, and Grieg. In Delibes's *Jean de Nivelle,* first performed in 1880, 'Tempo rubato' emphasizes the text: 'to hear his voice for the last time!'. Each syllable of 'pour la dernière fois' is marked with an accent sign and the passage involves a modulation from A minor to B flat major.[22] Three years later his opera *Lakmé* contains two similar passages marked 'Tempo rubato' for the phrase 'caresses with passionate kisses'. In this case the orchestral accompaniment is marked *suivez,* and the rubato is cancelled after two measures by the word 'Tempo'.[23]

In Italy we find both *rubato* and *rubando* in operas by Mascagni, Leoncavallo, and Puccini from 1890 to 1926. *Rubando* had been used in 1784, as we noted in Chapter 5, to describe Clementi's manner of playing the piano. With the Italian opera composers it is apparently synonymous with *rubato.* Pietro Mascagni writes *a tempo rubato* once

[20] See e.g. the Manfred Symphony, as well as the following piano works: the 4th piece in *The Seasons, Aveu passionné,* and the 5th of the *Dix-huit morceaux.*

[21] Balakirev, Glazunov, and Rimsky-Korsakov, for example.

[22] Piano/vocal score (Paris: Henri Heugel, 1880?), Act III, Air No. 17: 'Ah! Malgré les douleurs'.

[23] Piano/vocal score with English trans. (Ditson, 1883) and Italian (Paris: Heugel, 1884), Act I, Duet No. 6a: 'D'où viens tu?' Tchaikovsky admired the works of Delibes and thus might have been influenced by him in his own operatic rubatos; see David Brown, *Tchaikovsky,* ii: *The Crisis Years 1874–1878* (Norton, 1983), 59.

in his *verismo* opera *Cavalleria rusticana*, first performed in 1890.[24] In *L'amico Fritz* from 1891, however, he uses both terms: in one scene *un poco rubando* occurs fourteen measures after *un poco rubato*; later both *rubando un poco il tempo* and *rubato* appear simultaneously with *col canto* in the accompaniment.[25] His opera *Iris* includes five rubandos, whereas *Parisina* has four rubatos, nine rubandos, and one *rub.*; *Il piccolo Marat* from 1921 twice indicates *rubando appena*.[26]

Mascagni's rubatos and rubandos involve very brief passages, sometimes limited by 'a tempo' to one or two notes. Most are controlled by a singer, but a few occur in the orchestra. Almost all involve the sudden introduction of faster notes or an unusual group such as a triplet, quintuplet, or sextuplet. Other words also appear in the form of both past and present participles: thus, *legato* and *legando*, *sostenuto* and *sostenendo*, *ritenuto* and *ritenendo*, *trattenuto* and *trattenendo*. In addition, there are many tempo changes, as well as superlative expressions such as *tenutissimo*, *sostenutissimo*, *estremamente lento*, or *con estrema dolcezza*. Rubato and rubando seem to indicate ordinarily some sort of emphatic articulation of text or motive, with alteration of tempo perhaps incidental or very slight. For tempo changing there are many terms in Italian operas of this period, for *col canto* in the orchestra, followed by *a tempo*, coincides not only with *rit.*, *rall.*, *accel.*, *affrettando*, and *trattenuto*, but also with *a piacere*, *ad libitum*, *senza tempo*, *senza misura*, *senza rigore di tempo*, *allargando*, *recitativo*, *stentato*, even with *tenuto* and *sostenuto*, and with fermatas and accent marks.

Rubato and rubando also occur in operas by Ruggero Leoncavallo from the turn of the century[27] and in works by Giacomo Puccini published between 1899 and 1926. With Puccini, as in the operas of Tchaikovsky and Delibes, the subject matter involves love or death. Both are combined when Cavaradossi, in prison awaiting death, writes his farewell letter to Tosca. A clarinet plays a melody twice: first in 3/2 metre and marked *dolcissimo vagamente*, then as shown in Ex. 9.2*a*. In each case the four notes under the slur are marked *rubando*. When Cavaradossi sings the melody, the same four notes are rhyth-

[24] Full score with German and Italian texts (New York: Broude Bros., n.d.), Finale: a 16th-note quintuplet for voice.

[25] Piano/vocal score (Milan: Casa Musicale Sonzogno, 1891; repr. 1984), Act I, sc. iii, and Act II, sc. ii and iv.

[26] Full score of *Iris* (Ricordi, 1925), Acts II and III; *Parisina*, piano/vocal score (Milan: Edizioni Curci, 1974), Acts II, III, and IV; and *Il piccolo Marat*, piano/vocal score (Milan: Casa Musicale Sonzogno, 1921), Act III.

[27] See *Zazà*, piano/vocal score (Milan: Edoardo Sonzogno, 1900), Acts I and III; and *Maia*, piano/vocal score (Sonzogno, 1908), Act II.

Ex. 9.2 Puccini, *Tosca*, Act III (1899)

a Cl., sounding pitch

b Cavaradossi
   con grande sentimento vagamente

mically transformed as in Ex. 9.2*b*, but without the word *rubando*.[28] A rubando passage in the orchestra is associated with Suor Angelica when she recalls the death of another nun, and again as she kills herself after learning that her child is dead. The music is similar both times and the extent of each *rubando* is indicated by dotted lines that lead to the *a tempo* which marks its conclusion. The first time it lasts for almost two measures in 2/4, the second time for about one and one half measures in 4/4 and one in 2/4.[29] Suicide is again involved when Turandot tries to force Liù to divulge the identity of the Unknown Prince; when asked the source of her courage, Liù replies 'tanto amore segreto', with the four notes on '-mo-re se-' marked 'un po' rubato' and followed immediately by 'a tempo' on '-gre-to'.[30] Happier rubandos appear in *Il tabarro*: when Giorgetta recalls her childhood in Paris, and later when she and her lover Luigi remember it together. Violins play the melody in Ex. 9.3*a*, with *rubando* carefully marked for the last three notes in the 3/8 measure and cancelled by the following *a tempo*. The last two rubando notes are marked with tenuto dashes, and the accentuation on the first is emphasized by the staccato note that precedes it.[31]

When a composer writes *rubato* or *rubando* in a score, he leaves to the performer, of course, considerable freedom in the exact manner of execution. A number of composers around 1900 attempted in various

---

[28] Full score (Ricordi, 1924), piano/vocal score with English (New York: G. Schirmer, 1956), Act III.

[29] Full score (Ricordi, 1918; repr. 1980), 21 and 91; piano/vocal score (Ricordi, 1930), 23 and 92–3.

[30] Full score (Ricordi, 1926; repr. 1977), Act III, 379; piano/vocal score with English (Milan: Ricordi, and New York: Franco Colombo, 1929), 334.

[31] Full score (Ricordi, 1917; repr. 1980), 56 and 65; piano/vocal score with English (Ricordi, 1956), 52 and 59.

Ex. 9.3 Puccini, *Il tabarro* (1918)

a 1st vns.

rubando   *a tempo*

p

b Theoretical notation in 3/4

ways to incorporate fluctuating tempo in more precise ways. On the first page of his Second Symphony, Mahler describes in prose the tempo changes for the opening bass theme and gives metronome numbers for the melodic figures and the pauses.[32] On other occasions Mahler changes time signatures to indicate the precise duration of prolongations or breath pauses that could have been marked less precisely by fermatas or commas. At the beginning of his Eighth Symphony, for example, he lengthens the third note of the melody to a half note and changes the duple metre to 3/4 for a single measure. It sounds to a listener as though this note in the later version of the melody were being prolonged in response to a fermata or tenuto mark, or to some instruction such as *Zeit lassen*, *ritardando*, *sostenuto*, *allargando*, or *tempo rubato*.[33] Mahler frequently employs the expression *Zeit lassen* (allow time), presumably to indicate a momentarily flexible tempo.[34] Richard Strauss also changed time signatures in order to specify the precise prolongation of certain notes. In *Till Eulenspiegel*, however, he wrote the theme completely in 6/8, as in Ex. 9.4*a*. This has the accentual effect of the changing signatures shown in (*b*), which, in turn, seems like one possible solution to the prolonged notes shown, in 3/4 metre, in (*c*).

Puccini changes time signatures in Ex. 9.3*a*. The same results could conceivably have been achieved by continuing the 3/4 metre as in Ex.

---

[32] *Sämtliche Werke: kritische Gesamtausgabe* (Vienna, 1960–), ii, 3 n.; trans. William Bolcom in George Rochberg, *The Aesthetics of Survival: A Composer's View of Twentieth-Century Music* (UMP, 1984), 91. Rochberg refers to this as 'the attempt to provide clear directions for producing rubato, i.e. elastic durational values'.

[33] *Sämtliche Werke*, viii, 3, m. 3 (cf. m. 9). Edward Kravitt refers to this sort of notation as 'measured tempo-modification' in 'Tempo as an Expressive Element in the Late Romantic Lied', *MQ* 59 (1973), 508–14; see his Ex. 4*a* on 510.

[34] According to Thomas S. Wotton, in *A Dictionary of Foreign Musical Terms* (Br. & H, 1907), 224, the term means 'allow time; do not hurry'. In Christine Ammer's *Musician's Handbook of Foreign Terms* (New York: G. Schirmer, 1971), 70, it means 'perform freely (with regard to tempo)' and 'allow for a pause'.

Ex. 9.4 R. Strauss, *Till Eulenspiegel*, 1895

a Theme written

b With changing time signatures

c In 3/4 with notes prolonged

d Rhythmic transformation (transposed)

e Another transformation

9.3*b* and relying on the effect of the *rubando* to lengthen the final two notes of the measure. Twenty measures after Ex. 9.3*a*, the same melody recurs without rubando. In either case, however, the musical sense seems best revealed when one applies a conductor's beat in 3/4, as in Ex. 9.3*b*, but allows for rhythmic flexibility as one approaches cadence. In Ex. 9.2, (*b*) represents an expansion of (*a*) from 3/4 to 4/4 metre. Actually, the *rit.* at the beginning of (*a*) might have resulted in the longer note values shown in (*b*), and the *rubando* might have caused the distortion of note values later in the measure. Considering the great amount of rhythmic flexibility employed by performers during this period—whether indicated in the score or not—it is possible that there could be very little difference, as a matter of fact, between renditions of the different versions in Exx. 9.2 and 9.4. The melody in Ex. 9.2*b* apparently represents the composer's attempt to control the music more accurately by taking away from the performer the sort of freedom involved in the *rit.* and *rubando* of (*a*). He thus seems in (*b*) to have selected and frozen into notation one specific solution that might have resulted from (*a*).

The process of rhythmic transformation, which had been employed so successfully by Liszt, reaches a high point of intensity and complexity with the late romantic composers. In Ex. 9.4*d* and *e* Strauss transforms the theme in (*a*) in order to represent Till in different moods and situations. Notes are thus altered in their duration, their accentuation, and sometimes their metre, in order to produce a new rhythmic syntax. Such robbery is more complex than that of Liszt in Exx. 8.18 and 8.24, and it becomes an increasingly subtle means of manipulating melodic material for programmatic purposes. In spite of the extreme nature of this sort of rhythmic robbery, however, the procedure was not considered a type of rubato and was never identified in scores by this word.

Tempo rubato, on the contrary, was something rhythmic that the performer was expected to do, either in response to seeing the term in a score or on his own initiative. It is perhaps surprising that composers, during this very period when performers applied the greatest amount of tempo flexibility, should include the word *rubato* so seldom in their scores. With the exception of Liszt, Tchaikovsky, and Mascagni, no late romantic composer before Debussy marked *rubato* or *rubando* on more than ten occasions—and sometimes only once or twice. Some composers of the time never used the word, as far as I can tell, at all. Until very late in the century, there are very few rubatos in vocal music of any sort: almost none in songs and relatively few in operas. There are also none in the works of Richard Wagner, but he wrote in great detail concerning the tempo flexibility to be applied by conductors.

The Conductors

In 1852 Wagner wrote an essay for conductors and producers describing the proper performance of *Tannhäuser*. In order to avoid the sort of rhythmic freedom formerly associated with recitative, he advises the singer to observe first the strict values of the notes. After the music has been fully understood, however, he urges the addition of a new kind of flexibility, 'an almost entire abandonment of the rigour of the beat':

From the moment when the singer has taken into his fullest knowledge my intentions for the rendering, let him give the freest play to his natural sensibility, nay, even to the physical necessities of his breath in the more agitated phrases; and the more creative he can become, through the fullest freedom of Feeling, the more will he pledge me to delighted thanks.

The conductor is then to follow the singer, but the words and music of the vocal part should be copied into every orchestral part so that each player will understand the phrasing.[35] Although Wagner never uses

[35] 'Über die Aufführung des *Tannhäuser*: eine Mitteilung an die Dirigenten und Darsteller dieser Oper', in *Richard Wagner's Prose Works*, trans. William Ashton Ellis, iii (London, 1894; repr., New York: Broude Bros., 1966), 174–5.

the word *rubato*, Henry Finck describes in 1889 the 'dramatic *rubato*' of the bass singer Emil Scaria, produced by dwelling on important syllables. Scaria had studied with García in London and sang in many Wagnerian roles during the 1880s.[36]

In 1855 Wagner conducted a series of concerts in London. Most of the critics, accustomed to a stricter Mendelssohnian approach to rhythm, complained about 'retardations and accelerations of time' and 'ill-measured rallentandi'. Henry Smart describes the tempo changes more precisely:

he prefaces the entry of an important point, or the return of a theme— especially in a slow movement—by an exaggerated ritardando; and . . . he reduces the speed of an allegro—say in an overture or the first movement— fully one-third, immediately on the entrance of its cantabile phrases.[37]

In a letter from the same year, Berlioz mentions 'Wagner's delight at leaving London' and the 'renewal of the critics' fury against him . . . , for he conducts in a *free style* too, as Klindworth plays the piano'.[38] This was the letter we previously noted in which Berlioz relates that Klindworth's free rhythm provoked a member of the orchestra to yell 'sempre tempo rubato!'

Wagner explains his point of view, finally, in the book from 1869 entitled *On Conducting: A Treatise on Style in the Execution of Classical Music*. The central idea is that each theme, even within a single movement, has its own personality and hence requires its own proper tempo. This involves not only changing the tempo to suit each melody, but also making a transition from one tempo to another, with the appropriate retards and accelerations occurring in the proper places. In addition, fast movements are taken faster and slow movements much slower than formerly. These concepts are applied, then, to the various themes in the sonata forms. Wagner uses the expression 'Tempo modification' (*Modifikation des Tempos*) and gives examples from Beethoven's 'Eroica' Symphony and *Egmont* Overture, Weber's *Freischütz* Overture, and his own Overture to *Die Meistersinger*. 'We

[36] Henry T. Finck, *Chopin and Other Musical Essays* (New York, 1889; repr., Freeport, NY: Books for Libraries Press, 1972, as *Essay Index Reprint Series*, xi), 221–2 (in the essay on 'Italian and German Vocal Styles'). Concerning Scaria, see *New Grove*, xvi, 549. He may have studied also with García's sister Pauline: see Angela F. Cofer, 'Pauline Viardot-Garcia: The Influence of the Performer on Nineteenth-Century Opera', D.M.A. thesis (Univ. of Cincinnati, 1988; UM 89–8,464), 172.

[37] *Sunday Times* (17 June 1855), 3; quoted by Elliott W. Galkin in *A History of Orchestral Conducting in Theory and Practice* (New York: Pendragon Press, 1988), 579. See also his quotations from *Athenaeum* and *Musical World*.

[38] *New Letters of Berlioz 1830–1868*, trans. Jacques Barzun, 2nd edn. (Greenwood, 1974), 142–3.

may consider it established', he states, 'that in classical music written in the later style, *modification of tempo* is a *sine qua non*.' He conceives of such modification, however, as enhancing the expression of individual parts of a movement in order to articulate the structure as a whole. He recognized the dangers of modifications added for less than purely musical reasons:

nothing can be more detrimental to a piece of music than *arbitrary nuances of tempo*, ... such as are likely to be introduced by this or that self-willed and conceited time-beater, for the sake of what he may deem 'effective'. In that way, certainly, the very existence of our classical music might, in course of time, be undermined.[39]

In 1872 Hanslick commented on Wagner's conducting of Beethoven's 'Eroica' Symphony. He accepted the tempo modifications in the last movement, but thought the idea was exaggerated elsewhere:

After a very fast beginning of the first movement, for example, he takes the second theme (forty-fifth measure) conspicuously slower, thus disturbing the listener's hardly confirmed establishment in the fundamental mood of the movement and diverting the 'heroic' character of the symphony toward the sentimental.[40]

Hanslick was concerned about establishing 'purely individual points of view as exclusively valid laws':

Were Wagner's principles of conducting universally adopted, his tempo changes would open the door to intolerable arbitrariness. . . . *Tempo rubato*, that musical seasickness which so afflicts the performances of many singers and instrumentalists, would soon infect our orchestras, and that would be the end of the last healthy element of our musical life.[41]

Wagner's theories flow naturally from the tempo fluctuations described in the last chapter for Liszt's conducting. They bear remarkable resemblance to Schindler's description in 1840 of the way Beethoven performed his own piano sonatas (see Chapter 6).[42] In any event, the sort of rhythmic freedom practised earlier by solo pianists finally reaches the medium of the orchestra and is codified into concrete rules by Wagner. In spite of Hanslick's fears, Wagner's ideas did

---

[39] *Über das Dirigieren*, trans. Edward Dannreuther in *On Conducting* (London: William Reeves, 1887, 4th edn., 1940; repr., St Clair Shores, Mich.: Scholarly Press, 1976), 43 and 67.

[40] *Eduard Hanslick: Vienna's Golden Years of Music 1850–1900*, trans. Henry Pleasants, 108. Wagner himself mentions the themes beginning in mm. 84 and 288 as appropriate for tempo change; see *On Conducting*, trans. Dannreuther, 42.

[41] *Vienna's Golden Years of Music*, trans. Henry Pleasants, 108.

[42] See Harold C. Schonberg, *The Great Conductors* (S & S, 1967), 136. On p. 139, he describes Wagner's ideas: 'We get an approach to rhythm and tempo that demands perpetual fluctuation, a kind of super rubato'.

become the accepted manner and lived on in the conducting styles of Bülow, Nikisch, Mahler, Richard Strauss (during his earlier years), Mengelberg, and Furtwängler. A reaction against them appeared in the work of Strauss, Muck, and Weingartner, which led ultimately to a reversal of these ideals in the twentieth century by conductors such as Toscanini.

Bülow, whom we have already noted as a pianist, a composer, and editor of the Beethoven sonatas, also worked closely with both Liszt and Wagner as a conductor. From 1880 to 1885 he conducted at the Meiningen court and made the orchestra one of the most excellent of its day. Franz Kullak, however, was offended by the 'interpretive nuances' in his performance of Beethoven's Seventh Symphony in 1882, specifically the tempo changes and breath pauses (*Luftpausen*, which he equates with Schindler's *caesuras*).[43] Hanslick, in a generally favourable review of a concert in 1884, expresses 'reservations about certain rubati' which he considers 'unmotivated and affected'. He admits, however, that 'metronomic evenness of tempo has, in any case, been disavowed by all modern conductors'. Yet he seems to have doubts about conductors who imitate Wagner: 'In his treatise *On Conducting*, Richard Wagner gave expression to a number of dangerous theories, but when he himself conducted, one readily accepted many liberties.'[44] As both pianist and conductor, Bülow was characterized by contemporaries as intellectual more than emotional, calculating rather than spontaneous.

He apparently became even more eccentric in his later years and his manner of conducting was vigorously attacked, shortly after his death, by Felix Weingartner in his book on conducting. He complains that Bülow's tempo changing was exaggerated, capricious, and unmotivated, and that it was imitated by a host of 'little Bülows', whom he describes as follows:

The tempo-rubato conductors . . . sought to make the clearest passages obscure by hunting out insignificant details. Now an inner part of minor importance would be given a significance that by no means belonged to it; . . . often a so-called 'breath-pause' would be inserted, particularly in the case of a *crescendo* immediately followed by a *piano*, as if the music were sprinkled with *fermate*.

---

[43] *Der Vortrag in der Musik am Ende des 19. Jahrhunderts* (Leipzig: Verlag von F. E. C. Leuckart, 1898), 15–22. In the 1st movt., for example, Bülow, according to Kullak, becomes gradually slower following m. 309, extremely slow by m. 316, then suddenly resumes the original tempo at m. 319; he slows also (pp. 16–17) for the passage beginning in m. 401. At the end of m. 155 he inserts a breath pause to mark a sudden change from $f\!f$ to pp (pp. 15–16).

[44] *Vienna's Golden Years of Music*, trans. Henry Pleasants, 273–4. See Edward F. Kravitt, 'Tempo as an Expressive Element in the Late Romantic Lied', *MQ* 59 (1973), 504 n. 27, concerning tempo modification even by a conductor admired by Brahms.

These little tricks were helped out by continual alterations and dislocations of the tempo. Where a gradual animation or a gentle and delicate slowing-off is required—often however without even that pretext—a violent spasmodic *accelerando* or *ritenuto* was made.

He concludes more optimistically, however, by noting that some of the younger conductors have turned away from this style and that 'the "tempo rubato" is not at such a premium as formerly, and that its unhealthy excrescences represent a fashion that is gradually dying out if not yet quite extinct'.[45]

Wilhelm Furtwängler attempted later to apply Wagner's concepts with greater musical integrity. He was aware of Weingartner's complaints, for he refers in his own writings of 1918 and 1934 to 'rubato-musicians' and 'rubato-pianists'.[46] He distinguished between the real and the false rubato in an interview from 1937:

it is possible to tell from the treatment of the so-called *rubato* . . . —which is a temporary relaxation of rhythm under the stress of emotion— . . . whether or not the impulses provoking it are in accordance with the real feeling of the passage or not, whether they are genuine or not. For as soon as this *rubato* is 'put on' and is intentional and calculated, it becomes, as it were, automatically exaggerated. This is less noticeable with an orchestra, . . . although a false *rubato* can frequently be heard even there. But amongst pianists . . . there is a tendency nowadays to use this trick unrestrainedly, with positively devastating results.[47]

The personal, subjective manner became the dominating style for conductors as well as pianists during the late nineteenth and early twentieth centuries. The manipulation of time and rhythm was the main expressive element. Bülow applied tempo changes, according to Hanslick, 'at those places where they would seem appropriate to him if he were playing the same piece on the piano'.[48] The same temptation that plagued the virtuoso pianists, of course, then lured the virtuoso conductor—the temptation to 'divert the attention of the audience from the music to himself', and to make 'not the work but the conductor . . . the chief thing'.[49]

---

[45] *Über das Dirigieren* (Leipzig, 1895); trans. of 3rd rev. edn. of 1905 by Ernest Newman as *On Conducting* (Scarsdale, NY: E. F. Kalmus, [1925]), 27–8, 40; see also 13–21, 30–4.

[46] *Ton und Wort: Aufsätze und Vorträge 1918 bis 1954* (Wiesbaden: F. A. Brockhaus, 1955), 13 (from 'Anmerkungen zu Beethovens Musik') and 76 (from 'Interpretation—eine musikalische Schicksalsfrage', which is also printed in *Das Atlantisbuch der Musik*, i, 610).

[47] *Gespräche über Musik* (Zurich: Atlantis-Verlag, 1948; 9th edn., Wiesbaden: F. A. Brockhaus, 1978), 61–2; trans. by L. J. Lawrence in *Concerning Music* (B & H, 1953; repr. Greenwood, 1977), 52. See p. 72 concerning 'false, calculated rubato' and p. 75 for the dates of the interviews.

[48] *Vienna's Golden Years of Music*, trans. Henry Pleasants, 273.

[49] Weingartner, *On Conducting*, trans. Ernest Newman, 22.

Wagner never, as far as I know, uses the word *rubato*, either in his scores or his writings. Liszt would not have employed the term, as we have seen, for Wagner's modification of tempo. Berlioz, Hanslick, and Weingartner, however, seem to use it for this purpose when speaking in a derogatory sense, and Finck states that 'Wagner's essay, *On Conducting*..., is chiefly a treatise on modifications of tempo, or what is usually called tempo rubato.'[50] Such general tempo modifications, however, seem to be operating on a somewhat broader formal level than the continuous type of later rubato marked in scores by composers such as Tchaikovsky. Furtwängler may be referring to an even more momentary type, as Finck certainly is when he speaks of significant notes prolonged by Scaria. The breath pauses mentioned by Kullak and Weingartner remind one of Schindler's caesuras, but they may in some cases have seemed like the 'sudden, light suspension of rhythm' which Liszt called *rubato*. During this same period, however, the word *rubato* had an even wider variety of meanings for the various categories of writers.

The Writers

Writers sometimes describe the later rubato of this period in terms of changes in tempo, in which case they employ words such as 'retarded and accelerated', 'slackened and quickened'. At other times they speak of changes in note values and refer to lengthening and shortening, prolonging and curtailing. The concept of compensation, initiated probably by the lexicographers, becomes attached to both points of view. Various essayists—critics, teachers, or performers—support or reject this concept, others ignore it altogether. Of the latter, some are generally supportive of rubato, others are violently opposed. Still other writers attempt to weave it into a general theory of rhythm and expressive performance. Historians, finally, trace rubato back in history and rediscover the earlier types.

Turning first to the lexicographers, we find *rubato* defined in a series of English-language dictionaries that span the entire period. In 1851 John Stowell Adams published at Boston his *5,000 Musical Terms*, which, as stated on the title page, includes previous works by Busby, Hamilton, and others.[51] Under the word *rubato* he quotes

---

[50] Henry T. Finck, *Success in Music* (New York: Charles Scribner's Sons, 1927; copyright 1909), 429. This concept of rubato lives on in later sources involving conducting. See Benjamin Grosbayne, *Techniques of Modern Orchestral Conducting* (HUP, 1973), 159–65.

[51] *Complete Dictionary of Latin, Greek, Hebrew, Italian, French, German, Spanish, English and Other Words, Phrases, Abbreviations, and Signs...the Whole including the Celebrated Dictionaries of Dr. Busby, Czerny, Grassineau, and Hamilton* (Ditson, 1851), 118 and 134.

Hamilton's definition (see Chapter 5), which involves the lengthening and shortening of notes in such a way that 'the aggregate value of the bar may not be disturbed'. He defines *Tempo rubato*, however, as follows:

TEMPO RUBATO . . . implies that the time is to be alternately quickened and retarded, but so that one process may compensate for the other.

The first part of this definition comes directly from Busby (quoted in Chapter 5), but Adams has replaced Busby's final phrase ('for the purpose of enforcing the expression') with Hamilton's idea of compensation. In the 1865 edition of the dictionary, Adams repeats this definition for *Tempo rubato*, but reduces *rubato* to simply 'robbed, borrowed', thus eliminating Hamilton's remarks about changing note values. He has now confined rubato to tempo changes, but linked them to the concept of compensation.[52]

When Hamilton stipulates that the value of the bar is not to be altered, it is natural to assume that he is describing the earlier type of rubato, even though he does not mention the strict accompaniment, or the displacement between it and the melody. One wonders, then, about Adams' point of view. He may have intended in 1851 to provide a definition for both the earlier and the later types of rubato. On the other hand, he may have thought that Hamilton and Busby were describing the same effect in different ways. His omission of note changing in 1865, however, tempts one to imagine that he was thereby deliberately omitting the earlier type for the reason that it had by that time become almost obsolete. One wonders also about his understanding of Busby's definition—whether compensation was merely a further explanation of Busby's 'time alternately accelerated and retarded' or a new limitation similar to that imposed by Hamilton on note changing. In any event, this is the first time in history, to my knowledge, that compensation is applied to tempo changing. I know of no earlier source that suggests it for the free forms—the recitative, the cadenza, and the preludial forms discussed in Chapter 1—or even for the subtle flexibility in the stricter forms (see the comments of Türk, Czerny, Schindler, and others at the end of the eighteenth and early nineteenth centuries in Chapters 5 and 6). The only place it had previously existed seems to be in a melody constrained by a strict accompaniment in the earlier type of rubato.

There is a strong tendency for lexicographers—as well as other authors, of course—to retain in their own works elements from their

---

[52] *Adams' New Musical Dictionary of Fifteen Thousand Technical Words, Phrases, Abbreviations, Initials, and Signs Employed in Musical and Rhythmical Art and Science* (New York, 1865; repr. in *MD*), 195 and 231.

predecessors as a sort of historical thread of authority. In the case of rubato, the idea of compensation provides one thread that leads back to Jousse and Hamilton. Another thread of continuity is note changing, which, although rejected finally by Adams, does continue throughout the period in works by others. The alteration of note values is especially attractive for lexicographers, because it enables them, more easily than with tempo changing, to describe exactly what is robbed and what is stolen. The third thread, of course, is tempo changing, coming from Busby. These three aspects are combined in various ways in succeeding dictionaries.

In 1854 Moore simply quotes Hamilton and Busby. Stainer and Barrett cite both note and tempo changing in 1876, but with compensation for neither. Niecks (1884) also mentions both, but with notes changing only 'at the expense' of others.[53] Clifford (1893), Bekker (1908), and Dunstan (1908) confine themselves to note changing without compensation; Baker (1895) and Matthews & Liebling (1896) mention only note changing with compensation.[54] Wotton, finally, speaks in 1907 only of compensated tempo changing.[55] Niecks, as well as several later authors, confines rubato to a few notes or measures. Matthews & Liebling use tempo changing nomenclature to refer to notes that are retarded or accelerated. Similarly, Baker speaks of the prolongation and 'equivalent acceleration' of notes, and Hubbard of 'a slight deviation from the tempo . . . gained by retarding one note and quickening another'.[56] Some of these sources mention the expressive purpose of rubato, and Clifford states that 'this style of performance is used with great effect in the modern intensely emotional school of music.'

[53] John W. Moore, *Complete Encyclopedia of Music* (Boston: J. P. Jewett, 1854; repr. Ditson, 1880, with appendix from 1875), 834 and 916; John Stainer and W. A. Barrett, *A Dictionary of Musical Terms* (Ditson, n.d.; originally London, 1876), 381 and 431; Frederick Niecks, *A Concise Dictionary of Musical Terms* (London: Augener, 1884), 209 and 237.

[54] John H. Clifford, *The Musiclover's Handbook* (New York: University Society, 1911; copyright 1893), 148 of the 'Dictionary of Terms'; *Stokes' Encyclopedia of Music and Musicians* (New York, 1908), 569, or *Black's Dictionary of Music and Musicians* (London, 1924), 578, both ed. L. J. de Bekker; Ralph Dunstan, *A Cyclopaedic Dictionary of Music* (London, 1908; 4th edn. 1925, repr. Da Capo, 1973), 438; Theodore Baker, *A Dictionary of Musical Terms*, 16th edn. (New York: G. Schirmer, n.d.; original edn. 1895), 169; W. S. B. Matthews and Emil Liebling, *Dictionary of Music* (Cincinnati, New York, Chicago, 1896; repr., New York: AMS Press, 1973), 192 and 217.

[55] Thomas S. Wotton, *A Dictionary of Foreign Musical Terms* (Br. & H, 1907; repr., St Clair Shores, Mich.: Scholarly Press, 1972), 5. According to Karen C. Rosenak, in 'Eighteenth- and Nineteenth-Century Concepts of Tempo Rubato', D. M. A. Final Project (Stanford University, 1978), p. 32, compensatory tempo modification is mentioned also by Heinrich Germer in *Wie spielt man Klavier?* (Leipzig, 1881) and by Horatio Palmer in *Palmer's Piano Primer* (New York, 1885).

[56] William Lines Hubbard, *Musical Dictionary* in *The American History and Encyclopedia of Music*, x (Toledo and New York: Irving Squire, 1908), 517.

Presumably, all these authors were attempting to describe the same effect. The great controversy that developed out of the diverse dictionary definitions, however, concerned the issue of compensation. The concept was stated vigorously in 1883 in the first edition of George Grove's *Dictionary of Music and Musicians*. In the article on *rubato*, John A. Fuller Maitland first explains that the word refers 'to the values of the notes, which are diminished in one place and increased in another'. The word indicates, he continues, 'a particular kind of licence' used mainly in instrumental music:

This consists of a slight *ad libitum* slackening or quickening of the time in any passage, in accordance with the unchangeable rule that in all such passages any bar in which this licence is taken must be of exactly the same length as the other bars in the movement, so that if the first part of the bar be played slowly, the other part must be taken quicker than the ordinary time of the movement to make up for it; and *vice versa*, if the bar be hurried at the beginning, there must be a *rallentando* at the end.[57]

Others made similar declarations, but not always limited to a single measure. In the music of Chopin, according to Louis Ehlert, 'the first rule in *rubato* playing' is that 'loss and gain of time must be evenly balanced'.[58] Such unequivocal statements led to substantial acceptance of the idea and to its advocacy in various books and articles by performers, teachers, and critics. It is described by Liszt's piano pupil William Mason[59] and by Josef Hofmann.[60] Constantin von Sternberg passionately defends this 'law of balance' in several essays.[61] The concept is ultimately systematized by Tobias Matthay in a lengthy section of his book on *Musical Interpretation* (1913). He identifies three types: (1) the usual rubato, in which retarding is followed by accelerating; (2) *inverted* rubato, where acceleration precedes retardation; and (3) the more frequent *compound* rubato, in which the other two types are combined in a single phrase.[62]

---

[57] *A Dictionary of Music and Musicians*, ed. George Grove (London: Macmillan & Co., 1879–89), iii (1883), 188; and see iv (1889), 85, for 'Tempo rubato' by Franklin Taylor. The same definitions recur in the 2nd edn. (1904–10; repr. Presser, 1918), iv (originally 1908), 176, and v (1910), 70.

[58] *Aus der Tonwelt* (Berlin, 1877–84); trans. by Helen D. Tretbar in *From the Tone World, A Series of Essays* [which appeared earlier as articles in periodicals] (New York: Charles F. Tretbar, 1885), 285–6. He also mentions some disturbing places in Chopin's works, but feels that 'an intelligent use of the *rubato* may skillfully conceal many a questionable passage.'

[59] *Memories of a Musical Life* (New York: The Century Co., 1901), 246–7.

[60] *Piano Questions* (Doubleday, 1909), 100–2.

[61] 'Tempo rubato', *The Musician*, 17 (1912), 524–5; and *Tempo Rubato and Other Essays* (New York: G. Schirmer, 1920), 3–12. According to Harold C. Schonberg, in *The Great Pianists*, rev. edn. (S & S, 1987), 291–2, and 323–4, Sternberg (1852–1924) was a Russian pianist and conductor who had studied with Moscheles and Liszt.

[62] (London: repr., Boston: Boston Music Co., 1980), 60–106.

Compensation appealed to the lexicographers because it formed a link with the past and with that previous type of rubato and its strict accompaniment which guaranteed a compensated melody. It was attractive to teachers like Matthay, since it imposed limitations on inexperienced students. It probably seemed logical to pianists such as Mason and Hofmann because phrases often do require some sort of combination of both acceleration and retard. For those who were naturally conservative or those who opposed excesses of rhythmic freedom, the idea must have seemed a welcome restraint.

Although the concept continues in a few isolated sources as late as the 1970s and 1980s,[63] it is emphatically rejected by Paderewski in 1909, at least for pianists:

The technical side of Tempo Rubato consists . . . of a more or less important slackening or quickening of the time or rate of movement. Some people, evidently led by laudable principles of equity, while insisting upon the fact of stolen time, pretend that what is stolen ought to be restored. . . . The making up of what has been lost is natural in the case of playing with the orchestra . . . With soloists it is quite different. The value of notes diminished in one period through an *accelerando*, cannot always be restored in another by a *ritardando*. What is lost is lost.[64]

Finck declares that during forty years as a music critic he never heard any pianist, including Hofmann, actually apply compensation, and calls the concept a 'pedagogic hoax'.[65] Niecks refers in 1913 to the modern type of tempo rubato—'the tempo modifications, the accelerandos and rallentandos, which do not apply to details only, and are not a system of compensation, but one of arbitrary dispensation'.[66]

---

[63] Jon W. Finson, in 'Performing Practice in the Late Nineteenth Century, with Special Reference to the Music of Brahms', *MQ* 70 (1984), 471–2, defines rubato as 'the robbing of time from certain parts of a phrase and a return of that time elsewhere within the confines of a steady beat'. See also Benjamin Suchoff, *Guide to Bartók's Mikrokosmos*, rev. edn. (B & H, 1971), 89; and Eric Heidsieck, 'Dynamics or Motion?', trans. Charles Timbrell in *PQ* 36/140 (Winter, 1987/8), 56–8.

[64] 'Paderewski on Tempo Rubato' in Finck's *Success in Music*, 459. In *Etude Music Magazine*, 44 (1926), 95, Paderewski describes the music of Chopin: 'music which eludes metrical discipline, rejects the fetters of rhythmic rule, and refuses submission to the metronome as if it were the yoke of some hated government; this music bids us hear, know, and realize that our nation, our land, the whole of Poland, lives, feels, and moves "in Tempo Rubato".' Those who believe that Chopin invented rubato tend to consider it a Polish trait, especially since so many of the marked rubatos occur in the Mazurkas. Liszt no doubt encouraged this point of view in his book on Chopin, where he states, immediately following his remarks on tempo rubato, that Chopin's Polish students, especially the ladies, comprehended his style of playing better than others; see *Frederic Chopin*, trans. Edward N. Waters (London: Free Press of Glencoe, 1963), 81, and *Life of Chopin*, trans. John Broadhouse, 2nd edn. (London: William Reeves, [1913]), 84.

[65] Henry T. Finck, *Musical Progress* (Presser, 1923), 90 (in the chapter 'The Disgraceful Tempo Rubato Muddle'). See also his *Success in Music*, 429.

[66] Frederick Niecks, 'Tempo Rubato', *MMR* 43/506 (1 Feb. 1913), 30.

In 1928 McEwen demonstrates, by measuring the durations of notes on piano rolls made by well-known pianists early in the century, that such compensation does not occur.[67] The third edition of Grove's Dictionary in 1927 finally includes a new article on rubato in which A. H. Fox Strangeways asserts that robbed time need not be paid back during the bar—partly because the barline 'is a notational not a musical matter' and partly because 'there is no necessity to pay back even within the phrase'.[68]

A number of writers mention rubato, of course, without involving the matter of compensation. Some are opposed to rubato altogether. In reviews of piano performances during the 1850s, Hanslick complains about 'the disagreeable [*leidige*] Tempo rubato'—'this morbid [*krankhafte*] unsteadiness of tempo' which Chopin loved so much.[69] In 1856 Carl Engel seems to relate rubato to ritenuto: 'A fault to be found . . . in most virtuosi, is the entire disregard of time—the horrible tempo rubato. We hear constant successions of ritenuto, and so on, without any reason or taste.'[70] Louis Köhler seems to distinguish in 1874 between the senseless accelerando and ritardando produced by the unsuccessful imitators of Liszt and their 'inappropriate rubato, with which one does not even know whether the pianist or the listener is crazy'. Perhaps the latter refers to an attempt to imitate Liszt's 'suspension of rhythm'.[71] The pianist and writer Ernst Pauer, on the other hand, speaks in 1877 of tempo changes:

The *tempo rubato*, . . . which actually means 'stolen time', is inadmissible; in short, any dragging or quickening of the time which does not originate in the absolute feeling or in the character of the piece, and which appears entirely as

[67] John Blackwood McEwen, *Tempo Rubato or Time-Variation in Musical Performance* (OUP, 1928). See especially 13–22, where he analyses the Aeolian Company's Duo-Art piano rolls of performances by Pachman, Busoni, and Carreño. According to Alice Jean Anderson, in 'A Study of Tempo Rubato', M.M. thesis (Eastman School of Music, University of Rochester, 1948), pp. 14–17, McEwen advocates compensation in his earlier book *The Thought of Music* (London: Macmillan, 1912).

[68] *Grove's Dictionary of Music and Musicians*, 3rd edn. (1927–8; reissue, New York: Macmillan, 1935), iv, 465.

[69] Eduard Hanslick, *Geschichte des Concertwesens in Wien*, ii: *Aus dem Concertsaal* (Vienna: Wilhelm Braumüller, 1870; repr. Gregg, 1971), 85 (concerning the playing of Wilhelmine Clauss in 1855) and 131 (Nanette Falk in 1857). The parts I have quoted from the review of Clauss are omitted in the 2nd edn.; see *Aus dem Concert-Saal* (Vienna and Leipzig: Wilhelm Braumüller, 1897), 87 and 143.

[70] *The Pianists' Hand-Book* (London: Gustav Scheurmann & Co., 1856), 7; quoted in Rosenak, 'Eighteenth- and Nineteenth-Century Concepts of Tempo Rubato', 30.

[71] 'Das Pianistenthum II' in *Neue Zeitschrift für Musik* (repr., Scarsdale, NY: Annemarie Schnase, 1964), 70/14 (3 Apr. 1874), 137: 'ein ungehöriges Rubato, bei welchem man nicht gleich weiss, ob's im Pianisten oder im Zuhörer rappele'. Schonberg states in *The Great Pianists* (1987), 180–1, that Köhler had himself studied with Liszt.

the result of the peculiar or individual taste of the performer, is devoid of artistic feeling, and destroys truth and correctness.[72]

As late as 1926 Wanda Landowska deplores the 'epileptic rubato' in some current performances of the works of Chopin.[73]

In 1914 J. Alfred Johnstone describes the false manner in which most contemporary pianists perform rubato—'the modern *tempo rubato* of the ultra-romantic school, which plays havoc with both form and time':

[It] is made an altogether stiff, arbitrary, and palpably artificial interference with the time of the composition at certain selected parts of its progress. It is a stereotyped process which the observant hearer soon learns to predict with accuracy ... This curious artificial device may be described thus: the first few notes of the passage selected for the operation are taken slowly, and then, during the remainder of the passage, the pace is gradually increased until the regulation speed is attained.

To this 'senseless and artificial device' he contrasts his own concept of the true rubato, 'the spontaneously recognized, the instinctively felt, delicate variations arising naturally from the changing moods of the music':

Flexibility ... is one of the essentials of eloquent playing. And *tempo rubato* is the soul of flexibility. Not the modern *tempo rubato* ..., but that delicate *tempo rubato* which recognizes the inadequacy of our fixed note values to give voice to the varying shades of impassioned eloquence which issue from the human heart. This device, which consists in the almost imperceptible lengthening and shortening of certain notes, this instrument of expressive beauty used invariably in the declamation of verse, this vitalising method of treating the otherwise too rigid and mechanical record of music in our notation, this essential medium for the attainment of eloquence, is astonishingly neglected by our pianists.[74]

Six years later the same author devoted an entire book to a systematic explanation of rubato. He distinguishes rubato on three levels: (1) on a single note (including 'metric rubato' for accenting first beats, and 'double rubato' for short notes that lead to a larger one, which, in turn, moves to a more accented tone); (2) on a motive, phrase, or period; and (3) within an entire movement or composition. He emphasizes that the fundamental principle is 'the feeling of hastening towards the important point'. He seems only incidentally to add that

---

[72] *The Elements of the Beautiful in Music* (London: Novello; New York: H. W. Gray, 1877), 31.

[73] 'Advice on the Interpretation of Chopin', *Etude Music Magazine*, 44 (1926), 107–8.

[74] *Essentials in Piano-Playing and Other Musical Studies* (London: W. Reeves, [1914]), 45–6 and 51–2.

this hastening must be balanced, after the climax is reached, by restoring to the music the exact time that was robbed. His rubato is based strictly on analysis, and he rejects the types 'introduced in obedience to the caprice of fancy or to accord with some conventional or mechanical rule'.[75]

Other writers also speak of rubato in a more positive manner. Most often it is considered a general sort of slowing or hastening. Only rarely is it described as unexpected or capricious. Finck considers the first element of Chopin's rubato to be 'the frequent unexpected changes of time and rhythm, together with the *ritardandos* and *accelerandos*',[76] and Adolphe Carpe states in 1893 that 'Chopin's original style of playing, *tempo rubato*, is a capricious robbing of time by accelerando or ritardando, more often by sudden changes in time, subject to no particular rule but the whim of the performer.'[77]

The authors of essays thus express their views *pro* or *con* compensation, *pro* good or *con* bad rubato, or, indeed, *pro* or *con* rubato itself. During the second half of the nineteenth century, however, a more analytical approach becomes evident in a series of books concerning the aesthetics of performance—books devoted largely to the rhythmic elements of music, especially in relation to piano playing. They form a link with the past, for they draw upon the previous works of C. P. E. Bach, Türk, Hummel, and Czerny. Many include long lists of situations in which it is appropriate for the performer to retard or accelerate in order to articulate an element of form, the shape of a phrase, or the significance of a single note.[78] In 1884 Hugo Riemann published his ideas on *Agogik*, his own term, which 'relates to the small modifications of tempo (also called *tempo rubato*) which are necessary to genuine expression'. In his *Musik-Lexikon* of 1887, he writes:

[75] *Rubato, or the Secret of Expression in Pianoforte Playing* (London: Joseph Williams, 1920); see especially 15, 21, 48, and 56.

[76] *Chopin and Other Musical Essays*, 38–9.

[77] *Pianist and the Art of Music* (Chicago: Lyon & Healy, 1893), 68; quoted in Rosenak, 'Eighteenth- and Nineteenth-Century Concepts of Tempo Rubato', 34.

[78] See Adolph Kullak, *Die Ästhetik des Klavierspiels* (Berlin, 1861), trans. from 3rd edn. of 1889 by Theodore Baker in *The Aesthetics of Pianoforte-Playing* (New York, 1893; repr. Da Capo, 1972), 280–94; Mathis Lussy, *Traité de l'expression musicale* (Paris, 1874), trans. from 4th edn. by M. E. von Glehn in *Musical Expression, Accents, Nuances, and Tempo, in Vocal and Instrumental Music* (Novello, [1885]), 163–96, 229; and Otto Klauwell, *Der Vortrag in der Musik* (Berlin and Leipzig, 1883), trans. as *On Musical Execution: An Attempt at a Systematic Exposition of the Same Primarily with Reference to Piano-Playing* (New York: G. Schirmer, 1890), 2, 9–58. See also Marion Louise Perkins, 'Changing Concepts of Rhythm in the Romantic Era: A Study of Rhythmic Structure, Theory, and Performance Practices Related to Piano Literature', Ph.D. diss. (Univ. of Southern California, 1961; UM 61–6,302); and Artis Stiffey Wodehouse, 'Evidence of Nineteenth-Century Piano Performance Practice Found in Recordings of Chopin's Nocturne, Op. 15 No. 2, Made by Pianists Born before 1900', D.M.A. Final Project (Stanford Univ., 1977).

Tempo rubato is the free treatment of passages of marked expression and passion, which forcibly brings out the stringendo–calando in the shading of phrases, a feature which, as a rule [that is, when rubato is not involved], remains unnoticed.[79]

One of the most comprehensive theoretical works was written in 1885 by Adolph Friedrich Christiani, who worked during his later years in the United States and whose book was first published in English, then translated into German, French, and Russian. *The Principles of Expression in Pianoforte Playing* is dedicated to Liszt 'in admiration of his matchless genius and in recognition of his sympathetic kindness'.[80] Christiani considers rhythmic freedom in five categories: Accelerando, Ritardando, Sudden tempo changes, Tenuto and fermata, and Rubato. He devotes to the first two a chapter in which he quotes extensively from the works of Czerny and later theorists. He deals briefly then with sudden changes and tenuto, including under the latter the prolongation of individual notes which others inaccurately consider a type of ritardando. He defines rubato (also 'rubamento di tempo', 'contra-tempo', and 'contre-temps') first of all as 'any temporary retardation or acceleration'. Secondly, it is an accent, as described by Türk, on a weak beat.

His third definition, however, reveals even greater historical insight, for it appears to describe both the compensating later type and the earlier rubato of Chopin:

That capricious and disorderly mode of performance by which some notes are protracted beyond their proper duration and others curtailed, without, however, changing the aggregate duration of each measure, is a *rubato*.

[This] mode, which is, in fact, the real *rubato*, as it is usually understood, will receive particular notice.

[It] is the *rubato* of Chopin; very beautiful and artistic when in its proper

---

[79] *Encyclopaedic Dictionary of Music*, trans. J. S. Shedlock (Presser, 1908), 13 and 673, trans. of articles from the 3rd edn. of the *Musik-Lexikon* (Leipzig: Max Hesse's Verlag, 1887). The main definition of rubato continues as the opening sentence of subsequent entries as late as the 11th edn. of 1929. Riemann first presented his ideas on agogics in the *Musikalische Dynamik und Agogik* (Hamburg).

The expression *agogic accent* is often used to refer to a delay in the expected appearance of a note, somewhat like Couperin's *suspension* in Ex. 2.9. Will Crutchfield speaks of an 'agogic accent of delay' in 'Brahms, by Those Who Knew Him', *Opus*, 2/5 (Aug. 1986), 15, and in 'Gerald Moore and the Art of Accompaniment', *Opus*, 3/5 (Aug. 1987), 22, mentions 'the little hiccup of delay so many pianists use to set off a note of destination'.

[80] According to Arne Jo Steinberg, in 'Franz Liszt's Approach to Piano Playing', D.M.A. diss. (Univ. of Maryland, 1971; UM 71–25,283), 63, Christiani was 'a relatively obscure student of Liszt'. Liszt's pupil William Sherwood reports in 'Musical Europe of Yesterday and Musical America of To-day', *The Etude*, 27 (1909), 160, that after studying the MS of Christiani's book, he felt that it contained 'just what I learned from Weitzmann [another of Liszt's students] and Liszt'. It appeared in German as *Das Verständnis im Klavierspiel* (Leipzig, 1886).

place and limitation, but very ugly and pernicious when out of place, or exaggerated.

It may be executed in two ways:

(1) both hands in sympathy with each other, *i.e.* both hands accelerating or retarding together.

(2) Or the two hands not in sympathy, *i.e.* the accompanying hand keeping strict time, while the other hand alone is playing *rubato*.

The latter way is the more beautiful of the two, and is the truly artistic *rubato*.

The first manner of execution may refer to Liszt's light suspension of rhythm on a significant note. The second, of course, is the earlier type of rubato that we have traced from vocal and violin music to the keyboard examples by Chopin and others. Crucial to Christiani's definition are the words 'capricious and disorderly', for this clearly distinguishes the concept from accelerando and ritardando, and even from expressive lingering on passages or notes.

Christiani was one of the few musicians of any sort from this period who not only understood earlier rubato, but enthusiastically advocated its use. He quotes Chopin's analogy between the strict accompaniment and the conductor. He mentions the rubato of the Italian singers and of the pianist Thalberg. This sort of rubato, he adds significantly, occurs mainly in homophonic music, especially those types, such as dances, which have a distinctly characteristic rhythm. He mentions the waltz and the mazurka, two dance forms which, like the earlier Italian opera aria, provided a conspicuous and continuous rhythmic beat in the accompaniment against which the melody could sound 'capricious and disorderly'.[81] Christiani just as vigorously opposes 'the old *ad libitum* style of interpretation . . . , which only too often resembled the unsteady gait of a drunken man'. Fortunately, 'that pernicious *rubato* nuisance, that slippery downward course in time-keeping, which Chopin's disciples and unripe admirers are greatly responsible for, is happily becoming rarer and scarcer'. He also complains of incompetent performers and their 'out-of-time playing which they believe to be rubato'.[82]

Two other theorists also investigated the history of the earlier rubato, but could not understand and accept it as fully as Christiani

---

[81] *The Principles of Expression in Pianoforte Playing* (New York: Harper & Bros., 1885), 213–303. See also Harriette Brower, 'The Marriage of Rhythm and Rubato', *Etude Music Magazine*, 44 (1926), 805–6.

[82] *The Principles of Expression*, 53 and 18. In connection with the 'out-of-time playing' of incompetent performers, see the facetious definition of rubato in Alan Raph and Bucky Milan, *"Les" Brass* (Candlewood Isle, Conn.: AR Publishing Co., 1984), 34: 'Fast on the easy parts, slow on the hard parts'. I have also been told that rubato is for the purpose of facilitating page turning.

did. Franz Kullak presents quotations on rubato from Türk, Czerny, Mikuli, and Karasowski. He finally realizes that Chopin's rubato consists of a 'metrically inexact working together of the two hands in which, since one hand remains strictly in time, the continuation of the tempo itself is not at all changed'. Although he seems on the brink of accepting this form of rubato, he finally cannot do so—even, apparently, for the music of Chopin—since this method of playing reminds him of the offensive current mannerism of letting the left hand play before the right.[83] A. J. Goodrich goes a step further in his work from 1899. He accepts the strict left hand and the *ad libitum* style of the right, and thus equates tempo rubato with the altering of note values. Although he mentions syncopation produced by the quadruplet in measure 44 of Chopin's Minute Waltz against the three beats of the accompaniment, his examples of rubato show the dotting or undotting of notes on the main beats in such a way that no displacement occurs with the left hand.[84]

These investigations of the earlier rubato by a few theorists may have been the inspiration for even more substantial studies by those one might call historians or musicologists. We must also keep in mind that all the quotations by former students or acquaintances of Chopin describing the strict left hand as a conductor were recorded between 1868 (beginning with Lenz) and 1910 (Planté and Saint-Saëns). Historical facts begin to appear in works on Chopin or in articles devoted to the subject of rubato. Niecks traces the idea back to Quantz and the Italian singers in his biography of Chopin in 1888. Finck mentions Caccini and Mozart in an essay of 1889. In an article from 1912 on 'Tempo Rubato', Reginald Gatty quotes from Tosi, Galliard, and Agricola, and the following year Niecks, in a series of articles, surveys the different types of rubato from Caccini and Frescobaldi to Tosi, Mozart, and Türk, and finally to Busby, Koch, Hummel, and Chopin.[85]

The process of rediscovering the earlier type of rubato led, however,

---

[83] *Der Vortrag in der Musik am Ende des 19. Jahrhunderts* (Leipzig: F. E. C. Leuckart, 1898), 118: 'Es besteht also eigentlich nur in einem taktlich ungenauen Zusammenwirken beider Hände, wobei, da die eine Hand streng im Takte bleibt, das Tempo selbst in seinem Fortgange gar nicht verändert wird. Diese Spielweise, die übrigens auch vielen Dilettanten zu eigen ist, insofern sie mit Vorliebe die linke Hand der rechten voranschlagen lassen, will ich nun *nicht* als das wahre Tempo rubato . . . betrachtet wissen'.

[84] *Theory of Interpretation Applied to Artistic Musical Performance* (Presser, 1899), 154–7, 229–32, and 281–2; also 236–7 and 253–6. His rubato reminds one of the dots which Schindler added to notes in Beethoven's Op. 14 No. 1 (see ch. 6).

[85] Niecks, *Frederick Chopin as a Man and Musician*, ii, 100–3; and MMR 43 (1913), 29–31, 58–9, 116–18, and 145–6; Finck, *Chopin and Other Essays*, 33–45; Gatty, 'Tempo Rubato', *The Musical Times*, 53 (1912), 160–2.

to some misunderstandings of the technique as well as some fanatical reactions against it. Josef Hofmann seems uncomfortable with the idea.[86] Finck is upset by the 'left-hand-in-strict-time nonsense':

A lamentable amount of confusion has been caused by the preposterous 'tradition' that in playing Chopin the left hand must always play in strict time. The absurdity of this dictum (which reduces the 'rubato' to a mere mechanical question of dotted notes in the right-hand part) will be further exposed in a later chapter . . . , in which an attempt will also be made to discover the secret of the *true* rubato in the changing emotional character of the melody.[87]

Little suspecting the vocal origin of rubato, Constantin von Sternberg demonstrates its absurdity by imagining a singer who is accompanied by a strict beat and who therefore 'must renounce all artistic expression'.[88] Raoul Koczalski, in his book on Chopin from 1936, seems at one point to grasp the true meaning of Chopin's rubato:

When the right hand plays embellishing notes or when it, carried away by the warmth of feeling which it expresses, seeks lingeringly or energetically to free itself from the left hand, only then can there be a difference in the playing together of the two hands.

He finally concludes, however, that the left hand should follow the right like a good conductor follows a singer. In his analyses of individual pieces, he uses the word *rubato* simply to indicate expressive tempo changes.[89]

In the process of historical discovery, other early types of rubato

---

[86] *Piano Questions*, 100–1: 'I find an explanation of *tempo rubato* which says that the hand which plays the melody may move with all possible freedom, while the accompanying hand must keep strict time. How can this be done?

'The explanation you found, while not absolutely wrong, is very misleading, for it can find application only in a very few isolated cases; only inside of one short phrase and then hardly satisfactorily. Besides, the words you quote are not an explanation, but a mere assertion or, rather, allegation. . . . I can see only very few cases to which you could apply such skill [independence of the hands], and still less do I see the advantage thereof.'

[87] *Success in Music*, 271; see also 269, as well as his *Musical Progress*, 87–9 and 92. As late as 1971, David Bollard writes, in 'What they Taught—Part Two', *Australian Journal of Music Education*, 8 (Apr. 1971), 15: 'To reduce *tempo rubato* to a series of quasi-dotted melody notes over a metronomic left hand is nothing short of musical idiocy.'

[88] *Tempo Rubato and Other Essays*, 4; and 'Tempo Rubato', *The Musician*, 17 (1912), 524.

[89] *Frederic Chopin: Betrachtungen, Skizzen, Analysen* (Cologne-Bayenthal: Verlag Tischer & Jagenberg, 1936), 58: 'Wenn die rechten Hand Verzierungsnoten spielt, oder, wenn sie, mitgerissen von der Wärme des Gefühls, das sie ausdrückt, zögernd oder energisch versucht, sich von der linken Hand zu befreien, dann allein kann es eine Differenz im Zusammenspiel der beiden Hände geben.' He suggests on pp. 130 and 203 that one play rubato in m. 25 of the Mazurka Op. 7 No. 1 and mm. 30–1 of the Nocturne Op. 9 No. 2. For neither of these 2 compositions, however, does he mention applying his type of rubato at those places where Chopin himself marked the word. See also the rubato suggested in the analyses on pp. 84 and 139.

were also uncovered—those that relate to syncopation: displaced accents and metres, as well as the displacement of notes described by C. P. E. Bach. We have noted Goodrich's mention of four against three in the Chopin Waltz. In 1889 Finck includes as the second component of Chopin's rubato the irregular groups of small notes, such as nineteen against four, or twenty-two against twelve. Franklin Taylor, who had earlier described rubato as a compensated quickening and slackening of tempo, seems to take an even stricter stand in 1897:

any independent accompaniment to a *rubato* phrase must always keep strict time, and it is, therefore, quite possible that no note of a *rubato* melody will fall exactly together with its corresponding note in the accompaniment, except, perhaps, the first note in each bar.

Yet, as 'a good example' he quotes measures 152 and 153 from Chopin's F minor Ballade, which contain the notated syncopation caused by eight notes against six, and seven against four.[90]

In Mendel's *Lexikon* of 1878 the second definition of tempo rubato involves the displacement of accent, either incidental or of a sort to change the metre. As an example, he adds accent signs in Chopin's Mazurka Op. 7 No. 1 (see Ex. 7.2) on the second beat of the first bar and the third beat of the second. Chopin himself, of course, marks accent signs elsewhere in the same piece, as well as in many other works. Finck (in 1889) considers these accentuated second and third beats as one of the three aspects of Chopin's rubato. Displaced accents and metre appear as a secondary definition in the dictionaries of Riemann (1882, 1884, and 1887) and Niecks (1884), displaced metre in Taylor's article on 'Tempo rubato' in the first edition of Grove's (1883). Christiani includes 'any negative grammatical accentuation' as the second of his three definitions of rubato. In a separate chapter he discusses the subject in detail, including syncopation, both by anticipation and retardation, and changes of metre or *imbroglio* caused by repeated syncopation or by the recurrence of melodic figures. Goodrich, however, gives an example in 1899 of a hemiola passage in which two measures notated in 3/4 actually bear the accentuation of three measures in 2/4, but this, he adds, 'is not a tempo rubato, but a rhythmic effect'. After 1889, displaced accents and metre are apparently no longer connected with rubato.[91]

The late romantic writers thus defined rubato in a multitude of ways. Historical concepts played a significant role, as the lexicogra-

---

[90] *Technique and Expression in Pianoforte Playing* (London: Novello, [1897]), 72–3.

[91] Finck, *Chopin and Other Musical Essays*, 39–40; Christiani, *Principles of Expression in Pianoforte Playing*, 69–93 and 299; and Goodrich, *Theory of Interpretation Applied to Artistic Musical Performance*, 157.

phers reached back to Busby and Hamilton, theorists to Czerny, and musicologists to Tosi. Most often, however, the rubato of the writers seems to refer to expressive tempo changes added at the initiative of the performer at the level of the phrase, but sometimes on a broader scale and sometimes on the more detailed level of the single note. Such changes are occasionally described as slight or almost imperceptible, so that they sound to a listener as though the written note values had been performed, but with special expression.[92] According to some authors, the quickening and slackening were generally supposed to balance. Seldom is rubato characterized as sudden or capricious.[93] It is no longer described in terms of anticipation or delay. The earlier type was mainly a curiosity; the displaced accents were finally known by other names. Two new types of rubato seem to have been invented during this period: the later type with compensation, and the earlier type without displacement between melody and accompaniment, involving mostly the dotting of notes.

A few writers include marked ritardandos and accelerandos as a type of notated rubato. It is remarkable, however, that not a single writer of the period, to my knowledge, ever mentions the word *rubato* when it is written in a score by a composer. Furthermore, whenever writers extract a passage from Chopin's works to illustrate rubato, they invariably select a place where Chopin himself did not mark the word. Of the composers between Liszt and Debussy included in the opening section of this chapter, Saint-Saëns is the only one, as far as I know, who mentions rubato in his writings, and he was either relating the views of Chopin's students, as we have seen, or complaining about contemporary performances of Chopin's music.

The Performers

Leopold von Sonnleithner objects as early as 1860 to the deviations in tempo which cause Schubert songs to be 'distorted, disfigured, and

---

[92] John B. McEwen, in *Tempo Rubato*, writes on p. 31: 'the prolongation or shortening of any sound must be within such limits as enable the mind to realize that the performance presents the written value inflected in the presentation, not an entirely new value commensurable with the original', otherwise 'the composer originally would have written this performed version.' On p. 37 he expands the idea to broader variations of time. See the reactions of Moscheles, Meyerbeer, and Hallé in Ch. 7 to Chopin's performance of his Mazurkas, and the comments earlier in this chapter concerning the more precise notation of rubato-like passages by composers.

[93] In *Chopin: The Man and His Music* (New York: Charles Scribner's Sons, 1923; copyright 1900), 344–9, the American writer James Huneker, on the other hand, describes 'the much exploited rubato' of his day, with rhythms distorted, absurd and vulgar haltings, exaggerated and jerky tempi, rhythmic anarchy and disorder. He himself prefers Chopin's music played 'in curves', 'in a flowing, waving manner'.

robbed of their greatest charm'.[94] Almost half a century later, Mason criticizes a famous pianist's 'excessive use of tempo rubato', in which 'about one measure was added to every section of four'.[95] It is difficult for us to evaluate Sonnleithner's comments, since, as we have noted before, we have no way of knowing how slight or great the tempo changes may have been. For the later performers, however, we have, for the first time in our history, the means of determining precisely how they played, for beginning late in the nineteenth century we have recordings on piano rolls and eventually on phonograph discs. There are recordings by composers such as Brahms, Grieg, and Debussy, and others by famous pianists, singers, and string quartets. During the past thirty years a number of scholars have begun to study these sources, especially those involving the music of Chopin and Brahms.

Rhythmic freedom is most extreme with the pianists up to around 1930 and involves two broad categories. First, as one would expect, are the frequent retardations and accelerations at all levels of structure. Second, and far more conspicuous and unexpected for those who are not familiar with these performances, are the mannerisms related to arpeggiation and to the earlier type of rubato. To the performers, however, the expression *tempo rubato* referred not to the mannerisms, but, according to Hofmann, to 'a wavering, a vacillating of time values'.[96]

The latter changes operate on three levels: that of the entire movement, the phrase, and the single note. String quartets of the period apply Wagner's *tempo modification* as a sort of 'audible analysis', to distinguish one theme from another and unstable from stable tonal areas.[97] Pianists sometimes change the tempo for parts of a work by Chopin, often speeding up the less melodious sections which have arpeggios in the right hand.[98] Grieg accelerates markedly for the developmental middle sections of two brief compositions that he

---

[94] *Schubert: Memoirs by His Friends*, ed. Otto Erich Deutsch (London: Adam and Charles Black, 1958), 336–8. See also William S. Newman, 'Freedom of Tempo in Schubert's Instrumental Music', *MQ* 61 (1975), 528–45.

[95] *Memories of a Musical Life*, 246.

[96] *Piano Questions*, 101.

[97] See Jon W. Finson, 'Performing Practice in the Late Nineteenth Century, with Special Reference to the Music of Brahms', *MQ* 70 (1984), 473.

[98] In the Waltz Op. 42 on RCA Victor–1550 recorded in 1935, Hofmann suddenly increases the tempo for the section which first appears in m. 41. A similar passage in m. 329 of the Scherzo Op. 31 he takes about twice as fast; see X–904 in the *Everest Archive of Piano Music*, a set of disc recordings from Duo-Art and Ampico piano rolls originally made between 1916 and 1925. In the Waltz Op. 64 No. 2 on Victor set M–338, Rosenthal plays the section beginning in m. 33 noticeably slower, then its repetition in m. 49 very fast; I am grateful to the Stanford Archive of Recorded Sound, Braun Music Center, Stanford University, for providing me with a copy of this recording.

recorded.[99] Theoretically, these tempo changes act to articulate the formal structure for the listener.

The next level concerns the expressive shaping of phrases. Sometimes groups of shorter notes are accelerated and longer notes retarded, thus strengthening the sense of forward movement as well as articulating motivic structure.[100] Crescendos are accelerated and decrescendos retarded, as Klauwell and other theorists explain.[101] Accompanimental figures that are notated in even notes are performed with great flexibility in order to accommodate expressive events in the melody.[102]

Melodic expression on the most detailed level, however, may involve only a single note. A dissonant appoggiatura in a reaching motive, for example, may be prolonged or given an agogic accent.[103] Accented notes may be lengthened, unaccented ones shortened.[104] In a group of small notes leading toward a cadence or harmonic change, the penultimate note is often lengthened and the last one shortened.[105] Such alterations may result in the dotting of the first of a group of two evenly notated notes, or the over-dotting of already dotted notes—a practice which emphasizes the anacrustic impulse toward the next

[99] The 2nd piece of the *Albumblätter für Pianoforte*, Op. 28, and 'Erotik', the 5th of the *Lyric Pieces*, iii, on Allegro Records LEG–9021: *Famous Composers Play Their Own Compositions*.

[100] Jon Finson, in 'Performing Practice in the Late Nineteenth Century', *MQ* 70 (1984), 471–3, describes this situation in a recording of Brahms's F minor Quintet, Op. 34, by the Flonzaley Quartet and Harold Bauer (see the excerpt on p. 472). This process, which occurs 'within the confines of a steady beat', he calls *rubato*; I presume he means a compensated form of the later type. He states that 'rubato was a device applied to the motivic and melodic structure of a piece in order to outline that structure for the audience. It was not merely a sentimental device applied haphazardly.'

[101] See *Performance Practice: Music after 1600*, ed. Howard Mayer Brown and Stanley Sadie (Norton, 1989), 379 and 453 for comments by Wayne Leupold concerning organists and Will Crutchfield on singers. In 'Brahms, by Those Who Knew Him', *Opus*, 2/5 (Aug. 1986), 21, Crutchfield describes the application of this practice in an interlude in Brahms's *Sapphische Ode* and compares this 'old-fashioned' approach to more recent performances.

[102] Note the middle section of Chopin's Polonaise Op. 26 No. 1 as played by Bauer on *Everest Archive of Piano Music* X–911; in the opening 2 measures, for example, the 1st 4 8th notes in the left hand are fast, the last 2 considerably slower, in order to produce very *dolce* anacrustic notes in the melody. See also Hofmann's performance of Chopin's *Berceuse* on RCA Victor–1550, where the ostinato figure in the left hand is extremely flexible. See Kravitt, 'Tempo as an Expressive Element in the Late Romantic Lied', *MQ* 59 (1973), 506.

[103] Paderewski, on *Everest Archive of Piano Music* X–902, considerably prolongs the half note at the beginning of m. 26 in Chopin's Waltz Op. 34 No. 1, thus emphasizing the difference between it and the corresponding m. 18.

[104] See Crutchfield, 'Brahms, by Those Who Knew Him', 15, for a description of a syncopated chord which Brahms accentuates with an agogic accent in his recording of the Hungarian Dance No. 1.

[105] See William Hunter Heiles, 'Rhythmic Nuance in Chopin Performances Recorded by Moriz Rosenthal, Ignaz Friedman, and Ignaz Jan Paderewski', D.M.A. thesis (Univ. of Illinois, 1964; UM 65–832), 12–15 and, for example, 38, where he describes two cases in the Nocturne Op. 9 No. 2: in the group approaching the 3rd beat of m. 8, the C is lengthened and the D shortened before moving on to the dotted quarter note E♭, and the same thing happens in m. 6 as the group approaches the harmonic change accompanying the dotted quarter note on B♭.

long note.[106] Short notes tend to be rushed and played in a light, offhanded way.[107] In addition, some pianists emphasize dance rhythms by lengthening the first or second beats.[108]

In addition to all of these manifestations of tempo rubato, performers of this period also employed a number of other expressive rhythmic devices. Pauses could create accents on notes or chords and could also separate motivic units.[109] More often, however, delays and occasionally anticipations of melody notes occur because vertically aligned notes are not played together. When the notes of such a vertical group are played successively, usually in order from the bottom to the top note, arpeggiation results. When such a vertical group is played so that one hand precedes the other, with each hand separately playing all its notes together, then a device occurs which has been called the 'breaking' or 'separation' of hands.

We noted in Chapter 7 that Chopin opposed arpeggiation not marked in the score, even though Corri, Czerny, and Thalberg identified situations where it could be applied. Late romantic performers often arpeggiate left-hand chords in order to enhance a *dolce* quality in the melody[110] or to emphasize an important note or chord.[111] Sometimes arpeggiation marks the return of a thematic idea.[112] Some-

---

[106] See Robert Philip's remarks in Brown and Sadie, *Performance Practice: Music after 1600*, 474–6. Rakhmaninov clearly dots the 1st F in the opening bar of Chopin's Nocturne Op. 9 No. 2 in a recording from 1927 included in *The Complete Rachmaninoff*, iii, RCA ARM3–0294.

[107] Philip, in Brown and Sadie, *Performance Practice: Music after 1600*, 474–6, calls this practice 'throwaway rhythmic lightness'.

[108] Most spectacular are the performances of Chopin Mazurkas by Friedman on Columbia Masterworks Set No. 159 (copy obtained from the Stanford Archive of Recorded Sound). In Op. 7 No. 1, the 1st beat is considerably lengthened each time the main theme appears. More frequently, however, he vigorously lengthens the 2nd beat as in a Viennese waltz: see Op. 24 No. 4, Op. 33 Nos. 2 and 4, and Op. 41 No. 1 (beginning in m. 17, especially). Friedman sometimes accentuates the 1st beat by adding an octave, 5th, or 10th above the bass tone. See Heiles, 'Rhythmic Nuance in Chopin Performances', 20–2.

[109] Ibid. 6–7 and 10–12 regarding the separation of motives—a process that tends to attract the listener's attention to tiny details of construction. Concerning the 'agogic accent of delay' used by Joachim in a performance accompanied by Brahms, see Crutchfield, 'Brahms, by Those Who Knew Him', 15. Robert Philip quotes J. A. Johnstone and Fuller Maitland regarding Joachim's agogic accents in Brown and Sadie, *Performance Practice: Music after 1600*, 473. Concerning Joachim's rubato playing, see Louis Ehlert, *Aus der Tonwelt* (Berlin, 1877–84), trans. by Helen D. Tretbar in *From the Tone World* (New York, 1885), 286; and Andreas Moser, *Joseph Joachim: Ein Lebensbild*, i (Berlin, 1908), 54.

[110] See Friedman's recording of Chopin's Nocturne Op. 37 No. 1 on *Everest Archive of Piano Music* X–919 and Paderewski's performance, in the Ballade Op. 47 on X–902 of the same series, of the melody beginning in m. 54.

[111] Friedman thus emphasizes a long, high note in m. 33 of the Mazurka Op. 50 No. 2 on Columbia Masterworks Set No. 159. Rosenthal emphasizes expressive chords at the beginning of m. 16 in the Prélude Op. 28 No. 3 (a secondary dominant to a IV chord, with the 7th in the melody) and in m. 61 of the Waltz Op. 64 No. 2 (a Neapolitan chord)—both from Victor Set M–338.

[112] In m. 76 of the Mazurka Op. 50 No. 2, Friedman emphasizes the *sforzando* and *a tempo* which mark the return of the melody from m. 60 (Columbia Masterworks Set No. 159).

times it acts to propel the rhythmic flow: when a chord is followed by a rest, or when there is a long note or rest in the melody. Chords are strummed rapidly on third beats in 3/4 metre or rolled slowly to enhance a rallentando.[113] Sometimes arpeggiation occurs regularly on certain beats for several measures in order to lighten and give momentum to a passage that is leading back to a main theme.[114] On rare occasions a chord is arpeggiated downward to emphasize the bass note.[115]

In 1937 Leroy Ninde Vernon published the results of a detailed study of chord synchronization in Duo-art piano rolls made by Paderewski, Hofmann, Bauer, and Bachaus playing works by Beethoven and Chopin. He concludes that with some pianists as many as half the chords are asynchronous, that most are probably intentional, that more occur at a slow tempo, and that most involve melody notes emphasized by being played early or late. He notes that when 'a clearly defined and continuous melody has an accompaniment of chords contrasted in rhythm and rather separate from the melody, the two are seldom played together'. However, he discovered few examples of Chopin's type of rubato, since the accompaniment usually did not hold a steady tempo. He quotes A. B. Marx concerning chords with melody notes:

Occasionally, when no other means are available, it even becomes necessary to move the important tones almost imperceptibly forward or backward; or perhaps to give them in strict time, and imperceptibly to touch the others a little after.[116]

---

[113] Friedman arpeggiates a chord before a rest in m. 6 of the Mazurka Op. 33 No. 4 and in m. 8 of the Mazurka Op. 50 No. 2 in Columbia Masterworks Set No. 159, and during a melodic rest and long note in mm. 2 and 4, respectively, of the Nocturne Op. 37 No. 1 in Everest Series X–919. Rosenthal gives an anacrustic lift to the rhythm by rapidly strumming third beats almost like a guitar in the Nocturne Op. 9 No. 2 and the Mazurka Op. 33 No. 4 (Victor Set M–338). Rakhmaninov also strums 3rd and sometimes also 2nd beats in Op. 9 No. 2 in *The Complete Rachmaninoff*, iii (RCA ARM3–0294). See Crutchfield, 'Brahms, by Those Who Knew Him', 18, where he describes the playing of Ilona Eibenschütz, who 'almost always uses broad arpeggios to fill out a rallentando, and often uses quicker, tighter ones to propel an upbeat back into tempo'. Florence May, in *American Brahms Society Newsletter*, 6/2 (Autumn 1988), 7–8, states that Brahms disliked chords spread unless so marked; yet Rosenthal reported, according to Rosen, that Brahms himself 'arpeggiated all chords'. See also Will Crutchfield, 'Gerald Moore and the Art of Accompaniment', *Opus*, 3/5 (Aug. 1987), 18, on the usefulness of arpeggiation to the accompanist, and 19 on Saint-Saëns' arpeggiation in a recording of one of his songs.

[114] Note Friedman's arpeggiation in the 3 measures beginning in m. 29 of the Nocturne Op. 37 No. 1 (Everest X–919)—a passage which leads to the dominant chord in m. 32 and the return of the opening theme in m. 33.

[115] According to Heiles, in 'Rhythmic Nuance in Chopin Performances', 135, Friedman 'sometimes uses downward arpeggiation to give very telling emphasis to an important event in the bass'.

[116] 'Synchronization of Chords in Artistic Piano Music', *University of Iowa Studies in the Psychology of Music*, iv (1937): *Objective Analysis of Musical Performance*, ed. Carl E. Seashore, 306–45. The 1st quotation comes from p. 331; that of Marx on p. 320 from the *Introduction to*

It was the breaking of hands, however, that became a special characteristic of the period. It was employed frequently by all the famous
pianists of the day—more boldly by pianists such as Paderewski,
Bauer, Friedman, and Pachman, more subtly by Hofmann and
Rosenthal. Often only a single note appears in each hand, sometimes a
chord in the left and a single note in the melody. Usually the left hand
sounds first, causing a delay in the melody. Sometimes the melody
note in the right hand comes first.[117] Breaking can be used for the
same expressive, rhythmic, and structural purposes we noted for
arpeggiation. The delay of a melody note acts to articulate the beginning of a new section, emphasize expressive passages, accent strong
beats, or strengthen the singing quality of portato touch. Anticipation
seems to occur, as an alternative to delay, as a means of articulating
the commencement of a formal section.[118] In a curious example from
Brahms's Ballade Op. 118 No. 3, the arpeggiated left hand and the

the Interpretation of the Beethoven Piano Works, trans. Fannie Louise Gwinner (Chicago:
Clayton F. Summy Co., 1895) from the 2nd edn. in 1875 of his Anleitung zum Vortrag
Beethovenscher Klavierwerke (Berlin, orig. 1863). See also M. T. Henderson, 'Rhythmic Organ-
ization in Artistic Piano Performance', 281–305; Harold Seashore, 'An Objective Analysis of
Artistic Singing' based on recordings of Tibbett, Crooks, and others, 98–146 ('continuous
deviation from strict time is the rule in artistic singing'); and Arnold M. Small, 'An Objective
Analysis of Artistic Violin Performance', a study of recordings by Elman, Kreisler, Menuhin,
Szigeti, and others, 220–9 ('The violinists deviated 80% of the time from exact note values . . .
Over-holding and under-holding were about equally prevalent.')

[117] For conspicuous examples of hand breaking, see Everest X–921 played by Pachman. In
the 'Raindrop' Prelude, Op. 28 No. 15, the opening melody note is delayed, whereas the same
note in the final return of the passage in m. 80 is anticipated; delays occur frequently elsewhere,
as well as a few anticipations (2nd beat of m. 5, 3rd beat of m. 58, 1st beat of m. 59). In his
performance of the F sharp major Nocturne, Op. 15 No. 2, prominent anticipations occur on the
1st beat of the 1st full measure and the 1st beat of the 5th, delays on the 1st beat of m. 4 and
the high note of m. 14. Concerning the performance of this piece by a number of other pianists
of the period, see Wodehouse, 'Evidence of Nineteenth-Century Piano Performance Practice
Found in Recordings of Chopin's Nocturne, Op. 15 No. 2, Made by Pianists Born before 1900',
64, 77, and 93–5 on breaking hands.

[118] On Columbia Masterworks Set No. 159, Friedman marks the return of a passage by a
delay of its opening melody note in m. 25 of the Mazurka Op. 7 No. 2 and in m. 92 of the
Mazurka Op. 50 No. 2 (the same note, however, is anticipated in mm. 9 and 84); he anticipates
the 1st note of a recurring theme in m. 25 of the Mazurka Op. 33 No. 4, but delays the
corresponding note in m. 65. He marks a contrasting section in a similar manner, as in the
Mazurka Op. 33 No. 2, m. 49, where the bass precedes the F$^\sharp$ in the inner voice (the melody
itself is tied over from the previous measure). Hofmann also marks a contrasting section with a
delay in the Scherzo Op. 31, m. 26 (Everest X–904). Rosenthal, in the Nocturne Op. 9 No. 2 on
Victor Set M–338, adds a singing quality by delaying single notes such as the 1st G and the high
C in the 2nd full measure and the 1st 3 portato notes in m. 10. Crutchfield, in 'Brahms, by Those
Who Knew Him', 14 and 18, describes Brahms himself breaking hands in the Hungarian Dance
No. 1 on 'just about all the accented first beats where the texture is melody/accompaniment', as
well as other pianists breaking on every 1st and 2nd beat of the opening section of Brahms's
Intermezzo Op. 116 No. 2 (a section in 3/4 based on the rhythm of a quarter or 2 8th notes
followed by a half).

melody are supposed, according to the notation, to sound the same note at the same time. Since the melody at this point begins to repeat material from eight bars earlier, pianists of the day broke hands at this spot, playing the note first in the left hand, then again, with a fuller tone, with the right.[119]

In some unusual cases of three or more notes, breaking occurs not exactly between the hands, but more precisely between one of the notes and the remaining notes played together by both hands. When the right hand has two notes and the left hand one, the middle note is sometimes struck first and followed by the other two notes together. This is used to call attention to a melodic figure in an inner voice or to accentuate the upper note.[120] Occasionally octaves in the left hand are broken in order to make the upper note sing out melodically.[121] In addition, a sense of finality is sometimes given to the last chord of a section or piece by playing first the bass note by itself or in octaves, followed later by the rest of the chord played by both hands.[122]

The breaking of hands occurs more frequently, and with a greater amount of anticipation or delay, in performances early in the twentieth century. For some pianists it becomes a constant mannerism. Later the device is applied more subtly and less often and apparently becomes a legitimate nuance of expression in performances by Bartók, Ravel, Ives, Brahms, and Cortot.[123] A few performers warn against excessive use. Hofmann considers such 'limping' to be 'the worst habit you can have in piano playing', and the accompanist Coenraad Valentyn Bos regretted later in life his earlier use of the 'faulty mannerism' and 'unforgivable musical sin of anticipating the right hand with the left'.[124]

---

[119] Last 8th beat of mm. 48 and 64; see Crutchfield, 'Brahms, by Those Who Knew Him', 18.

[120] Pachman, 6 measures from the end of the Nocturne Op. 27 No. 2 on Everest X–921, plays the B^{bb} on the 1st beat before the other 2 notes in order to emphasize the middle voice. Paderewski articulates the beginning of a new section or the return of a theme in the Waltz Op. 34 No. 1 (Everest X–902) by playing first the middle note of a 3-note chord (mm. 81, 113, and 175), or by playing the bass note first, followed by the other 2 together in the right hand (the regular method of breaking hands, as in m. 128). Rakhmaninov arpeggiates 2 or 3 notes in the right hand of the Waltz Op. 64 No. 2 (mm. 5 and 9, for example) in a recording from 1927 in *The Complete Rachmaninoff*, iii, RCA ARM3–0294.

[121] Crutchfield describes breaking left-hand octaves in 'Brahms, by Those Who Knew Him', 19 and 21.

[122] Paderewski employs this on Everest X–902 in the Polonaise Op. 40 No. 1, m. 24 (and again in the repeat of the section, and when it recurs at the end of the piece), and in the Ballade Op. 47 (the last chord of the piece). Concerning Paderewski's breaking of hands, see Schonberg, *The Great Pianists* (1987), 308.

[123] See Crutchfield, 'Brahms, by Those Who Knew Him', 18. Concerning Ives' arpeggiation and rubato, see Will Crutchfield, 'The Concord Sonata: An American Masterpiece', *Opus*, 1/4 (June 1985), 22.

[124] Hofmann, *Piano Questions*, 25, and Coenraad V. Bos, *The Well-Tempered Accompanist, as told to Ashley Pettis* (Presser, 1949), 34–5. Bos considers breaking to be an example of the

In spite of the widespread use of breaking by most of the acclaimed pianists over a rather considerable period of time, however, the theorists and other writers, as far as I know, never mention it as a valid means of expression. A few complain about its abuse. We saw in Chapter 7 the warnings of Thalberg in 1853 and Kleczyński in 1879. We also noted previously that two sources connect it with the earlier type of rubato: F. Kullak, who could not accept Chopin's rubato in 1898 because it reminded him of breaking hands, and Saint-Saëns, who got his information from Viardot and Grandval and who felt in 1910 that just playing 'the two hands one after another' was the worst way of imitating Chopin's rubato.[125] Although breaking shared with the earlier rubato a displacement of the hands, it did not necessarily involve strict rhythm in the left hand. Incessant tempo changing of the various later types was so pervasive in the late romantic style of performance that it usually operates in conjunction with breaking.

Breaking seems to be utilized, however, for some of the same purposes for which Chopin indicates rubato: to mark the return of a melody, to emphasize an expressive note, or to set a mood. Curiously, the recordings available to me of works in which Chopin wrote the word *rubato* reveal that the late romantic pianists usually do absolutely nothing when they see this word. Occasionally they make tempo changes, in accordance with their understanding of the term. Even in some performances that involve frequent breaking, the hands are precisely together during the moments marked *rubato*.[126] Of course, they were understandably mystified by Chopin's indications, since the kind of rubato they knew did not usually fit in the places Chopin had marked in the score. Their attitude, however, was also part of a general lack of respect for the score. In addition to all the rhythmic liberties, they sometimes add notes, usually in the left hand, and ignore signs for accents, dynamics, or pedalling.[127] This was a

degeneration of the 'sympathetic approach' into 'sloppiness and sentimentality'. See Crutchfield, 'Brahms, by Those Who Knew Him', 20.

[125] See Eigeldinger, *Chopin: Pianist and Teacher*, 49 and 118 n. 94; and Wayne Leupold in Brown and Sadie, *Performance Practice: Music after 1600*, 379: 'Melodic rubato also led to the tradition at the end of the Romantic era of not playing the hands together'.

[126] This seems to happen in Rosenthal's recording on Victor Set M–338 of the Nocturne Op. 9 No. 2, and Friedman's on Columbia Masterworks Set No. 159 of the Mazurkas Op. 7 No. 1 and Op. 67 No. 3. In his performance of the Mazurka Op. 7 No. 3 (see Ex. 7.15), Friedman makes a big retard and extends the trill in the preceding measure, and then prolongs the rest below the word *rubato*. Heiles, in 'Rhythmic Nuance in Chopin Performances', 74, writes in connection with Op. 7 No. 1: 'It should be noted that Friedman disregards Chopin's indicated *rubato* in m. 49.' Rakhmaninov accelerates and gets louder for the rubato in Op. 9 No. 2 in *The Complete Rachmaninoff*, iii, RCA ARM3–0294.

[127] In Friedman's performance of Op. 7 No. 1, for example, he does not hold the pedal down for 7 measures beginning 4 bars before the *rubato*, as marked by Chopin, nor does he accent the 2nd beat 2 bars after the *rubato*.

general and accepted attitude of the period. A performance must therefore have often been more an expression of a pianist's personality than a projection of the sense of the music.

To us many of these performances sound nervous, careless, trivial, or arrogant. During the 1950s and 60s, a period of considerably more rhythmic strictness and greater fidelity to the score, authors almost unanimously rejected the breaking of hands as 'bad playing' or an 'old maid mannerism', or as 'anathema to the modern listener'.[128] Stravinsky, in a description of performances of Beethoven sonatas, writes in 1970:

> Rare ... are ... pianists unaddicted to such stage business as the 'artful' delaying of the right hand and the deliberate ... non-synchronizing of the treble and bass parts, as if taking the Bible literally about not letting the right hand know what the left hand doeth.[129]

More recently, however, the practice has been viewed more objectively as part of authentic performance practice of the period. László Somfai describes Bartók's method:

> Bartók generally produces a break and separates the right and left hands in order to (1) better outline the linearity of parts, or (2) afford a closer look at the dissonances that are part of the chord. ... Bartók's odd arpeggios are not a performer's mannerisms ..., but one of those problems of notation which he was not able to solve in his scores, but which he precisely demonstrated in recordings ...[130]

According to Wayne Leupold, breaking hands acts to dissipate 'over-accented metric accents' and creates 'a feeling of suspension, a yearning, churning, plastic, indistinct effect'.[131] And Will Crutchfield feels that breaking is so much a part of performing style that 'playing Brahms with the hands together is just like playing Bach with lots of pedal and octave doublings.'[132]

---

[128] Thomas Fielden, 'Tempo Rubato', *Music and Letters*, 34 (1953), 152; Jan Holcman, 'Liszt in the Records of his Pupils', *Saturday Review*, 44/51 (23 Dec. 1961), 47; and Thomas Higgins, 'Chopin Interpretation: A Study of Performance Directions in Selected Autographs and Other Sources', Ph.D. diss. (Univ. of Iowa, 1966; UM 67–2,629), 361–2.

[129] Igor Stravinsky, 'On Beethoven's Piano Sonatas', *Harper's Magazine*, 240/1440 (May 1970), 37; also in *Themes and Conclusions* (Faber, 1972), 269.

[130] *The Centenary Edition of Bartók's Records*, i: *Bartók at the Piano 1920–1945*, Hungaroton LPX 12326–33, p. 31 of booklet.

[131] Brown and Sadie, *Performance Practice: Music after 1600*, 379.

[132] 'Brahms, by Those Who Knew Him', 60. See also Andrew Trechak, 'Pianists and Agogic Play: Rhythmic Patterning in the Performance of Chopin's Music in the Early Twentieth Century', D.M.A. diss. (Univ. of Texas at Austin, 1988; UM 89–9,777), which is a study of 16 recordings made between 1903 and 1946 of Chopin's Waltz Op. 64 No. 2. His purpose is 'to show that the playing of the early twentieth century, though quite free, could be guided by a highly refined sense of proportion and balance' (p. 2). He refers to breaking hands as 'asynchronism'.

Romantic music in general, in contrast to that of the Classic period, tends to draw the listener's attention more to the beauty of the moment than to the formal structure. Singing melodies, elaborate arpeggiation, slow moving chords, conspicuous dissonances, and the increasingly sensuous sounds of instruments all attract attention to themselves. When, in addition, the performer adds breaking, arpeggiation, and much tempo modification on a detailed level, the listener's attention is even more completely fastened on the moment, and he has even greater difficulty in cumulating a perception of the musical architecture.

One of the romantic pianists, who was equally famous as a composer and conductor, recorded compositions of his own that include the word *rubato*. As a pianist, Sergey Rakhmaninov was eventually known for his rhythmic restraint and sensitivity to overall structure.[133] In his performance of Chopin's Nocturne Op. 9 No. 2 in 1927, however, he surrounds high notes and other expressive moments with enormous retards and strums most of the left-hand chords, perhaps to keep the rhythm moving in spite of the delays.[134] At a later date he recorded his *Rhapsody on a Theme of Paganini* with Leopold Stokowski (in 1934, the same year it was composed) and the Fourth Piano Concerto with Eugene Ormandy (1941, the year he revised it). In the former, *rubato* is marked four times in Variation XVIII for an anacrusis consisting of four melodic sixteenth notes played against three accented chords in the accompaniment. In the performance an allargando effect, in company with the polyrhythm, produces a powerful and warm melodic thrust to the following long notes. The rubatos are strong rhetorical gestures, but do not seem to involve conspicuous tempo changes of the sort Rakhmaninov employs in the preceding statement of the melody for solo piano. The same general effect occurs for four measures marked 'a tempo rubato' at a climactic moment in the first movement of the Concerto, where gigantic *ff* triplet chords in the piano accompany a melody of accentuated quadruplets and triplets in the strings. The rubato recurs near the end of the last movement, but with triplets in the orchestra, quadruplets in the piano.[135]

In neither work does the rubato involve any jerky rhythm or con-

---

[133] Schonberg, *The Great Pianists* (1987), 390–9.

[134] In *The Complete Rachmaninoff*, iii, RCA ARM3–0264, he approaches the high notes in mm. 4, 6, and 8, for example, with much rallentando; similarly, he slows down greatly for the Nachschlag of the trill in m. 7 (thus destroying the flowing effect of the trill which I described in Ch. 7) and for the portato repeated notes at the beginning of m. 10 (where Chopin marks *poco rit.*).

[135] For the Rhapsody, see the full score (New York: Charles Foley, 1934), mm. 13, 15, 20, and 22 of Var. 18; and for the Concerto, the full score (Foley, 1944), 38 and 127. Both are recorded in *The Complete Rachmaninoff*, v, RCA ARM3–0296.

spicuous reconfiguration of note values. Both include cross-rhythms and tenuto signs and hence a special type of articulation (or what would be called 'touch' on the piano). Both seem more concerned with declamatory emphasis than with acceleration or retard. The result is intensely earnest and demands special attention from the listener. Rakhmaninov recorded the Grieg Violin Sonata Op. 45 with Fritz Kreisler in 1928, but the brief passages marked *rubato* seem to pass by without any special treatment by either artist.[136]

One wonders if the music in all three of these works might have been played the same whether or not the word *rubato* were in the score. It is difficult to know whether one can draw general conclusions from these few performances from a relatively late date. No other composer from this period, as far as I know, recorded an example of his own marked rubato, and few could have performed them, in any case, as well as Rakhmaninov.[137] Considering that performers 'interpreted' rather than followed the notated score, however, it would not be surprising if their unnotated breakings, arpeggiations, and tempo changes were far more assertive and conspicuous than anything they did in response to the word *rubato*. When they saw the word in a score, they might well have felt that they were already performing this way—almost continually, in fact. Such interpretive performances were not only characteristic of conductors, as we have seen, but were brought to a high point of creativity, and sometimes eccentricity,[138] by solo performers, including not only pianists,[139] but also singers and other instrumentalists as well.[140] In addition, singers were still

[136] *The Complete Rachmaninoff*, iv, RCA ARM3–0295.

[137] Other instructive examples may well be uncovered by future studies of the recorded material of the period. Hanslick, however, writes in 1862 that as a pianist, Brahms 'neglects—especially in the playing of his own pieces—much that the player should rightly do for the composer'; see *Aus dem Concertsaal* (1870), 257, trans. Henry Pleasants in *Vienna's Golden Years of Music*, 84–5. In his book *On Conducting*, Wagner describes Brahms's playing as 'painfully dry, inflexible, and wooden'; see the trans. by Edward Dannreuther, 84. Tchaikovsky's piano playing was likewise considered cold by Laroche and others: 'the very opposite of what one might have expected'; see Brown, *Tchaikovsky*, i: *The Early Years 1840–1874*, 67 n. 28.

[138] Robin H. Legge, in Henry T. Finck's *Musical Laughs* (New York: Funk & Wagnalls, 1924), 65–6, relates a story about the pianist Anton Rubinstein (1829–94), who, when asked to play a waltz so that guests could dance, deliberately played so fast or with 'so many rubatos that dancing was practically impossible'.

[139] It is not surprising, then, that most pianists seldom include the word *rubato* in their own compositions. There are a few in Paderewski's Sonata for Violin and Piano, Op. 13, and in the 2nd of Friedman's *Six Mazourkas*, Op. 85, but I have found none in the pieces available to me by Rosenthal or Hofmann.

[140] See the references to rubato in the reviews repr. in *Herman Klein and the Gramophone*, ed. William R. Moran (Portland, Ore.: Amadeus Press, 1990), 88, 97, 140, 174–5, 212, 388, 532, and 544. In the chapter on 'The Teaching of Manuel García' in his book *The Bel Canto* (OUP, 1923), however, he does not mention rubato. He had studied singing with García and helped him in 1894 prepare his *Hints on Singing*, which included, as we have seen, a clear description of the earlier type of rubato.

employing, at least in performances of works from Verdi's middle period, the anticipations and delays of the earlier type of rubato, but now no longer known by this name.[141]

For performers there seem thus to have been mainly two kinds of rubato: the tempo changes that they added themselves according to their own personal feelings, and the rubato marked in scores, which was apparently either optional or was treated, like the word *espressivo*, in a very general way. At the same time, they indulged in breaking, dotting, arpeggiation, prolongation, and delay—devices they inherited from the earlier rubato, which by that time had been mostly forgotten.

Debussy

The composers, conductors, writers, and performers just discussed in this chapter represent, for the most part, the mainstream development of Austro-German romanticism and its adaptation in other countries. Famous pianists around the turn of the century played music mainly by Beethoven, Chopin, and Liszt, and occasionally by Schumann, Brahms, Tchaikovsky, Grieg, Saint-Saëns, and others in the same romantic tradition. They rarely included pieces by Mozart or Schubert, and they played works of Bach only in transcription. They also seldom performed the music of Claude Debussy—music which required a different approach to the instrument and to music making in general. Although musical impressionism, it seems to me, is essentially a late manifestation of romanticism, it pointed in many ways to the future, and it set Debussy apart from other musicians of his time.

The rhythm of Debussy's music results from his special way of using melody, harmony, and structure. He reduces to a minimum the sort of forward momentum that had been traditionally generated by the expectancy that a note, a chord, or a brief passage would be followed

---

[141] Concerning recordings by singers of the period, see Kravitt, 'Tempo as an Expressive Element in the Late Romantic Lied', *MQ* 59 (1973), 505; and Will Crutchfield in Brown and Sadie, *Performance Practice: Music after 1600*, 452–7. Note especially the excerpt from *La traviata* in his Ex. 11*b*, in which he notates the embellishments and earlier rubato used by Gemma Bellincioni, a singer whom Verdi admired. In 'An Open Ear', *Opera News*, 52/2. (Aug. 1987), 22, he states that 'she is quite often not together with her accompanist . . . , often leaping to notes well in advance of (but sometimes delaying them until after) their literal place.' He adds on p. 23: 'Her intention was that the accompanist would maintain a basically steady tempo and that she would come and go around it.' In 'Twin Glories II', *Opera News*, 52/12 (27 Feb. 1988), 13, Crutchfield notates similar earlier rubato in a recording by Fernando de Lucia of arias from Verdi's *Rigoletto* and Bizet's *Les Pêcheurs de perles*. See also his article 'Vocal Ornamentation in Verdi: The Phonographic Evidence', *19th Century Music*, 7 (1983), 3–54, and Robert Philip's comments on pp. 477–8 of Brown and Sadie, *Performance Practice: Music after 1600*.

by a particular consequent event. This he accomplishes by using scales without semitones (such as pentatonic or whole-tone), by freeing scale degrees 6 and 7 from a need to resolve even within a tonic chord, by employing chords in nonfunctional ways (such as parallel streams) or prolonging them so long or in such a sensuous way that one no longer cares where they resolve, and by juxtaposing brief fragments of music which simply contrast rather than grow organically from each other. The listener must be prepared to hear melodies only suggested and units of construction hazy and blurred. He must be willing at any moment to accept the unexpected and the incomplete. This includes the special sense of time suggested by the programmatic titles and their images of rain and fountains, moon and perfume, exotic and faraway places.

Rhythm tends to move like the natural, effortless flow of water. Except for brief passages, metrical accents are avoided by syncopation and changes of metre. Debussy frequently includes such instructions as 'librement rythmé', 'sans rigueur le rythme', 'dans un rythme un peu abandonné', 'dans un rythme très souple', or 'dans un rythme nonchalamment gracieux'. He sometimes marks 'à l'aise', 'volubile', or 'ad libitum'. Still other terms may also involve a rhythmic response: *ondoyant, mormorando, dolce sostenuto, lusingando, caressant, expressif*, and perhaps even *doux, délicat,* or *con morbidezza*.

Within the generally free and relaxed flow of rhythm, however, the character of each contrasting fragment is precisely defined—and more exactly than ever before in music. Because of its brevity and the fact that it does not necessarily develop out of anything previously heard, each unit is carefully crafted not only for the pitches it contains, but now also for other elements that may be equally or even more important: register, sonority, tone colour, dynamics, touch, articulation, pedalling, and, of course, all the dimensions of rhythm. All of these matters are marked in great detail by Debussy, with the expectation that they will be carefully followed by the performer. He marks the frequent changes of rhythmic flow by *cédez* (ritardando), *serrez* (accelerando), or *retenu* (ritenuto), and often indicates their exact duration by dashes and their termination by a double slash.

Within the rhythm of the precisely determined musical fragment Debussy sometimes also includes rubato. In his earlier works before 1902 he follows the practice of Liszt and Tchaikovsky by including *tempo rubato* as part of a main tempo marking or as a tempo change in effect for an entire section of up to thirty-two bars. He had met Liszt and heard him play at Rome in 1885, and had become familiar with Tchaikovsky's music during his summer travels in 1880–2 as one of Madame von Meck's musicians.[142] After 1902, however, he more often marks the single word *rubato* for a brief fragment, whose length (between one half and four bars) he sometimes indicates by dashes and a double slash. Such precise notation emphasizes the importance of rubato in the music, and it is therefore not surprising that Debussy includes the word far more frequently than any composer before him. I have found one hundred and two rubatos in fifty-seven of his pieces or movements composed between around 1880 and 1917 (not counting transcriptions and piano reductions he made from the same works): fifty-two for piano, eighteen for orchestra, ten for solo voice and piano, and twenty-two for various chamber combinations. Thirty-six of the total are in late works composed between 1915 and 1917.

For the longer rubatos included in the main tempo markings of the early works, Debussy usually separates the expression *tempo rubato* from the actual tempo terms by a comma, by parentheses, or by the word *et*: 'Andantino moderato, tempo rubato'; 'Moderato (tempo rubato)'; 'Très modéré et tempo rubato'.[143] Sometimes, however, he simply sets them side by side, as in 'And.^te tempo rubato', and on one occasion he writes 'Tempo quasi andante rubato'.[144] In some cases it is not completely clear how far the rubato should extend.[145] On the tempo changing level, *tempo rubato* or *a tempo rubato* may appear alone, or *tempo rubato* may coincide with a tempo change in Italian or French, as in 'Meno mosso tempo rubato' or 'Tempo rubato (un peu moins vite)'.[146]

After 1902 Debussy occasionally uses the full expression *tempo*

[142] See Edward Lockspeiser, *Debussy: His Life and Mind* (London: Cassell and Co. Ltd., 1962 and 1966; repr. with corrections, CUP, 1978), i, 40–51 and 81–3; also *Portrait of Liszt by Himself and His Contemporaries*, ed. Adrian Williams (OUP, 1990), 657–8.

[143] 'Fleur de blés', *Frühe Lieder nach verschiedenen Dichtern*, ed. Reiner Zimmermann (Leipzig: Peters, 1976), 6; *Suite bergamasque* (Paris: E. Fromont, 1905), repr. in *Claude Debussy, Piano Music 1888–1905*, 2nd edn. (Dover, 1973), Prélude; and *Les Chansons de Bilitis* (Paris: Société des Éditions Jobert, 1971).

[144] 'Harmonie du soir' from *Cinq poèmes de Baudelaire* (Paris: Durand, 1902–4), repr. in *Claude Debussy, Songs 1880–1904*, ed. Rita Benton (Dover, 1981), 41; *L'Enfant prodigue* (Durand, 1908, repr. 1949), 'Air de danse'. Durand often reprinted later its own original editions of Debussy's works; I have indicated this by following the original copyright date with 'repr.' and, when known, the date of reprint.

[145] In a somewhat later work, 'Reflets dans l'eau' from *Images I* (Durand, 1905), repr. in *Piano Music 1888–1905*, 153–9, the main tempo marking is 'Andantino molto (Tempo rubato)'. 'A tempo' appears in m. 16, 'Quasi cadenza' in m. 20, and 'Mesuré' in m. 24, one of which may signal the cessation of rubato. With the return of the opening motive in m. 35, however, 'au Mouvt.' is the only marking, but one might expect a return of the rubato as well as the tempo.

[146] 'Clair de lune' from *Suite bergamasque*, repr. in *Piano Music 1888–1905*, 26; *Cinq poèmes de Baudelaire*, repr. in *Songs 1880–1904*, 49, 52, and 56; and *Deux arabesques* (Durand, 1904), repr. in *Piano Music 1888–1905*, No. 1, 3.

rubato, but now more often writes simply *rubato*, sometimes *molto rubato*, occasionally *poco rubato*, and once *un pochettino rubato*.[147] The effect may still be quite lengthy, but now is apt to be brief.[148] The rubatos in these later works often alternate in rapid succession with other brief, self-contained units marked *cédez* or *serrez*. All these rhythmic effects, including rubato, are cancelled by *a tempo*, *in tempo*, *(in) tempo I°*, *mouvt.*, *au mouvt.*, *(in) 1er mouvt.*, or *tempo giusto*. In one work, 'Retenu' or 'En retenant' is followed by 'Mouvt. rubato' or '1er mouvt. Rubato'; on another occasion 'rallentando' precedes 'I° Tempo rubato', followed ten bars later by 'I° Tempo giusto'.[149] This implies, it seems to me, that rubato by itself does not essentially involve retardation or acceleration, at least not on the same level as *cédez* or *serrez*, and that Debussy conceives of a basically stable tempo either with or without rubato.

In the twelfth Prélude of Book II, composed in 1912–13, furthermore, the expression 'Tempo (Rubato)' follows a passage marked *Quasi cadenza* in order to indicate, presumably, that a free rhythm is to be followed by the original stable tempo in company with rubato.[150] In succeeding works Debussy seems simply to remove the parentheses from *Rubato* and use the expression *Tempo rubato* to mean a combination of *a tempo* and *rubato*. Several later works include both the expression *tempo rubato* and the single word *rubato*. In such cases, the single word must indicate that rubato continues in whatever tempo preceded it, whereas *Tempo rubato* involves a return to the basic tempo of the piece. Thus, when *rubato* follows *cédez* or *serrez*, it proceeds presumably at the pace attained at the end of the tempo change. At the same time, one can still find *rubato* in one bar followed by simply *mouvt.* in the next, even though both are basically *a tempo*. In the fourth movement of his *Épigraphes antiques* for piano four hands from 1914, for example, the following sequence of terms occurs over the course of twenty-five bars: 'au Mouv^t, Poco rubato, Mouv^t, Retenu, Molto rubato, au Mouv^t (sans rigueur), en serrant, Tempo rubato'. In this case 'Mouv^t' cancels 'Poco rubato', but does not alter its basic tempo; 'Molto rubato' continues the slower tempo of 'Retenu', but 'Tempo rubato' slows down, following 'en serrant', to 'a tempo'.[151]

[147] *Douze études II* (Durand, 1916), No. 9, 15.

[148] For longer ones, see the 10-measure middle section of *Syrinx* for solo flute (Paris: J. Jobert, 1927; repr., Paris: Société des Éditions Jobert, 1954); and 12 measures marked 'Lento. Molto rubato con morbidezza' in the Finale of the Cello Sonata (Durand, 1915 repr.).

[149] *La Plus que lente* for orchestra (Durand, 1912 repr.), 8 and 11; Sonata for Flute, Viola, and Harp (Durand, 1916 repr.), last movt., 32–3.

[150] *Préludes, deuxième livre* (Durand, 1913 repr.), No. 12: 'Feux d'artifice', m. 68.

[151] *Six épigraphes antiques* (Durand, 1915 repr.), No. 4: 'Pour la danseuse aux crotales', mm.

Before 1912 Debussy indicates the combination of *a tempo* and rubato in a different manner. The main tempo marking of the Valse *La plus que lente* for piano from 1910 is 'Lent (Molto rubato con morbidezza)'. Following a 'Retenu' of two bars, he marks the second half of the theme 'Mouv^t', referring presumably to the return of both 'Lent' and 'Molto rubato'. Then, following 'Animez un peu', 'En serrant', and 'Retenu', the word 'Rubato' occurs twice, implying either that 'Animez un peu' replaced 'Molto rubato' as well as 'Lent', or that the intensity of rubato should now be less than 'Molto'. For a later return to the opening idea, however, he marks 'Mouv^t (Rubato)' to indicate both the rubato and the original tempo. In the orchestral version two years later, he writes simply 'Mouv^t rubato' at this point, and also adds, for the final appearance of the theme: 'Ier Mouv^t Rubato', which is missing in the original piano piece. Again, it appears that tempo changing terms at some point cancelled the rubato as well as the tempo of the preceding 'Mouv^t rubato'.[152]

Debussy's transcriptions of his own works sometimes reveal changes in rubato notation. These may reflect a change of mind, a response to the new medium, or simply careless omission. In his 1907 orchestration of the piano part of his early song 'Le jet d'eau' from 1889, he replaces 'Meno mosso tempo rubato' the first time it occurs with 'Animez légèrement', and marks its second appearance one measure earlier.[153] In the 'Air de danse' from *L'Enfant prodigue* he marks the ornate flute melody 'Tempo quasi andante rubato' in the first score of 1908, but 'Andante—dans un rythme un peu abandonné' in the piano/vocal score from the same year and in the extract for piano four hands in 1905.[154] In the ballet *Jeux* two rubatos as well as other markings that appear in the orchestral score of 1914 are missing in the piano reduction of 1912.[155]

Debussy's rubatos usually appear in connection with terms such as *doux*, *gracieux*, *expressif*, *ondoyant*, *passionné*, *avec charme*, *lusingando*, *harmonieux*, and *dolcissimo*. An orchestral rubato accom-

---

25–49. For other works that include both 'Rubato' and 'Tempo rubato', see *Jeux* (Durand, 1914 repr.), *Sonate pour flûte, alto, et harpe* (Durand, 1916 repr.), and *Sonate pour violon et piano* (Durand, 1917 repr.).

[152] *La Plus que lente, valse pour piano* (Durand, 1910, repr. 1962), mm. 1–45, 59, and 103; *Valse pour orchestre*, 8 and 11.

[153] *Cinq poèmes de Baudelaire*, No. 3, repr. in *Songs 1880–1904*, 49 and 52 (mm. 35 and 63); for *chant et orchestre* (Durand, 1907), 10 and 18.

[154] Full score (Durand, 1908, repr. 1949), 31; *Partition chant et piano* (Durand, 1908, repr. 1960), 13; and *Extrait de la cantate* (Durand, 1905), 4.

[155] Full score, 39 and 43; *Partition pour piano réduite par l'auteur* (Durand, 1912), 15 and 16. The same 2 rubatos are missing also in the *Transcription pour piano 4 mains* (Durand, 1914), 14 and 15.

panies the passionate embrace of Pelléas and Mélisande,[156] and in *La Boîte à joujoux* it marks the dropping of a flower and the awakening of love between a doll and a toy soldier.[157] In the songs, rubato most often emphasizes love, either gentle (a bouquet gathered for a lover) or passionate (the soul 'set aflame by the burning flash of pleasures'), or the memory of love ('sounds and perfumes turn in the evening air': 'Is there no longer a perfume that remains of the celestial sweetness ... of faithful love?').[158] Other moods involve a fluttering fan, the gloomy atmosphere of evening, pessimism, and even the comic.[159]

The earlier continuous rubatos of Debussy probably involve, as they did with previous composers, expressive lingerings on important notes, as well as subtle accelerations and retards to project the shapes of phrases. Clues for the execution of his more typical brief rubatos reside in two aspects which we have also noted with some earlier composers: polyrhythms and specially accentuated touches or articulations. Cross rhythms are evident in sixteen of Debussy's rubatos, involving two beats against three, three against four, and two or three against five. The rubato in one song has two beats in the voice against three in the right hand of the pianist and four in the left. Another displays not only the polyrhythm of two against three, but also the polymetric effect of a melody in syncopated 3/4 metre against an accompanying ostinato figure in 2/4.[160] Twenty-five of his rubatos include accentuations marked by dashes, or portato touch indicated by various combinations of dots and dashes under a slur. In one case both a dot and a dash appear above each of five notes under a slur.[161] Only rarely does he describe the touch by terms such as *soutenu*, *lusingando*, or *murmurando*.[162]

A number of works contain rubatos with both cross rhythms and

[156] Full score (Paris: E. Fromont, 1904; repr. Durand, 1950) and piano/vocal score (Paris: E. Fromont, 1902; repr. Durand, 1907), Act IV, sc. iv.

[157] During the 'Poco rubato' in the 1st tableau of *La Boîte à joujoux* the doll drops a flower in front of the little soldier, who picks it up and kisses it; the doll rejects the soldier. In the rubato near the end of the 2nd tableau, however, the doll is gently caring for the soldier, who has been wounded in battle and still carries the flower—the symbol of his love.

[158] 'Fleur des blés' (Flower of the Wheat Fields), 'Le Jet d'eau' (The Fountain), 'Harmonie du soir' (Debussy used the line 'Les sons et les parfums tournent dans l'air du soir' as the title for the 4th Prélude in Book I), and the 1st of the *Deux romances* on poems by Paul Bourget (Durand, 1906), repr. in *Songs 1880–1904*, 72.

[159] 'Éventail' (The Fan) from *Trois poèmes de Mallarmé* (Durand, 1913, repr. 1971); 'Recueillement' (Meditation) from *Cinq poèmes de Baudelaire*; 'Le Faune' from *Fêtes galantes* (2nd collection) on poems of Verlaine (Durand, 1904), repr. in *Songs 1880–1904*, 155; and 'Des femmes de Paris' from *Trois ballades de François Villon*.

[160] 'Des femmes de Paris', m. 43; and 'Recueillement', m. 16.

[161] *Douze études*, No. 2, mm. 13 and 14.

[162] *Ibéria*, *Images pour orchestre*, No. 2 (Durand, 1910, repr. 1956), 37; Sonata for Flute, Viola, and Harp, 32, or Violin Sonata, 4; and the 8th Prélude of Book II ('Ondine'), m. 44.

accentuated articulation. Fortunately, Debussy himself recorded two such pieces for piano.[163] *D'un cahier d'esquisses*, composed in 1903, includes three appearances of the single rubato measure shown in Ex. 9.5. The first half of the bar produces two beats against three; the

Ex. 9.5 Debussy, *D'un cahier d'esquisses*, 1903

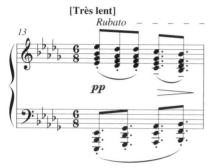

dots, dashes, and slurs indicate accentuated touches. Debussy verifies in his own playing what we deduced previously from the notation: that rubato does not necessarily involve noticeable changes of tempo, either internally or in relation to the surrounding music. He plays the rubato chords deliberately and with a declamatory emphasis that commands the listener's concentrated attention. The first time he makes the opening chord sharply staccato and follows it with a substantial silence of articulation, thus carefully articulating the polyrhythmic displacement. If the tempo is broadened somewhat in the process, he does it in an exceedingly subtle manner. He precedes the second rubato, in fact, with an unmarked retard, so that the rubato itself, by comparison, seems to be *in tempo*. The first and third times the measure in Ex. 9.5 leads to a single chord, the second time to a new idea. It also appears a fourth time, but without rubato and without dashes under the last two chords. Although Debussy plays it much like the others, the stream of chords continues this time on into the following measure.[164]

Debussy also recorded the 'Soirée dans Grenade' from his *Estampes* of 1903. A six-bar passage marked 'Tempo rubato' begins with the two measures in Ex. 9.6, continues the whole-tone scale with the same

---

[163] *Welte–Mignon Digital: Berühmte Komponisten spielen eigene Werke* (Intercord INT 860.855, recorded 1985). Debussy performed in 1912 and 1913.

[164] *D'un cahier d'esquisses* (Brussels: Schott Frères, 1904), repr. in *Piano Music 1888–1905*, 89 and 91 (mm. 13, 15, and 38).

Ex. 9.6 Debussy, *Soirée dans Grenade*, 1903

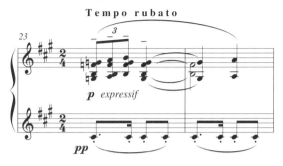

melody up a third in the right hand, then repeats the opening two measures marked 'Retenu'. The same section recurs later, but slightly more animated by additional notes. Debussy again emphasizes the accented chords in an almost parlando fashion, and clearly enunciates the three against four polyrhythm, with the sixteenth note in the left hand's habanera pattern clearly falling after the third triplet chord. Very little tempo changing takes place within the rubato or between it and the preceding music. Debussy actually changes tempo more conspicuously when he rushes the sixteenth-note chords in the following section, which, curiously enough, is marked 'Tempo giusto'. Debussy generally follows his own markings fairly faithfully, however, and displays, especially in the melody beginning in measure 7 and marked *expressif*, a warm and sensitive, but subtle and refined application of the performer's type of rhythmic freedom—the expressive lingerings and slight accelerations and retards.[165]

Jacques Durand, from the firm which originally published most of Debussy's works, lists the characteristics of his playing: 'a full sonority, a remarkable delicacy, a perfect mastery of *nuance*, an impeccable finish, an imperceptible rubato always framed within the beat, an astounding use of pedal'.[166] Marguerite Long, who studied Debussy's piano works with him and taught at the Paris Conservatoire, states that 'this delicate rubato . . . is confined by a rigorous precision, in almost the same way as a stream is the captive of its banks', and that it 'does not mean alteration of line or measure, but of nuance or élan'. In her description of the rubato in the warm, singing melody in

---

[165] *Estampes* (Durand, 1903), repr. in *Piano Music 1888–1905*, 102–3 (mm. 23 and 61). Pierre Boulez suggests a subtle rubato for the performance of Debussy's works in his 'Reflections on Pelléas et Mélisande', notes with the recording on Columbia M3 30119 from 1970, also in *Orientations*, ed. Jean-Jacques Nattiez, trans. Martin Cooper (HUP, 1986), 315.

[166] *Quelques souvenirs d'un éditeur de musique*, 2e *Série 1910–1924* (Durand, 1925), 21; trans. from Eigeldinger, *Chopin: Pianist and Teacher*, 128 n. 120.

Ex. 9.7 Debussy, *L'Isle joyeuse*, 1904

Un peu cédé. Molto rubato

L'Isle joyeuse from 1904 (see the opening measure in Ex. 9.7), she seems to be emphasizing the special touch involved: 'At the *Molto rubato*, I still seem to feel the fingers of the master pressing on my shoulder, and his fingers commanding mine to be nearer still to the keyboard, as if going into it.' The left hand plays five sixteenth notes per measure while the right has three eighth notes, an eighth followed by a dotted eighth and sixteenth, or two dotted eighth notes, some marked by dashes of accentuation.[167]

According to Long, Debussy admired the works of Bach and Liszt, but felt a special affinity for the music of Chopin. In addition, his own playing seems to resemble that of Chopin: in the flowing manner ('one must forget that the piano has hammers'), the imaginative use of the pedal, the preference for soft sounds, the sincere rather than theatrical approach, and the generally disciplined, elegant, and refined style.[168] Debussy had studied piano at the Paris Conservatoire with Antoine François Marmontel, who had met Chopin in 1832 and often heard him perform. Even earlier Debussy had studied with Antoinette Mauté de Fleurville, whom he described as knowing many things about Chopin, and who, according to Long and others, had herself been a pupil of Chopin.[169] Debussy composed mazurkas, ballades, nocturnes, préludes, études, and other pieces with titles familiar from Chopin. In 1915 he dedicated his *Douze études* 'to the memory of Frédéric Chopin'.[170]

[167] *Au piano avec Debussy* (Paris, 1960), trans. Olive Senior Ellis as *At the Piano with Debussy* (London: J. M. Dent & Sons, 1972), 25 and 39.

[168] Ibid. 12–13. See also Eigeldinger, *Chopin: Pianist and Teacher*, 128 n. 120, and Schonberg, *The Great Pianists* (1987), 363–4.

[169] In a letter of 27 Jan. 1915, in *Lettres de Claude Debussy à son éditeur*, 131, he mentions Madame Mauté de Fleurville, 'à laquelle je dois le peu que je sais de piano...Elle savait beaucoup de choses sur Chopin'. See Long, *At the Piano with Debussy*, 13, and Eigeldinger, *Chopin: Pianist and Teacher*, 129 n. 121. Also see Lockspeiser, *Debussy: His Life and Mind*, i, 19–23 and 25–30.

[170] Debussy had a difficult time deciding between Chopin and F. Couperin. UCLA ML

During the same year he was preparing an edition of Chopin's piano works, which Durand published between 1915 and 1917 as part of his series *Édition classique*. The volumes containing the Mazurkas, the Nocturnes, and some miscellaneous pieces appeared in 1915, with some rubatos added by Debussy and with special treatment of those marked by Chopin. Debussy adds the word 'Rubato' four times in the Nocturne Op. 9 No. 1, a piece in which Chopin marked no rubatos. Plate XX shows, for comparison, roughly corresponding measures from Kistner's publication (Leipzig, originally in 1832) and Debussy's in 1915. The rounded binary form in the middle of the piece has the following design: AB–AB–CB–CB, where each letter represents four bars and B, repeated four times, contains an astonishing modulation from D flat major to D major and back— precisely the sort of harmonic movement bound to catch Debussy's attention.

On Plate XX, the original excerpt contains the first three appearances of the modulating phrase B, with the first and third marked *poco rallent.*; the excerpt from Debussy's edition includes the second and third occurrences of B, both with 'Rubato' in parentheses to indicate an editorial addition. Debussy's page commences seven bars after the other one, which, in turn, begins with the third bar of the binary section. Debussy thus substitutes 'Rubato' for *poco rallent.* three times and adds it where Chopin has no marking, thus treating each of the four appearances of B the same. He seems to have felt that an intensified touch (perhaps more like a ritenuto) was more effective than a rallentando or perhaps needed to be added to it. Each rubato is apparently in effect until the following 'a Tempo'. In any case, effective rendition of the passage requires strict observance of the carefully marked dynamics.[171]

Ex. 9.8 shows the other rubato added by Debussy, this time for a transition back to a cantabile theme in the *Allegro de concert*. He carefully marks its extent with dashes, as he often does in his own works. He adds a decrescendo sign in measure 199, an accent sign also on the second beat of measure 198, and releases the pedal at the

possesses 2 versions of Durand's 1916 publication of the Études: one has no dedication at all, whereas the other, reduced in size and 'published in the United States' by Elkan–Vogel Co., Philadelphia, is dedicated 'à la mémoire de Frédéric Chopin (1810–1849)—Claude Debussy— Été 1915'.

[171] The excerpts in Pl. XX come from BL, the Kistner from h.471.j.(11.), the other from g.553.h: Chopin, *Œuvres complètes pour piano, révision par Claude Debussy, Édition classique*, No. 9704 (Durand, 1915). For M. Schlesinger's edition, first published in 1833, see *Stirling*, 20–1.

PLATE XX. Frédéric Chopin, excerpt from Nocturne Op. 9 No. 1 in two versions:

(a) from *Trois nocturnes pour le pianoforte* (Leipzig: Kistner, originally 1832)

(b) from *Nocturnes, révision par Claude Debussy* (Paris: A. Durand & Fils, 1915)

XX*b*

Ex. 9.8 Chopin/Debussy, *Allegro de concert*, Op. 46 (1915)

beginning of measure 199 rather than after the first beat of measure 198. Judging from his performances of his own rubatos, the sixteenth notes, or perhaps the entire rubato passage, should be played as an intensely flowing anacrusis to the opening melody note in measure 200. Conceived as a single impulse and as a transition in dynamics from *ff* to *p*, such a rubato would not necessarily involve any considerable changes of tempo to be effective. Debussy wanted an effect, I believe, quite different from *ad libitum, quasi cadenza,* or *senza tempo.* He seems to indicate a careful and emphatic enunciation of separate notes within a powerful anacrustic sweep of rhythm.[172]

It is clear from this last example that Debussy has his own special type of later rubato in mind. Chopin, of course, would not have marked rubato here, for there is no possibility for displacement between a melody and its accompaniment. Evidently Debussy did not learn about the earlier rubato and its strict left hand from Marmontel or Mauté and therefore attempted, not without some difficulty, to interpret Chopin's marked rubatos from his own point of view. In his own works published by Durand around the same time, the word 'Rubato' is almost always capitalized and printed in large, dark letters above the top staff, as in Plate XX. This is the way he marks the three rubatos of Chopin that seem to fit best with his concept of the term: in the Mazurka Op. 6 No. 2 (see Ex. 7.16) he no doubt intends that his kind of rubato will give a sense of finality to the last eight bars of the piece, and in Mazurkas Op. 7 No. 1 and Op. 24 No. 2 (Exx. 7.17 and 7.18), he probably feels that rubato will intensify the repeated four or eight measures preceding 'a Tempo'.

Debussy prints all of the other Chopin rubatos, however, in small uncapitalized italics like the words *poco stretto* in the third line of his

---

[172] Cf. the edn. of Br. & H first published in 1841, copy in the William Mason Collection, v, at UCLA ML.

excerpt on Plate XX. He marks the Mazurkas Op. 24 No. 1 (Ex. 7.20) and the Nocturne Op. 15 No. 3 (Ex. 7.28) in this manner, placing *languido e rubato* above the top staff of the latter; he probably intends the rubato in both pieces to apply to an entire section. This leaves three Chopin pieces which must have been puzzling to Debussy, for their rubatos seem to interrupt an idea in midstream rather than identifying an entire unit. He prints the *poco rubato* in the Nocturne Op. 9 No. 2 (Ex. 7.21) exactly as Schlesinger did in 1832, including even the period after *rubato*. He would no doubt have preferred to see *poco rubato* written a half or one and a half bars earlier, but perhaps he realized that his rubato of emphatic enunciation might be effective in this momentary situation for expressive reasons.

The two remaining pieces, however, must have baffled him completely, for in the Mazurkas Op. 6 No. 1 (Ex. 7.14 and Plate XVI) and Op. 7 No. 3 (Ex. 7.15) Chopin articulates the parallel second half of a theme with rubato. In a letter to Durand, Debussy calls these rubatos 'assez fantaisiste', meaning unusual and whimsical, I think, as well as rather fanciful.[173] He could not, of course, imagine applying his rubato in either case to the entire second half of the theme, and he could not, at the same time, identify an expressive purpose, as in the Nocturne Op. 9 No. 2, for a purely momentary rubato. He simply marked them, for the most part, as they were in the original editions.[174]

His type of rubato spanned entire blocks or units of music, however brief they might be, and could not, as Chopin's earlier type and Liszt's suspension of rhythm could, articulate—with exact precision—a repetition, an approaching cadence, a dissonant note, or a chromatic chord. He would have been even more astonished at the asterisk Liszt placed in Weber's sonata (Plate XIX). His edition of Chopin's works

---

[173] Letter dated 14 July 1915, in *Lettres de Claude Debussy à son éditeur*, 137: 'Voyez donc: page 5 (mazurkas), un *Rubato*, assez fantaisiste et aussi: page 18, même observation.' Roger Nichols translates this in *Debussy Letters*, 298, as: 'Please note: on page 5 of the Mazurkas, a rather imaginative *rubato*, and the same on page 18.' In the published work, however, the 2nd rubato of Op. 7 No. 3 appears on p. 18 and no rubato at all on p. 5. The only 2 rubatos that are 13 pages apart are those on pp. 2 and 15, which occur in Op. 6 No. 1 and Op. 7 No. 3 (the 1st one). Therefore, the proofs that Debussy was working on when he wrote the letter must have been paged differently, counting the 2 sides of the title page and the 2 pages of the table of contents as pp. 1–4 and beginning the music on p. 5. In the published version the 2 pages of contents are numbered II and III and the music starts on p. 2.

[174] Debussy makes a few other alterations in the Mazurkas. Like Liszt in his edn., Debussy ties the 2 F#s between mm. 8 and 9 in Op. 6 No. 1 (see Pl. XVI). In Op. 7 No. 1 (Ex. 7.17) the rubato commences over the 1st instead of the 2nd note. The rubato at the beginning of Op. 67 No. 3 is missing, but does appear in the incipit in the table of contents; here the *p* (cf. Ex. 7.27) occurs under the anacrusis and *rubato* under the D#. He probably considered *rubato* to apply, however, to the whole piece, as though it were part of the main tempo marking.

thus reveals and puts into perspective his own concept of the device. His rubato gives a special character to a musical passage rather than accentuating an instantaneous event. It is perhaps similar, although achieved by different means, to the singing effect produced by earlier rubato in Chopin's Nocturne Op. 9 No. 2 (Ex. 7.21). Debussy's rubato was sensual and smooth and not jerky or capricious, and operated over a carefully specified length of time. It simply acted, like dynamics, touch, pedalling, and other elements in the performance of music, to characterize one passage in contrast to another.[175]

Debussy's intense involvement with rubato was thus manifested through his activities as an editor as well as through his compositions and his performances. The unusually frequent use of the word in his scores is not equalled by any of his contemporaries. The only other composer from this period with a substantial number of rubatos is Alexander Skryabin, who seems to follow, for the most part, the practice of Liszt. Those most heavily influenced by Debussy use rubato far less than either Debussy or Skryabin.[176] Only a few examples occur in works by early twentieth-century composers, such as Sibelius, who continue the mainstream romantic tradition. The striking contrast between Debussy and that tradition is emphasized by Edward Elgar, who marks an oboe melody in his Second Symphony 'espress. (molto rubato, quasi ad lib.)', and who himself conducted, according to a contemporary critic, with a 'very personal rubato'. His recordings reveal 'a rubato . . . characterized by the subtle use of tenutos and agogic accents, accelerandos and rallentandos' in company with the formal tempo modifications, rushing of short notes, over-dotting of dotted notes, and offhand sense of rhythm typical of most performances at that time.[177]

The period from 1850 to the early years of the next century represents unprecedented diversity in the interpretation of rubato. The writers of various sorts differ among themselves. They, in turn, differ from the conductors and performers as well as from most of the composers.

---

[175] Concerning the performer's type of rubato, Rosen writes: 'Rubato in Debussy always, I think, involves the lengthening of one or more beats. The *tempo* [of the beat] remains constant but the bar is lengthened, the rhythm deformed.' He cites *Images*, 2ᵉ *Série* No. 3: 'Poissons d'or' (Durand, 1908 repr.), m. 46, where the 1st beat in 3/4 must be doubled in length. Debussy does not mark the word *rubato* here, but does in 2 places later in the piece.

[176] There are examples by Albéniz, also Ravel, Joseph Marx, Delius, and Griffes.

[177] Robert Philip, 'The Recordings of Edward Elgar (1857–1934)', *EM* 12 (1984), 483–4 and 487–9 (quotation from *The Times*, 27 Apr. 1926, p. 14). He describes the oboe melody in Elgar's recording as 'entirely without vibrato' and mentions the frequent lack of dynamic nuances in oboe playing of the period. In this case, the main element remaining to project the *espressivo* would be the tempo changes involved in the *molto rubato*.

The increasingly free rhythmic treatment of traditional romantic composers and performers, however, finally finds a complete contrast in the innovative style of Debussy—a disciplined style in which rubato is carefully restrained and confined.

10 THE MODERN RUBATO

THE dominating musical style of the first part of the twentieth century involves, in some essential respects, a complete reversal of the ideals of late Austro-German romanticism. Debussy represents a substantial but still romantic reaction to this tradition. The new neo-classical style emerges when composers turn from a subjective to an objective point of view,[1] and when performers reject excessive personal interpretation in favour of fidelity to the musical score. Along with the new spirit come new concepts of time and rhythm which involve a far greater adherence to strict tempo. Influence comes occasionally from the rhythms of folk-music, the music of other cultures, or Western music of the past. Some of the popular music of this period incorporates the earlier type of rubato. In art music, performers make far less use of the later type, while composers indicate it in their scores far more. All of these practices are documented with unprecedented fidelity by increasingly excellent recordings and carefully notated scores.

Before turning to the neo-classic rubato of Bartók, Stravinsky, and others, we will first consider the expressionism of Schoenberg and Berg, which, in spite of startling dissonance and lack of tonality, continues the central romantic ideal of subjective expression.

Schoenberg, Webern, and Berg

We have already noted the general tendency during the course of the Romantic period for the composer to increase the number of tempo directions, and we saw in Ex. 9.4 how later composers like Strauss could notate, in effect, an exact degree of tempo alteration. Arnold Schoenberg's expressionistic compositions from around 1908 to 1923 represent a high point in this sort of rhythmic control. Elliott Carter, whose own rhythmic concepts we will consider later, mentions Schoenberg's *Erwartung* from 1909 as a superlative example of rubato, using the word to refer to 'irregularity of speed, . . . a constantly fluc-

---

[1] See Aaron Copland, *The New Music 1900–1960* (Norton, 1968), 17–18.

tuating tempo'.[2] This work is a monodrama in which a woman wanders through a forest at night and discovers her murdered lover. The tempo changes frequently in order to give expression to the powerful text. This is achieved in a number of ways: by terms that alter the tempo gradually ('rall.', 'rit.') or at once ('langsamer', 'rasch'); by numerous metronome markings (as often as twice in one measure); by changes of time signature, with equivalence of old and new note values; triplets alternating with duplets and general changes of note values, especially in the voice part. In addition, Schoenberg states in the instructions preceding the score that the numerous metronome numbers must not be taken literally, for they only indicate a basic figure from which the tempo is to be freely formed.[3] Occasionally the increase of movement caused by shorter note values (as in measure 15) seems to contradict the tempo marking ('viel langsamer' in this case). When to all of this notated fluctuation is added the tempo flexibility of the singer and the vigorous pronunciation of words, a highly dramatic declamation results which follows with great intensity the rapidly changing emotions in the text.

Schoenberg uses the word *rubato* in two of his twelve-tone works. In the third movement ('Largo') of the Fourth String Quartet from 1936, the first violin has brief unaccompanied cadenza-like passages marked 'dolce', 'ad libitum', or, in the case of a motive consisting of two thirty-second-note triplets falling to an eighth note, 'rubato (presto)'. It is not clear whether the rubato applies only to this motive or continues until the 'Grave' in the following measure. In the Piano Concerto of 1942, the marking 'poco più mosso (rubato)', which appears in measures 160–1, is cancelled in the following measure by 'a tempo'. The rubato in this case seems to involve the vigorous and very clear enunciation of a four-note rhythmic motive which occurs twice in succession in the piano part, each time marked separately with the sign for *Hauptstimme*. The last note of the first statement of the motive has a hairpin accent sign. Schoenberg no doubt wanted the motive played with special emphasis so that the listener would not only notice it, but would remember it when repeated later many times throughout the rest of the work. In an extensive cadenza for piano alone later in the Concerto, the word 'rubato' by itself occurs twice, each time cancelled a half measure later by 'a tempo'. In both cases the rubato spans three rhythmic figures in the right hand: the first two

---

[2] Carter's comments appear on a postcard, dated 12 July 1989, in which he replies to my questions concerning rubato.

[3] *Erwartung* (Universal, 1950): 'Die Metronomzahlen dürfen nicht wörtlich genommen werden, sondern sollen bloss die Zähleinheit des Grundtempos andeuten, aus dem das Tempo frei zu gestalten ist.'

like the first two notes of the earlier motive in measure 160, the other like its first three notes. Over the first note of each figure is a caret, which, according to the explanatory notes preceding the score, means that the note thus marked 'should be given a certain degree of importance'. In the left-hand triplets played against these three figures, moreover, each note is marked with a caret, apparently with the intent of clearly articulating the polymetric effect of three against four and the consequent displacement of the hands. The rubatos in this work seem to have the specific function of declaiming important rhythmic motives with special emphasis. On the other hand, the work contains very few tempo changing instructions of the sort so frequent in *Erwartung*.[4]

Schoenberg's student Anton Webern includes rubato only in his orchestration in 1934–5 of the six-part Ricercar from Bach's *Musical Offering*. He splits the nineteen notes of the subject (see Ex. 2.19 for the first eight) into seven segments, each played by an instrument different from that used for the preceding segment. This *Klangfarben* technique, which occurs often in his own works, is intended 'to reveal the interrelation of motifs'.[5] In nine of the entrances of the subject, notes 6 to 14—the chromatic portion of the theme—are marked 'poco rubato', followed in most cases by 'poco allargando' or 'poco rit.' for the remaining notes. In addition, two brief chromatic passages within episodes are to be 'sehr fliessend, rubato'. In this unusual environment, the rubato probably involves, as it seems to in Schoenberg's works, emphatic motivic articulation rather than conspicuous alterations of tempo.

It is Schoenberg's more conservative and more romantic student Alban Berg, however, who is more deeply involved with rubato. The

---

[4] Schoenberg, *Sämtliche Werke* (Schott and Universal, 1966–), VI: A/xxi, p. 100 (m. 625) for the String Quartet, and IV: A/xv, pp. 188 (mm. 160–1) and 218 (mm. 291 and 292) for the Concerto. The motive in mm. 160–1 of the Concerto recurs often in mm. 165–95, 216–18, 233–5, and 313–16. For an English trans. of the 'Explanatory Notes', see *Concerto for Piano and Orchestra, Op. 42* (Los Angeles: Belmont Music Publishers, 1972). Concerning Schoenberg's use of the musical motive as a neo-classical element, see Charles Rosen, *Arnold Schoenberg* (New York: The Viking Press, 1975), 100–1; see also 77, where he mentions 'the occasional 4/8 measure' which Schoenberg includes in the *Walzer*, Op. 23 No. 5 'in homage to the traditionally exaggerated *rubato*' in the Viennese waltz.

[5] J. S. Bach, *Fuga (Ricercata) a 6 voci, No. 2 aus dem 'Musikalischen Opfer' für Orchester gesetzt* (Universal, 1935 and 1963); quotation comes from a letter from Webern to Hermann Scherchen in 1938, trans. on p. iv of the Preface by F.S. Concerning the motivic aspects, see Paul Griffiths' article in *New Grove*, xx, 276, and Carl Dahlhaus, 'Analytical Instrumentation: Bach's Six-Part Ricercar as Orchestrated by Anton Webern' in *Schoenberg and the New Music*, trans. Derrick Puffett and Alfred Clayton (CUP, 1987), 185; the article was originally printed in *Bach-Interpretationen*, ed. Martin Geck (Göttingen, 1969), repr. in *Schönberg und andere: Gesammelte Aufsätze zur Neuen Musik* (Mainz, 1978). Concerning the *Klangfarben* style, see the comments by Robert Craft on pp. 12 and 13 of the booklet accompanying the recording Columbia K4L–232: *Anton Webern, the Complete Music Recorded under the Direction of Robert Craft*.

word appears thirty-eight times, in a positive sense, in nine pieces or movements composed between around 1922 and 1935 (the year of his death); he uses it fourteen more times in order to prohibit the device. Berg's interest in rubato increased over the years, so that the greatest number are in the uncompleted opera *Lulu*, with seventeen applications and eleven prohibitions. He first employs it, however, in *Wozzeck*. He marks 'bedeutend langsamer und molto rubato' when the Captain, in the midst of taunting Wozzeck about Marie's affair with the Drum Major, sings 'ich habe auch einmal die Liebe gefühlt!' (I too once felt love!); the last four notes are *frei* (*ohne Rücksicht auf den Takt*). Later in the same Act, a polymetric 'Rubato in 9/8' accompanies music in 4/4 as the First Apprentice sings 'My own immortal soul stinketh of brandywine'. Tempo changing terms appear below the word 'Rubato' in smaller italics: *a tempo* for four eighth beats, *accel.* for three, and *rit.* for two. He repeats this pattern of tempo fluctuation for eight repetitions of an ostinato rhythm which continues for nine measure of 4/4. After Wozzeck says in the next Act 'All Heaven would I give . . . if I still could sometimes kiss you so', a 'rubato' with *poco accel.* and *rit.* three times in parentheses accompanies the text 'But yet I may not!' The parenthetical tempo changes seem to constitute a sort of explanation of the rubato, but in notation more like that in *Erwartung*.[6]

In addition, the final scene in 12/8 (the invention on a persistent movement in triplets) bears the tempo marking 'Fliessende Achtel, aber mit viel Rubato'. When one of the children asks 'Where is she?' (referring to the murdered Marie), the music is marked 'poco rit.'; the reply 'Out there, on the path by the pond' is 'a tempo, ma sempre rubato'. When the dialogue ceases and Marie's child, left all alone, 'hesitates a moment, then rides off after the other children', the music becomes 'etwas zögernd [hesitating] und molto rubato' for two measures. In parentheses below this marking, *rit.* and *acc.* alternate nine times, occurring in the first measure on the first note of each triplet and in the second on the last note of each triplet. The preceding rubato, however, occurs simultaneously with 'a tempo' and is therefore perhaps involved more with emphatic articulation than with noticeable changes of tempo.[7] On the other hand, Berg marks 'A

---

[6] *Georg Buechners Wozzeck*, rev. by H. E. Apostel according to the final corrections and amendments left by the composer, English trans. by Eric Blackall and Vida Harford (Universal, 1955), Act II, sc. ii, mm. 326–9, and sc. iv, mm. 456–64; Act III, sc. ii, mm. 91–2. Concerning the polymetric rubato, see George Perle, *The Operas of Alban Berg*, i: *Wozzeck* (UCP, 1980), 173 (and Ex. 168) and 144 (and Ex. 111); see also 183 (and Ex. 184 on 179), where he describes as 'quasi-rubato' a statement of the *Hauptrhythmus* from Act III, sc. ii ('Invention on a Rhythm') in which the metronome figure changes from 60 to 80 and back again to 60.

[7] Act III, sc. v, mm. 372–88.

tempo, ma molto rubato' following a 'Rit.' in 'Sommertage' from his
Sieben frühe Lieder, but below 'A tempo' the terms *rit.* and *accel.*
alternate five times in parentheses. The parenthetical *rit.* must surely
represent a less substantial level of slowing than the 'Rit.' which
precedes 'A tempo'. Such seemingly contradictory notation must be an
attempt to control the rubato, yet allow it, at the same time, to sound
rhythmically free.[8]

It is in the twelve-tone concert aria *Der Wein* from 1929 that Berg
first specifically prohibits rubato. In one case, 'a tempo senza rubato'
follows 'Noch langsamer und rubato' and 'poco accel.' On two other
occasions, however, 'Quasi Tempo I, ma senza rubato' and 'Subito
senza rubato' occur when no previously marked rubato is in effect.
This seems to imply that performers are presumed to play ordinarily
with some degree of rubato, and that they can at any particular time
increase the effect when they see the word *rubato* as well as decrease
or eliminate it when so instructed.[9] None of the fourteen prohibitions
in *Lulu* follows a marked rubato. Sometimes they are followed,
however, by 'a tempo', a term which must imply in this case an
increase in rubato.[10] Three of the four times the marking 'Subito senza
rubato' appears, its exact duration is shown by dashes.[11] At other
times Berg writes 'Tempo senza rubato', 'subito *p* e senza rubato',
'*Senza rubato*' by itself,[12] or simply '*Non rubato*' (an expression
which is used much earlier, as we will see, by Bartók). During
one twenty-three-bar passage, '*Senza rubato*' appears over the first
measure, after which '*Rubato*' and '*Non rubato*' alternate three times
at three- or four-bar intervals.[13]

For two of the rubatos in *Lulu* Berg indicates the extent by dashes
(following Debussy's practice) and the specific nature by *poco rit.* and
poco accel. in parentheses below the dashes. On another occasion a
two-bar *molto rubato* is followed only by dashes. At another spot,
'(*rubato*)' appears midway in a 'molto riten.' which extends over two

---

[8] Piano/vocal score (Universal, 1928), full score (Universal, 1959), No. 7, m. 26.

[9] (Universal, 1966), mm. 24, 27, 29, and 178.

[10] See George Perle, *The Operas of Alban Berg*, ii: *Lulu* (UCP, 1985), 222 and 232. Rosen, on the other hand, comments that '*senza rubato* in Berg may not imply that rubato is ordinarily used elsewhere, but that the temptation here is to be resisted.'

[11] *Lulu*, *Akt 1–2* (completed by Berg), ed. H. E. Apostel, 1963, rev. Friedrich Cerha (Universal, 1985), Act I, m. 976, and Act II, m. 1073; and *Symphonische Stücke aus der Oper Lulu* (also completed by Berg) (Universal, 1935), m. 143 of the Rondo (corresponding to Act II, m. 1073) and m. 62 of the Adagio. 'Subito senza rubato' appears without dashes in *Akt 1–2*, Act II, m. 285 (*Symphonische Stücke*, m. 38 of the Rondo).

[12] *Akt 1–2*, Act I, m. 1113, and Act II, m. 302 (*Symphonische Stücke*, m. 44 of the Rondo); Act II, m. 427; *Symphonische Stücke*, m. 1 of the Adagio, which is printed also in the *Anhang* to *Akt 1–2* and in *Akt III*, reconstructed by Friedrich Cerha (Universal, 1985).

[13] *Symphonische Stücke*, mm. 1–24 of the Adagio.

thirds of a measure.[14] Nine times the word *rubato* appears alone and capitalized, as Debussy wrote it in his later works. It sometimes occurs with other terms such as *affettuoso, animando,* or *tumultuoso,* and is often coupled with tempo changing words or with 'a tempo'. Berg sometimes writes the expression 'a tempo rubato' following 'riten.' and 'Zeit lassen', presumably to mean, like Debussy, both 'a tempo' and 'rubato'.[15] In his Violin Concerto from the same period Berg marks the second movement 'Allegro, ma sempre rubato, frei wie eine Kadenz', and later refers back to Tempo I as '*Allegro rubato*'.[16]

Berg's general style seems to be influenced mainly by Schoenberg and, beyond him, by the sensuous orchestral music of Wagner, Brahms, Mahler, and Richard Strauss. These composers, however, seldom wrote *rubato* in their scores—Wagner never, Brahms, Mahler, and Strauss only a few times, and Schoenberg only in two works composed after Berg's death. Rubato influence seems to come to Berg most directly from Debussy. Some similarities have already been noted: the single, capitalized word 'Rubato', the dashes that may follow it, and the expression 'a tempo rubato'. The idea of *non rubato*, however, seems to come from Bartók. It may be significant that Berg was closely involved from 1918 to 1921 in the Verein für Musikalische Privatauf-führungen, a society formed by Schoenberg and his students to arrange performances of works by modern composers, the most frequently represented of which were Reger, Debussy, Bartók, and Schoenberg.[17]

Artur Schnabel was also influenced by Schoenberg, but developed his own atonal style before turning to the serial technique. In the few compositions available to me, Schnabel includes *non rubato* more than twice as often as *rubato*. In his *Piece in Seven Movements* from 1936 for piano, the rubatos are sometimes connected with special accentuation or with polyrhythm, and in one case he indicates the notes to be lengthened by *ten.* and a fermata in parentheses. His rubatos seem to apply to carefully selected notes, even when a marking such as *sempre un poco liberamente (rubato)* refers to a passage of at least ten bars. He often writes *liberamente* by itself, to indicate, pre-sumably, a quite different sort of rhythmic freedom. His *non rubato* may refer to the entire music or specifically to a left-hand accompani-ment. It is combined with terms such as *semplice* or *egualmente* and often occurs with *senza pedal.*[18]

---

[14] *Akt 1–2,* Act II, mm. 501 and 506; *Symphonische Stücke,* mm. 11 and 16 of the 'Lied der Lulu', mm. 28 and 89 of the Adagio.

[15] *Akt 1–2,* Prolog, m. 17.

[16] *Violinkonzert* (Universal, 1936), 2nd movt., mm. 1 and 96.

[17] *New Grove,* xix, 733; and William W. Austin, *Music in the 20th Century* (Norton, 1966), 222.

[18] (Marks, 1947), Nos. 4 and 6.

Further insight into Schnabel's rubato comes from his edition of the Beethoven piano sonatas in 1935. In the earlier sonatas (through Op. 31 No. 2) he adds the word *rubato* three times by itself, five times with *con poco*, and once with *molto*. *Senza rubato* appears once and *non rubato* fifteen times, mainly in the late sonatas from Op. 106 on. *Non rubato* is often coupled with *egualmente* or *tranquillo*, and seems to be used for three purposes: to make syncopation or other complex rhythm clear; to keep sixteenth- or thirty-second-note accompanimental figuration steady, especially in the left hand; and to prevent rubato where the musical flow should move onward in spite of some local event which might invite excessive expression. Schnabel obviously understands the natural tendency of late romantic pianists to be preoccupied with the beauty of the moment rather than the broader aspects of form. Most of his *non rubatos* therefore seem to be a warning that the continuity of the music will be injured if the performer responds in too great detail to a particular momentary event.[19]

Schnabel himself recorded all of the sonatas from 1932 to 1935, and on three occasions noticeably applies rubato approximately where he marks the word in his edition.[20] Each case involves declamatory emphasis on only a few notes in order to intensify an effect which Beethoven has already indicated in the score. In one example, rubato reinforces a three-note motive which is suddenly played forte and leaps up to a sforzando note in a transition back to a main theme.[21] In another, rubato articulates a slightly expanded anacrusis that marks the third successive appearance of a melodic phrase.[22]

In Ex. 10.1 it enhances the powerful reaching gesture from the final appearance of the main theme in the last movement of Op. 2 No. 2.[23] In most of the earlier statements of the phrase, the ascent ends with two quarter-note Es on beats 3 and 4 in bars corresponding to measure 177, with E once more at the beginning of the next measure. In measure 177, however, Beethoven ornaments the repetition of the Es with lower neighbour notes, and Schnabel indicates rubato which is to extend, as shown by the dashes, through half of the following

[19] Ludwig van Beethoven, *32 Sonatas for the Pianoforte, Memorial Edition*, ed. Artur Schnabel (S & S, 1935), i and ii (with continuous paging). For rhythms that gain clarity by *non rubato*, see 289, 837, and 839. For accompanimental figures in even notes in the left hand, see 721, 731, 795, 798, and 844; for those in the right hand, 66, 68, and 845; and in both hands, 722, 732, and 786. For unusual harmonic movement, see 853, and for a rhythmically activated pedal point, 774. Schnabel now uses the expression *con rubato* as the opposite of *senza rubato*.

[20] *The Beethoven Society Edition*, first issued as a whole by RCA Victor in a limited edn. in 1956, later released again by Victor and other companies.

[21] 2nd movt. of Op. 10 No. 3, m. 43; Schnabel edn., 165, and RCA Victor LM–2151.

[22] Op. 14 No. 1, 1st movt., m. 69; Schnabel edn., 200, and RCA Victor LCT 1110.

[23] Schnabel edn., 52, and Seraphim IC 6063–1. I have omitted fingering and some of the broader slurs.

Ex. 10.1 Beethoven/Schnabel, Piano Sonata Op. 2 No. 2, last movt. (1935)

Ex. 10.2 Mozart, Fantasia K. 397/385g

measure. He confines his rubato performance, it seems to me, to a vigorous articulation of the two-note motives (D♯ to E). Schnabel's pupil Konrad Wolff explains this as 'the deliberate separation, by *rubato* playing, of groups of slurred notes'—a separation achieved by playing each group a little too fast and by inserting a *Luftpause* before the next group. Wolff also states that Schnabel, 'like Mozart and Chopin... demanded an unbroken line in the accompanying hand'. The melodic D♯s in Ex. 10.1, however, sound precisely with the corresponding notes in the accompaniment, so that the alteration of their length takes place between left-hand notes. According to Wolff, Schnabel's pupil Leon Fleisher teaches that 'any new bass note must be secured before the *rubato* begins'; thus, in the Mozart excerpt in Ex. 10.2 the E in the left hand on the first beat of measure 13 must be in time, but the second eighth note of the bar 'can then be a trifle late'. Such a delay of course—like an anticipation in Ex. 10.1—carefully avoids the possibility of displacement between the two hands.[24]

---

[24] Konrad Wolff, *Schnabel's Interpretation of Piano Music*, 2nd edn. (Norton, 1979), 20, 68–71. On the recording Schnabel actually plays the rubato in Ex. 10.1 so impetuously that it is difficult to perceive the 2-note groupings. Played on a Viennese fortepiano of Beethoven's day, of course, such small phrase units could be rendered with quite a different effect. Ex. 10.2 comes from the Mozart *Neue Ausgabe*, IX: 27/ii, 30.

Schnabel's performances of his own marked rubatos seem restrained, and his *non* rubatos seem to urge rhythmic restraint for the sake of a larger flow of music. His own playing, however, also includes other retards, accelerations, and pauses, often seeming very personal in nature.[25] In a review from 1944 the music critic and neo-classical composer Virgil Thomson speaks of Schnabel's romanticism, 'which seems to be based on Richard Wagner's theories of expressivity'. He complains that Schnabel plays both themes and passage work with equally intense emotion and thus does not project a sense of formal structure.[26] Charles Rosen later writes that 'we cannot ever be certain exactly how liberties were taken with music in the early nineteenth century, but we can be sure that the rhythmic distortions with which Schnabel presented the sonatas of Beethoven are more closely related to the expressive *rubato* indicated for the music of Schoenberg.'[27]

The transition to a more objective style thus occurred slowly among the keyboard performers. Schnabel in the Beethoven sonatas, and Landowska with the music of Bach on the harpsichord, felt that they were following the composer's intentions and both influenced the general direction toward a more objective style. Yet both still displayed in their own playing a considerable amount of subjective rhythmic flexibility. Josef Hofmann and Arthur Rubinstein similarly move toward objectivity, yet Thomson notes in 1940 that Rubinstein's rubato 'is of the Paderewski tradition'—that is, it involves 'a flexible distortion of strict rhythm'.[28] In a similar manner, Schoenberg and Berg utilize some motivic and contrapuntal techniques and formal structures that remind of earlier periods, but their central expressive intentions are still romantic. Their ideas, however, lead to a more objective style in the music of Webern.

At the same time, objectivity was achieved in quite different ways by a number of composers throughout the world. We shall turn now to the rubato of the neo-classicists, the principal composers before mid-century: Bartók and Kodály in Hungary, Stravinsky from Russia, Hindemith, Orff, and Toch from Germany and Austria, Copland and others in the United States.

---

[25] Note, for example, the acceleration (which recurs in the repeat) in m. 21 from the Trio in the 3rd movt. of Op. 2 No. 1, or the lingerings, pauses, and generally unstable tempo in the 2nd movt. of Op. 10 No. 3.

[26] *New York Herald Tribune*, 28 Mar. 1944, printed in Virgil Thomson, *The Musical Scene* (Knopf, 1947), 192–3, and in *A Virgil Thomson Reader*, with Introduction by John Rockwell (Boston: Houghton Mifflin Co., 1981), 248–9.

[27] 'The Proper Study of Music', *PNM* 1/1 (1962), 84.

[28] *New York Herald Tribune*, 26 Oct. 1940, in *The Musical Scene*, 187, and in *A Virgil Thomson Reader*, 194. See also Andrew Trechak, 'Pianists and Agogic Play: Rhythmic Patterning in the Performance of Chopin's Music in the Early Twentieth Century', D.M.A. diss. (Univ. of Texas at Austin, 1988; UM 89–9,777), 42–63.

Bartók and Kodály

Like Liszt, Béla Bartók was productive in many fields, for he was an ethnomusicologist, a writer, an editor, and a concert pianist as well as a composer. In all of these areas he was intensely involved with rubato. The main influences came from the folk-music of Eastern Europe and from the art music of Liszt and perhaps also Debussy. Bartók studied piano with István Thomán, one of Liszt's most talented students, and often played Liszt's works on his piano recitals. He helped prepare the scores and critical notes of Liszt's Hungarian music for the *Gesamtausgabe*. In his writings he discussed Liszt's involvement with gypsy music and his influence on future composers—'of far greater importance than that of Strauss or even Wagner'. In 1907, according to his autobiography, he thoroughly studied the works of Debussy.[29]

In several articles Bartók explains the nature of the Hungarian music described in Liszt's book and included in the rhapsodies and dances of Liszt and Brahms. This was not generally folk-music, but contemporary popular music written by amateur composers and played by the gypsies in their own style—a style which included 'excessive *rubato*' and florid embellishment. Bartók reports that 'they made use of the *rubato* in a special way in melodies with strict rhythm: certain small melodic portions of equal length (for instance, each couple of measures) remain equally long temporally, while inside these measures the value of the quarter-notes, for example, is variable.'[30]

Bartók himself, however, was interested in quite a different sort of Hungarian music: the authentic rural folk-music of the peasants. Such music was 'the ideal starting point for a musical renaissance' by composers seeking a new objective style, since it was simple, sincere,

---

[29] László Somfai, 'Liszt's Influence on Bartók Reconsidered', *New Hungarian Quarterly*, 27/102 (1986), 210–19. See also *Béla Bartók Essays*, ed. Benjamin Suchoff (St Martin's, 1976), 409–10 from Bartók's 'Autobiography' (1921) and 362 from 'Harvard Lectures' of 1943. See *Franz Liszts musikalische Werke* (Br. & H, 1907–36; repr. Gregg, 1966), ii/xii, *Einleitung* to the Hungarian Rhapsodies.

[30] *Béla Bartók Essays*, 68–70 from 'Hungarian Folk Music' (1921); also 71 from 'Hungarian Folk Music' (1933) and 361 from 'Harvard Lectures' of 1943. See also Béla Bartók, *The Hungarian Folk Song*, ed. Benjamin Suchoff, trans. M. D. Calvocoressi, annotations by Zoltán Kodály, in *The New York Bartók Archive Studies in Musicology*, xiii (SUNYP, 1981), 99 (part of an appendix added in 1931 to the English trans. of *A magyar népdal*, which was published at Budapest in 1924. For a repr. of the latter and the German trans. of 1925, see *Béla Bartók, Ethnomusikologische Schriften, Faksimile-Nachdrucke*, i: *Das ungarische Volkslied*, ed. D. Dille (Schott, 1965); on 435–6 is a fac. of Bartók's handwritten appendix in German. For the tunes used by Brahms and Liszt, see Zoltán Kodály, *Ungarische Volksliedtypen* (Schott, 1964), iii: *Volkstümliche Lieder*.

austere, and devoid of any 'sentimentality or exaggeration of expression'.[31] In 1905, along with Zoltán Kodály, he began his study of 'the simple and non-romantic beauties of folk music'.[32] They collected, recorded, transcribed, and analysed thousands of folk-melodies from Hungary, and later from Romania, Czechoslovakia, Yugoslavia, and Turkey. Particularly striking were the pentatonic scales of the melodies and the two types of rhythm identified as 'parlando-rubato' and 'tempo giusto'.

The parlando-rubato type occurs in the oldest Hungarian songs, and Bartók describes it as

free, declamatory rhythm without regular bars or regular time signatures. Its nearest equivalent in Western European art music may be found in recitative music; Gregorian music probably had a similar rhythm. . . .

This kind of musical recitation is in a certain relation to that created by Debussy in his *Pelléas et Mélisande* and in some of his songs which were based on the old French *recitativo*. This recitation is in the sharpest possible contrast to the Schoenbergian treatment of vocal parts in which the most exaggerated jumps, leaps, and restlessness appear.[33]

In the performance of parlando-rubato rhythm, 'single notes are often diminished or augmented by rational or, still more often, by irrational values'. Prolongation frequently occurs on the first and on the last one or two syllables of a line of text. The resulting rhythmic flexibility conveys a feeling of improvisation and spontaneity.[34]

'Tempo giusto', on the other hand, refers to the relatively strict dance rhythm of the newer Hungarian folk-melodies. Even this type, however, is not metronomic, but is enlivened by subtle modifications of tempo.[35] It occurs with notes of equal value, as well as with more complex patterns that seem like a sort of 'hardened rubato'—the result, that is, of vocal melodies employed for dance music, or of notes originally flexible in value acquiring, during the course of time, a definite duration.[36]

---

[31] *Béla Bartók Essays*, 341 from 'The Influence of Peasant Music on Modern Music' (1931) and 395 from 'Hungarian Music' (1944).

[32] *Béla Bartók Essays*, 427 from 'Contemporary Music in Piano Teaching', 1940.

[33] Ibid. 383 and 386 from 'Harvard Lectures' of 1943.

[34] Ibid. 51 from 'The Melodies of the Hungarian Soldiers' Songs' (1918) and 86 from 'Hungarian Peasant Music' (1933); and Bartók, *The Hungarian Folk Song*, 14.

[35] See Somfai's comments in *The Centenary Edition of Bartók's Records*, i: *Bartók at the Piano 1920–1945*, Hungaroton LPX 12326–33, pp. 29–30 of booklet.

[36] Bartók, *The Hungarian Folk Song*, 9, 13, and 14. See also Judit Frigyesi, 'Between Rubato and Rigid Rhythm: A Particular Type of Rhythmical Asymmetry as Reflected in Bartók's Writings on Folk Music', *Studia musicologica*, 24 (1982), 327–37; and László Somfai, 'Die *Allegro barbaro*—Aufnahme von Bartók textkritisch bewertet', *Documenta bartókiana*, 6 (1981), 275.

Bartók and Kodály transcribe both types of rhythm with barlines to indicate accentuation and with small grace-notes slurred forward or back to a large note to depict ornaments of indefinite duration. They use time signatures for tempo giusto, but not for parlando-rubato. They indicate an alteration in the length of a regular note in parlando-rubato by a special sign above or below: an arc curving toward the note (like a fermata without a dot) means a slight lengthening, whereas the fermata itself indicates a prolongation at least twice as long; an arc curving away means shortening.[37]

In addition, they include a tempo marking for each transcribed melody. In his collection of Romanian melodies published in 1913 Bartók includes a great variety of markings: sometimes just 'Rubato', 'Poco rubato', or 'Molto rubato', occasionally 'Parlando rubato', 'Parlando, rubato', 'Rubato (parlando)', or 'Poco rubato, parlando'. Sometimes he combines *sostenuto* with *rubato* in various ways, with either word modified by *poco* or *molto*: 'Sostenuto, rubato', 'Sostenuto (rubato)', 'Rubato sostenuto', or 'Rubato, sostenuto'. Similarly, he often joins *rubato* with tempo terms, and again each can be preceded by *poco* or *molto*: 'Moderato, rubato', 'Andante rubato', 'Andante (rubato)', 'Rubato, largo', 'Rubato (lento)'. Usually he links rubato with moderate tempos, which may explain the marking 'Poco vivo (ma rubato)'. *Non rubato* occurs ten times, either joined with *sostenuto*, as in 'Sostenuto, non rubato' or 'Sostenuto (non rubato)', or with tempo terms, as in 'Non rubato, lento', 'Non rubato (Andante)', or 'Allegretto, non rubato'. In addition, some melodies are marked simply 'Allegro' or 'Vivo', with the implication, which we noted also in the music of Berg, that they involve neither *non* rubato nor conspicuous rubato.[38]

In a letter from 1926 Bartók suggests some changes for a new edition of this work. Four tempo markings—'Molto rubato', 'Poco lento, rubato', 'Andante rubato', and 'Quasi recitativo'—become simply 'Parlando', while the marking 'Molto largo, non rubato' becomes 'Tempo giusto'.[39] Bartók simplifies the markings considerably

[37] Bartók, *The Hungarian Folk Song*, 195, and *The Selected Writings of Zoltán Kodály* (Budapest: Corvina Press, 1974), 15 from 'The Pentatonic Scale in Hungarian Folk Music', 1917–29. See also Béla Bartók and Albert B. Lord, *Yugoslav Folk Music*, i, originally published by Columbia University Press in 1951 as No. 7 of *Studies in Musicology*, new edn. by Benjamin Suchoff in *The New York Bartók Archive Studies in Musicology*, ix (SUNYP, 1978), 7 and 90.

[38] Bartók, *Cântece poporale românești din comitatul Bihor (Ungaria)* (Bucharest, 1913), repr. in *Béla Bartók, Ethnomusikologische Schriften, Faksimile-Nachdrucke*, iii: *Rumänische Volkslieder aus dem Komitat Bihar*, ed. D. Dille (Schott, 1967).

[39] *Béla Bartók Letters*, ed. János Demény, trans. Péter Balabán and István Farkas, rev. Elisabeth West and Colin Mason (Budapest: Corvina Press, 1971), 170–7. Concerning parlando-rubato rhythm in Romanian folk-music, see *Béla Bartók Essays*, 106–9 from 'The Folk Music Dialect of the Hunedoara Rumanians' (1914) and 115–25 from 'Rumanian Folk Music' (1931).

in his book *A magyar népdal* (Hungarian Folksong) of 1924. Most of the markings now are either 'Parlando' or 'Tempo giusto'; only a few are 'Poco rubato' or 'Rubato'.[40] A host of new variant markings, however, appear later in Bartók's collection of Slovak folksongs and in the Hungarian collection of Bartók and Kodály published after Bartók's death. These include modifications of the word *parlando* ('Quasi parlando', 'Poco parlando', 'Parlando, quasi recitando'), of *rubato* ('Rubato molto', 'Moderato [or Lento] poco rubato'), and of *giusto* ('Quasi giusto'), or combinations of *parlando* and *rubato* ('Parlando rubato', 'Rubato, poco parlando', 'Poco parlando, rubato') or even *rubato* and *giusto* ('Tempo giusto, poco rubato').[41] In a book on Yugoslav folk-music written with Albert B. Lord, the authors refer to 'free rhythm, called *parlando-rubato* or, in short, "*parlando*"'. Yet in an earlier article Bartók describes a particular rhythm as 'neither *rubato* nor a strict dance-step, ... it often comes nearer to a *parlando* performance'.[42]

This diverse and evolving nomenclature must somehow indicate nuances of rhythm between the two extremes of tempo giusto and parlando-rubato. It is significant, however, that the rubato involved is not the subjective type of flexibility added by the performers of art music early in the century, but an objective, speech-like declamation coming out of essentially non-romantic folk-music. Bartók does describe a practice of 'rhythmic compensation' in certain Hungarian, Romanian, and Turkish melodies, in which some notes are shortened and others lengthened in such a way that the total time within one or two measures does not change.[43] In most melodies, however, the parlando-rubato seems to be free.

Bartók and Kodály used folk-melodies in their own compositions in various ways: in settings, either simple or more complex, for voice and piano, or for piano or other instruments alone; as a model for the creation of new melodies in the same style; and as an often subconscious influence on their music in general. In the settings of actual folk-tunes, the pentatonic melodies inspired innovative and often highly

[40] For *rubato*, see his musical Ex. 295*a*, and for *poco rubato* Exx. 259*e* and 263 in *A magyar népdal* or its trans. in *The Hungarian Folk Song* or *Das ungarische Volkslied*.

[41] Bartók, *Slowakische Volkslieder (Slovenské l'udové piesne)* (Bratislava, 1959); and Bartók and Kodály, *A magyar népzene tára*, in *Academia scientiarum hungarica, Corpus musicae popularis hungaricae* (Budapest, 1953).

[42] Bartók and Lord, *Yugoslav Folk Music*, ed. Suchoff, 15; and *Béla Bartók Essays*, 273 from 'The Folklore of Instruments and Their Music in Eastern Europe' (from essays published between 1911 and 1931).

[43] Bartók, *Turkish Folk Music from Asia Minor*, in *The New York Bartók Archive Studies in Musicology*, vii (PUP, 1976), 43; and *The Hungarian Folk Song*, 33. On 309 of the latter source, Kodály adds this note: 'The *rubato* of Chopin was of this type'. As we have seen, this would be true if an accompaniment simultaneously provided the strict beats against which the melody notes could be displaced.

dissonant accompaniments, as well as instrumental preludes, inter-
ludes, and postludes. The settings include time signatures, although
they often change in the course of a piece, and the declamation is
often indicated by fermatas or various signs of accentuation.

Bartók recorded Kodály's vocal setting of 'Szomoru fűzfának' (The
Weeping Willow) with the contralto Mária Basilides. Kodály marked
the original tune 'Rubato' in his collection of Hungarian folksongs;
the setting itself is marked 'Lento, poco rubato', and includes three
stanzas of the tune, each with a different piano accompaniment and
each including some notated alterations of note values as a part of the
rubato response to the text. The singer, of course, adds her own 'poco
rubato' as part of the pronunciation and projection of the words.
Their performance of Kodály's 'Meghalok' (Woe is Me), which is
marked 'Lento doloroso rubato', demonstrates the extreme rhythmic
flexibility sometimes involved in parlando-rubato rhythm.[44] Bartók
also recorded Kodály's setting of 'Székely keserves' for piano alone. It
is marked 'Rubato, parlando' and the tune contains many repeated
notes. Bartók's performance of the three contrasting statements of the
melody clearly demonstrates the considerable alterations applied to
the notated note values.[45]

Bartók himself has settings—for solo voice and piano, for two
violins, and for piano alone—which involve rubato. He sometimes
includes *rubato* in the main tempo marking, as Kodály usually does
and as they both did in collections of the monophonic tunes. In such a
case, of course, rubato simply applies in general to the entire piece.
More often, however, Bartók indicates a briefer or even momentary
rubato during the course of a piece. The first of his *Eight Hungarian
Folksongs* for voice and piano has the main marking of 'Adagio', but
poco rubato appears at the beginning of both the piano prelude and a
later interlude. The second song of the collection is 'Andante', with
rubato over three notes at the high point of the brief piano introduction and *parlando* for the voice when it enters.[46] In a drinking song

[44] Kodály, *Hungarian Folk Music (Magyar népzene), The Songs for Medium Voice* (Universal,
n.d.; originally published in 11 volumes between 1925 and 1964, with songs for high and
medium voice range mixed), 42 and 49 (No. 9 from the original vol. ii and No. 13 from iii). For
the tune of 'The Weeping Willow', see Kodály's *Magyar népdaltípusok* (Budapest, 1961),
published with German texts in *Ungarische Volksliedtypen*, ed. Pál Járdanyi (Schott, 1964), i,
171. Bartók's recordings from 1928 are in *The Centenary Edition of Bartók's Records*,
Hungaroton LPX 12326–12338, vol. i, sides 3 and 4.

[45] Kodály, *Zongoramuzsika* (Seven Pieces for Piano), Op. 11, composed 1910–18 (Universal,
1952), No. 2. Bartók's recording from 1936 of this piece is in *The Centenary Edition*, ii, side 2.

[46] *Nyolc magyar népdal* (Hawkes, 1955). For the original melodies and their sources, see
Vera Lampert, 'Quellenkatalog der Volksliedbearbeitungen von Bartók', *Documenta bartókiana*,
Neue Folge, 6 (1981), 15–149; the tune used in the 1st song is marked 'Poco rubato' (Lampert,
No. 166), the 2nd 'simple recitation' (Lampert, No. 167).

from *Twenty Hungarian Folksongs*, the main marking is 'Non troppo vivo', but the first statement of the tune in the voice is *poco rubato*, the second 'Più vivo' and *non rubato*, and the third, which begins with the piano alone, 'Largamente, molto rubato'; Bartók had previously labelled the monophonic tune 'Tempo giusto'.[47] Two of the forty-four duets for two violins have *rubato* in the main marking, another has *poco rubato* for the last four measures.[48]

Bartók has recorded four of his piano settings of folksongs. In one of them from *Three Hungarian Folk-Tunes* the 'poco rubato' becomes especially evident during the steady eighth-note movement in the second half of the piece.[49] In the sixth of his *Improvisations on Hungarian Peasant Songs* Bartók marks the first appearance of the melody 'poco rubato fin al segno ✵ ' (which appears six bars later), and partially notates the rubato by using quintuplets which sound against duplets in the accompaniment. The simpler statements of the melody that follow are thus specifically not to be played rubato, and once again Bartók presents a contrast between two rhythmic styles within a single piece. In the main marking of the seventh improvisation Bartók links *sostenuto* and *rubato* together, as he often does in the monophonic melodies.[50]

His performance of this and other pieces based on folk-melodies shows that, in spite of numerous tenuto marks and other accentuation signs, the touch is extremely legato. His rubato alterations are thus only seldom emphasized by silences between the notes, and his *parlando*, perhaps surprisingly, seldom includes the sort of vigorous articulation that could be provided by a singer passionately pronouncing the consonants in a text. Only a gentle articulation is sometimes produced by repeated notes in the melodies. In addition to the parlando-rubato rhythm of folksong, Bartók sometimes includes in some piano settings which he did not record, internal rubatos that mark the

---

[47] *Húsz magyar népdal* (Hawkes, 1939), iii, No. 15: *Bordal*, the tune in Bartók's *Hungarian Folk Song*, No. 74*b*, and in Lampert, No. 252.

[48] *44 Duos* (Universal, 1933; Hawkes, 1939), i, No. 23, and ii, No. 28. Bartók transcribed the latter for solo piano in the movement 'Lassú' (Slow Tune) in his *Petite Suite* (Universal, 1938), which he recorded *c*.1941 (*The Centenary Edition*, i, side 15). The Duo that ends with *poco rubato* is No. 25 in vol. i.

[49] *Three Hungarian Folk-Tunes* (B & H, 1942), No. 2. The piece appears also in *Homage to Paderewski* (B & H, 1942). Bartók recorded it *c*.1941 (*The Centenary Edition*, i, side 16).

[50] Op. 20 (Universal, 1922; Hawkes, 1939), Nos. 6 and 7, recorded by Bartók *c*.1941 (*Centenary Edition*, i, side 16). Concerning No. 6, see Lampert, 30–1, as well as the simple version of the theme in No. 205; the tune for No. 7 is No. 206. No. 7 also includes markings such as 'Sempre più sostenuto' which, together with changing metronome figures, show that sostenuto implies slowing as well as sustaining when used during the course of a piece. This work was published also in *Tombeau de Claude Debussy*, supplement to *La Revue musicale*, 1/2 (1 Dec. 1920).

commencement of a new accompanimental figure and thus seem to require the momentary and seductive rubato of Liszt.[51]

Inspired by these folksong settings for voice, violins, and piano, Bartók sometimes composed original melodies in the same style. The most notable example is 'Este a székelyeknél' (Evening in Transylvania) from his *Ten Easy Piano Pieces* of 1908. Bartók describes the piece in a radio interview in English in 1944:

'Evening in Transylvania' is an original composition, that is, with themes of my own invention, but the themes are in the style of the Hungarian-Transylvanian folk tunes. There are two themes. The first one is [in] a parlando-rubato rhythm and the second one is more in a dance-like rhythm. The second one is more or less the imitation of a peasant flute playing. And the first one, the parlando-rubato, is an imitation of song, vocal melody. The form of it is ABABA.

The A section is 'Lento, rubato', the B 'Vivo, non rubato', and each is varied each time it appears. Bartók made four recordings of the piece between around 1920 and 1945, and thus demonstrates not only the difference between rubato and *non* rubato, but also the manner in which each can change during different performances. All the performances reveal considerable declamatory flexibility during the rubato sections—different, of course, for each—along with a very legato, sostenuto touch; the *non* rubato sections sometimes contain substantial unnotated accelerations and retards, as well as a crisp shortening, especially in the second B section, of the sixteenth notes. This is no doubt one of the most studied of all Bartók's works, especially by László Somfai, who has transcribed portions of each recording for comparison.[52] Bartók felt that recordings could illuminate certain

[51] See *Gyermekeknek* (For Children) (1910–12), ii: *Slovak Melodies*, repr. from 'original editions' in *Piano Music of Béla Bartók: The Archive Edition*, ed. Benjamin Suchoff (Dover, 1981), ii, Nos. 30 and 43. The markings remain unchanged in the revision of 1945 (B & H, 1946), although the numbers change to 28 and 39.

[52] *Ten Easy Piano Pieces*, No. 5, in *Piano Music of Béla Bartók*, i, 110–11 (see also pp. ix and x). The recordings are in *The Centenary Edition*, i, side 1 (HMV Budapest, 1929), side 8 (Welte piano roll, c.1920), side 15 (Vox U.S.A., 1945); and ii, side 1 (Pleyel piano roll, c.1922). See Somfai's comments in the booklet for i, 29–30, and ii, 27.

Somfai presents more detailed analysis in 'Bartók rubato játékstílusáról' (Bartók's Rubato Performing Style), *Magyar zenetörténeti tanulmányok Mosonyi Mihály és Bartók Béla* (Essays in the History of Hungarian Music, in Memory of Mosonyi and Bartók), 225–35 (Budapest, 1973), see English summary by Vera Lampert in *RILM* (1974), No. 3834 on 266; and in 'Über Bartók's Rubato-Stil', *Documenta bartókiana*, 5 (1977), 193–201. A summary of his *Centenary Edition* comments, as well as his 4 transcribed excerpts, are in *International Council for Traditional Music, UK Chapter, Bulletin*, 19 (1988), 48–9, in a report by Harold Dennis-Jones. See also Márta Conrad, 'Eine dritte Autorenaufnahme von "Abend am Lande"'. *Studia musicologica*, 24 (1982), 295–302. Bartók arranged the piece in 1931 as the 1st movt. of *Magyar képek* (Hungarian Sketches) for orchestra (Budapest: Editio Musica, 1953), but replaced 'Vivo, non rubato' with 'Allegretto'.

aspects of performance difficult to notate in a score, but cautioned that even 'the composer himself . . . does not always perform his work in exactly the same way'.[53]

Bartók also recorded three of his pieces not connected directly with folk-music. Each includes the contrast, which we have noted in many of his folksong pieces, between *rubato* and *non rubato* or *tempo giusto*. In the second movement of the Suite Op. 14 from 1916 a passage marked 'Meno mosso (poco rubato)' and *espressivo* is followed seventeen bars later by 'Tempo I' and '(*tempo giusto*)'. 'Slightly Tipsy', the second of *Three Burlesques* from 1911, commences with *molto rubato*. The transition to the return of the main theme involves four measures of even, staccato eighth notes rising chromatically and marked *non rubato*. The effect ends presumably when the theme returns, for the last few measures of the piece are marked *comodo* and, again, *non rubato*. In the original piano version of the Rhapsody Op. 1 from 1904, 'Poco più allegro (non rubato)' follows 'Poco maestoso' and *rubato*, and later in the work dashes lead three and a half bars from *poco rubato* to *tempo giusto*.[54]

The presence of both *non rubato* and *tempo giusto* in the Rhapsody suggests that the terms may not always be synonymous. As we have seen, Bartók used *non rubato* in his Romanian collection of 1913 to distinguish dance rhythm from *parlando*, but by the time of his book on Hungarian folksong in 1924 had changed to *tempo giusto*. These two categories were apparently sufficient to characterize the two extreme types of rhythm in folk-music. In his own compositions *non rubato* occurs eight times in six works mostly between 1904 and 1911 (but one in 1929), and *tempo giusto* six times in five works between 1904 and 1921.[55] Unlike Berg and Schnabel, whose *senza* and *non rubatos* come later from 1929 to 1940, Bartók never seems to write *non rubato* or *tempo giusto* unless there is a previously marked *rubato*.

Having heard Bartók's recordings, I suspect that Liszt's unexpected,

---

[53] From 'Mechanical Music' (1937), quoted by Somfai in *The Centenary Edition*, booklet for vol. i, 19.

[54] *Suite*, Op. 14 (Universal, 1918; Hawkes, 1939), 'Scherzo'; *Three Burlesques*, No. 2, in *Piano Music of Béla Bartók*, ii, 88–91; and *Rhapsody*, Op. 1, in *Piano Music of Béla Bartók*, i, 56 and 58. One performance of the *Suite* and two of 'Slightly Tipsy' are in *The Centenary Edition*, i, side 1, both recorded in 1929; the *Rhapsody* for piano and orchestra is in ii, side 6, recorded in 1939.

[55] In addition to works already cited, 'non rubato' appears also in the *Suite for Two Pianos*, Op. 4b, rev. edn. of 1941 (Hawkes, 1960), 3rd movt., 63. In the original *Second Suite*, Op. 4, rev. edn. 1943 (Hawkes, 1948), the same spot is marked simply 'a tempo'. 'Tempo giusto' occurs also in the ballet *The Wooden Prince* (Universal, 1951), 217, 15 bars after 'Allegretto capriccioso (rubato)'; it is missing in the version for solo piano (Universal, 1921). It appears twice in the First Sonata for Violin and Piano of 1921 (Universal, 1923), 24–5 in the 2nd movt., each time 1 to 4 bars following a cadenza-like passage marked *poco rubato* or *rubato*.

deceptive, and seductive type of momentary rubato was transmitted fairly faithfully to Bartók by Thomán. Extended over a span of time, *sempre rubato* probably consists of a series of such momentary rubatos executed at selected spots. Bartók no doubt perceived the parlando of folksong as a sort of declamatory version of Liszt's *sempre rubato*. Bartók's rubato is therefore essentially a brief, but noticeable distortion of rhythm on a few notes at a time, involving the lengthening and shortening of note values in a deceptive, jarring manner, as though instantaneously threatening the onward flow of music. Conceived in this manner, rubato contrasts with the smooth tempo changes that extend over a longer span of time—the notated accelerandos and rallentandos, or the hastening and lingering added by a performer as he shapes a phrase.

If rubato is different from general tempo flexibility, then four performing styles seem to be indicated by *rubato, non rubato, tempo giusto*, and the tempo changing terms. The basically separate nature of rubato and tempo changing becomes apparent when *non rubato* occurs simultaneously with *poco accel.*, or *rubato* with *rallentando*,[56] or when Bartók performs the *non rubato* section of 'Evening in Transylvania' with unnotated accelerations and retards. The terms *rubato* and *non rubato* apparently impose no restrictions on tempo flexibility. *Tempo giusto*, in contrast to *non rubato*, then seems to mean that neither rubato nor conspicuous tempo changing should be added by the performer.[57] When none of these terms appears, the composer must expect the performer to add whatever degree of tempo flexibility and rubato is usual for performers at that particular time and place. In the fifth of the *Fifteen Hungarian Peasant Songs* for piano, the main marking is 'Allegro', but four measures in the middle are 'Sostenuto, poco rubato' followed by 'Tempo I *(tempo giusto)*'. It therefore appears that the piece includes three different rhythmic styles.[58]

At the same time, Bartók sometimes seems to be attempting to

---

[56] '*Scherzando, non rubato*' and *poco accel.* appear together in the second of *Three Folk Songs from the Csík District*, in *Piano Music of Béla Bartók*, i, 67; *rallentando* and *rubato* in *Cantata profana*, 3rd movt., full score (B & H, 1955), 91, and piano/vocal score (Universal, 1951), 53, both with English and German texts.

[57] Below the words 'Tempo giusto' in *The Wooden Prince*, full score, 217–18, however, appears the instruction '*comminciandolo meno mosso* — — —' (for 3 measures), followed by '*poi poco a poco accelerando il tempo sino al Vivace*'. It is not clear whether 'Tempo giusto' refers only to the 1st 3 measures in a steady *meno mosso*, or whether Bartók conceives of it operating also during the accelerando.

[58] *Tizenöt magyar parasztdal* (Universal, 1920; B & H, 1948), 'Scherzo'. Arthur Benjamin, in the *Homage to Paderewski* which contains Bartók's *Three Hungarian Folk-Tunes*, begins his *Elegiac Mazurka* with 'Mesto. Appassionato', but marks 'Tempo I' and '*giusto*' in m. 33, *rubato* in m. 35. Here we are even more clearly dealing with three rhythmic styles.

specify the nature of a particular rubato by applying tempo changing terms to very brief spans of time. During the course of the third song of Op. 16, 'Molto rubato' occurs as a main marking, whereas '(*accel.*)' in smaller type and italics appears twice above two sixteenth-note chords. During the following one and a third measures, instructions in larger roman type either cancel or explain the rubato further by specifying 'rit. molto', 'rit.', and 'accel.', each time with metronome numbers. In this case the tempo changing terms, unlike those we noted with Berg, do not appear in parentheses.[59] In the Suite for two pianos, 'accel. rubato' occurs twice above a group of six sixteenth notes followed in the next measure by 'a tempo'. This might mean that the first three notes should be accelerando and the next three rubato, but it also might indicate that all six notes should be played with an accelerating sort of rubato.[60] A similar ambiguity occurs in the First Violin Sonata, where Bartók writes 'poco rubato, stringendo' above a passage of one and a third measures.[61] In the Dance-Suite, Bartók marks a five-bar passage 'poco accel. (quasi rubato)'.[62] In addition, Bartók adds to his orchestral transcription of 'Slightly Tipsy' numerous markings of *accel.*, *rall.*, and metronome figures to match the way he actually performed the piano version. Since an accelerando and a rallentando both occur frequently within a single measure of eight eighth notes, these rapid changes of tempo may, once again, be part of the 'rubato' in the main tempo marking.[63] When linked closely with rubato, such tempo changing seems to operate on a detailed level quite different from ordinary accelerations and retards.

In general Bartók's metronome markings probably do not indicate the nature of a rubato, for the word occurs as often with a single figure as with a range. In the First Quartet, however, *molto appassionato, rubato* coincides with 'a tempo' and an eighth-note speed of '70 (76)'. In a letter Bartók explains that the tempo is to be 70, but 'here and there also 76'—perhaps an indication of the nature and extent of this passionate rubato.[64]

Bartók occasionally uses rubato in unusual ways. In his second

[59] *Öt dal* (*Five Songs*) (Universal, 1923), No. 3: 'Az ágyam hivogat' (Lost Content), mm. 29–31.

[60] Op. 4*b*, 1st movt.; 'Serenata', mm. 185 and 193.

[61] 1st Sonata for Violin and Piano (1923), 1st movt., 16.

[62] *Táncszvit* (Universal, 1924; B & H, 1951), 1st movt., 56; see also the version for piano (Universal, 1925; B & H, 1952).

[63] Compare the 4th movt. of *Magyar képek* with Bartók's performance in *The Centenary Edition*, i, side 1.

[64] 1st String Quartet (B & H, 1945), in *The String Quartets of Béla Bartók* (B & H, n.d.), 1st movt., 25. See *Béla Bartók Letters*, 218: a letter of 6 Nov. 1931 to Max Rostal, a violinist and member of a string quartet which was rehearsing this work.

Elegy for piano the main marking is 'Molto adagio, sempre rubato (quasi improvisando)'. Certain sections are enclosed with special signs (tiny circles with a dash to one side or the other) which 'indicate that the accompaniment figures [in the right hand] . . . need not be fixed in number but may be varied according to the *rubato* [in the left-hand melody], especially in the middle part (upper staff 4/4, lower one 3/4), where the accompaniment should be an even roll'. In two other pieces, passages marked *quasi cadenza* and *rubato* include brief figures for solo violin or flute to be played *molte* or *più volte ad libitum*.[65] The twelfth Bagatelle, marked 'Rubato', commences with a measure in which a single pitch moves progressively from an eighth note to a one-hundred-and-twenty-eighth note. This is accompanied by *sosten. accel.* and *molto espress.* and a footnote which explains the execution as 'a gradual acceleration in which there is no definite number of notes (and similarly in subsequent measures [three others] with the same figuration)'. The same effect, which reminds one of the accelerating roll in Asian music, occurs also in the Sixth Quartet.[66]

Like other composers before him, Bartók often indicates a continuous rubato for solo passages marked *quasi cadenza*.[67] A momentary rubato may occur within a solo passage to articulate the commencement of an idea or the repetition of a figure.[68] Occasionally, especially in later works, Bartók seems to distinguish within a single work between a momentary rubato (with the word in small type, uncapitalized, and perhaps in italics and sometimes in parentheses) and the continuous, but often brief rubato (capitalized and in larger roman type). In the pantomime *The Miraculous Mandarin*, for example, he indicates a brief continuous 'Rubato' for a five-bar clarinet melody played when the young girl reluctantly takes a seductive pose at the window in order to lure men who can be robbed by the three tramps. Later in the

[65] *Two Elegies*, Op. 8*b*, in *Piano Music of Béla Bartók*, i, 127, 130–1. Repetition of figures occurs in the *Divertimento for String Orchestra* (Hawkes, 1940), 3nd movt., m. 256; and *Concerto for Orchestra* (Hawkes, 1946), 4th movt., m. 143.

[66] *14 Bagatelles*, Op. 6, in *Piano Music of Béla Bartók*, i, 95; and 6th String Quartet (Hawkes, 1941), 2nd movt., mm. 80–2. I am indebted to Paul Humphreys for information on the 'accelerating roll'—the accelerated repetition of a beat on a wood block or drum. He presented a paper entitled 'Time, Rhythm, and Silence: A Phenomenology of the Buddhist Accelerating Roll' to the Fourth Symposium of the International Musicological Society, July 1990 in Osaka, Japan.

[67] In the *Rhapsody No. 1 for Violin and Piano* (Universal, 1929), 20, also arranged as *First Rhapsody for Violin and Orchestra* (Universal, 1931; Hawkes, 1939), 2nd part ('Friss'), 41; *Divertimento for String Orchestra*, 3rd movt., m. 256; 6th String Quartet, 2nd movt., m. 115; and *Concerto for Orchestra*, 4th movt., m. 143.

[68] See the Viola Concerto (B & H, 1950), 1st movt., m. 11: 'poco rubato' to emphasize the repetition of a syncopated figure; and 2nd movt., m. 8: *poco rubato* to articulate the beginning of a brief solo passage.

same work, a momentary *rubato* occurs at the beginning of a phrase of five notes for solo oboe when the frightened girl hesitatingly begins a dance to impress the Mandarin.[69] On two occasions Bartók specifies the length of a three- or four-bar rubato, like Debussy, with dashes.[70] Occasionally rubato involves unusual groups of notes, sometimes causing polyrhythmic displacement.[71] The moods vary from the soft and gentle ones favoured by Liszt to the passionate and capricious ones sometimes used by Debussy and Tchaikovsky. Thus we find Bartók's rubato linked with *dolce, espressivo, tranquillo,* and *quieto,* as well as with *appassionato, agitato, capriccioso,* and *scherzando.* Terms such as *slargando, largamente,* and *sostenuto* suggest special emphasis as well as broadening, sustaining, and perhaps slowing of the tempo.

Additional insight into Bartók's rhythm comes from his performances and editions of the works of earlier composers. In his recording of Liszt's *Sursum corda* the evenly notated eighth-note chords in the accompaniment allow one to notice very clearly his flexibility: often first beats are delayed and measures commence slowly and accelerate later. In a Brahms Capriccio and a Chopin Nocturne he dwells on high notes or on colourful chords. He plays nineteenth-century music with romantic flexibility, and even in an excerpt from Beethoven one can hear the lengthening and shortening of notes. At all times, however, Bartók's playing, it seems to me, is superlatively musical, and whatever he does is done for valid musical reasons.[72]

From 1907 to 1934 Bartók was a professor of piano at the Budapest Academy of Music and during this period prepared instructional editions of the works of a number of composers. Even for music as early as Couperin and Bach, he adds numerous indications for mood, dynamics, tempo changing, articulation, fingering, pedalling, and the execution of ornaments. In his edition of the *Well-Tempered Clavier*

---

[69] Full score (Universal, 1955), 38 and 114; keyboard 4 hands (Universal, 1925), 13 and 33. The 5-note phrase involved in the momentary rubato is notated in the full score as 5 8th notes in a quintuplet filling a measure of 3/4; in the earlier keyboard version, the same pitches (some changed enharmonically) occur, along with an 8th rest, as a sextuplet followed by a quarter rest.

[70] *44 Duos,* i, No. 25; 5th String Quartet (B & H, n.d.), 5th movt., m. 516.

[71] Violin Sonata of 1903, ed. D. Dille (Budapest: Editio Musica, 1968), also in *Documenta bartókiana,* 1 and 2 (1964–5), 2nd movt., mm. 34–40.

[72] The recordings are in *The Centenary Edition*: Liszt's *Sursum corda,* recorded in 1936, is in vol. i, side 2, Brahms's Capriccio Op 76 No. 2 and Chopin's Nocturne Op. 27 No. 1 (1939) in ii, side 3. Concerning Beethoven's *Variations on an Original Theme,* Op. 34, a portion of which Bartók plays in ii, side 3, see George Robert Barth, 'The Fortepianist as Orator: Beethoven and the Transformation of the Declamatory Style', D.M.A. diss. (Cornell Univ., 1988; UM 88–21,213), 97–100; he describes Bartók's 'expressive *tempo giusto*' and the ' "temporal dissonance" between mechanical time and rhetorical time' by indicating on musical excerpts precisely which notes are lengthened and which are shortened (the 'good' and 'bad' notes).

he states that rubato is scarcely used with Bach, but that a completely rigid rhythm is also not proper; he recommends a retard at decisive cadences.[73] He makes the distinction between rubato and tempo changing even clearer, however, in a footnote to a Mozart sonata: 'the rapid notes [in long ascending or descending series of eighteen to twenty-four notes] . . . must start slowly and then be executed with an accelerando—avoiding any rubato'.[74]

Bartók actually adds *poco rubato* in two of Mozart's sonatas, in both cases at the beginning of a free section leading back to a rondo theme.[75] In addition, he includes *rubato* in toccatas by Della Ciaia and Michelangelo Rossi,[76] as well as in the two different places in a sonata by Domenico Scarlatti shown in Ex. 10.3.[77] The double bar marks the end of an internal section within the first half of this sonata in B flat major. The original sources show only the notes, one slur, and the first fermata. The first *rubato* indicates a rhetorical execution for the soloistic ornamental conclusion to the opening section; the second enunciates the sense of the descending scale, which provides a modulation to the key of D flat major. In both cases Bartók adds the same signs of articulation and accentuation that he uses in abundance in his own compositions.

Bartók is the first composer in our history to explain systematically the precise meaning of these signs. In a preface to his edition of pieces from the *Clavierbüchlein* for Anna Magdalena Bach, he describes the various types of articulation and accentuation.[78] The signs in his list

[73] *Das wohltemperierte Klavier* (Budapest: Editio Musica, 1964), *Anhang* at end of i: 'Das Rubato ist bei Bach kaum gebräuchlich, hingegen soll ein ganz strammer Rhythmus im Vortrag auch nicht überwiegen.' Modification of a 'completely rigid rhythm' presumably involves the subtle flexibility that operates in the shaping of a motive or phrase—the type of rubato which Rosen has in mind, I believe, when he comments on the opening fugue of Bach's *Kunst der Fuge*: 'the climax at bars 34–37 will bring itself out when played straightforwardly (but with the *rubato* constantly necessary in Bach to inflect the line)'; see *Oxford Keyboard Classics: Bach, The Fugue*, ed. and annotated by Charles Rosen (OUP, 1975), 60.

[74] Mozart, *Twenty Sonatas for Piano*, ed. Béla Bartók (New York: Edwin F. Kalmus, 1950), 269, with reference to mm. 29 and 30 in the Adagio movt. of the Sonata in C minor, K. 457.

[75] Ibid. 99 and 179 from the Sonatas in B flat major, K. 281/189*f*, and D major, K. 311/284*c*.

[76] *XVII and XVIII Century Italian Cembalo and Organ Music Transcribed for Piano* (New York: Carl Fischer, 1990; originally published by Fischer in 1930 in 11 vols.), 22: 'Moderato, un poco rubato' for the opening section of Rossi's Toccata in A Minor; 23: 'Molto moderato, non rubato' for the following imitative passage; 34: 'Largamente rubato (*quasi improvvisando*)' for the beginning of the Toccata movement in Della Ciaia's Sonata in G major); and 39: '(*rubato*)' in parentheses for the commencement of a long ascending run in the same toccata.

[77] *Scarlatti*, ed. Bartók (in *Masterworks of Keyboard Literature*) (Budapest: Editio Musica, 1958), ii (preface dated 1921), 3–4. The 2nd and 3rd notes of m. 13 appear with different values in different sources.

[78] J. S. Bach/Bartók, *Tizenhárom könnyű kis zongoradarab* (Budapest: Editio Musica, 1950), copy in LC. The preface is in Hungarian and German. Earlier, Czerny and Hummel describe portato. Thalberg, in the English preface of *L'Art du chant appliqué au piano*, Ser. 1–2

Ex. 10.3 D. Scarlatti/Bartók, Sonata K. 332

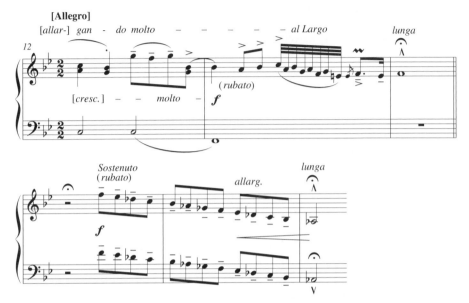

of touches involve a decreasing amount of silence between the notes: a series of wedges indicates a sharp staccato (or *staccatissimo*) combined with a certain accentuation and a sharper tone colour; a series of dots represents the usual staccato in which notes are held between an instant and almost half value; a series of dots under a slur (which he calls *portamento* and which we today call *portato*) means that notes with a certain special colouring sound almost half their value; with a series of dashes with dots below, notes will not be shorter than half value; with a series of dashes (the tenuto sign) the notes sound their full value, if possible, without binding them together; and finally, the slur by itself indicates legato. Next he names the types of dynamic accent, progressing from the most emphatic to the least: *sf* is the strongest; the caret is less vigorous but still strong; the small hairpin is weak; and the tenuto dash on a single note within a legato phrase means a slight emphasis attained by a different tone colouring.

Bartók emphasizes the fact that touch, accentuation, and tone

(London: Cramer Beale & Co., [1853]), states that a series of dashes above dots or a series of dots under a slur should be neither legato nor staccato, but carried over [*portées* in the Parisian edn. of Heugel] 'as by the human voice', with 'the former pressed somewhat heavier than the latter'. He also mentions that a note marked by the sign of a caret should be 'more firmly pressed [*d'autant plus vigoureusement enfoncée*], especially in slow melodies, that it may be of long duration'.

colour are not mutually exclusive. Staccatissimo involves some sort of accentuation, and a special tone colour occurs with staccatissimo, portato, and tenuto. Furthermore, he combines signs such as the dot and the dash and distinguishes between tenuto as an accent on a single note and tenuto as a type of touch on a series of consecutive notes. In the latter case, slight articulation must be present even though one theoretically holds each note its full value. Thus signs which originally were applied as a purely dynamic accent to single notes or chords seem to require articulative separation when applied to several notes in succession. In the first *rubato* of Ex. 10.3 Bartók writes a hairpin accent over three consecutive notes. Although he does not explain this situation in his preface, such notes may involve some brief silences of articulation in addition to their basic dynamic accentuation. On the last quarter beat of the same measure he combines the hairpin accent with the touch required for two consecutive tenutos. More modern sources carry Bartók's definitions a few steps further by showing silences of articulation between notes in a consecutive series of hairpin accents as well as dashes or dots, and a different amount of silence when the signs occur under a slur.[79]

We have often noted the concurrence of rubato with special marks of accentuation or touch in composers before Bartók. Chopin uses the small hairpin sign on single notes or several in succession. In Ex. 7.18 it indicates an accent on the third beat of the mazurka rhythm. In Ex. 7.24 it follows a staccato beat, so that the dynamic accent is strengthened by an agogic accent caused by the preceding silence. Ex. 7.26a shows a series of six in a row. Clara Wieck includes four in the *rubato stretto* of Ex. 8.6a. In Ex. 8.3 Viardot marks the sign on six consecutive notes in the vocal rubato of her transcription of a Chopin Mazurka, and Fauré uses the sign on four notes to indicate stentato in Ex. 3.19. Chopin marks a caret in Ex. 7.23. Gottschalk has a succession of wedges in Exx. 8.26 and 8.28, Puccini two tenuto dashes in Ex. 9.3a. Debussy has a series of dashes in Exx. 9.5 and 9.6, as well as a series of hairpin signs in Ex. 9.8. Even more frequent than any of these signs, however, is the combining of rubato, as we have often noted, with portato.

Accent and articulation signs have different meanings within different musical styles. In the staccato style of Mozart they would have a different effect than in the later legato style. By the time of Bartók the number of such signs, as well as the variety and complexity of

---

[79] See Gardner Read, *Music Notation: A Manual of Modern Practice*, 2nd edn. (New York: Taplinger Publishing Co., 1969), 260–71, and especially Exx. 15.21–15.22 on 270, which show the amount of silence between accented notes with or without slurs.

their execution, increases enormously and seems to be a visible indication of new twentieth-century methods of declamation. The signs may be part of the Klangfarbe concept, or part of the new percussive rhythms of neo-classicism. In the case of Bartók they seem to come, at least to a great extent, from folk-music. He does not use the signs in his monophonic collections of songs or in the voice part of his vocal settings of folk-tunes. They occur in great abundance, however, in his piano settings of the folk-melodies—apparently to indicate the same sort of declamation required by the voice in singing the Hungarian texts. Bartók then applies this special rhetorical style, originating in the Hungarian language, in his own independent compositions, which, like the instructional editions, are filled with a rich supply of accent and articulation signs. Although they may appear anywhere in his compositions or editions, they are especially important, I think, in understanding the special declamatory nature of his rubatos.

Bartók was thus involved in music education in a number of ways: through these instructional editions of works by Mozart, Scarlatti, Chopin, Schumann, Heller, and others; through his teaching at the Budapest Academy of Music; through his own compositions such as *Mikrokosmos*; and through writings such as his article from 1940 on 'Contemporary Music in Piano Teaching'. His educational purpose in the collection *For Children* was to acquaint young piano students with folk-music.[80] Kodály was even more intensely involved with music education. He taught subjects in music theory for many years at the Academy of Music, but took special interest finally in the education of young children. For this purpose he wrote many books of exercises and songs and eventually developed the 'Kodály Method' based on group singing of folk-melodies.

The concept of parlando-rubato developed by Bartók and Kodály influenced later composers and performers. In her analysis of the E minor fugue from the second book of Bach's *Well-Tempered Clavier*, Wanda Landowska writes that the fantasy-like section near the end 'reminds us of Béla Bartók's indication *parlando-rubato* for a folk song. This rubato is indispensable.' The passage interrupts the flow of counterpoint with brief cadenzas for a single line—first in the left hand, then the right—which lead to a fermata. In her recording, Landowska dramatically alters the rhythm, moving at times capriciously from note to note, but maintaining, as Bartók does in his own performances of folk-melodies, an essentially legato style.[81]

---

[80] See *Béla Bartók Essays*, 427.

[81] *Landowska on Music*, ed. Denise Restout and Robert Hawkins (New York: Stein & Day, 1964), 201, and a slightly different version in the notes accompanying her recording, made in

Parlando-rubato also influenced later writers and collectors of folk-material. The word *rubato* continues in more recent works devoted to Hungarian and Yugoslavian melodies.[82] It also occurs in connection with folk-music from other Eastern European countries such as Bulgaria, Czechoslovakia, and the Ukraine,[83] as well as from locations as distant as Spain and the United States. It is used to describe the folk singers of the Maine woods,[84] and Harvey Worthington Loomis indicates *rubando* for descending passages of eight sixteenth notes in a piano setting of an original American Indian melody.[85] Percy Grainger translates 'rubato il tempo' as 'wayward in time' in settings of folk-tunes from England, Ireland, and Denmark.[86] The free parlando-rubato mode of performance occurs also, according to George Herzog, in Spanish songs influenced by Arabic music and in the heroic epic poetry of Russia, Ukrainia, and Yugoslavia.[87]

In addition, Bartók felt that the influence of Russian folk-music could be traced in the works of Stravinsky—especially in their mosaic-like character and the extensive use of brief ostinato figures.[88] Although he does not mention a declamatory influence, this may, indeed, have been one of the sources of Stravinsky's style of rubato.

Stravinsky

In 1963 Stravinsky states in a list of differences between Schoenberg and himself that the former made 'much use of rubato', whereas his

May 1950, of the complete *Well-Tempered Clavier* released by RCA Victor on LM 6801 (1958). Although she refers to 'the phrase in the 82nd bar', she actually begins the rubato in the preceding measure. Her capricious rubato contrasts with the steady but substantial retard with which she plays mm. 68 and 69, which also lead to a fermata.

[82] See Vargyas Lajos, *A magyarság népzenéje*, dedicated to Bartók and Kodály (Budapest: Zeneműkiadó, 1981); *MGG* vii, 367 and 370; and *New Grove*, xx, 596.

[83] See Nikolai Kaufman, *New Grove*, iii, 431–8, and the folksong collections listed there. On the 'frequent use of rubato' in East Moravian music, see *New Grove*, v, 129.

[84] See *British Ballads from Maine*, ed. Phillips Barry, Fannie Hardy Eckstorm, and Mary Winslow Smyth (YUP, 1929), p. xi; and Phillips Barry, 'American Folk Music', *Southern Folklore Quarterly*, 1/2 (June 1937), 42–3. George Herzog, in *Funk & Wagnalls Standard Dictionary of Folklore, Mythology, and Legend*, ed. Maria Leach (New York, 1972), 1041, mentions that the parlando-rubato style of Eastern Europe has been observed also in the southern and north-eastern parts of the United States as well as in French Canadian melodies.

[85] *Music of the Calumet* (Newton Center, Mass.: The Wa-Wan Press, 1903), repr. in *Nineteenth-Century American Piano Music*, ed. John Gillespie (Dover, 1978), 206–8.

[86] See *British Folk-Music Settings*, No. 6, repr. in *Percy Grainger Piano Album* (New York: G. Schirmer, 1982), 113; and No. 18 (G. Schirmer, 1918). See also *Danish Folk-Music Settings*, No. 8, repr. in the *Piano Album*, 126.

[87] *Funk & Wagnalls Standard Dictionary of Folklore, Mythology, and Legend*, 1043.

[88] 'The Influence of Peasant Music on Modern Music', *Béla Bartók Essays*, 341 and 343; also p. 410 from his autobiography.

own music is characterized by 'metronomic strictness, no rubato, ...
mechanical regularity'.[89] His attitude toward performance has become
well known, chiefly due to remarks by himself and others during his
neo-classical period from around 1920 to 1951. He advocated, in
reaction to late romantic practice, the subjugation and control of
emotion, citing as example the sonatas of Weber, 'which are of an
instrumental bearing so formal that the few *rubati* which they permit
themselves on occasion do not manage to conceal the constant and
alert control of the subjugator'.[90] He reacted against the late romantic
concept of performance, for he wanted his music simply executed and
transmitted, and not interpreted through the personality of a performer.[91] As both a pianist and a conductor, he was 'often regarded as
a life-size ... metronome' and is quoted as saying that 'if the conductor
beats the measures precisely and at the indicated speed, the music will
take care of itself'.[92]

Such an objective method of performance was necessary, as it was
in the Classic period, in order to project the sense of the musical
architecture and to render properly music 'based on objective elements
which are sufficient in themselves'. One could not add a performer's
personal 'interpretation' in such a piece 'without risking the complete
loss of its meaning'.[93] Such music cannot admit 'wrong or uncertain
tempo', and the main 'stylistic performance problem ... is one of
articulation and rhythmic diction'.[94] Elliott Carter describes the 'telling
quality of attack' which he noticed in Stravinsky's piano playing in
1934: 'incisive but not brutal, rhythmically highly controlled yet filled
with intensity so that each note was made to seem weighty and
important'; every time he heard Stravinsky play he got 'the strong

[89] Igor Stravinsky and Robert Craft, *Dialogues and a Diary* (Doubleday, 1968), 108, and
quoted by Donald Harris in 'Stravinsky and Schoenberg: A Retrospective View', *PNM*, 9/2–10/1
(1971), 118. The composer Vittorio Rieti, in 'The Composer's Debt', *Stravinsky in the Theatre*,
ed. Minna Lederman (New York: Pellegrini & Cudahy, 1949), 134, writes: 'Special thanks are
due him [Stravinsky] for not asking us to swallow crescendo porridge, pedal sauce and rubato
marmalade'.

[90] *The Poetics of Music* (HUP, 1947), 79–80. These essays were first given in 1939–40 as a
series of Charles Eliot Norton lectures at Harvard and published in French in 1942. Weber did
not mark the word *rubato* in his scores, although Liszt, as we have noted in Pl. XIX, included it
once in his edn. of 1870.

[91] *Chronicle of My Life* (London: Victor Gollancz, 1936), 60 and 126; later published as *An
Autobiography* (Norton, 1962), 34 and 75.

[92] *Bravo Stravinsky*, photographs by Arnold Newman, text by Robert Craft (Cleveland and
New York: World Publishing Co., 1967), 41. See also *Themes and Conclusions*, written with
Robert Craft (Faber, 1972), 226–7.

[93] 'Some Ideas about my Octuor', *The Arts*, Jan. 1924, printed also by Eric Walter White in
Stravinsky: The Composer and his Works, 2nd edn. (Faber, 1979), 574–7.

[94] Robert Craft, *Conversations with Igor Stravinsky* (Doubleday, 1959), 135–6. See Pieter C.
van den Toorn, *The Music of Igor Stravinsky* (YUP, 1983), 206, 214, 239–42; and White,
Stravinsky: The Composer and his Works (1979), 562.

impression of highly individualized, usually detached notes filled with extraordinary dynamism . . . and this was true in soft passages as well as loud'.[95] Robert Craft writes that Stravinsky's 'keyboard style in any music was marked by a *staccato-sforzando* touch, a *secco* tone, an avoidance of the pedal—all in the interest of . . . clarity of articulation'.[96]

Stravinsky's rhythm often involves a constantly recurring eighth- or sixteenth-beat pulse, sometimes combined with changing time signatures. According to Carter, Stravinsky 'substituted for the rubato, so to speak, of the Romantic period an irregular rhythm which in a certain way is a kind of stiffening of this rubato into patterns which still give a certain sense of intensity' (perhaps somewhat like the procedure we noted in Exx. 9.2 and 9.3, in which the composer has notated what sounds like one possible rubato alteration of a more regular pattern).[97] The choreographer George Balanchine writes:

Since his rhythms are so clear, so exact, to extemporize with them is improper. There is no place for effects. With Stravinsky, a fermata is always counted out in beats. If he intends a rubato, it will be notated precisely, in unequal measures. (Elsewhere, of course, a good instrumentalist, . . . or a resourceful dancer, can give the feeling of rubato in Stravinsky's music without blurring the beat.)[98]

Stravinsky seems to have been more flexible rhythmically during both his early and late periods. The original piano reduction of *The Firebird* (1910), for example, contained expressions such as *passionato* and *con tenerezza*, which were deleted in the full score. He included very few indications in later works, in fact, for fear of evoking too emotional a response from the performer.[99] According to Craft, Stravinsky's suggestions in 1916 to the conductor Gabriel Pierné for some unmarked rallentandos in *Petrushka* may indicate a preference for 'more rubato' during his earlier years.[100] By 1953, on the other

---

[95] 'Igor Stravinsky, 1882–1971', PNM 9/2–10/1, pp. 1–2; also in *The Writings of Elliott Carter: An American Composer Looks at Modern Music*, ed. Else and Kurt Stone (IUP, 1977), 302.

[96] Vera Stravinsky and Robert Craft, *Stravinsky in Pictures and Documents* (S & S, 1978), 215.

[97] *Igor Stravinsky: The Man and his Music*, a documentary radio series by Educational Broadcasting Associates, made in 1976 by Frederick Maroth and William Malloch (Berkeley, Calif.: Educational Media Associates, 1977), EMA–103, side 4, Program ii: *The Character of the Music*. See *Stravinsky in Pictures and Documents*, 564–5.

[98] 'The Dance Element in Stravinsky's Music' in *Stravinsky in the Theatre*, 75.

[99] Stravinsky is quoted by his son Soulima, in an interview with Ben Johnston in PNM 9/2–10/1, p. 16: 'If I put in a crescendo, they will give me too much, so I'd better put nothing.' Concerning the *Firebird*, see White, *Stravinsky: The Composer and his Works* (1979), 188.

[100] Stravinsky, *Selected Correspondence*, ed. Robert Craft, i (Knopf, 1982), 393 n. 4 in app. C on 'Petrushka: Revisions, Early and Late'.

hand, he seems to modify the severity of his neo-classical position by writing that 'too much rigidity and an excessive overcaution not to go astray create a general uneasiness and dullness which is as detrimental as too much liberty or "interpretation"'.[101] In 1959 he felt that certain stylistic questions in his music did, in fact, require some interpretation and for this reason regarded his recordings as 'indispensable supplements to the printed music'.[102] Stravinsky's son Soulima considers the orchestral recordings authoritative, whereas in those for piano one 'could add something more', since they were recorded when the composer 'still held very strict opinions as to rhythm'.[103]

In his writings and remarks, Stravinsky seems to link rubato with romanticism, with romantic performance practice, and generally with everything which he is most against. Hence it is surprising to discover that he himself marked the word *rubato* thirty times in sixteen pieces between 1908 and 1963 (not counting alternate versions, arrangements, or suites drawn from the same works). Furthermore, twenty-two of these rubatos occur in eight works from the neo-classical period when his rhythmic attitude was at its strictest. Nine of them, however, come from *The Rake's Progress*, and none at all appears in works composed from 1931 to 1942. Five are from the early years and three are in late twelve-tone works. The early rubatos occur in piano pieces or in orchestral music for ballet or opera. Those from the middle period are in works for two pianos or piano with orchestra, or in dramatic works for solo voice or chorus. Those from the late years involve non-dramatic orchestral, solo vocal, or choral works.

Rubato influence may have come to Stravinsky from earlier Russian composers: his teacher Rimsky-Korsakov, Glazunov, Balakirev, or Skryabin, although only the latter (see Chapter 9) employed the device with any frequency. Far stronger influence no doubt came during his early years from Debussy's numerous rubatos. Equally if not more influential, not only in the early period but especially in the middle years, may have been the many examples by Tchaikovsky, his favourite

[101] Ibid. iii (1985), 379: letter to Erwin Stein 27 Nov. 1953.
[102] Craft, *Conversations with Igor Stravinsky*, 135.
[103] *PNM* 9/2–10/1, p. 18. See also White, *Stravinsky: The Composer and his Works*, 564: 'His own recorded performances occasionally showed that a greater degree of latitude in tempo and interpretation was permissible than might have been expected after reading his intransigent public utterances.' Concerning his interest in the player piano, see *Chronicle of My Life*, 166–7, or *Autobiography*, 101, and Van den Toorn, *The Music of Igor Stravinsky*, 245–6. Even the pianola, however, did not prevent unsatisfactory results and personal interpretation: after hearing an 'expert' play one of Stravinsky's rolls, Ansermet wrote in 1919 that 'unfortunately I heard the piece on a very bad instrument, and to make matters worse, the expert made *rubati*.' See *Stravinsky in Pictures and Documents*, 164, and Charles M. Joseph, *Stravinsky and the Piano*, in *Russian Music Studies*, viii (UMI, 1983), 93–4.

Russian composer and the one most completely Westernized.[104] There may even be some influence from Liszt.[105] In addition, he may have been attracted, as Bartók was, by the declamatory rubato in folk-music. In his compositions before World War I he frequently used Russian folk-tunes, folk-stories, and methods of text setting related to folk-singing. He drew upon the folksong collections of Rimsky-Korsakov, Balakirev, Tchaikovsky and others. He also knew the collections and writings of Evgeniya Linyova, the first to record Russian folk-tunes on the phonograph. She describes the rhythmic robbery that is apt to occur when attempting to notate the irregular rhythm of folksong:

When taking a song down by hand little rhythmic compromises are possible—one can steal an eighth note here, a quarter note there, and in this way smooth over the apparent rough spots and bring the recalcitrant, capricious tune into conformity with a general mold. But . . . the phonograph insistently claims its due and will not admit such errors.[106]

Stravinsky's earliest rubato appears in the first of his *Quatre études* for piano composed in 1908 and influenced by Skryabin. The piece moves with an almost continual accompanimental figure in quintuplets against a melody in triplets and duplets. This agitated motion is interrupted only once—by three measures marked *Tempo rubato*, which act as a transition to the repetition of the opening material. The rubato commences with a half-measure rest, followed by an arpeggiated chord and then a brief melodic passage in which the notes become progressively slower in a sort of notated retard. The passage concludes with *rall.* and *diluendo* (dying or fading away). The rubato is thus a declamatory statement concerning a sensitive moment in the musical structure. It is made dramatic by the sudden cessation of motion during the rest. Its declamation is controlled by the arpeggiated chord as well as by the partially notated slowing of the rhythm.[107]

---

[104] See Stravinsky's *Poetics of Music*, 79–80, and Claudio Spies's comments in 'Conundrums, Conjectures, Construals; or, 5 vs. 3: The Influence of Russian Composers on Stravinsky' in *Stravinsky Retrospectives*, ed. Ethan Haimo and Paul Johnson (UNP, 1987), 106–40.

[105] See 'Chronological Progress in Musical Art: An Interview . . . with . . . Stravinsky', *Etude Music Magazine*, 44 (1926), 559, where Stravinsky compares Chopin and Liszt: 'I have higher honor and admiration for the great Liszt, whose immense talent in composition is often underrated.'

[106] Trans. from Richard Taruskin, 'Stravinsky's "Rejoicing Discovery" and What it Meant: In Defense of his Notorious Text Setting', *Stravinsky Retrospectives*, 173–9. See also Eugenie Lineff, *The Peasant Songs of Great Russia as They are in the Folk's Harmonization, The First Series* (St Petersburg: Imperial Academy of Science, 1905), p. xvi.

[107] Op. 7 (Moscow: P. Jurgenson; Leipzig: Rob. Forberg, [1910]), another edn. marked 'Rev. M. Frey' (Hamburg: Anton J. Benjamin, 1925). Soulima Stravinsky, in his edn. in *The Short Piano Pieces* (B & H, 1977), replaces *Tempo rubato* with *poco rubato*, a reflection, perhaps, of the way he was taught by his father to play the passage. See *Memories and Commentaries*, with Robert Craft (UCP, 1981), 64–6; and Joseph, *Stravinsky and the Piano*, 43–54.

Ex. 10.4 Stravinsky, the Magician's motive from *Petrushka*

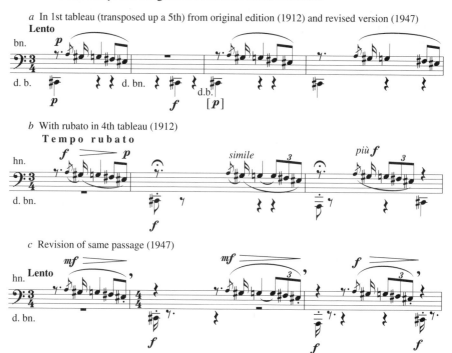

a In 1st tableau (transposed up a 5th) from original edition (1912) and revised version (1947)

b With rubato in 4th tableau (1912)

c Revision of same passage (1947)

An even more carefully notated rubato occurs near the end of *Petrushka*. When the Magician appears in the first tableau, he is accompanied by a three-fold repetition of the chromatically descending motive shown in Ex. 10.4*a*. In the original version, the fourth tableau contains a similar passage, shown in Ex. 10.4*b*, which is marked 'Tempo rubato'. During the first measure of this excerpt and the preceding two bars, 'the Magician arrives', and during the following two bars 'he picks up the corpse of Petrushka'. Horns now replace the bassoons for a darker colour, and decrescendos replace *p* for a more intense, sigh-like quality. The rhythmic aspects of the rubato are indicated by the two fermatas and by alterations in the notation of the second and third statements of the motive. Comparison of (*a*) and (*b*) in Ex. 10.4 shows that the motive first appears unchanged; then in its second appearance the G♮ is extended by robbing the last two notes, in the manner of the earlier type of rubato, and in the third the last three notes are all robbed of some portion of their value. Such precise notation leaves little room for additional alteration of note values. In the revised edition of 1947, as a matter of fact, Stravinsky completes the process by changing time signature to indicate the precise values of the fermatas (see Ex. 10.4c). In this case, either because other conductors interpreted the rubato too broadly, or perhaps because the rubato was now completely determined by the notation, Stravinsky replaces 'Tempo rubato' with simply 'Lento'. A performance by Stravinsky in 1960 of the revised version reveals only slight rhythmic freedom added to the notation of Ex. 10.4c. The music sounds like rubato, for some notes are robbed of their value by other notes. The precise nature of the robbery has not been left to the discretion of the performer, however, but seems to have been almost completely notated by the composer.[108]

Although *Petrushka* contains a number of folk-tunes elsewhere in the score, there is no influence from folk-music, as far as I know, on the Magician's motive. *The Rite of Spring*, however, begins with a melody adapted from a Lithuanian folksong and marked 'tempo rubato'. Comparing the folk-tune in Ex. 10.5a with Stravinsky's theme beginning in (b) and continuing in (c) and (d), it is clear that he treats the borrowed material not as Bartók and Kodály ordinarily would—as a cantus firmus to which new voices are added—but in a rhythmic transformation.[109] The folk-melody in Ex. 10.5a shows accentuation on the three first beats of its opening phrase: on C, E, and A. Stravinsky's phrase in (b) creates a new accent on the first B, partly because it now begins a five-note group and partly because it is emphasized by small ornamental notes. At the same time, the accentuation on E has been reduced considerably. The repetitions of this phrase—at the beginning of Ex. 10.5c and then again at (d)—retain the ornamental accentuation of the first B, but sound like rubato

---

[108] Full score (ERM, 1912); also ed. Charles Hamm with English stage directions in the Norton Critical Score (Norton, 1967), and repr. with English trans. (Dover, 1988). For the revised 1947 version, see the full score (B & H, 1948). The passages in Ex. 10.4 are essentially the same in the composer's reduction for 4 hands of the original version (ERM, 1913; repr. Dover 1990) and the revision (B & H, 1948). In Ex. 10.4 I have omitted other voices as well as octave doublings. For the recording in which Stravinsky conducts the Columbia Symphony Orchestra, see, among other releases, the series *Igor Stravinsky: The Recorded Legacy* (CBS Records: GM31 US–LXX 36940, 1981), v.

[109] André Schaeffner, to whom Stravinsky initially disclosed the origin of the theme, prints the folk-tune, together with the opening measures of *The Rite of Spring* from the autograph MS belonging to ERM, in *Stravinsky* (Paris: Les Éditions Rieder, 1931), pl. XXI (see also pp. 43 and 125). Lawrence Morton, in 'Footnotes to Stravinsky Studies: "Le Sacre du printemps"', *Tempo*, Ser. 2, No. 128 (Mar. 1979), gives the theme again on p. 12, but with the corrections he lists on p. 16 n. 24. See also Richard Taruskin, 'Russian Folk Melodies in *The Rite of Spring*', *JAMS* 33 (1980), 502. The melody comes from *Melodje ludowe litewskie* (Lithuanian Folk Melodies) compiled by Anton Juszkiewicz (Kraków, 1900), No. 157. In the edition available to me from Pennsylvania State Univ., the tune is numbered 157 (158) and the last pitch in the repetition of the opening phrase is the same as the preceding note. See Stravinsky's comments in *Memories and Commentaries*, 98. See also Richard Taruskin, 'From *Firebird* to *The Rite*: Folk Elements in Stravinsky's Scores', *Ballet Review*, 10 (1982), 72–87.

Ex. 10.5 Stravinsky, opening rubato of *The Rite of Spring*, 1911–13

a Lithuanian folksong 'Tu, manu seserėlė' (transposed down a major 3rd)

versions of (*b*) in which note values have been altered. In his own copy of the 1921 score, Stravinsky 'marked (and always wanted)', according to Craft, 'a break (short silence) after the second fermata in the very first measure of the piece'.[110]

Carl Dahlhaus describes the freedom he feels is appropriate in this passage:

> The bassoon solo . . . is rhythmically imprecise. One hardly needs the indication *Tempo rubato* for it to be clear that the notation should not be taken too literally . . . The presentation of the E minor broken chord in semiquavers [sixteenth notes], quaver [eighth-note] triplets and semiquaver quintuplets is not meant as rational augmentation and diminution—the proportion 1/4 : 1/3 : 1/5—but as an uncertain, vague broadening out and acceleration. And the apparent shift of accent in the first bar, the displacement of the stress from the first to the second note of the E minor broken chord, is nothing other than a deception inherent in the notation. In the irrational, melismatic rhythm that Stravinsky has in mind, only the turn [the grace-notes] before the figure's first note acts as an accentuating factor.[111]

In 1964 Stravinsky wrote a review of three recordings of *The Rite*. Concerning a performance conducted by Pierre Boulez, he states that 'the second fermata in the first measure is too long' and objects to retards added elsewhere and places where the 'tempo is unsteady' or

[110] *Stravinsky in Pictures and Documents*, 530.
[111] 'Problems of Rhythm in the New Music', *Schoenberg and the New Music*, trans. Derrick Puffett and Alfred Clayton (CUP, 1987), 47. The essay first appeared as 'Probleme des Rhythmus in der Neuen Musik' in *Terminologie der Neuen Musik* in the series *Veröffentlichungen des Instituts für Neue Musik und Musikerziehung Darmstadt*, v (Berlin, 1965), and was printed also in *Schönberg und andere: Gesammelte Aufsätze zur Neuen Musik* (Mainz, 1978).

the 'rhythm is askew'. He similarly complains about retards and accelerations added by Herbert von Karajan and mentions that his rubato three measures before the 'Danse sacrale' is 'unnecessary and debilitating'.[112] Three years later he describes a performance by Leonard Bernstein: 'Even the peculiarities of *tempi* and *rubati* in the *Sacre*—a more handwringing reading than I am accustomed to—were derived, as he told me, from Stokowski's performances in the thirties.'[113] In a review from 1970 he criticizes a performance by Zubin Mehta: 'The second note of the horn [notated to sound between the middle two eighth notes in the second measure: see Ex. 10.5*c*] comes *with*, instead of before, the bassoon, a mistake that changes the whole character of the beginning.' Stravinsky no doubt felt that when performed strictly as notated, the displacement between the two instruments would emphasize the free and improvisatory nature of the bassoon's rubato.[114]

Stravinsky's original manuscript contains the main tempo marking 'Lento, tempo rubato'. This recurs also in the full scores revised and published in 1921 and in the 1940s. Although rubato seems most appropriate in the opening bassoon solo, it is possible that it applies to the opening thirteen measures that precede the tempo change of 'Più mosso'. In addition, it may return at 'Tempo I' when the bassoon melody recurs, especially since the 'Augurs of Spring' eleven measures later is marked 'Tempo giusto'. Curiously, however, the reductions for piano four hands made by Stravinsky for publication in both 1913 and 1947 omit 'tempo rubato' altogether; 'Lento' remains as the main marking, whereas *a piacere* appears in smaller italics below the melody. Perhaps Stravinsky did not trust pianists as much as conductors to react properly to the word *rubato*.[115]

The same sort of changes occur in the various versions of the Chinese March from *Le Rossignol*, where Stravinky writes 'rubatis-

---

[112] *Dialogues and a Diary*, 82–90: 'Stravinsky Reviews *The Rite*'. In the original article in *HiFi/Stereo Review*, 14/2 (Feb. 1965), 60–3, Stravinsky does not mention Boulez's 2nd fermata, and concerning Karajan's approach to the 'Danse sacrale' briefly states: 'The third measure before 142 should not be played, as here, *rubato*.'

[113] *Themes and Conclusions*, 106–7, originally in 'Stravinsky at Eighty-Five', *The New York Review of Books*, 8/10 (1 June 1967), 14.

[114] *Themes and Conclusions*, 234: 'Spring Fever: A Review of Three Recent Recordings of *The Rite of Spring* (June 1970)'.

[115] For the full scores, see the original edn. (ERM, 1921; repr. B & H) and the revision of 1947 (B & H, 1967). For a repr. of the draft of the beginning, before *Lento* and *tempo rubato* were added, see *Stravinsky in Pictures and Documents*, 78. For the piano reductions, see the 1947 revision (B & H, 1947) and the original 1914 edn. repr. in the *Kalmus Piano Series* (Melville, NY: Belwin-Mills, 1970), and in *Petrushka and The Rite of Spring for Piano Four Hands or Two Pianos* (Dover, 1990). It was part of the latter version that Stravinsky played with Debussy in 1912; see *Expositions and Developments*, 162, and *Stravinsky in Pictures and Documents*, 87, 90, and 613 nn. 153 and 154.

simo' over a single measure in the orchestral score, but 'Allarg.' or *molto allargando* in the piano arrangements. The March occurs in Act II of the opera, composed in 1913–14. When the stage directions read 'Servants bear in with pomp the Emperor of China, seated in his canopied chair', the music commences a series of five upward and downward waves through a pentatonic scale in 5/16 and 7/16 metre. This leads to a final rise in a measure of 6/16 in which the last five sixteenth notes are marked with accent signs, followed by a measure in 4/16 metre in which four accented *fff* sixteenth notes sustain the summit of this climactic point. This last measure is the one marked 'rubatissimo' (an expression which had previously been used, as we have seen, by both Ernst and Tchaikovsky) in the full scores of both the opera and the symphonic poem *Chant du rossignol*. In a recording of the opera in 1960, Stravinsky makes a substantial unmarked retard on the accented notes in the bar preceding 'rubatissimo', and then, with heavy articulation, sustains each of the sixteenth notes marked 'rubatissimo' for about double their usual value. He notates this manner of performance more exactly in an arrangement for violin and piano in 1934: here the last three sixteenth notes of the preceding measure are 'allarg.', whereas the measure originally with 'rubatissimo' is now 'a tempo' and 'marcatissimo' in 4/8 metre with four accented eighth notes.[116]

Stravinsky demonstrates a similarly exaggerated rubato performance in the Trio of the 'Valse' from his Suite No. 2 for chamber orchestra. In 1914–15 he wrote the piece originally as one of *Three Easy Pieces* for piano duet. The Trio consists of three statements of a two-bar unit in which three quarter notes move like an anacrusis to a quarter note at the beginning of the following bar; two additional measures then complete the eight-bar Trio, which is immediately repeated. In the piano duet 'poco rubato' appears above the opening measure of the melody, with *col parte* for the accompaniment. In the arrangement of 1921 for orchestra, however, Stravinsky has marked 'Rubato' above each of the three measures containing three quarter notes and 'Tempo' above the final two thirds of each succeeding measure. In his recording of this version in 1963 he adds about a quarter rest before beginning

---

[116] See the full score of the *Chant du Rossignol* (ERM, 1921; repr. B & H) and the rev. edn. of the opera (ERM and B & H, 1962). The rubatissimo measure in the reduction for piano solo of the original version of the opera (ERM and B & H, 1961) is marked *molto allargando* and the 1st 16th beat has an arpeggiated chord. The same spot in the piano reduction of the *Chant* (ERM, B & H, 1947) has 'Allarg.'. A copy of the *Airs du Rossignol et Marche Chinoise* (ERM, 1934) for violin and piano is in BL: h.3992.s.(7.). CBS's *Recorded Legacy* includes a performance of the *Chant du Rossignol* conducted by Robert Craft in 1967; he makes no noticeable retard in the preceding measure and sustains each note of the 'rubatissimo' far less than Stravinsky does.

the Trio, then doubles in length each of the three quarter notes in the three measures marked 'Rubato'. He could thus have notated the exact values of the notes by alternating metres of 3/2 and 3/4.[117]

Another example of substantial note lengthening occurs in *Les Noces*, composed between 1914 and 1917. During the fourth scene of the second part ('The Wedding Feast'), the tenors in the English version shout 'black her brows and beautiful'. The single measure of 2/4 is marked *clamando*, and the rhythm is indicated without pitches. It includes four sixteenth notes marked 'poco rubato' plus three triplet eighth notes with 'tempo' above the last two. In Stravinsky's recording of 1959, the sixteenth notes are performed evenly, but each with a value equal to about a dotted eighth. On the word *beautiful* two short notes move to a longer one. Although no accentuation is marked, the tenors yell with heavy articulation during the rubato on the first half of the measure. No doubt the phrase takes on a somewhat different rhythm in each language, and perhaps the original Russian words fit the notated values more closely.[118]

Stravinsky thus seems to employ rubato during his early period in two ways: by altering the relationship between note values without indicating any special articulation, or by heavy articulation accompanied by ritenuto but without any substantial change in the relative note values. The first method occurs in the earliest examples, where he notates the rhythmic changes in such a detailed manner that any further flexibility by the performer would seem inappropriate. In the case of the Étude, the alterations are perceived in relation to the different rhythm of the surrounding music; in *Petrushka* and *The Rite* the passage involved appears previously in unaltered form. In *Le Rossignol*, the Suite No. 2, and *Les Noces*, on the other hand, rubato seems to consist, at least judging from his recordings from a later period, of substantially slowing down a series of three or four evenly notated notes with a heavy articulation, while the notes, at the same

[117] See the original *Waltz*, dedicated to Erik Satie (London: J. & W. Chester, 1917) and with fingering added by Gerard Alphenaar (New York: Omega Music Edition, 1949); and the *Suite No. 2* (J. & W. Chester, 1925). In an arrangement for a single pianist in *The Short Piano Pieces*, Soulima Stravinsky marks *rit.* instead of 'Rubato' above the 3 measures involved and writes out the repeat of the section. In *Selected Correspondence*, i, 413, Craft states, after consulting the original MSS, that 'the Trio is marked simply "rubato" (not "rit.")'. *The Recorded Legacy* contains Stravinsky's performance of the Suite in vol. v and a recording of the piano duet in 1961 by Gold and Fizdale in vol. ii. In the latter case, the rubato seems confined mostly to the opening measure, which is deliberate, weighty, and considerably slower.

[118] See the full score with Russian and French texts (London: J. & W. Chester, 1923), 118, as well as the piano/vocal scores in Russian and French (London: J. & W. Chester, 1922) and in German and English (London: J. & W. Chester, 1922), 157. The shouted Russian and German texts appear with the same rhythmic notation as the English, whereas for the French trans. the triplet occurs first under 'Poco rubato', followed by the 4 16th notes.

time, do not alter their relationship to each other, but continue to have equal value. It is the unexpectedly large degree of slowing that is surprising in Stravinsky's performances.

The rubatos in the neo-classical works differ in several respects from those of the early period. Now Stravinsky does not usually notate alterations in the relationship between note values as he did in the Étude, *Petrushka*, and *The Rite*, but sometimes adds them in performance. He does not change the tempo, as he did in *Le Rossignol*, the Suite, and *Les Noces*, but often includes marked or unmarked articulations of a special nature. He continues to use rubato for some of the same purposes: to articulate musical structure, as in the Étude; to declaim the text with intensity, as in *Les Noces*; to strengthen a humorous mood, as in Suite No. 2. Although Stravinsky considered that music was 'essentially powerless to *express* anything at all',[119] his rubatos in *Petrushka*, and probably also in *The Rite* and *Le Rossignol*, seem to have a programmatic significance which is to be realized also by the choreography. In addition, Stravinsky uses rubato during the neo-classical period, as Tchaikovsky did, for cadenza-like passages for solo instruments.

When Stravinsky compared Schoenberg's abundant use of rubato to his own metronomic strictness without rubato, he gave as examples his Octet of 1922–3 and his Concerto for Piano and Wind Instruments of 1923–4. In 1927 he wrote that 'the Concerto . . . risks being compromised if incompetent or romantic hands begin to "interpret" it before undiscriminating audiences'.[120] It was the interpretive rubato of the romantic pianists that he wanted to eliminate, however, and not his own type of rubato. There are, in fact, two cadenzas in the second movement, each marked 'poco rubato'. Each contains two statements of an ascending passage which begins with vigorously accented notes punctuated by low notes in the trombone, tuba, and timpani. An acceleration occurs in the notation itself, which moves from four sixteenth notes to a septuplet in the time of a quarter note and finally a quintuplet in place of a sixteenth. In the revised version of 1950 'poco accel.' appears on three of these passages. The unusual groups also create polyrhythms of three, five, and seven beats against four. Elsewhere in the cadenzas are long groups of notes marked *staccato*. The rubato thus seems largely concerned with conspicuous articulation, with the 'extraordinary dynamism' on each note which Carter noticed in Stravinsky's own playing. This dynamic articulation in-

---

[119] *Chronicle of My Life*, 91, or *An Autobiography*, 53.
[120] Letter of 14 Aug. 1927 to Gavril Païchadze, quoted in *Stravinsky in Pictures and Documents*, 629 n. 34.

volves a percussive touch with little or no use of the sustaining pedal. Stravinsky commented concerning the Concerto: 'I have endeavored to restore the piano to its rightful place as a percussion instrument.'[121]

The *Capriccio* of 1928–9 is in effect another piano concerto. The second movement ('Andante rapsodico') contains a cadenza-like solo section with two marked rubatos which Stravinsky, in a recording from 1930, plays with considerable flexibility. He pauses before the first rubato, starts four beamed sixty-fourth notes hesitatingly with the value of thirty-second notes, then accelerates on the second group of four sixty-fourth notes and on the quintuplet marked 'accelerando'; he slows down the last half of the measure more than the word 'tempo' would indicate. At the beginning of the next measure there is a group of four accented and unbeamed sixty-fourth notes marked *rubato*: he pauses first, then plays them with a sharp staccato touch and in a rhythm more like slightly uneven sixteenth notes. There are two places in the third movement ('Allegro capriccioso ma tempo giusto') which apparently were originally marked *poco rubato* and later changed to *poco rall.* in the revised orchestral score. Both involve a lively melody that falls impishly through three octaves to the E above middle C during a single measure of dotted rhythm. For the first rubato Stravinsky slightly delays the chord on the second beat as well as the E; the second rubato is even less noticeable, with the E delayed somewhat less and played with a special colour.[122] These two subtle rubatos enhance the capricious mood, whereas the bolder ones in the second movement strengthen the aggressive improvisatory character appropriate in a solo cadenza.

[121] 'Chronological Progress in Musical Art: An Interview', *Etude Music Magazine*, 44 (1926), 559. See Joseph, *Stravinsky and the Piano*, 8, 152–3, and 158; and *Memories and Commentaries*, 25–6. For the original 2-piano reduction published in 1924 see *Concerto pour piano suivi d'orchestre d'harmonie* (B & H, 1947), 35 and 40. For the rev. version, see the full score (B & H, 1960) and the reduction for 2 pianos (B & H, 1968). The revisions are summarized in a chart by White in *Stravinsky: The Composer and his Works* (1979), 319. See also Barbara Oakley Allen, 'A Comparison and Critique of the Recorded Performances of Stravinsky's *Concerto for Piano and Wind Instruments*', D. M. A. diss. (Stanford Univ., 1980; UM 80–24,768).

[122] At 1 bar after rehearsal no. 66 in the 3rd movt., as well as 4 bars later, there are no markings in the original reduction for 2 pianos (ERM, 1929). In the original full score (New York: Kalmus, 1930; repr. later) the 1st spot is unmarked, the 2nd is *poco rubato*; in the reduction for 2 pianos of the revised version (ERM and B & H, 1949) both are *poco rubato*; and in the revised full score (B & H, 1952), both are *poco rall.* For a sketch including the cadenzas before the word *rubato* was added, see *Stravinsky in Pictures and Documents*, 292; see also Joseph, *Stravinsky and the Piano*, 192–3. A performance from 1930 with Stravinsky as pianist and Ernest Ansermet conducting the Orchestre des Concerts Straram appears on *Stravinsky Playing Piano* (Seraphim Mono 60183, released 1972). For a recording with Maria Bergman, pianist, and Stravinsky conducting the Orchestra of the Southwest German Radio at Baden-Baden in 1954, see *Igor Stravinsky: The Man and his Music*, EMA–103, side 5 (Educational Media Associates, 1977).

One final neo-classical example of rubato in a cadenza occurs in the ballet *Le Baiser de la fée* (The Fairy's Kiss), based on music by Tchaikovsky. In the third scene the Young Man, who had been marked with the Fairy's kiss while still a child, meets his Fiancée and her friends at a mill and they all dance. The Adagio for the Young Man and his Fiancée from the 'Pas de deux' commences with a cantabile melody played by a solo cello, which is interrupted by a three-bar cadenza written by Stravinsky for solo flute. 'Poco rubato' appeared at the beginning of the cadenza in the original score published in 1928, but in the *Divertimento* (1938) and later versions of both the ballet and the orchestral suite 'Poco rubato' occurs only when the flute has staccato triplets during the second half of the passage, and now with tempo changes indicated, in the manner we noted in some rubatos by Bartók, in smaller print below 'Poco rubato': 'poco rit.' at the end of the second bar of the cadenza, 'poco a poco accel.' through most of the third, but ending with 'poco rit.' again. The rubato thus employs articulation as well as carefully defined rhythmic flexibility to mark the change from duple to triplet movement.[123]

Stravinsky also uses rubato during this period for purely structural purposes—to mark the beginning of a new musical unit or the transition to a new one. Following the Adagio that includes the flute cadenza in *Le Baiser de la fée*, Stravinsky modulates from E flat major to G minor, the key of the Variation for the Fiancée. In the 1950 revision he marks the first of the two modulating measures 'Poco rubato' and the second one 'acceler.', then, following a double bar, the main tempo indication 'Allegretto grazioso'. In the earlier versions, there is neither rubato nor acceleration, and 'Allegretto grazioso' occurs two measures sooner. In recordings of both the *Divertimento* and the ballet in 1955 and 1965, Stravinsky begins the first measure very slowly and reluctantly, with heavy articulation of the dotted rhythm; in the second measure he makes a rapid and exhilarating acceleration to the main tempo of the Variation. For comparison, one can hear the modulating measures played strictly in the tempo of the

---

[123] Compare the original reduction for solo piano of the ballet *Le Baiser de la fée* (ERM, 1928) with the 1950 revised full score (B & H, 1952) and piano reduction (ERM and B & H, 1954) as well as with the full score of the *Divertimento*, both in the original edn. (ERM, B & H, 1938) and the revised (B & H, 1950). In the transcription by Stravinsky and Samuel Dushkin of the *Divertimento* for violin and piano (ERM, 1934), 32–3, there are no markings at all for rubato or tempo changing during the cadenza, and most of the staccato dots are missing or replaced by tenuto signs. Stravinsky conducts a recording of the ballet in 1965 in *The Recorded Legacy*, vii, and of the *Divertimento* in 1955 in *Igor Stravinsky: The Man and his Music*, EMA–103, side 14 (Educational Media Associates, 1977).

main piece in a recording of the original version by Ernest Ansermet.[124]

A similar two-bar transition occurs also in the comic opera *Mavra* from 1921–2. At one point the Mother exits, whereupon Parasha and her lover, who is masquerading as a cook in order to be near his beloved, look around, presumably to see if the coast is clear (during the measure marked *Rubato*) and then rush into each other's arms (the accelerando in the succeeding measure). The transition consists only of 'oom-pah' chords in the orchestra, so that in Stravinsky's performance recorded in 1964 the rubato, although perhaps some-what deliberate, is not very conspicuous.[125] In the Variation for the male dancer in *Scènes de ballet* of 1944, a solo violin imitates the preceding dotted-note rhythm in a measure marked 'poco rubato' which becomes part of a transition that leads to a return of the open-ing theme. In his recording of 1963 Stravinsky slows down noticeably so that the solo instrument can enunciate the rhythm with great clarity.[126] The final example of a structural rubato appears in the third movement of the Sonata for Two Pianos of 1943–4. Here it acts to articulate the commencement of the middle section of a ternary form. The opening section in G minor ends with a full cadence. The contrasting middle section is in G major, and the change of mood and manner is marked by *poco rubato* in the opening measure and can-celled by *a tempo* in the second. The melody, a Russian folk-tune, opens with repeated notes marked with tenuto dashes and accom-panied by motives consisting of an eighth and two sixteenth notes played staccato by the second piano. For the rubato to be effective, the second pianist would have to feel a declamatory emphasis on the accompanimental motives.[127]

In addition to the instrumental rubatos in solo cadenzas or at sen-sitive structural points, Stravinsky also employs declamatory rubatos

[124] Rubato appears in both the full score of the ballet (1952) and the piano reduction (1954), but not in the earlier editions and not in any versions of the *Divertimento* or the arrangement for violin and piano. Ansermet's recording with L'Orchestre de la Suisse Romande is on London CM 9368 (London: Decca Record Co., 1963).

[125] Full score with English trans. (B & H, 1969) and piano/vocal score with French, German, and English (B & H, 1947). Stravinsky conducts the CBC Symphony Orchestra in *The Recorded Legacy*, x.

[126] Full score (AMP, 1945). For the performance in which Stravinsky conducts the CBC Symphony Orchestra, see *The Recorded Legacy*, vii.

[127] (AMP, 1945). Concerning the folk-melody, see Joseph, *Stravinsky and the Piano*, 220–5. In 'Music for the Ballet', *Stravinsky in the Theatre*, 49, Arthur Berger writes: 'In romantic music, the device of rubato (either specified or implied) allows retards at the performer's whim. But Stravinsky's beat is metronomic, and when we come upon the indication "rubato" in the charming *Sonata for Two Pianos* (1944), leaving a slight metrical license to the performers themselves, it is unusual indeed.'

in the vocal music of this middle period. The vocal rubatos, however, are far more concerned with special articulation and voice quality than with alterations in tempo. In *Mavra*, for example, a large section consisting of twelve measures of instrumental introduction and thirty-two more of aria is marked 'Lento (poco rubato)', with the melody to be *tranquillo e molto rythmico*. The disguised Vasili sings of the agony of waiting alone for his beloved. The accompaniment marches with quarter notes in 2/4 like a steady drumbeat, while first the trumpet and then the tenor sing heavily articulated triplet figures marked with dots, dashes, and accent signs. This heavy and sombre articulation ceases only when Vasili imagines himself finally alone with Parasha, at which time the music moves with 'Più mosso' and 'tempo giusto', and sounds as though it had been suddenly freed from a powerful constraint. The rubato therefore consists of special articulation and voice quality within a relentlessly strict tempo.[128]

In Stravinsky's own recording of *The Rake's Progress* of 1948–51, most of the nine rubatos are performed fairly strictly according to the notated rhythm, but with conspicuous and special declamatory emphasis. They generally vary from a half to one and a half measures in length and are usually indicated by directions in the voice part and cancelled by 'a tempo' or a new tempo marking. When Tom Rakewell discovers that life is no better in the city than in the country, he sings a rubato of disdain on 'City! City!', with the first syllable delayed, somewhat in the manner of the earlier type of rubato, until after a sforzando chord. When Shadow advises him to 'take Baba the Turk to wife', the words 'Baba the Turk' are sung rubato, with an accent sign on each of the four notes. Sellem, the auctioneer, has two rubatos: a boastful one on 'ne plus ultra of auctions', with an accent on each note of 'ne plus ultra', and a sarcastic one when he suggests that those who take from those who lose are acting as 'nature's missionaries' (the rubato includes 'missionaries' and 'you are her instruments', with the four notes on the first word accented). When Shadow is about to claim Tom's soul, there are two more rubatos: one when the patronizing devil suggests 'a game of chance to finally decide your fate', and another when the frightened Tom, asked if he has a pack of cards, answers, 'all that remains me of this world, and for the next'. In Bedlam Tom, thinking he is Adonis, sings: 'Come quickly, Venus, or I die', where the last three words are unaccompanied and the last two rubato. When Anne arrives, finally, he greets her with *poco rubato* on 'My queen, my bride, at last'.

---

[128] Stravinsky provides a vigorous performance in his 1964 recording with the CBC Symphony Orchestra in *The Recorded Legacy*, x.

Most of these rubatos are performed in Stravinsky's recording with striking emphasis and tone colour, but with strict rhythm or, at the most, with slight allargando. The only noticeable alteration of note value occurs when the quarter note on the first word of the phrase 'I die' is about doubled in length. There is, in addition, one more rubato indicated when Anne, preparing to leave home in order to find the lover who deserted her, beseeches the moon to guide her 'and warmly be the same he watches without grief or shame'. *Poco rubato* appears above the word *and* in the voice part and is apparently not cancelled until the next tempo change eight bars later. In this case, however, I suspect that the rubato is, in fact, extremely brief—perhaps mainly involving the syllable '*warm*' of *warmly*. In the recording Stravinsky keeps the orchestra strictly playing its accompanimental figures in sixteenth notes, so that one notices only a slight leaning on '*warm*' and a softer quality on '*ly*' rather than any alteration of note values.[129]

These declamatory rubatos of the neo-classical period, being less concerned with tempo flexibility than those at cadenzas or structural points, seem, in general, more subtle and restrained. Coming at the end of the period, the rubatos in *The Rake's Progress* thus seem to point toward the even more austere examples from Stravinsky's last period. Two of the late twelve-tone vocal works involve *quasi rubato* cadenzas. In *Canticum sacrum* from 1955, the fourth movement is a setting of a portion of the text from the Gospel of St Mark concerning Jesus' casting out of an unclean spirit from a father's son. After the chorus sings in Latin the text: 'And straightway the father of the child cried out, and said with tears', the baritone soloist repeats and thus emphasizes the last three words in a single unaccompanied and unmetred bar consisting of the equivalent of thirteen quarter beats. This measure is marked *quasi rubato discreto* in the full score, *quasi rubato, con discrezione* in the piano/vocal score. Although the word *discrezione*, as we saw in Chapter 1, referred in the Baroque period to rhythmic freedom in the performance of preludial works by composers such as Froberger, it seems more likely here to mean moderation. In the recording conducted by Stravinsky in 1957, the baritone seems to sing the passage strictly in time and without any striking articulation or tone colour—thus like a very moderate rubato indeed.[130]

---

[129] Full and piano/vocal scores (B & H, 1951), Act I sc. iii; Act II, sc. i; Act III, scs. i, ii, and iii. Stravinsky conducts the Royal Philharmonic Orchestra in *The Recorded Legacy*, xii. The line sung in the recording as 'a game of chance to finally decide your fate' appears in the piano/vocal score with the 5th word missing and in the full score with *finally* replaced by *thoroughly*.

[130] Full and piano/vocal scores (B & H, 1956). In the miniature full score the half-note C$^\sharp$ in m. 281 looks like a quarter note and is so printed by White in the excerpt of this 'unaccompanied cadenza' in *Stravinsky: The Composer and his Works* (1979), 488. In *The Recorded Legacy*, xiv, Stravinsky conducts the Los Angeles Festival Symphony Orchestra.

This passage also illustrates the manner in which rhythmic transformation is applied to melodic lines in twelve-tone music. A transposed version of the tone-row used for the twelve notes in the *quasi rubato* solo also occurs fifteen measures earlier, but with some notes repeated or moved to a new octave level, and with the rhythmic values completely changed. An even better demonstration of rhythmic transformation in serial music appears in the 'Elegia prima' of Stravinsky's *Threni*, where the alto bugle, following the rhythmic reading by the chorus, plays the same series of twelve tones five times in succession, each time with a different rhythm. In this case the transformation is easily perceived by the listener.[131] Stravinsky rhythmically transformed themes occasionally even in earlier works which do not employ the twelve-tone technique.[132]

In *Abraham and Isaac* from 1962–3, a sixteen-measure instrumental interlude separates the preliminary portion of the narrative from the dramatic part in which Abraham offers to sacrifice his son. The interlude commences with a two-bar *Cadenza quasi rubato* for solo flute, divided into three segments by string chords. The first measure of the cadenza is in 3/4, with '5 for 4' sixteenth notes on the first beat and '7 for 4' on the second. The following meaure is unmetred, with *colla parte* in the strings and with trill-like figures for the flute notated in great detail by means of sixteenth- and thirty-second-note triplets and a sixty-fourth-note septuplet. The only freedom allowed to the performer by such specific notation seems to be provided by the breath pauses before and after the group of two string chords in the second measure and the fermata at the end. *Quasi rubato* in this piece, as well as in *Canticum sacrum*, seems to imply that a precisely notated solo passage should be played rather strictly, but with the same feeling of improvisatory freedom as though it had resulted from the flexible performance of a simpler version. The *quasi rubato discreto* of *Canticum sacrum* mainly involves only quarter and

[131] Full and piano/vocal scores (B & H, 1958), mm. 43–61, 89–107, and 143–65.

[132] We have already seen the transformation in Ex. 10.5. See Louis Andriessen and Elmer Schönberger, *The Apollonian Clockwork: On Stravinsky*, trans. from Dutch by Jeff Hamburg (OUP, 1989), 122, where the 1st 5 bars of the theme from the 2nd movt. of the *Sonata for Two Pianos* are compared with a transformed version at the beginning of Variation 3. On pp. 120–1 the technique of rhythmic transformation is illustrated by showing the opening pitches of 'Happy Birthday to You' notated in different rhythmic values by Stravinsky, Boccherini, Donizetti, Michael Haydn, and Rakhmaninov. See also p. 175, where 'the romantic rubato on "stolen time"' is distinguished from the baroque 'rubato on "borrowed time" (in general, not equal semiquavers, but a slightly longer first semiquaver compensated by the other three being shorter)'. Then on 177, the authors state: 'Stravinsky's rhythms are the authenticated interpretation of the baroque rubato on "borrowed time". Everything is conceived proceeding from the down-beat, but the beat itself has become *inégal*: down becomes up and an up (sometimes a few beats later) turns out to be a down-beat.'

half notes, whereas the florid *quasi rubato* of *Abraham and Isaac* includes far smaller note values disposed in unusual groups.[133]

One more rubato occurs, finally, in the second of the three madrigals which Stravinsky arranged for instruments in 1960 in the *Monumentum pro Gesualdo di Venosa ad CD annum*. The original Gesualdo madrigal is the *seconda parte* of a larger work and begins with a setting of the three words 'Ma tu, cagion', which are then repeated as part of the longer phrase 'Ma tu, cagion di quella atroce pena' (But you, cause of this terrible pain). Sensing the introductory nature of the opening three words, Stravinsky writes 'poco rubato (rit.)' over the same music scored for oboes, bassoons, and a trumpet; he dramatically rearranges the pitches within the individual lines and even moves some to a different octave level. The rubato includes a whole-note chord in a measure of 2/2 followed by a few quarter and half notes on the opening two half beats in a measure of 3/2. The last half beat of this measure is marked 'a tempo', whereupon the instrumentation changes to trumpets and trombones and each instrument suddenly follows the original pitches and voice leading. Furthermore, Stravinsky shortens the final chord of the rubato to half its original length so that it does not overlap with the following music as it did in the original madrigal. Since the 'poco rubato' begins with a sustained chord, the listener does not have much sense of metre and tempo before the 'rit.', which presumably refers mainly to the continuation of the rubato on the more active notes of the next measure. In a recording of the piece, Stravinsky performs the rubato subtly, for the ritardando is very slight, but with extremely eloquent declamation. At the beginning of the second bar, three of the instruments have two slurred quarter notes, the top voice falling the interval of a ninth. Stravinsky executes the slur elegantly, with the second quarter beat shortened to provide a silence of articulation, which, in turn, throws a special tone quality onto the final chord. The marking 'poco rubato (rit.)' thus seems to require a very sensitive declamation of this introductory gesture, which, in turn, depends upon a very clear understanding of the way it relates to the following music.[134]

Most of Stravinsky's rubatos are declamatory in nature, whether they occur in vocal music, in an instrumental cadenza, or at a sensitive spot in the musical structure. The declamation is projected mainly by

---

[133] Full and piano/vocal scores (B & H, 1965), mm. 89–90. See White, *Stravinsky: The Composer and his Works* (1979), 531.

[134] Full score (B & H, 1960). For the original piece from *Madrigali libro quinto* (1611), see C. Gesualdo: *Sämtliche Werke*, v, ed. Wilhelm Weismann (Hamburg: Ugrino Verlag, 1958), 69–71 (the 2nd part of 'Poichè l'avida sete' for 5 voices). See *New Grove*, vii, 318 and 322. In *The Recorded Legacy*, xi, Stravinsky conducts the Columbia Symphony Orchestra (1960).

intense articulation—an articulation sometimes marked by signs and sometimes causing lesser or greater alterations of the note values. Stravinsky cultivated a superlatively articulate style. Most instruments were eventually used in a percussive manner, even those, like the violin or piano, which had formerly been associated with singing legato lines. He indicates the pedal only for his first youthful rubato in the Op. 7 Étude. His later style became so dry and crisp that it was unnecessary for him to mark rubatos, as Schnabel did, *senza pedal*. Within such a style, subtle details of articulation become enormously conspicuous, and in this context rubato generally requires even more intense articulation in order to contrast with its surroundings. Often Stravinsky's rubatos involve a solo singer or solo instrument with the accompaniment absent or minimal, which makes the articulation even more apparent. Sometimes they include the animated articulation of a dotted rhythm, with the rests carefully notated between the notes. Like a number of preceding composers, Stravinsky associates rubato with a variety of accent signs, with unusual groups of notes, and with polyrhythms which cause displacement between different parts.

Stravinsky obviously felt ambivalence toward rubato. Trained as a pianist, he saw the word in the scores of composers he liked and composers he was reacting against. He opposed the rubato added by interpretive performers and felt it as a threat in the performance of his own compositions. As we have seen, he sometimes replaced *rubato* by other words when making a piano reduction or when revising an earlier work. In spite of all his fears, however, he continued indicating rubato in his scores until 1963. Perhaps he felt that by marking the word in a few specific places he was, in effect, prohibiting its use elsewhere. He may also have felt that his general style was so obviously unsuited for rhythmic flexibility that unlike Bartók, Berg, or Schnabel, he had no need to mark *non rubato* or *senza rubato*. He attempted in both the early and late periods to notate a sort of frozen and completely determined rubato. At other times he himself performed his own rubatos with a flexibility which he would surely have roundly condemned in another performer. His scores and his recordings, however, show that his own kind of rubato, which evolved during his lifetime along with his changing attitudes toward music making in general, constituted a significant way in which he bestowed prominence upon certain special and intense moments in his music.

Other Composers from the First Half of the Century

Many other composers contemporary with Schoenberg, Berg, Bartók, and Stravinsky also indicate rubato in their scores, but generally in the more conventional ways inherited from Liszt and Tchaikovsky. Most write in an objective neo-classical style which involves a relatively strict sense of tempo. Within this context, rubato and other tempo alterations are ordinarily applied in an extremely restrained manner.

Paul Hindemith employs rubato in a number of works composed in Germany between 1916 and 1934. A relatively long rubato seems clearly intended when the word occurs in the marking for a main tempo or a tempo change, or a section in 'Quasi recitativo'. A momentary rubato seems appropriate when the word is carefully placed on specific internal notes of a phrase, or when it recurs with an immediately repeated motive. Several of Hindemith's rubatos seem to be brief—less than a measure in length, yet continuing longer than the momentary type. As usual, there are a few cases in which it is difficult to determine whether the rubato is brief or momentary. Like many composers before him, Hindemith associates rubato with notes of short value, sometimes arranged in unusual groups, sometimes involving polyrhythmic displacement, and sometimes bearing signs for special articulation.

Carl Orff, strongly influenced by works of Stravinsky such as *Les Noces*, includes rubato in his scores with unusual frequency. Although he occasionally applies *sempre rubato* to an entire section, most of his rubatos are momentary or brief. Within his extraordinarily austere style, many occur when a solo voice or solo instrument is singing alone or accompanied only by a long sustained note or chord. Sometimes they occur with rhythmic speaking. Sometimes they involve a chorus. Many utilize repeated notes, like the reciting tone in chant, in a declamatory manner. Others involve a melismatic or cadenza-like flourish of small notes. Most, like Orff's music generally, display an abundance of articulation signs. Some of the moods are indicated by espressions such as *rubato e affetato*, *rubato con slancio*, *agitato e sempre rubato*, *dolce (rubato)*, *rubato con molto entusiasmo*, *tranquillo e rubato*, *molto infiammato e rubato*, *leggiero e rubato*, and *pesante e rubato*. Orff also uses other terms that imply free and expressive changes in the tempo: *libero*, *frei*, *flessibile*, *stentato*, *(quasi) senza misura*, and *(quasi) parlando*. These terms, however, are apparently different from rubato, for they sometimes occur together in the same indication, such as *rubato e flessibile* and *libero e rubato*, or in succession, such as *rubato* followed immediately by *libero*. *Rubato*, *libero*, or *frei* occurs in a few cases simultaneously with 'a tempo'. The latter expression, on the other hand, often follows *rubato* and in this case apparently marks its termination.

Almost all of Orff's rubatos occur in texted works. Although some are performed only by instruments, most are vocal and thus reflect the declamatory accentuations and emotional nuances of the text. In addition to the many rubatos in his serious compositions, there are also a few in the short instructive pieces in his *Klavier-Übung* of 1934 and *Music for Children* from 1950–4. As with the Kodály Method, the *Orff-Schulwerk* has become internationally known, so that school children in many countries have been exposed to its materials and to its rubatos.[135]

Some more unusual applications of rubato occur in pieces by Kurt Weill and the Austrian composer Ernst Toch. In his Cello Sonata from 1920 Weill indicates *rubato quasi 3/4* for the four-fold repetition of a musical motive notated, somewhat like the *imbroglio* in Ex. 5.8, in three measures of 4/4. The rubato is apparently intended to emphasize the triple quality even more intensely than would have been indicated by an actual change of metre.[136] In his *Circus Overture* of 1953 Toch includes a 'clown episode' which is to be 'flexible—tempo rubato along the lines indicated here'. He indicates first a scene in which one clown slowly climbs a ladder, then falls down as he attempts to rescue another clown. Then the clowns quarrel, alternating quintuplet motives marked 'poco rubato'.[137]

Among Soviet composers, Prokofiev and Shostakovich indicate rubato very seldom. Kabalevsky uses it somewhat more often, especially for continuous sections in a reciting, improvisatory, or cadenza style. It occurs far more frequently and for a greater variety of situations in the works of Khachaturian. Bacewicz from Poland writes *poco meno rubato* one bar after *rubato*,[138] and Janáček from Czechoslovakia has some brief rubatos whose duration is precisely indicated by dashes. From France come a few by Poulenc and Milhaud, from England several by Vaughan Williams and Britten.

In the Western hemisphere a few rubatos appear in works by Villa-Lobos from Brazil and Ginastera from Argentina. In the United States they occur occasionally in works by Hanson, Barber, Gershwin, Menotti, and others. John Cage sometimes indicates within his metronome marking a tempo range for the rubato.[139] Roy Harris's *David Slew Goliah* begins 'Quasi rubato', referring perhaps to the syncopations and triplet figures already notated in the melody for tenor

---

[135] For Orff's concept of rubato performance, compare the piano/vocal score of *Die Kluge* (Schott, 1942) with the recording, conducted by Wolfgang Sawallisch, which he supervised on Angel Records, Album 3551 B/L (35389–90).

[136] *Sonata for Violoncello and Piano* (European American Music Corporation, 1976, 1982), 1st movt., mm. 113 and 162.

[137] Full score (New York: Mills Music, 1954), mm. 90–142.

[138] *Sonata II* from 1953, repr. in *Historical Anthology of Music by Women*, ed. James R. Briscoe (IUP, 1987), 301, mm. 33–5.

[139] See, for example, his String Quartet from 1949–50 (New York: Henmar Press, 1960), beginning: half note is '54 (rubato: 48–60)'.

soloist. Later in the piece a chorus accompanies a solo speaker, whose rhythm is notated and described as follows:

To be spoken freely, as a preacher exhorting his congregation. Not to be spoken metrically—meter is used only to keep the chorus with the speaker. A good deal of rubato must be allowed the speaker, and the conductor must move the chorus accordingly—as undulating waves of sound.[140]

Aaron Copland sometimes uses rubato in more innovative ways. The second movement of his Piano Concerto from 1926 begins with two measures for solo piano marked *molto rubato*. The passage is notated in 3/4 metre, but the motives, consisting of two eighth notes in the right hand, suggest 6/8, for they begin on eighth beats 1 and 4 with an accent sign on the first note. In addition, there is a comma between the two notes of each motive to indicate a breath pause, which is emphasized in the two-piano arrangement of the work by a squared fermata above each comma. While the right hand plays this partially notated rubato on eighth beats 1, 2, 4, and 5 of each measure, the left hand provides accented sforzando notes on beats 3 and 6. Copland seems preoccupied here with the pulse of the eighth note as it struggles to resolve the tension between 3/4 and 6/8.[141]

In the 'Hoe-down' from *Rodeo* Copland presents a section of seventeen measures in a sort of 'oom-pah' rhythm, with bass notes on beats 1 and 3 of 4/4 metre and, for the first eight measures, two sixteenth notes on beats 2 and 4. Pairs of measures alternate between major triads on the first and flatted seventh degrees. The passage, which sounds like an introduction to a melody which never actually appears, is marked in the piano arrangement *rubato—unsteady rhythm*. Copland thus combines two contrary elements: an accompanimental pattern associated with a strict dance rhythm, and the freedom of rubato—here used to mean a deliberately flexible tempo.[142]

A more unusual type of rubato occurs, however, in the instrumental introduction to Act III of *The Tender Land*. As the curtain opens the timpani commences a basso-ostinato pattern spanning two sixteenth-note triplets with the pitches DBD—BBB. After three repetitions, the cellos begin a melody marked *rubato, espress., ardente*, and the timpani ostinato is instructed to 'continue in tempo without regard to conductor's beat' or, in the piano score, to 'continue in tempo without

---

[140] *Folk Fantasy for Festivals*, No. 3 (AMP, 1957).

[141] *Concerto for Piano and Orchestra*, full score (Cos-Cob Press, 1929; repr. later by Arrow Music Press) and the arrangement for 2 pianos by John Kirkpatrick (New York: Cos-Cob Press, 1929). According to Gardner Read, *Music Notation* (1979), 108, the squared fermata indicates a pause of moderate duration, briefer than the regular fermata.

[142] *Four Dance Episodes from Rodeo* for orchestra, 1942 (B & H, 1946), 89; and *Rodeo*, arranged by the composer in 1962 for piano solo (B & H, 1973), 35.

regard to rubato of melody'. In the piano score the rubato is partially determined by '(*accel.*)' written twice in parentheses. In both scores the passage ends with an unparenthetical 'accel.', then 'rit.' The melody lasts for twelve measures of 4/4, each with four notated ostinato patterns. Because of the accelerations in the melody, however, the timpani could presumably play fewer repetitions than notated, and in any case they would not be co-ordinated precisely with the notes of the melody.[143]

This work composed between 1952 and 1954 thus illustrates a special interest in contrasting two different types of rhythm—one strict, the other free. This concept becomes important in the works of a number of composers, especially during the second half of the century. It is part of a broader concept of contrast achieved also on occasion by other musical elements. It is also a part of an approach that seems to characterize American music in particular. Virgil Thomson writes in 1944 that 'the basis of American musical thought is a special approach to rhythm. Underneath everything is a continuity of short quantities all equal in length and in percussive articulation.'[144] In 1946 he states that 'American music ... requires a high degree of metrical exactitude, emphasized by merely momentary metrical liberties.' He speaks of the American type of crescendo, which is not accompanied by acceleration or retard, and adds that 'the separation of these devices is as characteristic of American musical thought as is our simultaneous use of free meter with strict meter and free with strict pitch. These dichotomies are basic to our musical speech.'[145]

In 1952 Copland, after speaking of the rhythmic counterpoint in Elizabethan madrigals, contrasts the American approach as follows:

Our polyrhythms are more characteristically the deliberate setting, one against the other, of a steady pulse with a free pulse. Its most familiar manifestation is in the small jazz band combination, where the so-called rhythm section provides the ground metrics around which the melody instruments can freely invent rhythms of their own.[146]

As early as 1927 Copland had quoted Thomson as saying that 'jazz is a certain way of sounding two rhythms at once, ... a counterpoint of

[143] *The Tender Land*, suite from opera (B & H, 1960), 1st movt.: 'Introduction and Love Music', 6; and piano/vocal score of the opera by the composer (B & H, 1956), 158 in Act III, where 'Martin, restless and unable to sleep, comes out of the shed; he walks uneasily about the yard, then over to the gate.'

[144] *A Virgil Thomson Reader*, with introduction by John Rockwell (Boston: Houghton Mifflin Company, 1981), 224, from the *New York Herald Tribune*, 25 Jan. 1942.

[145] *A Virgil Thomson Reader*, 281, from the *New York Herald Tribune*, 27 Jan. 1946.

[146] *Music and Imagination* (HUP, 1952), 86–7.

regular against irregular beats'. He illustrates by showing eighth notes grouped as 3 + 3 + 2 over four even quarter notes in a strict accompaniment.[147] More characteristic, however, is the rhythmic freedom added by solo performers. Copland describes it thus:

In the Americas . . . the typical feature of our own rhythms was this juxta-position of steadiness, either implied or actually heard, as against freedom of rhythmic invention. Take, for example, the stylistic device of 'swinging' a tune. This simply means that over a steady ground rhythm the singer or instrumentalists toy with the beat, never being exactly *on* it, but either anticipating it or lagging behind it in gradations of metrical units so subtle that our notational system has no way of indicating it. Of course you cannot stay off the beat unless you know where the beat is. Here again freedom is interesting only in relation to regularity.[148]

Copland made use of this concept, as we have seen, by contrasting a rubato melody with a strict ostinato bass in *The Tender Land*. Before exploring this concept further in art music, however, we shall first turn to the manifestations of rubato in American popular music—manifestations which Copland, Thomson, and others felt as a strong influence on their own music.

American Popular Music

The readers of this book will immediately recognize the striking resemblance between the popular performing style described by Copland and the earlier type of rubato defined by Tosi, North, and others. This popular style, which I attempted to demonstrate roughly in Ex. 1 of the Introduction, probably originated in the expressive declamation of text by singers, was then imitated by solo instrumentalists, and reached a climax during the 1930s in the jazz movement known as *swing*.

Those who were familiar with the earlier rubato in classical music were quick to note the similarity and to apply the word to swing. As early as 1936 Percy Scholes writes in the article 'Rubato' in *The Oxford Companion to Music*: 'For an undoubted application in later times of the alleged Chopin principle of rubato, see "Ragtime and Jazz".' In the latter article he describes swing music, current around 1935, as follows:

All such music consists apparently of a *simple harmonic basis* supplied largely by guitars, piano, percussion instruments, etc. (what is called the

---

[147] 'Jazz Structure and Influence', *Modern Music*, 4/2 (Jan.–Feb. 1927), 9 and 11.
[148] *Music and Imagination*, 87.

rhythm section . . .), *with a melodic thread superposed*, this last being assigned to some one instrument (occasionally more)—saxophone, trumpet, etc. The accompanimental-harmonic part is played in strong rhythm, rigid, unvarying; the melodic part (often improvised or so much 'decorated' in performance as to take on an improvisatory character) uses a free rubato. The contrast between the two is piquant . . .

Obviously two conditions necessary to this style are: (a) an accompanimental part using such simple and rarely changed chords that a melody note may anticipate or outlast its legal accompaniment chord without producing a clash . . . , and (b) an assignment of the melody (usually) to a single instrument so that the rubato can be carried out with . . . unanimity . . .[149]

Virgil Thomson writes enthusiastically in 1946 about the 'expressive devices of absolute originality' which 'American ensemble playing on the popular level has given to the world':

One is a new form of tempo rubato, a way of articulating a melody so loosely that its metrical scansion concords at almost no point with that of its accompaniment, the former enjoying the greatest rhythmic freedom while the latter continues in strictly measured time.[150]

Actually, he knew that such a rubato was not precisely *new*, since he had written six years earlier:

Chopin's prescription for rubato playing, which is almost word for word Mozart's prescription for playing an accompanied melody, is that the right hand should take liberties with the time values, while the left hand remains rhythmically unaltered. This is exactly the effect you get when a good blues singer is accompanied by a good swing band.[151]

The word *rubato* continues to be applied to the swing style in some more recent writings. Bradford Robinson speaks of the 'rhythmic phenomenon resulting from the conflict between a fixed pulse and the wide variety of accent and rubato that a jazz performer plays against it'.[152] Hans-Peter Schmitz identifies tempo rubato as one of many elements shared by both jazz and the art music of the Baroque period.[153] Henry Pleasants describes in detail how the appoggiatura,

---

[149] *The Oxford Companion to Music* (OUP, 1938), 779 and 817. He first wrote the article in 1936.

[150] *A Virgil Thomson Reader*, 280, from the *New York Herald Tribune*, 27 Jan. 1946.

[151] *A Virgil Thomson Reader*, 194, from the *New York Herald Tribune*, 26 Oct. 1940. See also Thomas Higgins, 'Chopin Interpretation: A Study of Performance Directions in Selected Autographs and Other Sources', Ph.D. diss. (Univ. of Iowa, 1966; UM 67–2,629), 100: 'it [Chopin's rubato] remains in its healthiest state in certain kinds of jazz.'

[152] *New Grove*, xviii, 416. See also *Harvard Dictionary of Music*, 1st edn. (HUP, 1953), 378, where Lloyd Hibberd describes 'Swing (a word which seems to be of largely subjective import referring to subtle and desirable rubato)'.

[153] 'Jazz und Alte-Musik', *Die Stimmen*, 18 (1949), 498.

mordent, turn, slur, portamento, and rubato appear in popular singing 'exactly as they existed in early seventeenth- and eighteenth-century practice, employed in the same way and for the same purpose: to heighten and elaborate the expression of oral, vocal, linguistic communication'.[154] 'Rubato in particular', he writes, 'relies on a steady accompanying rhythm, enabling the singer to displace notes according to the prosody, or to achieve an idiosyncratic style.'[155] He mentions the application of rubato by Al Jolson, Bessie Smith, Mildred Bailey, Judy Garland, Ethel Merman, and especially Frank Sinatra, whose style he describes by quoting from Tosi.[156] He defines rubato finally as 'a device, now more widely employed by popular than by classical singers, by which time is stolen from one note and given to another, producing in the listener a sensation of suspense as he waits for restitution to be accomplished'.[157] Thurston Dart names still other singers when he describes Chopin's 'true rubato (heard today only in the singing of such artists as Hoagy Carmichael and Dinah Shore), in which the rhythm of the tune sways about the fixed and unchanging metre of its accompaniment'.[158]

This same style was applied also by the instrumental soloists. Louis Armstrong played 'whole phrases that seemed at first to contradict the underlying pulse only to merge with it again'[159]—a practice described on Plate V, as we noted earlier, for Paganini and the elder García. Earl Hines the pianist would similarly '"lose" both rhythm and chord changes within a song only to bring them together perfectly at the end of the phrase or chorus'.[160] Tenor saxophonist Coleman Hawkins would place notes 'ahead of or behind the beat'.[161] 'By playing imperceptibly behind the beat', Pee Wee Russell 'often gave a weighty quality to individual notes' on the clarinet.[162] Lester Young, a per-

---

[154] *The Great American Popular Singers* (S & S, 1974), 35.

[155] *New Grove*, xvii, 346. He also describes how the popular singing style became different from the classical during the 1920s due to the use of the microphone.

[156] *The Great American Popular Singers*, 58, 75, 150, 288, 338–40, and 189–92, respectively.

[157] Ibid. 374.

[158] *The Interpretation of Music* (London: Hutchinson University Library, 1964), 39.

[159] James Dapogny in *New Grove*, i, 600–1. Mark C. Gridley, in *Jazz Styles: History and Analysis*, 3rd edn. (Prentice–Hall, 1988), 69, speaks of Armstrong 'staggering the placement of an entire phrase, as though he were playing behind the beat, a technique called "rhythmic displacement"'; see p. 44 for a description of this device used by earlier singers. Concerning Armstrong, see also Winthrop Sargeant, *Jazz: Hot and Hybrid*, 3rd edn. enlarged (Da Capo, 1975), 57–8; and *New Grove DJ*, i, 29, and ii, 457.

[160] Len Lyons, *The Great Jazz Pianists* (New York: Quill, 1983), 27.

[161] Barry Kernfeld, *New Grove DJ*, i, 557; see 556–7 for examples of the earlier type of rubato incorporated into the ornamental styles of Hawkins, Benny Goodman (clarinet), Roy Eldridge (trumpet), and Teddy Wilson (piano).

[162] James Dapogny, *New Grove*, xvi, 336, and *New Grove DJ*, ii, 404.

former on the saxophone and clarinet, placed accents 'almost imperceptibly before or behind the pulse' and anticipated or prolonged chords by several beats.[163] The phrasing of saxophonist John Coltrane was 'consistently behind the pulse and then for one dramatic instant squarely on top of it'.[164] The trumpet player Miles Davis sometimes allowed 'his melodic line to vary in pulsation—going a little slower at one moment, then making up for lost time by going a little faster immediately afterwards—but ending at precisely the same moment as he would have done by keeping in strict time; this "rubato" is used particularly in slow blues-like improvisations'.[165] With the pianist Erroll Garner 'the two hands did not seem to keep time with each other',[166] and his right hand might 'delay the melody line by almost a full eighth note'.[167]

Most of the jazz experts who describe the instrumental performers do not refer to this rhythmic practice as rubato. Instead, they employ the nomenclature used by the players themselves to describe a player or singer who 'places notes slightly before [or after] the beat, as articulated by the rhythm section or implied by the playing of the rest of the ensemble'. Thus to 'lay back' or play 'behind' or 'after the beat' means to delay intentionally, whereas to 'run away' or play 'ahead of the beat' is to anticipate. To play 'on top of the beat' refers to deliberately or unintentionally placing notes 'slightly before or too precisely on the beat'. 'Drag' also means delay, but often refers simply to improperly controlled rhythm.[168]

The nomenclature of the jazz performer thus depicts rubato exclusively in terms of anticipation and delay. It is only the historian, who sees a connection with the past, and the lexicographer, who seeks a logical etymology, who sometimes describe this rubato as time stolen from one note and given to another. Two recent and influential sources from 1988, however, employ the word *rubato* for a form of the later type which is used in jazz on rare occasions. For Mark

---

[163] Martin Williams, *New Grove*, xx, 580.

[164] Charles M. H. Keil, 'Motion and Feeling through Music', *The Journal of Aesthetics and Art Criticism*, 24 (1966), 346. See 341–8 concerning drummers who play 'on top' of the pulse or 'lay back' behind it.

[165] Avril Dankworth, *Jazz: An Introduction to its Musical Basis* (OUP, 1968), 33 and 69.

[166] Paul Tanner and Maurice Gerow, *A Study of Jazz*, 4th edn. (Dubuque, Ia.: William C. Brown, 1964), 134.

[167] Leonard Feather, *Inside Be-Bop* (New York: J. J. Robbins & Sons, 1949), 57.

[168] See especially the section on tempo in Barry Kernfeld's article 'Beat' in *New Grove DJ*, i, 85–6, as well as i, 8 (ahead of the beat), 92 (behind the beat), and 305 (drag), and ii, 16 (lay back) and 541 (top of the beat, on). See also Robert S. Gold, *Jazz Talk* (Indianapolis: The Bobbs–Merrill Co., 1975), 77 (drag), 161 (lay back), and 228 (run away), or 90, 183, and 258 in the earlier edn. entitled *A Jazz Lexicon* (Knopf, 1964); and Mark Gridley's *Jazz Styles* (1988), 7 and 403 (laid back).

Gridley, author of the widely used textbook *Jazz Styles*, rubato is 'the absence of steadiness in tempo' which is applied sometimes for introductions or endings. Barry Kernfeld defines rubato as 'the stretching or broadening of tempo'. Both authors mention the rubato that may occur in the freer and slower opening statement of a series of instrumental variations on the chorus of a ballad.[169] This is similar to the sort of declamatory freedom employed by the popular singer in rendering the verse which precedes a chorus—the same sort of freedom we have already seen in the preludial forms or recitatives of art music.[170] In addition, Gridley identifies as 'pseudo rubato' an alteration in the density, but not the tempo, of the music due to an increase or decrease of rhythmic activity.[171]

Thus the different types of rubato are almost as confused in the terminology of jazz as they are in descriptions of classical music. It is the earlier type, however—whatever its name—that is a special trait of jazz. Certain jazz elements occur from time to time in works by Stravinsky, Bartók, Hindemith, or Milhaud. But it was the American composer who was surrounded by jazz in his daily life and who intimately experienced its special kind of rubato and the idea of combining two opposite kinds of rhythm, the strict and the free. He heard this rubato in his dance music, popular songs, jazz concerts, and

[169] *New Grove DJ*, i, 85 (section 2 of 'Beat'). Although Henry Pleasants uses *rubato* in the earlier sense in his article on Sinatra in the same dictionary (ii, 454), I presume that the word used elsewhere in this work ordinarily means the later type defined by Kernfeld. See, for example, the *rubato* in connection with John Coltrane (i, 237), Duke Ellington (i, 333: *sempre un poco rubato* on Gunther Schuller's transcription in Ex. 2), 'free jazz' (i, 405), Thelonious Monk (ii, 122), Stan Kenton (ii, 312), Chick Corea (ii, 316), and Bobby Stark (ii, 488).

For Gridley's ideas, see *Jazz Styles* (1988), 33–4, 264–5 (in connection with Keith Jarrett), and 404 (definition). On 33 Gridley unexpectedly states parenthetically: 'Classical musicians ... reserve the term rubato for designating the situation in which the tempo remains steady but part of the time originally occupied by some notes is robbed from them and used to extend the sound of other notes. In other words, classical musicians use the term rubato to indicate that a passage is not played exactly as written and that the relative durations of the notes are rearranged at the discretion of the performer.'

[170] I am indebted to Prof. Alden Ashforth at UCLA for pointing out this use of rubato to me. He is himself preparing a book on *Traditional Jazz in New Orleans* to be published by OUP. He provided me with the following recorded examples of various types of jazz rubato. For the later rubato at the beginning and end of a set of variations on the chorus of a popular song, see Art Tatum's version of Gershwin's 'Someone to Watch over Me' recorded in 1949, the original LP: *Capitol Jazz Classics*, iii (M–11028), and the later CD: *The Complete Art Tatum*, i (Capitol C21K–92866). See also his variations on Gross's 'Tenderly' from 1956, original LP: *Art Tatum Piano Discoveries*, i (20th Century Fox 3029) and the later LP: *The Complete Art Tatum Piano Discoveries*, i (French RCA T607); and Arlen's 'Over the Rainbow', original LP: *Art Tatum Piano Discoveries*, ii (20th Century Fox 3033), later LP: *The Complete Art Tatum Piano Discoveries*, ii (French RCA T608). For the later rubato in the verse of a song and conspicuous application of the earlier rubato in the chorus, see 'Someone to Watch over Me' in *Marni Nixon Sings Gershwin*, recorded in 1985 by Reference Recordings (LP RR19 and CD RR1900).

[171] *Jazz Styles* (1988), 33–4.

musical comedies, and it is therefore not surprising that it finally influenced the composition of his art music.

The New Concertato Style of the 1950s and 1960s

A number of American composers, and eventually some from other countries as well, developed the concept of contrast into a complex aesthetic philosophy involving many other musical elements in addition to rhythm. This development, which began slowly in the early years of the twentieth century, reached a climax after around 1950. It is most clearly seen in the works of Elliott Carter.

Although influence came, as we have seen, from the rhythms of jazz, it may also have come from other sources such as the neo-classical interest in baroque music. The baroque *concertato* style involved the contrast between different sounds, manifested either by alternating relatively brief passages of each or by performing them simultaneously. The verb *concertare*—to join together (in Italian) or to contend (Latin)—implies the interaction between contrasting entities. The instrumental concerto of the later Baroque period is the best known example of the alternation of contrasting sounds. Earlier in the period *concertato* or *concerto* referred to the simultaneous sounding of instruments and voices, especially the solo voice with continuo instruments.[172] This sort of simultaneous contrast involves, it seems to me, a contrapuntal aspect and hence might be called concertato *counterpoint* as distinct from concertato *alternation*.

The contrast itself may involve not only the tone quality of one solo voice or instrument in relation to another, but also the solo performer versus several or many, one group versus another, high register versus low, soft versus loud, and one physical location versus another. It might also involve quite different types of music, as in the aria, where the voice differs markedly from the continuo, or the solo instrumental concerto, where the solo sections differ from those for orchestra. It is the rhythmic aspects of such contrasting types of music, of course, that concern us here. We have already seen that in the aria of the late seventeenth and early eighteenth centuries, the strict, motoristic rhythm of instruments combined in concertato counterpoint with the free flow of ornamentation and tempo rubato in the voice.

Strict and free rhythm may also appear in the new concertato style of the twentieth century, either in counterpoint or in alternation. We

---

[172] Manfred F. Bukofzer, *Music in the Baroque Era* (Norton, 1947), especially ch. 2 on 20–70; also *New Grove*, iv, 627–8, and *NHD* 186.

have already seen both in early pieces by Bartók: the 'Evening in Transylvania' from 1908, in which sections in 'Lento, rubato' alternate with others in 'Vivo, non rubato', and the Second Elegy of 1909, in which a rubato melody in the left hand is accompanied by a variable number of rhythmically strict arpeggio figures in the right. We have also noted Copland's rubato melody against a strict timpani ostinato in *The Tender Land*. In the case of the Elegy and *The Tender Land*, a rubato melody contrasts with an accompaniment in which rhythmic strictness is given an audible presence by the repetition of brief ostinato figures involving equal notes of short value. Such a strict beat becomes even more conspicuous when highly articulated, as it probably is in Copland's timpani ostinato.

Groups of even thirty-second notes recur in Toch's Third Symphony of 1955, where 'the Organ player should repeat figures ad lib. at a steady, even speed, like varying trill figures, unmindful of measured bars, the conductor's beat, or any change of tempo.' The percussion ostinato in Toch's *Chinese Flute* from 1921 involves notes of different value: 'While the percussion section continues playing strictly in time, the Solo Violin and Flute do not adhere to the beat but play freely, approximating the time values as indicated.'[173]

Lukas Foss includes a section in his *Time Cycle* for soprano and orchestra in which a temple block is struck with even beats; in spite of a retard for the orchestra, 'the clock-pulse, as if "outside" the score, continues at its established pace...unperturbed by the course the "music" is taking'. The strict and free elements are finally co-ordinated by means of a fermata-marked rest in the other parts during the last five strokes on the block, as well as a footnote to 'repeat ad lib. until vocal entrance'.[174] He also uses fermatas in a setting of a synagogue chant for the purpose of accommodating concertato counterpoint between strict and free parts. Here the alto section begins each line *non rubato*, and two solo altos 'enter at random' and without exact co-ordination: 'Solo Alto I imitates the rhythm [and pitches] of the Altos, but rubato. The Solo Alto II is also rubato but there is no rhythmic imitation [the pitches occur as even eighth notes in a rhythmic transformation of the melody].' A squared fermata at the end of

[173] *Third Symphony* (New York: Mills Music, 1957), 32 in the 1st movt.; *The Chinese Flute* (London: Schott & Co., 1949), 53.

[174] (New York: Carl Fischer, 1962), No. 3: 'Sechzehnter Januar', footnotes to mm. 86 and 95. In m. 88 'senza rall.' occurs in the temple block simultaneously with 'Poco rall.' in the other parts. In the piano/vocal score of the arrangement for soprano and chamber orchestra (Carl Fischer, 1964), the footnotes are missing and the percussion rhythm is completely notated: beginning with 'poco rall.' in m. 88, the strokes of the temple block have progressively shorter values, which, when combined with the 'poco rall.', presumably produce the same audible strict beat specified in the earlier version.

each *non rubato* line allows the two solo altos to complete their rubato versions before the alto section moves on to the next line.[175]

A different situation, however, occurs in Orff's *Oedipus der Tyrann*, where a solo voice sings *quasi rubato* against repeated percussion patterns marked *sempre senza rub*. In this case, neither a fermata nor a variable number of strict patterns facilitates co-ordination between the strict and the free. Thus, any flexibilities in the tempo of the melody must occur in such a way that they ultimately produce a duration exactly equal to that spanned by the notated accompaniment. The strict, insistent beat of the percussion thus imposes a restriction upon the rubato of the melody, requiring a compensation for every deviation in tempo. Perhaps Orff included the word *quasi* in recognition of this limitation.[176] A somewhat similar situation seems to occur in Hindemith's Concerto for Woodwinds, Harp, and Orchestra, where a passage in imitative counterpoint played by the woodwind quartet provides an insistent triplet eighth beat; at the same time, the solo harpist is to play a long series of unmetred half notes 'without any regularity' and to 'disregard the meter of the other Players'.[177] In his *Circus Overture* of 1953 Toch depicts the roaring of ferocious animals with horns, trumpets, and trombones which ignore the conductor, who 'keeps beating time with utmost precision [for the rest of the orchestra]', and 'play "wild", as it were, against his beat'. An even stream of descending and ascending sixteenth notes by the woodwinds and piano provides very little sense of pulse at first; later, however, when the trombones continue growling by themselves, the other instruments make their strict rhythm more conspicuous by repeating staccato chords in the rhythm of a quarter and two eighth notes.[178]

The examples discussed so far illustrate some of the ways that strict and free rhythm can be united in concertato counterpoint within relatively brief sections of larger movements or works. Steady tempo often accompanies a rubato melody. Sometimes programmatic purpose is involved. Co-ordination may be achieved by fermatas in the free part, a variable number of repetitions in the strict part, or it may, on the other hand, require a balance between acceleration and retard. The projected sense of strictness may vary considerably, but often it is made apparent by a steady reiterated pulse, repeated notes and figures, or staccato articulation.

---

[175] *Lamdeni* (Teach Me) (Paris: Éditions Salabert, 1975), No. 2, 'Wa-eda mah'. Rounded and squared fermatas indicate that one or both alto soloists should 'catch up before continuing'.

[176] *Oedipus der Tyrann*, piano/vocal score (Schott, 1959), 99.

[177] Hindemith, *Sämtliche Werke* (Schott, 1975–), iii/viii, 52–4 in 3rd movt.

[178] Footnote to m. 41; ostinato rhythm commences in m. 60 and 'complete exactness is . . . restored' in m. 68.

Many of these possibilities occur on a more extensive scale in Foss's *Echoi* from 1961–3 for four soloists (clarinet, cello, percussion, and piano). First of all, he uses a special notation in which small eighth notes are faster and unblackened beamed notes are slower than the usual eighth note, and the speed of ordinary eighth notes varies depending on their spacing on the page. At one point he explains in a footnote that 'when a group of small notes begins with a regular size note, a *rubato-accelerando* is implied.' As a matter of fact, some groups include white notes, ordinary eighth notes, and smaller eighth notes under a single beam, with the notes written progressively closer together.[179] Furthermore, black notes without stems and accompanied by a wavy line 'indicate an area of "no coordination": notes should be played once through (rather fast), then in fragments starting at any point...'[180] He summarizes the relationship between the parts as follows:

Exact simultaneity between players is desired only when indicated (dotted line). Elsewhere the music is so composed as to allow each entrance to occur an eighth note earlier than notated. This will make a free delivery possible, and render all coordination via counting impossible.[181]

Lengthy sections, especially in the first and fourth movements, present a conspicuous alternation or combination of strict and free elements. At one point in *Echoi I*, the vibraphone has a long series of even notes marked *non rubato* and not co-ordinated with the other parts. A footnote reads: 'Whenever the player finds himself behind..., play two [eighth notes] simultaneously in order to catch up. If ahead, repeat a group of two notes.' This is followed by the sound of strict staccato eighth notes co-ordinated in all the parts. Next, *tempo rubato* appears in each part, followed by 'even eighths' in the clarinet, cello, and percussion, with the piano part marked 'do not coordinate'. Then the clarinet and cello have '*string.* (do not coordinate)' while the percussion is marked '*senza string.* (As others accelerate, remain in strict time; stop all coordination)'. The movement then ends with the chimes sounding a strict, even pulse on a single pitch, with the instruction: 'Do not count; the number [of written notes] is arbitrary, a mere approximation.'[182]

---

[179] *Echoi* (New York: Carl Fischer, 1964): see the instructions preceding the 1st movt., p. 34 n. 2, and, for a good example combining most of these innovative devices, the *rubato* passages for clarinet and cello on p. 39.

[180] Ibid., instructions preceding the music. I have changed the punctuation in the quotation slightly so that the brackets in the original will not be mistaken for an editorial addition.

[181] Ibid. 3 n. 1.

[182] Ibid., *Echoi I*, 9–11.

In the fourth *Echoi* the clarinet, then the cello have a melody marked *'rubato* (do not coordinate)' to be played against what amounts to a third rubato in the piano indicated by regular and small eighth notes. Later, the cello, then the clarinet have an arpeggiated melody with a 'rubato-accelerando' while the piano repeats an ostinato pattern in strict eighth notes. After two bars 'in tempo (*non rubato*)', the same rubato melody recurs in both clarinet and cello. During each statement of this melody by the clarinet, one of the notes is marked with a fermata. Shortly after this, *poco più rubato* appears in the piano part, with this explanation:

A staggering (drunken) unevenness of eighth notes is particularly apropos when a phrase is repeated. If the identical *poco rubato* is then applied to each repeat, the pianist will achieve the desired 'obsessive' effect, which is best likened to a 'stuck' phonograph needle and which should be made increasingly obvious as the piece progresses.

Later in the movement, strict and free rhythms are co-ordinated as follows:

As the clarinet and cello parts are here following their own measured course, independent of the non-measured piano, it will be necessary for the clarinetist and cellist to either *add* (by repeating the last phrase or a part thereof) or to *subtract* (by omitting a number of notes) in accordance with the piano's progress, in order to stop abruptly when the moment indicated by the arrow has arrived.[183]

Foss thus demonstrates a variety of ways that strict and free rhythm can be combined in concertato alternation and concertato counterpoint. In the recording supervised by the composer, the strict parts are projected very clearly and often emphasized by articulate even notes on the piano. This provides a steady pulse with which the rubatos can either combine in counterpoint or contrast in alternation.[184] Foss discusses his philosophy of free rhythm in an article from 1963. In comments weighing the freedom of the performer against precision of notation, he presents an example of rubato which consists of a series of eight sixteenth notes with 'poco accel.' above the opening notes and 'rit.' over the last two. In a second example he shows what 'the accelerando, ritardando written out would produce': some notes sound sooner, some are grouped differently in a triplet or quadruplet, and this compensating type of rubato occupies exactly the same span

[183] Ibid., *Echoi IV*, 36, 38–9, 40, and 44. See also *Echoi III*, 24, for a series of 8th notes marked '*rubato* (uneven)'.
[184] Epic Record LC 3886, performed by the Group for Contemporary Music at Columbia Univ. and produced by Paul Myers and Lukas Foss.

of time as the notated example without any tempo changes. He comments from one point of view:

This seemingly precise notation puts the performer in a strait jacket. It is a translation of the supple into the realm of the rigid. A rigid rubato: contradiction in terms. Imagine asking the performer to feel a moment 'out of time', as it were, when it is notated slavishly 'in time'.

On the other hand, he feels, as a composer, that 'it is our traditional notation's ability to translate subtleties like a rubato into measured exactitude which makes it a highly developed tool.' This leads him to consider music in which 'the instruments or voices either individually or in groups, act and react to and against one another like characters in a play'. This concept, which he himself employed in ensemble improvisation, can be developed into 'a veritable polyphony of musics, with each music independent of the tempo and pulse of the other'.[185]

When concertato counterpoint involves an entire work, perhaps with performing forces spatially separated, such a 'polyphony of musics' creates a more complex relationship than we have noted in the examples above. As early as 1906 Charles Ives had experimented with such multiple musical forces. *Central Park in the Dark* contrasts the strict adagio tempo of a string orchestra, which represents 'night sounds and silent darkness', with the simultaneously changing tempos of the other instruments, which depict sounds made by human beings. *The Unanswered Question* contrasts three entities: strings, representing silence or the 'harmony of the spheres', which play with no change of tempo or dynamics; a solo trumpet, which asks the question always 'in the same tone of voice'; and a flute quartet, to represent human beings attempting, but ultimately in vain, to answer the question. The quartet, which becomes faster, louder, and more animated with each entrance, 'need not be played in the exact time position indicated; it is played in somewhat of an impromptu way'. These three forces are heard sometimes in alternation and sometimes together in pairs (the trumpet or the quartet with the strings).[186] Henry Brant writes that 'there is complete contrast between the three elements: in tone quality, tempo (which includes speedups, retards and rubato), meter, range, harmonic, melodic, and contrapuntal material.' In spite of the variability in the location of entrances, Brant does not consider this

---

[185] 'The Changing Composer–Performer Relationship: A Monologue and a Dialogue', *PNM* 1/2 (Spring 1963), 48–50; also in *Contemporary Composers on Contemporary Music*, ed. Elliott Schwartz and Barney Childs (New York: Rinehart & Winston, 1967), 329–30.

[186] *Two Contemplations for Small Orchestra*, No. 1: *The Unanswered Question* (New York: Southern Publishing Co., 1953), and No. 2: *Central Park in the Dark* (Hillsdale, NY: Boelke-Bomart, 1973). Both contain notes by Ives concerning the performance.

aleatoric or chance music: 'When non-co-ordinated rhythm is combined with spatial distribution, accident is no more a factor than it is in the performance of rubato in a complex Chopin ratio [he gives an example of a twenty-note group in the right hand against six notes in the left from the Nocturne Op. 9 No. 1].'[187] He feels that 'spatial separation is essentially a contrapuntal device; it makes counterpoint more distinct. It destroys harmony and ensemble, and it enhances the polyphonic contrast between widely separated groups.'[188] As a composer, Brant himself favours strict tempos, even when two independent forces are combined in concertato polyphony.[189]

In 1930 Henry Cowell described mathematical methods for combining parts in different tempos, or combining a part that accelerates or retards with one in steady tempo, or a part that accelerates with one that retards.[190] The composer Chou Wen-Chung later likened this procedure to the Indian practice of *talas*, in which 'the soloist and the drummer carry on an improvised rhythmic interplay, synchronizing only at the . . . initial beats'.[191] Cowell also mentioned that simpler cross-rhythms had formerly resulted from different note values and that 'Chopin in notating his *rubato* improvisations [presumably as in the Nocturne Brant quoted] hit on some extraordinary combinations'.[192] Unlike Brant, however, Cowell does not avoid rubato. He includes it as one element of contrast in his *26 Simultaneous Mosaics*. Here five solo performers play different pieces at the same time, and

---

[187] *Contemporary Composers on Contemporary Music*, 234–5. Brant's example comes from m. 73 of Chopin's Nocturne, and he adds *senza rigore* for the last 8 notes in the right hand.

[188] Cole Gagne and Tracy Caras, *Soundpieces: Interviews with American Composers* (Metuchen, NJ: Scarecrow Press, 1982), 60 (from a conversation in 1980); see also 57–8 concerning the 'polyphonic complexity' in Ives's *Unanswered Question*.

[189] He specifically rules out rubato in some of his early works. In the preface to *Variations for Four Instruments* (1931), he states concerning 'rubatos, dynamics and the like': 'None of these things have any particular relation to the character of the music; the detailed consideration of them which is customary might be omitted entirely in performance.' Preceding the 1st of his *Two Sarabandes for Keyboard Instrument*, he writes: 'The player should be careful to avoid any such "expressive nuances" as gradations of tone, rubatos, etc. There is no "phrasing" or accentuation of any type; the playing should be an even, cool, unrelenting legato throughout.' In his spatial music after 1953 Brant still indicates strict tempos, even though he contrasts metres (as in *Verticals Ascending* from 1967, with the 2 groups in 3/4 and 4/4) or degrees of co-ordination (as in *Ice Age* of 1954, where exactly co-ordinated passages alternate with 'independent' ones in which parts enter at variable moments, but maintain a steady tempo).

[190] *New Musical Resources* (Knopf, 1930; repr. 1969), 90–8.

[191] *Contemporary Composers on Contemporary Music*, 313–14 and n. 16. See his composition *Cursive* for flute and piano (New York: C. F. Peters, 1965), mm. 56, 68, and 98, where the 2 instruments have different metronome markings and one accelerates while the other retards; arrows show where the parts should coincide. See 'Variable Tempo' on the 2nd page of the instructions preceding the score.

[192] *New Musical Resources*, 108. See Cowell's own composition *Set of Five* for violin, piano, and percussion (New York: C. F. Peters, 1968), 3rd movt., m. 48, where *rubato* appears with a passage in which the right hand of the piano part has 13 notes against 6 in the left.

one of them is marked 'Adagio rubato'. Thus free rhythm heard simultaneously with strict becomes one of the many elements of contrast in this complex polyphony of many musics.[193]

Carter and the Concertato Style

The rhythmic aspects of the twentieth-century concertato style thus take many forms. Strict and free rhythms contrast for brief periods, as we have seen, in works by Bartók, Copland, Toch, Hindemith, and others. This concept occurs even more extensively in works by Foss and combines with still other elements of rhythmic contrast in the multiple musics of Ives and Cowell. No composer, however, draws these rhythmic aspects all together with a more consistent aesthetic philosophy than Elliott Carter—and no one describes them more clearly. Carter began to develop his ideas of concertato rhythm around 1945. He was concerned with many elements of contrast, of course— the sounds of different instruments, their location in relation to each other, the notes they play, and the way they play them. He had a special interest, however, in rhythm and describes the influences coming to him from jazz, from non-Western sources (the Indian *talas*, as well as Arabian, Balinese, and African music), from early fifteenth-century Western practice,[194] and from later composers such as Skryabin, Ives, Brant, and Cowell.[195]

The listener perceives two essentially different rhythmic textures in Carter's music: the pulsed and the unpulsed. In the first, pulses are articulated as a series of down-beats, either steadily (*sempre giusto* or *meccanico*) or with an acceleration or retard. Although the music is notated with barlines and time signatures, the metre is not to be

---

[193] (New York: C. F. Peters, 1963).

[194] Referring, no doubt, to the 'manneristic' style of the late 14th and early 15th centuries, characterized by a superlative degree of rhythmic contrast between voices. Concerning the complexities of rhythm and notation during this period, see Richard H. Hoppin, *Medieval Music* (Norton, 1978), 477–87, as well as the modern transcription of Baude Cordier's rondeau 'Amans ames' in *HAM* i, 51 (No. 48). Regarding the special interest during the Medieval, Baroque, and Modern periods in differentiating polyphonic voices in various ways, both rhythmic and otherwise, see my article 'A Theory of Alternating Attitudes in the Construction of Western Polyphony', *The Music Review*, 34 (1973), 221–30.

[195] *The Writings of Elliott Carter: An American Composer Looks at Modern Music*, ed. Else and Kurt Stone (IUP, 1977), 162–6 from 'The Rhythmic Basis of American Music', *The Score*, 12 (June 1955); 270 from the programme notes for the Nonesuch recording in 1969 of the *Sonata for Cello and Piano* and *Sonata for Flute, Oboe, Cello, and Harpsichord*; and 347 from 'Music and the Time Screen' in *Current Thought in Musicology*, ed. John W. Grubbs (UTP, 1976). Concerning non-Western influence, see also Allen Edwards, *Flawed Words and Stubborn Sounds: A Conversation with Elliott Carter* (Norton, 1971), 41–2 and 91.

heard. There are frequently different pulse tempos in different voices, and their relationship is varied through a process of 'tempo modulation' in which the value of a pulse is renotated as a new value with a changed metronome marking. In the opposite texture, indicated by *espressivo*, *rubato*, or *quasi rubato*, 'the player must stretch the implied pulse expressively' so that the music sounds 'freer and warmer'. In this case, one hears a smoothly flowing line rather than accentuated pulses. Both textures can occur in concertato alternation when an improvisatory style is appropriate, as in a solo cadenza which moves 'in and out of a pulsed feeling with flamboyant abandon'.[196]

Although such alternation of rhythmic textures occurs often in Carter's music, it is the 'simultaneous oppositions' or 'superimposed contrasts'—what I am calling 'concertato counterpoint'—that is most characteristic and most complex.[197] Most important for us, of course, is the contrapuntal contrast between the strict and the free—the steady pulses in *tempo giusto* versus free and flowing rubato. Carter first employs this technique in 1948 in his *Sonata for Violoncello and Piano*. The opening movement establishes 'two simultaneous, but differently characterized, planes of music':

Here the clock-like regularity of the piano is contrasted with the singing, expressive line of the cello, which, although accurately written out, sounds as free from the underlying beat as the jazz improvisor from his rhythm section—the musical situation which suggested this passage.[198]

Commencing in the third measure, the piano provides a sharp, even, staccato pulse on every quarter beat without regard to the 4/4 metre. In counterpoint with this steady ticking, the cello plays an *espressivo* melody, which begins as in Ex. 10.6*a*, and which is to be played *quasi rubato*—that is, following the notated rhythm, but sounding as though free. Comparison of this notated rubato with my imagined unrobbed version in Ex. 10.6*b* reveals the same sort of anticipations and delays of melody notes that we saw in the earlier type of rubato.

Later in the piece, the metre changes to 5/4, but with the quarter-

---

[196] See David Schiff, *The Music of Elliott Carter* (London: Eulenburg Books, and Da Capo, 1983), the book recommended to me by Carter himself. Concerning rhythm, see ch. 3 on 'Musical Time: Rhythm and Form', especially 25–32, from which I have drawn the quoted passages. He explains tempo modulation on 26–8. See Carter's explanation in *The Writings of Elliott Carter*, 349–50 from 'Music and the Time Screen'.

[197] See Schiff, *The Music of Elliott Carter*, 13; the book begins with the sentence: 'Elliott Carter makes music out of simultaneous oppositions.' For the other term, see David Schiff, 'Elliott Carter at 80' in The Grove Music Society's *Encore*, 3/3 (1988), 1.

[198] *The Writings of Elliott Carter*, 196 from a radio lecture commissioned by the Fromm Foundation. See also 245, 271–2, and 349, as well as Benjamin Boretz, 'Conversation with Elliott Carter', *PNM* 8/2 (1970), 18–19, and Schiff, *The Music of Elliott Carter*, 25–9: 'The strung-out cello line at the opening of the Sonata sounds like Lester Young.'

Ex. 10.6 Delays and anticipations in Carter's Cello Sonata, 1st movt., 1948

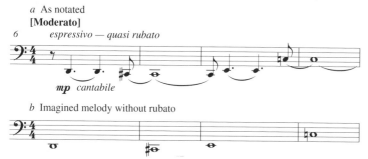

Ex. 10.7 Rhythmic relationships in Carter's Cello Sonata, 1st movt., 1948

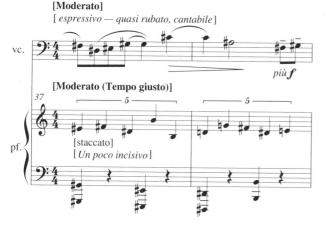

beat pulse remaining the same; then there is a tempo modulation back to 4/4, but with the new quarter-note quintuplet equal to the previous five quarter beats. Now, as in Ex. 10.7, each note in the top staff of the piano part still has the same pulse rate as each quarter note in the 5/4 measures or in the 4/4 measures that began the piece. The left hand produces displacement and polyrhythm with the right—the sort of rhythmic counterpoint which Brant and Cowell noted in the unusual groups of Chopin and which C. P. E. Bach, as we recall, considered rubato. The cello melody, however, relates in quite a different way with the piano rhythms, for its notes do not coincide with the piano's at the barlines or anywhere else. This relationship is polytextural in the rhythmic sense, for it juxtaposes two rhythmic textures—the strict and the free—somewhat in the manner of the early vocal rubato, but spanning a far greater length of time.[199] In the Cello Sonata, however, the two instruments are equally important and the exposition of their differences constitutes the central idea of the piece.

The Second String Quartet includes a similar procedure: at the opening of the fourth movement the first violin is marked *sempre quasi rubato*, and its notes never coincide with the strict pulses in the *Tempo giusto* of the viola and cello. Elsewhere in the same work Carter combines free and strict rhythm in other ways. 'In the Cello part', according to the Prefatory Note, 'as well as in the coda of the Fourth Movement ... various kinds of *rubati* are indicated.' The indications in the coda consist of '(*sempre accel.*)' or '(*sempre accel. molto*)', a metronome number for every measure, and tempo modulations. Rubato in the cello part involves 'various gruppetti of [from four to around twenty] accelerating and retarding notes' indicated by a dotted arrow curved like a slur:

In all cases ... the first note-value, over which the arrow starts, and the last, to which it goes, are to be played in the metrical scheme in which they occur. The intervening notes are to be played as a continuous *accelerando* ([or] ... *ritardando*), the notation indicating approximately whether the *accelerando* (or *ritardando*) is regular, or more active at the beginning or end of the passage.

Such accelerating groups are particularly evident in the 'Cadenza for Cello' in the second movement, where they both alternate and occur simultaneously with steady, articulated pulses marked *molto giusto* in the first violin. At other times, the strict, accelerating, retarding, or free rhythm is written into the note values themselves.[200]

The *Piano Concerto* of 1965 contrasts a *ripieno* orchestra with a *concertino* chamber group. In the second movement 'the orchestra should play its lines of short, accented notes very regularly and

---

[199] Charles Rosen, in *The Musical Languages of Elliott Carter* (Washington, D.C.: Library of Congress, 1984), 11, suggests the word 'polytempo' to describe the rhythmic relationship between cello and piano, in distinction to 'cross-rhythm'; see also 36–7 and 43. See Schiff, *The Music of Elliott Carter*, 28 (2nd paragraph) and 26 (at top): 'Its [the cello's] line must be played not as a series of syncopated off-beats but as an independent rubato shape which irregularly speeds up and slows down.' The excerpts in Exx. 10.6 and 10.7 are the same in both the original version (New York: G. Schirmer for the Society for the Publication of American Music, 1953) and the corrected edn. (AMP, 1966). For a recording of the work, see Nonesuch H–71234 (1969); Carter's notes are included also in *The Writings of Elliott Carter*, 269–73.

[200] *String Quartet No. 2*, 1959 (AMP, 1961), pp. ii–iii and mm. 243–73, 427–40, and 563–87. I have reversed the order of the two sentences in the quotation. See Schiff, *The Music of Elliott Carter*, 29–36 (including Exx. 3–9) and 198–9; and *The Writings of Elliott Carter*, 247. The work was recorded on Nonesuch H–71249 in 1970 by The Composers Quartet, with notes by Carter (printed also in *The Writings of Elliott Carter*, 274–9), and on Columbia M32738 in 1974 by the Juilliard Quartet.

emphatically' while 'the concertino, including the piano, should play their parts with an expressive rubato as suggested by the notation, staying, however, within the general metric scheme'. The orchestra opens the movement 'MOLTO GIUSTO *sempre*' and is joined in the fourth bar by the concertino playing 'QUASI RUBATO *sempre*'; in this case the simultaneous 'contrast between the *tempo giusto* of the ripieno...and the *quasi rubato* playing of the concertino and piano...should be as defined as possible'. In a later section, 'the concertino (including the piano) should play with a slight rhythmic liberty', a liberty sometimes suggested by dotted arrows preceded by 'A' or 'R' (for accelerando or ritardando). Sometimes different accelerating groups sound simultaneously in different instruments, and sometimes an accelerating group coincides with a group that is retarding. At one place where this sort of counterpoint is most complex, there is a note for the pianist to 'play each of the three or four contrapuntal lines with a different type of rubato'. Later another note reads: ' "QUASI CADENZA" applies only to Bass Cl., orchestra remains in TEMPO GIUSTO'. Still later, 'the soloist should accelerate gradually while the orchestra maintains a strict, regular tempo'; at the end of the passage, the length of a tremolo in the piano 'is determined entirely by the time it takes the orchestra to reach this point, playing in strict tempo'. This work thus includes an impressive variety of elements contrasted in concertato counterpoint: *tempo giusto* versus *quasi rubato*, *quasi cadenza*, or *accelerando*; an acceleration versus a retard; one acceleration versus another in a different instrument; and one type of rubato versus other types played by a single performer.[201]

He treats the four instruments in his Third String Quartet in two

---

[201] *Piano Concerto*, full and 2-piano scores (AMP, 1967), pp. iv–v and mm. 352, 375, 379–86, 438, and 522. See *The Writings of Elliott Carter*, 297–8 and 356–7; Schiff, *The Music of Elliott Carter*, 236–7; and Rosen, *The Musical Languages of Elliott Carter*, 37.

For a more extensive example of a simultaneous acceleration and retard, see his *Double Concerto for Harpsichord and Piano with Two Chamber Orchestras*, 1961 (AMP, 1964), 108: 'Beginning with m. 453, the Piano alone observes an *accelerando poco a poco* and, consequently, is not coordinated with any other instrument. Simultaneously, the remainder of the orchestra, including the Harpsichord, observes a *ritardando poco a poco*. All instruments are coordinated again at m. 466 (*Tempo giusto*).' During a portion of the slow movement, the wind instruments are notated in such a way that they *sound* in even notes while other instruments are accelerating. Elsewhere long retards (mm. 359–74) and accelerations (mm. 379–402) achieve a vivid presence by means of evenly notated note values that provide an audible and articulated pulse. See the description of this work by Rosen, who played the piano part in the first performance in 1961, in *The Musical Languages of Elliott Carter*, 11 and 23–31.

The *Variations for Orchestra* of 1954–5 (AMP, 1957) also include a long ritardando (Variation 4, mm. 213–58) and accelerando (Variation 6, mm. 289–360), both made clearly audible by notated even notes. Concerning both the *Double Concerto* and the *Variations*, see *The Writings of Elliott Carter*, 246–7 and 356–7. For a recording of both, see Columbia MS7191 from 1968, with programme notes by Carter.

pairs, with Duo I (violin and cello) 'playing quasi rubato throughout' and Duo II (violin and viola) 'playing in quite strict rhythm throughout'. According to the Performance Notes, 'suggestions . . . as to how to interpret the indication *quasi rubato* of Duo I have been given in parentheses in various places where Duo I plays alone [(*poco esitando*) and (*a tempo*) in one section, (*poco allarg.*), (*poco accel.*), and (*a tempo*) in another]. A similar flexibility can be observed in the passages when both Duos are playing, provided the strict tempo of Duo II is not disturbed and the harmonic structure relating the two Duos is not destroyed'—a restriction also applying, as we have seen, to the earlier type of rubato and requiring a compensating give-and-take in the free part. Duo I has four 'movements' and Duo II six, all of which are broken into segments and shuffled together in one continuous piece. Strict pulse is most evident in Duo II's two movements marked 'Pizzicato giusto, meccanico'. The entire work concludes with several alternations between 'Tempo giusto' in all the parts followed by 'Tempo rubato'.[202]

In another composition, the *Duo* from 1974, the violin plays in a 'rubato, rhythmically irregular style, while the piano constantly plays regular beats, sometimes fast, sometimes slow'. Near the end of the work the pianist plays even, staccato pulses in the left hand and the violin plays according to these instructions: '*quasi rubato*—approximate note-values; passages [of three or four notes] in brackets should be coordinated with the piano as written.' Later Carter writes: 'From here to the end the violin should play somewhat freely, but the regular attacks of the piano should come as indicated.'[203]

In some works concertato contrast consists mainly in the alternation rather than the simultaneity of free and strict rhythm. In the first movement of the *Sonata for Flute, Oboe, Cello, and Harpsichord* passages with a clock-like ticking of a steady pulse alternate with others marked *rubato* or *sempre rubato*.[204] In the *Concerto for*

---

[202] *String Quartet No. 3*, 1971 (AMP, 1973), Performance Notes preceding p. 1, as well as m. 1: 'Maestoso (giusto sempre)' in Duo II and 'Furioso (quasi rubato sempre)' in Duo I; mm. 28–38 and 96–105: parenthetical suggestions for quasi rubato; also mm. 363, 462, 477, 479, and 480. See Schiff, *The Music of Elliott Carter*, 260–1; Rosen, *The Musical Languages of Elliott Carter*, 14; and the recording by the Juilliard Quartet on Columbia M32738 (1974), with programme notes by Carter (also in *The Writings of Elliott Carter*, 320–2). For a repr. of the 1st page of the composer's MS, with 'quasi rubato sempre' at the beginning, see the cover of *Elliott Carter: Sketches and Scores in Manuscript* (New York: The New York Public Library and Readex Books, 1973).

[203] *Duo for Violin and Piano* (AMP, 1976), mm. 340 and 358. See *The Writings of Elliott Carter*, 329–30 from the programme notes for the recording on Nonesuch H–71314 (1975) by Paul Zukofsky (violin) and Gilbert Kalish (piano).

[204] (AMP, 1962), cf. mm. 1, 16–20, and 32–4 with mm. 9–15 and 30. See Schiff, *The Music of Elliott Carter*, 166.

Orchestra 'certain sections, especially the cello . . . , timpani, tuba, and bass solos . . . , should be played with enough rubato to bring out their character. Since a considerable amount of this is written into the parts, the interpretive rubato should not be exaggerated.'[205] The *Brass Quintet* of 1974 includes brief alternations between *rubato* and *a tempo* in the solo sections.[206] This, however, is simply the usual sort of rubato found in the works of many other composers and in Carter's own compositions from an earlier period.[207]

Carter's music presents a veritable catalogue of concertato rhythmic possibilities. Rubato, signifying to Carter irregularity of speed or fluctuating tempo, may contrast with strict pulses, or one rubato may contrast with another. Accelerations and retards—also a part of Carter's rubato—may contrast with each other or with steady tempos. All of these contrasts may occur simultaneously or in alternation. Rosen comments that 'these contrasting temporal perspectives are not simply opposed but are combined in ways that rarely allow one of them to become dominant'.[208] Carter has used a variety of methods to notate these rhythmic effects. Sometimes he simply writes the word *rubato*, in which case the performers add an irregularity to the notated rhythm. At other times he indicates *quasi rubato* and writes into the note values most of the desired flexibility. He sometimes uses the words *accelerando, ritardando, esitando,* and *allargando,* often accompanied by exact metronome numbers for every measure or so. Occasionally dotted arrows will indicate acceleration or retard. Sometimes brackets show where parts synchronize together. Often he includes a prose explanation within the score in a footnote, or in the performance instructions preceding the score.

In the later works that are available to me, however, terms such as *rubato, ritardando,* and *accelerando* are less frequent and sometimes missing altogether. One can still recognize the *quasi rubato* type of notation in Ex. 10.6*a*, however, even when not labelled as such.[209] In

---

[205] (AMP, 1970), mm. 29–125 and 289–418.

[206] (AMP, 1976), mm. 246–7 and 280.

[207] See his *Elegy*, originally composed in 1943 for viola or cello and piano. The arrangements in 1946 for string quartet (New York: Peer International Corporation, 1958) and in 1952 for string orchestra (Peer, 1957) include '*rubato* (*a tempo*)' in m. 44 for a melody ascending by leaps over a held chord; the previous bar is marked *ritenuto*. In the later revision in 1961 for viola and piano, both markings are missing. Similarly, compare the rubatos in his *Piano Sonata* from 1945–6 (Bryn Mawr, Pa.: Mercury Music, original edn. 1948, rev. edn. 1982), with the holograph score with corrections from 1946 (photocopy in the UCLA ML). See Rosen, *The Musical Languages of Elliott Carter*, 7–11, and his performance and programme notes on Epic LC 3850.

[208] 'Happy Birthday, Elliott Carter', *New York Review of Books*, 35/19 (8 Dec. 1988), 24.

[209] See, for example, *A Mirror on Which to Dwell* (AMP, 1977). Schiff, in *The Music of Elliott Carter*, 291, describes the contrasting rhythms in 'Sandpiper': the oboe has the 'same

one case he writes 'quasi accel.' over a passage in which the note values themselves get progressively shorter.[210] Even in these later works in which tempo fluctuation has apparently been completely incorporated into the note values, Carter continues his systematic exploration of concertato contrast. Titles such as *Triple Duo* and *Penthode* reveal the number of performing forces in the concertato drama.[211]

Curiously, the new concertato style of the twentieth century, especially in its contrapuntal form, has not yet acquired a generally accepted name. Various authors refer to 'superimposition', 'stratification', 'differentiation', 'multi-layered', 'simultaneous oppositions', 'planes of music', or, as we have seen, 'a polyphony of musics'. The concept has become widespread, especially during the second half of the century. It is applied in essentially the same ways it was during the Baroque period. It bears similarity also with the stratification and differentiation of voices in medieval music, and hence becomes one of those aesthetic ideas that recur from time to time in music history.

Boulez and Stockhausen

Some composers of the next generation also explored the concertato combination of free and strict rhythm. Rosen contrasts Carter's concept of time with that of some other composers:

It has been said that just as Schoenberg destroyed the sense of a unique harmonic center, so Carter has done away with the idea of a single rhythmic framework: in his music, time is no longer measured by a single yardstick or by a uniform beat. However, he has not destroyed the sense of uniform objective time by a continuously fluctuating rubato, as in Boulez, or by recourse to the play of chance, as Cage and Stockhausen preferred to do, but by organizing conflicting subjective perceptions of time into a dramatic form in which they each simultaneously find free and individual play. Carter's works embody the multiform experience of time in modern life.[212]

---

note-lengths throughout', while the strings and piano are 'rubato—many changes in tempo'. Although Carter never writes the word *rubato* in this score, one can also recognize his *quasi rubato* in the many tied notes that carefully avoid coinciding with pulsed notes: see the voice part at the end of the 1st song at mm. 61–5, for example, or the beginning of the 5th.

[210] *A Mirror on Which to Dwell*, 4th song: 'Insomnia', m. 16.

[211] *Triple Duo* (New York: Hendon Music, 1983) contrasts 3 different pairs of instruments; *Penthode* (Hendon, 1985) involves 5 groups, each with 4 players and each with its own combination of instruments. In both of these works 'rit.' and 'accel.' are completely missing, although one still finds 'espr.' or even 'molto espr.' occasionally.

[212] 'Happy Birthday, Elliott Carter', *New York Review of Books*, 35/19 (8 Dec. 1988), 24.

Comparing Carter specifically with Boulez, Rosen writes:

Carter's music demands a small-scale freedom and rubato even within the most complicated works, but the large scale is always held within a tight frame. A sense of nonmeasured time in his work results from a sense of conflict among various concepts of time that are all realized simultaneously. The music of Pierre Boulez . . . achieves a sense of nonmeasured time, but in a way that is less demanding. Boulez's music generally demands to be played with a floating beat: the rhythms must be fairly exact on a small-scale, but the music is most often conceived with a floating tempo within which there are continuous ritards and accelerandos.[213]

Pierre Boulez contrasts this 'floating' effect with strict or pulsed rhythm. In an article from 1957 on the role of chance in music, he speaks of the 'rigorous' versus the 'interpreted' aspects of a composition, the latter involving a new sort of rubato—one no longer mainly 'suppleness of articulation', but one in which the performer is deliberately made to lose control of the tempo in an attempt to include grace-notes, ornaments, or rapid changes in register, dynamics, attack, or number of simultaneous notes.[214] Boulez eventually defines more precisely his two categories of musical time: the free or floating type is *temps amorphe*—amorphous or smooth time in which 'time is filled without counting'; the strict, pulsed time is *temps strié*—striated time in which the pulse may be regular or irregular, metrical or not, and in which 'time is filled by counting'.[215]

Many of his compositions present a concertato juxtaposition of these two types of rhythm. He describes the oscillation in his *Livre pour quatuor* of 1949 between 'rigid' or 'austere' passages and movements and others which are 'relatively supple and flexible' and in a quasi-improvisatory style. These two rhythmic styles make a contrast which he feels is fundamental to his music.[216] The second movement of this work includes three brief sections marked 'Vivo rubato' and one marked 'Rubato'; in each case he specifies the range of tempo by

---

[213] Rosen, *The Musical Languages of Elliott Carter*, 14. I have reversed the order of the sentences; Rosen's comments on Boulez immediately precede those on Carter. Concerning *Le Marteau sans maître*, see Stravinsky's remark in Craft, *Conversations with Igor Stravinsky* (1959), 122: 'You are never in a tempo but always going to one'; and Boulez's comments from 1972–4 in *Conversations with Célestin Deliège*, original French version (Paris: Les Éditions du Seuil, 1975), English trans. (London: Eulenburg Books, 1976), 68.

[214] *Notes of an Apprenticeship*, ed. Paule Thévenin, trans. Herbert Weinstock (Knopf, 1968), 41–2 from 'Alea', *Nouvelle revue française*, 59 (1 Nov. 1957); and 49 on performing and conducting strict and free music simultaneously. See also *Conversations with Célestin Deliège*, 69.

[215] *Boulez on Music Today*, trans. of the original French and German edns. of 1963 by Susan Bradshaw and Richard Rodney Bennett (Faber, 1971), 88–9 and 94.

[216] *Conversations with Célestin Deliège*, 53–4; see also 13.

showing a half note preceded by a downward arrow pointing to the metronome number 88 and followed by an upward arrow and the figure 108. For other sections of the work marked *très libre et mobile*, he gives three metronome numbers separated by a downward and upward arrow.[217]

In the first of his *Notations* for orchestra a passage marked 'Tempo rubato' shows for each of four consecutive measures an upward arrow labelled 'poco acc.' followed by one pointing downward and marked 'poco rall.'[218] In other works he distinguishes free and pulsed rhythm by instructions such as 'il faut comme improviser' on the one hand, and 'strict dans les relations rythmiques' on the other.[219] Although he seldom writes the word *rubato* in his scores, his own association of the term with his floating *temps amorphe* has encouraged analysts to employ it in descriptions of his works. Thus, in the second movement of his First Piano Sonata from 1946, sections of strictly pulsed rhythm alternate with others in a 'continuous written-out rubato', and 'there are effects of rubato throughout almost every movement of *Marteau*'.[220] In addition, Boulez sometimes presents contrasting rhythmic styles in counterpoint. Groups of spatially separated instruments may play similar music but be out of phase with each other,[221] or one group in strict time may sound simultaneously with another in flexible tempo.[222]

Karlheinz Stockhausen also contrasts different rhythmic effects. In his *Zeitmasse* (Tempos) from 1955−6 for five woodwinds, one instrument sometimes retards while another accelerates; still another

---

[217] *Livre pour quatuor* (Paris: Heugel, 1982), II, mm. 247, 259, 264, and 270; III−a, m. 34.

[218] *Notations pour orchestre*, the recomposition of earlier piano pieces, No. 1 (London: Universal Edition, 1978), 5.

[219] *Conversations with Célestin Deliège*, 54. Concerning the notation of free and strict rhythm, see *Orientations: Collected Writings by Pierre Boulez*, ed. Jean-Jacques Nattiez, trans. Martin Cooper (HUP, 1986), 86−7 from 'Temps, notation et code', a lecture at Darmstadt in 1960.

[220] David Gable, 'Boulez's Two Cultures: The Post-War European Synthesis and Tradition', *JAMS* 43 (1990), 436−9, also 435 n. 18. On p. 437 he writes: 'Rubato in Chopin allows both a measure of freedom to the melody, which "robs" from the relatively strict time values marked by the bass [the earlier type], and a degree of flexibility in the projection of meter [presumably the later type]. . . . Where Chopin's flexible surface still implied an essentially stable underlying meter, Boulez's rhythms often float on an unstable flux.'

[221] Ibid. 439−40 and Ex. 5 on 442−3 from *Rituel*. Boulez greatly expands the concept of heterophony, which presumably would include, on the simplest level, those examples we have already noted in which both a robbed and unrobbed melody sound simultaneously. For a detailed explanation of his concept, see *Boulez on Music Today*, 120−9; on 121 he mentions the heterophony used by Beethoven 'for ornamental purposes' in the Adagio of the 9th Symphony: beginning at m. 99 the winds play the plain melody while the first violins simultaneously play an ornate version which includes added notes and the occasional shifting of the main melody notes in the manner of the earlier type of rubato.

[222] Gable, 'Boulez's Two Cultures', 441, referring to Part I of Boulez's *Répons* from 1981−4.

may at the same time be playing as fast as possible. In some places each player has a different metronome mark. For an acceleration the notes are printed closer together, for a retard they get further apart. Sometimes a performer is to play as slowly as he can in a single breath; sometimes he plays with or after another performer without counting beats himself.[223]

Around the same time, Stockhausen became interested in spatial contrast and in contrasts between live performers, electronic sounds, and *musique concrète*. In *Gruppen*, three orchestras, 'each under its own conductor, sometimes play independently and in different tempi'. *Gesang der Jünglinge* contrasts five loudspeaker groups in space, as well as electronic sounds versus manipulations of recorded singing.[224] *Mixtur* from 1964 is for five orchestral groups, sine-wave generators, and four ring modulators, thus contrasting live sounds with their own electronic transformation; the section called 'Stufen' is marked 'SEHR SCHNELL MIT RUBATO' and, when interchanged with another section, 'there must be enough rubato' to give a particular total duration.[225] *Mantra*, for two pianos, modulators, and percussion, contains a number of sections marked 'RUBATO' or 'mit RUBATO'.[226]

The use of electronic sounds, in which performers do not participate, required a re-evaluation of the time elements in music.[227] With Stockhausen, all musical elements are eventually serialized, including rhythms, durations, and tempos. Often he achieves a sort of static quality—a sense of time associated with Eastern religions in which one moment does not lead to the next but simply exists in a continuous present. Such a concept, like Boulez's *temps amorphe*, seems to be an almost inevitable consequence of rhythmic develop-

---

[223] Work No. 5: *Zeitmasse für fünf Holzbläser* (London: Universal Edition, 1957). See, for example, mm. 161 and 203. Stockhausen discusses the instruction 'as fast as possible' in a section dealing with 'degrees of freedom' for the performer in his article 'How Time Passes', written in 1956 and published in *Die Reihe* (English version), 3 (1959), 37–8. Robert Craft describes the innovative features of the work in his notes for Columbia Record ML 5275, printed also in 'Boulez and Stockhausen', *The Score*, 24 (Nov. 1958), 61–2; and in Elliott W. Galkin, *A History of Orchestral Conducting in Theory and Practice* (New York: Pendragon Press, 1988), 779–80. Concerning irregular note spacing, see Gardner Read, *Modern Rhythmic Notation* (IUP, 1978), 115–16.

[224] See Stockhausen's notes on his works in Karl H. Wörner, *Stockhausen: Life and Work*, ed. and trans. Bill Hopkins (UCP, 1973), 37–9 and 40–1. See also his lecture 'Music in Space', *Die Reihe* (English), 5 (1961), 67–71, where he traces spatial contrasts back to Willaert, G. Gabrieli, Mozart, and Berlioz.

[225] Work No. 16 (London: Universal Edition, 1966), p. 13 of the explanatory notes, and p. 16A (1st page of 'Stufen').

[226] Work No. 32 (Kürten: Stockhausen-Verlag, 1975), mm. 146, 158, 185, and 269.

[227] See Stockhausen, 'How Time Passes', *Die Reihe* (English), 3 (1959), 10–40; also Boulez, 'At the Ends of Fruitful Land...', *Die Reihe* (English), 1 (1958), 23–6 (note on 23 his reference to 'irrational values' [my 'unusual groups'] as more than 'only a written out rubato').

ments earlier in the century, yet it poses, of course, enormous difficulties for the Western listener.[228]

In addition to the concertato works which emphasize the contrast between different spatial positions and different methods of sound production, there are several later works by Stockhausen for a single solo instrument that also employ rubato. In *Der kleine Harlekin* for solo clarinet, a section of even sixteenth notes with changing metres is marked 'con rubato', with the explanation that 'the first note of each measure is slightly emphasized and somewhat longer.'[229] 'Die Schmetterlinge spielen' from *Amour* for solo clarinet is marked 'con poco rubato', with the instruction that 'at each bar line the change of duration [in the melodic line] should be made clearly audible (rubato).'[230] In the different versions of *In Freundschaft* for a single solo instrument, large groups of beamed small notes commence with a slash through the first two or three stems. Although some composers employ this notation to indicate that they should be played as fast as possible, Stockhausen combines it in the bassoon score with 'frei' and 'nicht zu schnell' on one occasion, and '*schnell—con rubato*' on another.[231]

The New Expression in the 1970s and 1980s

The music of Carter, Boulez, and Stockhausen represents a high point in rhythmic complexity. Performers of the time seemed as eager to conquer formidable challenges as composers were to provide them. The electronic medium finally extended the complexities even beyond those possible from a human performer. The new concertato style, in which rubato played such a vital role, provided the framework for the display of the rhythmic drama—most complex, of course, in concertato counterpoint. More recently, however, there has been a reaction to this development, leading to a simpler texture and a renewed interest in some familiar elements from the past. Thus there is new concern for performed music, for melody, for tonality and repetition, and for pulsed and metrical rhythm and its expressive modification by a more traditional sort of rubato.

---

[228] See George Rochberg, *The Aesthetics of Survival: A Composer's View of Twentieth-Century Music*, ed. William Bolcom (UMP, 1984), 116 and 132–3.

[229] Work No. 42½ (Kürten, 1978), 4 and trans. on p. iv. Since the passage consists of a long series of even notes, the rubato would be particularly conspicuous to the listener.

[230] Work No. 44 (Kürten, 1978), 4 and trans. on p. i.

[231] Work No. 46¾ (Kürten, 1983), 2, lines 6 and 7. The same spots in the clarinet version, Work No. 46 (Kürten, 1979) are marked 'frei' and 'äusserst schnell—poco rubato'.

In much music from the 1970s and 1980s, then, rubato occurs, at least for the listener, in a more familiar manner and in a more familiar environment. At the same time, brief passages of free time may contrast in alternation or in counterpoint with strict rhythm. Groups of beamed notes are frequently notated with a grace-note slash or with fanned beams to indicate various types of freedom in performance. The slash, mentioned above in connection with the music of Stockhausen, may occur on a beamed group of a few or many notes and for notes both large and small. For Toru Takemitsu this notation means 'jouer très rapidement' in a work from 1970; for Daniel Pinkham in 1973 it is 'as rapidly as possible'.[232] John Corigliano, for whom it means 'out of tempo' or 'as fast as possible, do not synchronize', includes such groups simultaneously with strictly metrical voices or one such group simultaneously with another.[233] In a composition from 1981 Donald Martino explains that grace-notes with a slash are 'as fast as possible', whereas small notes without a slash 'are notes of unspecified duration which may or may not be played "as fast as possible"'.[234] In a piece from 1968 Gardner Read defines the slashed notes as 'short, as fast as possible', and in a book ten years later writes that the 'grace-note format' is favoured 'for note groups to be played as fast as possible, a specification that allows for considerable latitude, depending upon the abilities of the individual performer and the nature of the instrument played'. He includes this among the techniques whereby 'the composer specifies a free or highly variable durational scheme, allowing performers to adjust both tempo and rhythm according to their spontaneous reactions'.[235]

In addition, Read describes the fanned beams as the 'notational device by which rubato is given graphic representation'. For an accelerando, two or three beams begin at a single point on the stem of the first note and fan out so that they are not parallel but at an angle with each other and therefore at some distance apart on the final note of the group. The reverse indicates ritardando, with the beams separated on the opening note and gradually reaching a single point on the stem

---

[232] Takemitsu, *Eucalypts* (Paris: Éditions Salabert, 1970), 1; and Pinkham, *Mourn for the Eclipse of His Light* for violin, organ, and electronic tape (Ione Press, 1980), title page.

[233] *Poem on His Birthday* (New York: G. Schirmer, 1981), 2 n. See the free groups on pp. 61–3 which the piano, xylophone, and flute play against the regular rhythm of the chorus, and the footnote on p. 44: 'R.H. [right hand of the pianist in this reduction of the orchestral parts] continues same tempo while L.H. and chorus slow to new tempo.' See also his *Concerto for Oboe and Orchestra* of 1975 (New York: G. Schirmer, 1983), the footnote to p. 21 for the definition, and p. 22 in the 3rd movt. for an example of 2 slashed groups in counterpoint.

[234] *Fantasies and Impromptus* for solo piano (Newton, Mass.: Dantalion, 1982), Notes.

[235] *Sonoric Fantasia No. 3*, Op. 125 (New York: Seesaw Music Corp., 1971), 3; and *Modern Rhythmic Notation*, 115–16.

of the last. Although both, according to Read, 'are intended to be nonmetrical and hence only approximate in durational relativity', many composers also employ them in 'metrically organized music'. Alfred Schnittke, for example, includes groups of seven notes with fanned beams on the last beats of some measures in his Second String Quartet; other instruments, however, play with metric exactness at the same time, so that the total duration of the accelerating group is fixed.[236] Martino adds a thinner slanting line closer to the note heads than the usual parallel beams to indicate such an 'acceleration within the ligature without alteration of the ligature's total duration'.[237] In addition, Read uses fanned beams to indicate an accelerating trill, and Pinkham employs them to signify 'an expressive and gradual accelerando' in trill-like figures within free fantasy sections of music.[238] Corigliano, finally, sometimes combines the slash and the fanned beam on the same group, signifying, presumably, an acceleration which eventually becomes as fast as possible.[239]

Although the word *rubato* seldom appears in connection with slashed or fanned beams, the same composers often do write the term elsewhere in their scores. Martino states in his *Fantasies and Impromptus*: 'Except where terms like "alla misura" appear, metronome marks are not meant to impose rigidity of pulse. On the contrary, "Tempo Rubato" is the norm.' Within his scores he writes *tempo rubato* or *con rubato* for sections of one to nineteen measures in length and indicates either the desired range of metronome tempos or specifies an exact tempo for each group of two or three notes. In addition, he sometimes marks single notes with two parallel dashes to indicate ritenuto.[240] The word *rubato* occurs in even more conventional ways in the tempo markings and smaller italic instructions in works by Corigliano, Pinkham, Takemitsu, and Crumb.

Gunther Schuller weaves two kinds of rhythm together in a passage in his *Episodes* for solo clarinet which he explains as follows: 'In this section the effect should be that of two separate lines. Dynamics must be sharply delineated. The line with stems down is to be played very

---

[236] (Universal, 1981), p. 25 in 3rd movt.

[237] *Fantasies and Impromptus*, Notes. See the groups of 5, 7, 8, or 14 notes in the final Fantasy. In m. 124 the accelerating group is part of a *ritardando*; in m. 151 3 groups are marked *animando*. Also note the passage beginning in m. 135, which is marked 'Due tempi: the melody is to be played slower . . . than the accompaniment.'

[238] Read, *Sonoric Fantasia No. 3*, 3 and 17; Pinkham, *Lessons for the Harpsichord* (New York: C. F. Peters, 1973), 2 for the definition, 8 and 9 in the fantasy sections of the 3rd movt.

[239] *Concerto for Oboe and Orchestra*, 5th movt., m. 198 (in a brief section marked 'Quasi Cadenza').

[240] *Fantasies and Impromptus*, No. 2, m. 5; No. 9, m. 81; and the markings at the beginning of Nos. 4 and 6.

strictly [and generally loud]; the lines with stems up at all times with a degree of rubato [and soft].'[241] György Ligeti has a piece for wind quintet marked 'Presto bizzaro e rubato, so schnell wie möglich'. A note explains that 'the tempo is gauged to the Bassoon's technical capabilities; free changes of tempo (accel., rall.) are permissible at any time, the other instruments suiting their tempo to the Bassoon (colla parte).' In a piece for organ marked 'Rubato, sempre legatissimo', he changes one note each measure in a sustained ten-note chord and requests that 'nowhere in the piece should the chord successions create an impression of metre or periodicity.'[242]

Krzysztof Penderecki sometimes employs a sine wave to indicate *quasi senza misura*: 'The rhythmic values need not be strictly observed.' This sign may occur alone or with the word *rubato*.[243] He sometimes depicts changes of tempo by varied spacing between the notes; in his *Dies irae* a section marked *tempo rubato* contains eight simultaneous vocal lines, each with differently spaced notes on the same text.[244] He often indicates rubato for a long series of evenly notated pitches or for passages in which the composite notated rhythm produces a steady pulse. In such an environment, of course, the effects of the rubato become especially conspicuous.[245] His rubatos may occur also in passages with varied note values and in a single line or in several parts at once. In addition, he occasionally includes groups with fanned beams, played either alone or against a strict rhythm in another part.

Peter Maxwell Davies, in a section marked 'Lento, rubato' in *Stedman Caters* of 1968, uses fanned beams for trill-like tremolo figures heard against strictly notated lines in other parts.[246] In *The Blind Fiddler* from 1975 a voice sings accelerating groups, as well as accelerating−retarding groups in which the beams fan out toward the middle stem and back again to a single point on the last one; at the same time, evenly notated tones in the celesta are to be played *rubato*,

---

[241] (AMP, 1979), 2 n.

[242] *Zehn Stücke für Bläserquintett* (Schott, 1969), No. 10, main tempo marking and note following m. 4; *Zwei Etüden für Orgel* (Schott, 1969), No. 1: 'Harmonies', 4.

[243] For a definition, see, for example, the *Passio et mors Domini Nostri Iesu Christi secundum Lucam* (Celle: Moeck Verlag; Kraków: Polskie Wydawnictwo Muzyczne, 1967), 6. It occurs with *tempo rubato* on 42, and by itself 1 and 2 bars earlier.

[244] *Dies irae* (Celle and Kraków, 1967), 24. Witold Lutosławski describes a similar effect in his *Concerto for Cello and Orchestra* (London: Edition Wilhelm Hansen, 1971), 61, where the solo cello plays 2 notes *rubato* and *dolente*, and a footnote for the 5-part string accompaniment states that 'the tempo should be slightly different with each instrument (even within the same part).'

[245] *Dies irae*, Part II, 26−7; *De natura sonoris* (Celle and Kraków, 1967), starting on p. 22; and *Capriccio* for violin and orchestra (Celle and Kraków, 1968), 16.

[246] (B & H, 1980), 4th movt.

which is explained in a footnote as 'freely and irregularly within the indicated bars'.[247] In addition, Davies sometimes indicates free rhythm by a group of black, stemless notes. They may occur simultaneously with exactly notated rhythms or with slashed groups; they are sometimes marked *rubato*, and sometimes closely resemble the notation used in the unmeasured preludes of the seventeenth century.[248] In *The Blind Fiddler* he superimposes free and strict rhythm by indicating 'allegretto rubato (*independent*)' for the unmetred but rhythmically notated violin part, which sounds simultaneously with 'adagio' and changing metres in the other instruments.[249] In the *Hymn to St Magnus* a variety of rhythms in the viola and cello, marked '*allegro moderato, rubato*', sounds against a long series of eighth notes in '*fluctuating tempo*' played by the flute, clarinet, and harpsichord, each independent of the other.[250] His rubatos sometimes involve a series of evenly notated tones: in the *Symphony* a succession of fourteen half notes marked *rubato molto* are to be repeated many times by the tubular bells; in *Blind Man's Buff* a brief passage of eighth notes is marked '*rubato—not equal values*'.[251] He sometimes includes displacement between the parts due to polyrhythm or notated syncopation.[252]

Many of the composers from this period use rubato for increasingly expressive purposes within contexts familiar from late romantic music. The most prolific composer of this type is George Rochberg, who noticeably increased his use of rubato around 1970. He employs rubato sometimes in a main tempo marking, sometimes in italics for a brief passage of a few notes or bars. He uses the expressions *sempre rubato, con rubato*, or *sempre con rubato*, and often writes *molto rubato* and occasionally *a tempo ma rubato* or *quasi rubato*.[253] His rubato sometimes occurs with signs of accentuation,[254] unusual groups

[247] (B & H, 1981), beginning of 6th movt.

[248] See his *Missa super l'Homme armé* (B & H, 1980), 395; *Hymn to St Magnus* (B & H, 1980), 2nd movt., 10; and *Fiddlers at the Wedding* (B & H, 1980), No. 7, 13 and 15.

[249] 9th movt., 30.

[250] 2nd movt., 4.

[251] *Symphony* (B & H, 1978), last movt., 178 ff.; *Blind Man's Buff* (London: J. & W. Chester, Edition Wilhelm Hansen, 1981), 35.

[252] See his *Sinfonia* for chamber orchestra (London: Schott & Co., 1968), 2nd movt., 9; and *Ecce manus tradentis* (B & H, 1982), 14.

[253] See *To the Dark Wood* for woodwind quintet (Presser, 1986), where *quasi rubato* at the beginning seems to refer to rhythmic flexibility written in by syncopations and triplet groups.

[254] For a series of accent signs, see *Duo concertante* for violin and cello (Presser, 1960), m. 187; for dashes, *Quintet for Piano and String Quartet* (Presser, 1984), 2nd movt., m. 7, 3rd movt., m. 318, and 4th movt., beginning; for dots, *Quintet for Two Violins, Viola, and Two Violoncellos* (Presser, 1983), 5th movt., mm. 1, 17, 51, and 80; for portato touch on the piano, *Songs in Praise of Krishna* (Presser, 1981), Nos. 5 and 8.

(mostly triplets or quintuplets),[255] or polyrhythm,[256] and sometimes in cadenza-like passages for a solo instrument.[257] Occasionally it accompanies groups of notes with fanned or slashed beams, and sometimes its duration is marked with dashes.[258] He often employs rubato with a series of notated even notes,[259] and sometimes indicates a declamatory style by marking *parlando* or *recitando*.[260] The essentially expressive purposes of his rubatos are most clearly revealed, however, by the large number of descriptive terms which accompany them: often *molto espressivo* or *con gran espressione*, as well as *freely, loose, fluid, senza rigore, libero, liberamente, flessibile*, or *come una improvvisazione*;[261] frequently *dolce* or *dolcissimo, cantando, cantabile*, or *lyrical*; and moods from *amoroso, grazioso, romantico, tranquillo*, or *teneramente* to *passionatamente, intenso, doloroso, wry, puckish, misterioso, restless, threatening, animando, agitato, violent, ominous*, and *foreboding*. Such a list reminds us of the terminology employed so abundantly by Liszt and later romantic composers and describes, for those who feel strangers in the musical worlds of Carter, Boulez, and Stockhausen, a more familiar environment in which to experience rubato.

David Del Tredici, in his works on Alice in Wonderland, combines rubato with terms such as *grazioso, intimissimo, passionato, agitato*, and *con amore*, as well as *molto espressivo* and *dolce* or *dolcissimo*. He sometimes explains the rubato in greater detail by including also *rit.* and *accel.*[262] Roger Bourland includes rubato within a passage to be performed 'with cosmic resignation'.[263] Stephen Albert has a rubato 'with feeling'; elsewhere he shows an *accel.* and *sub. rit.* on two septuplets, with the footnote: 'Start slower than tempo; tempo rubato for two beats only.'[264] William Bolcom includes the instructions

---

[255] *Partita-Variations* for piano solo (Presser, 1977), No. 10; *Trio for Piano, Violin, and Cello* (Presser, 1986), 2nd movt.

[256] *String Quartet No. 3* (New York: Galaxy Music Corp., 1973), Part B: III, mm. 33, 41, 65, and 69.

[257] *Sonata for Viola and Piano* (Presser, 1979), 1st movt., mm. 121, 128, and 245, 3rd movt., mm. 27 and 32; *Concerto for Oboe and Orchestra*, piano reduction (Presser, 1985), mm. 60, 72.

[258] Accelerating groups with fanned beams are in *Four Short Sonatas* for piano solo (Presser, 1986), No. 2, beginning; slashed beams in *Slow Fires of Autumn* for flute and harp (Presser, 1980), 9, 10, 11, and 20; dashes to show duration in the *Sonata for Violin and Piano* (Presser, 1989), 4th movt., mm. 49 and 52.

[259] *Caprice Variations* for unaccompanied violin, Nos. 5, 6, 20, and 50; *Ricordanza* for cello and piano (Presser, 1974), beginning.

[260] *Twelve Bagatelles* (Presser, 1955), No. 5.

[261] For rubatos in free fantasy-like music, see *Nach Bach, Fantasy for Harpsichord or Piano* (Presser, 1967), 5, 7, 15, and 16; and *Between Two Worlds* for flute and piano (Presser, 1983), 1st movt.: 'Fantasia', mm. 12 and 26.

[262] *Final Alice* (B & H, 1978), m. 75.

[263] *Cantilena* for string orchestra (Ione Press, 1983), m. 43.

[264] *RiverRun* (New York: G. Schirmer, 1985), 2nd movt., 41 and 65.

'Molto espressivo, rubato'; 'Slower, quizzically, rubato'; 'Swoopy, with rubato: sentimental waltz-tempo'; and also has a rubato that is shouted as well as a *non rubato* marked 'simply'.[265] William Helps also has some *non rubatos*, one even marked '*non rubato ma espressivo*'.[266]

Composers of this period also include terms, such as *free, improvisatory, liberamente, senza misura, parlando,* or *declamando*, that suggest some sort of rhythmic freedom. Occasionally these terms are also linked with rubato. Del Tredici combines *a piacere* and *sempre rubato*, and Albert joins *poco rubato* with *slowly and freely with feeling*. Thea Musgrave links rubato with *ad libitum, recitativo,* or *senza misura*: in *The Voice of Ariadne* she marks 'ad lib. T.^{po} rubato' and 'quasi recit.' on one occasion and states in her prefatory comments that '*ad lib.* passages or those in an unmeasured bar should continue in the same tempo, unless otherwise indicated, though they should be played with some rubato.'[267] Frederic Rzewski indicates '*con rubato e flessibile; quasi una fantasia*', and on another occasion writes: 'The groups of 10 [notes] may be stretched by lengthening the first notes and accelerating the rest, but without destroying the sense of the quintuplet: literally "tempo rubato".'[268] Ralph Shapey marks '*rubato—quasi cadenza*', 'Cadenza: rubato-bravura', or 'rubato, cantabile', the latter 'to be played with great freedom; fit the rhythm into the approximate metronome tempo'.[269] Henri Lazarof writes 'un po' rubato-quasi cadenza', 'Molto rubato (*quasi improvisato*)', or 'libero e molto rubato',[270] and Friedrich Cerha 'rubato, flexibel' and 'Rubato (Zeit lassen)'.[271]

A number of other composers also associate rubato with rhythmic freedom or with explicitly expressive purposes. This more conventional type of rubato occasionally occurs even before 1970. It seems to become part of a more general trend, however, during the 1970s and 1980s. The composers involved are too recent to be categorized easily

---

[265] *Commedia for (almost) 18th-Century Orchestra* (Marks, 1977), m. 356; *Seasons* for solo guitar (Marks, 1978), No. 3; *Cabaret Songs*, i (Marks, 1979), 4 and 18; and *Three Gospel Preludes for Organ* (Marks, 1980), No. 3.

[266] *The Running Sun* (New York: C. F. Peters, 1976), II and V.

[267] Composer's fac. of vocal score (Borough Green, Sevenoaks, Kent: Novello, 1977), 13, and Notes on Performance.

[268] *North American Ballads* (Zen-on Music Co., 1982), No. 3, m. 19; and *Four Pieces for Piano* (Zen-on, 1981), No. 3, 29.

[269] *Evocation No. 3 for Viola and Piano* (Presser, 1985), 1st movt., m. 4; and *Evocation No. 2 for Violoncello, Piano, and Percussion* (Presser, 1985), 3, 6, 19, and 33.

[270] *Three Pieces for Harpsichord* (AMP, 1972), 'Notturno'; *Trio for Wind Instruments* (Bryn Mawr, Pa.: Merion Music, 1982), 2nd movt., Var. 6; and *Violin Concerto* (Merion Music, 1987), 1st movt., m. 146.

[271] *Baal Gesänge* (Universal, 1982), No. 7, 90 and 118.

or for their significance in music history to be properly assessed. For our purposes, however, it is sufficient to note that they prefer a kind of rhythmic flexibility which contrasts with the strict, even pulses of neo-classicism from the first half of the century and that this flexibility, along with rubato, seems to serve now the purpose of a more intense and personal sort of musical expression.

The twentieth century has thus produced many kinds of rubato. We have noted the expressive rubatos of Schoenberg, Berg, and Schnabel, which continue the tradition of late nineteenth-century romanticism. The declamatory and structural rubatos of Bartók, Stravinsky, and many other neo-classical composers from the first half of the century, reflect the mainstream development of the new twentieth-century aesthetic ideals. After mid-century rubato becomes an important element in the new concertato style—a style which is again involved with expressive purposes. Carter feels that in all his works 'the primary intention is expressive and the entire musical vocabulary, instrumentation and form have been chosen to further this'.[272] According to Rosen, 'modernism in music is not confined to the hard-edged neo-classicism of Stravinsky, but has its neo-Romantic side in Schoenberg and Berg which reaches into the work of Elliott Carter and Karl-Heinz Stockhausen and even into much of Pierre Boulez.'[273] During the 1970s and 1980s rubato is increasingly involved with expression, feeling, and interpretation—all significant elements in the new attitude which has finally become so prevalent that many now refer to it as 'neo-Romanticism' or 'The New Romanticism'.

[272] Notes accompanying the recording of his *Concerto for Orchestra* on Columbia M 30112 from 1970.

[273] Charles Rosen, 'The Shock of the Old', a review of *Authenticity and Early Music*, ed. Nicholas Kenyon (OUP, 1988), in *The New York Review of Books*, 37/12 (19 July 1990), 46.

11 THE FUTURE OF RUBATO

THE overlapping arcs in Figure 1 depict my own conception of music history. Each arc represents a unified system of musical ideas which gradually evolve toward a moment of fulfilment, then decline. As one cycle declines, it overlaps with the beginning of the next. The process is essentially unrelated to artistic quality, since a great or trivial work may be created at almost any point in a cycle. Furthermore, the evolving musical elements themselves bear no absolute value, for any particular idea such as imitation or counterpoint may be cultivated during one cycle and carefully avoided or restricted during another. The systems of ideas are generated by two opposing factors: the need for change and for something new, and the desire to retain sufficient features from the past to assure a sense of continuity. The first half of each cycle is thus occupied with the fashioning of a new musical language, the second half with the increasingly expressive application of this language. Hence, the first half tends to be objective and intellectual and the second half more subjective and emotional. We can thus speak of a classic and romantic phase in each cycle.

The historical periods, which are considerably less significant than the cycles themselves, can only loosely be defined as that span of time during which one particular cycle dominates over others that may be in a state of beginning or decline. Thus the points where the arcs of Figure 1 intersect set the boundaries for the traditional periods of music history. Each spans about one hundred and fifty years, and the dominating cycle of ideas attains the climax of its development about seventy-five years or halfway through the period. Thus, Flemish imitative polyphony is fully developed as a technique by 1525 and can be employed thereafter for increasingly expressive purposes in the madrigal and motet. Tonality crystallizes around 1675 and provides a language that is then used for more vivid expression of the affections. By 1825 the homophonic style and its sonata forms constitute a common language which can be used during the remainder of the nineteenth century for more intensely expressive purposes.[1]

[1] See my article 'A Theory of Alternating Attitudes in the Construction of Western Polyphony', *The Music Review*, 34 (1973), 221–30. I have quoted a few sentences from 221–2 in this article, and Fig. 1 is an expansion of the diagram on 222.

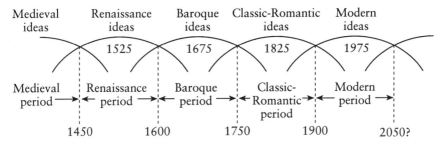

| Medieval | Renaissance | Baroque | Classic-Romantic | Modern |
| ideas | ideas | ideas | ideas | ideas |

FIG. 1. The cycles of ideas and periods in Western music history

Our own modern period might then be expected to extend from 1900 to around 2050. At some point around 1975 a change would become evident as the cycle moved from the classic phase to the romantic—from an objective, rational, intellectual point of view to one more subjective, irrational, and emotional. As early as 1909 Landowska wrote: 'Let us not say adieu, but au revoir to romanticism because soon it will come back adorned with new attractions and under a changed name.'[2] In order to return, however, romanticism requires a familiar language through which to speak, and it seems as though, for the first time in history, a new and commonly understood musical language has not, in fact, emerged. And yet, we do have such a language, but one provided less by newly composed music than by the music commonly listened to: that is, the music of the eighteenth and nineteenth centuries. This may indeed be the reason why many composers of the 1970s and 1980s, as we have seen, have returned to techniques familiar from the type of music heard extensively in concerts and recordings. Only by employing a well-known musical language can the modern composer achieve the more personal and subjective expression which is now desired as we turn into the second half of our historical period.

Signs of a new romanticism are widespread and come from diverse sources. Composers, as we have seen, now reflect this trend in their scores. Steve Reich refers to a New Classicism or New Romanticism which involves 'much more of a regular rhythmic profile' and 'a much clearer tonal center'.[3] In 1983 and 1984 the New York Philharmonic's Horizons Festivals were devoted to the New Romanticism, and the

[2] *Musique ancienne* (Paris: Mercure de France, 1909), trans. from *Landowska on Music*, ed. Denise Restout assisted by Robert Hawkins (New York: Stein & Day, 1964), 54–5.
[3] Cole Gagne and Tracy Caras, *Soundpieces: Interviews with American Composers* (Metuchen, NJ: The Scarecrow Press, 1982), 308 (from an interview in 1980).

accompanying booklet for 1983 speaks of 'renewed interest in the emotionally communicative qualities of music', 'the expression of feeling and emotion', and 'the re-emergence [during the mid-1960s] of those Dionysian qualities: sensuality, mystery, nostalgia, ecstasy, transcendency'.[4]

Music history texts are beginning to speak of neo-Romanticism or neo-Expressionism. According to Glenn Watkins, 'the wholesale recovery of tonality and the triad in combination with a newly released expressivity has led to the recognition of a lively trend that has, appropriately or not, been dubbed the New Romanticism'.[5] Robert P. Morgan describes the 'new simplicity' or 'new romanticism' in the movement, which began during the 1970s, toward 'a more traditional and conservative orientation'—a movement less interested in 'new possibilities' than 'in finding ways to incorporate what was already available into a more consistent and directly communicative musical language'. To this end, 'the more sensual aspects of music—sustained melodic development, richly colored textures, and the like—have received new and general emphasis.'[6]

A number of musicologists have been studying nineteenth-century romanticism in considerable detail for several decades now, and a few, as we have seen, have begun to investigate performance practices, such as the breaking of hands, in a more sympathetic manner. Some scholars are applying new methods of narrative analysis,[7] and some are ap-

---

[4] The New York Philharmonic, Commemorative Magazine for *Horizons '83: Since 1968, A New Romanticism?*, ed. Jacob Druckman; the quotations come, respectively, from the Introduction by Linda Sanders on 2, from Zubin Mehta's 'Message' on 5, and from Druckman's article on 7. The concert series itself included works by a number of composers mentioned in Ch. 10 in connection with rubato: Davies, Takemitsu, Del Tredici, Martino, Foss, Rochberg, Rzewski, and Schuller.

[5] *Soundings: Music in the Twentieth Century* (New York: Schirmer Books, 1988), 645–51, section on 'Nineteenth-Century Models and the "New Romanticism"'. He cites *The New Romanticism: A Broader View*, the magazine for the 1984 Horizons Concerts, in which the editor, Jacob Druckman, 'pointed to the presence of a new Dionysian sensibility, present in much music since 1970, that seemed to stem directly from a reaction to the Apollonian objectivity of the period of the 1950s and 60s'.

[6] *Twentieth-Century Music: A History of Musical Style in Modern Europe and America* (Norton, 1991), 482. Morgan includes in the movement Del Tredici, Rochberg, Davies, Penderecki, Henze, and others, and even later works by Stockhausen, Reich, and Glass. See also Eric Salzman, *Twentieth-Century Music: An Introduction*, 3rd edn. (Prentice-Hall, 1988), 207–15; K Marie Stolba, *The Development of Western Music: A History* (Dubuque, Ia.: Wm. C. Brown, 1990), 891; Douglass Seaton, *Ideas and Styles in the Western Musical Tradition* (Mountain View, Calif.: Mayfield Publishing Co., 1991), 393; and Bryan R. Simms, *Music of the Twentieth Century: Style and Structure* (New York: Schirmer Books, 1986), 428–30.

[7] See Anthony Newcomb's description of narratology in 'Once More "Between Absolute and Program Music": Schumann's Second Symphony', *19th Century Music*, 7 (1984), 233–50; and 'Schumann and late Eighteenth-Century Narrative Strategies', *19th Century Music*, 11 (1987), 164–74.

proaching criticism with the same sort of personal interpretation which they feel appropriate for performers.[8] Performers are likewise turning to a more subjective attitude: 'the pendulum has swung back and a strong personal taste is now accepted; expressive instincts can now be unleashed without any danger of their being proved unhistorical.'[9] This new attitude is evident even among organists, who generally tend to be more conservative,[10] and among the performers of early music.[11] At the same time there has been a renewed interest in the performance of lesser-known works of the nineteenth century and in works of composers such as Liszt.[12] Taken together, these signs seem to point without much doubt to a new romanticism which is sufficiently substantial that it must be reckoned with in one way or another by almost everyone involved in the business of music.

In addition, psychologists, especially during the 1980s, have been

---

[8] Joseph Kerman, *Contemplating Music: Challenges to Musicology* (HUP, 1985).

[9] Nicholas Kenyon, 'Introduction: Some Issues and Questions' in *Authenticity and Early Music: A Symposium*, ed. Nicholas Kenyon (OUP, 1988), 17. He adds: 'Perhaps it has already swung too far: only in the last year have we been able to hear period-instrument Mozart performances in neo-Furtwängler style or neo-Toscanini style, with voluptuous rhythmic rubati or driven like a rhythmic bulldozer'. See Richard Taruskin's comments in 'The Pastness of the Present and the Presence of the Past' on p. 204 of the same symposium: 'Harnoncourt's [performance of early music, such as the 1st movt. of Bach's Fifth Brandenburg Concerto] is a challenge [to 'the accepted canons of modernism'] . . ., not unlike the challenge lately issued by the so-called neo-romantics to modernist canons of composition. We are in the midst of what may yet be another major shift in aesthetic and cultural values'.

[10] See the recent pleas by organists for a more subjective approach. Scott Cantrell, in *The American Organist*, 24/8 (Aug. 1990), 45, complains of the typical organist who possesses 'technique buffed to a high gloss, rhythmic integrity, and a certain abstracted reserve—polish cultivated at the cost of personality'. In 25/6 (June 1991), 13–14 of the same source, Robert Noehren writes: 'Rubato has been inseparable from musical expression from the beginning . . . Musical style is ever changing, but the natural elements for musical expression, especially the use of rubato, have remained unchanged from 1200 to the present day'.

[11] See the *Lute Society of America Quarterly*, 25/4 (Nov. 1990), 24–6, report by Christopher M. Morrongiello on a class in 'Renaissance performance techniques with Paul O'Dette': 'Today we are learning to parrot others instead of thinking for ourselves. . . . We need to develop individual interpretations—with more personality'. He notes that 'fifteen years ago, it was fashionable not to bend any rhythms at any time. Bending rhythms was felt to be something unhistorical . . . There is, however, a lot of evidence for the use of rubato in Renaissance music, and in general, the contrast of tempo within the same piece.' He describes a basic pulse which can be expanded or contracted in 'a sort of rubber-band approach': 'If you stretch in one place you have to make up the time in order to get to the next downbeat in time. This is quite different from the 19th-century concept of *rubato* where you could take time wherever you want to and you don't have to make it up.'

[12] In *The Liszt Society Journal*, 15 (1990), 1, Dudley Newton speaks of 'the increased attention which is now being paid to Liszt the composer': 'We now have the opportunity of hearing works which previously would never have been performed'. This includes even works long considered of lowly status artistically. Rosen observes that 'the operatic paraphrases are coming back today with the revival of interest in nineteenth-century salon music and the neo-conservative antimodernism facetiously called the New Romanticism'; see 'The New Sound of Liszt', a review of Alan Walker's *Franz Liszt: The Virtuoso Years* (Knopf, 1983) in *The New York Review of Books*, 31/6 (12 Apr. 1984), 17.

studying the performance and perception of musical expression, particularly in regard to rhythm. Many of the studies deal with rubato—mainly the later type, but occasionally also the earlier. Carl Seashore had earlier observed that 'beauty in music lies largely in artistic deviation from the exact or rigid'. A number of psychologists have shown that performers cannot play in perfect metronomic time or play all the notes of a chord exactly together, even when they try or when a listener perceives the playing thus.[13] Therefore, the use of tempo changing or chord asynchronization for expressive purposes requires a level of deviation that can be perceived.[14] Modern psychologists have recently measured the shortening and lengthening of notes during performance.[15] Some studies relate tempo deviation to musical structure, leading to the conclusion that 'rubato [meaning variations in tempo] contains a great deal of information which the listener uses in the comprehension of music.'[16]

Electronically generated music makes very clear the difference between absolutely strict and performed rhythm. When a computer prints out pitches played supposedly 'in time' on a keyboard, and

[13] Carl E. Seashore, *Psychology of Music* (New York: McGraw-Hill, 1938), 249 (also 91–2, 244–53). See also Christoph Wagner, 'Experimentelle Untersuchungen über das Tempo', *Österreichische Musikzeitschrift*, 29 (1974), 601–4; L. H. Shaffer, 'Performances of Chopin, Bach, and Bartók: Studies in Motor Programming', *Cognitive Psychology*, 13 (1981), 348–55; and Michael Peters, 'Performance of a Rubato-like Task: When Two Things Cannot be Done at the Same Time', *Music Perception*, 2 (1985), 471–82.

[14] It may even be that tempo, like a Greek column which is curved so that it will appear straight, needs on occasion to be modified in order to project the impression of strictness. Moreover, other aspects of performance may influence a listener's perception of tempo, so that playing staccato with flexible rhythm, for example, might at times sound stricter in tempo than playing legato with a steady beat.

[15] See Alf Gabrielsson, 'Perception and Performance of Musical Rhythm', in *Music, Mind, and Brain: The Neuropsychology of Music*, ed. Manfred Clynes (New York: Plenum Press, 1982), 159–69; also pp. 174 and 178 in 'Neurobiologic Functions of Rhythm, Time, and Pulse in Music' by Manfred Clynes and Janice Walker. See also Ingmar Bengtsson and Alf Gabrielsson, 'Analysis and Synthesis of Musical Rhythm' in *Studies of Music Performance*, ed. J. Sundberg in *Publications Issued by the Royal Swedish Academy of Music*, xxxix (Stockholm, 1983), 27–59; and Alf Gabrielsson, 'Interplay between Analysis and Synthesis in Studies of Music Performance and Music Experience', *Music Perception*, 3/1 (1985), 62, 64, and 80.

[16] Neil Todd, 'Towards a Cognitive Theory of Expression: The Performance and Perception of Rubato', *Contemporary Music Review*, 4 (1989), 405–16; also 'A Computational Model of Rubato', *Contemporary Music Review*, 3 (1989), 69–88; 'A Model of Expressive Timing in Tonal Music', *Music Perception*, 3/1 (Fall 1985), 33–57; and, with L. H. Shaffer and E. F. Clarke, 'Metre and Rhythm in Piano Playing', *Cognition*, 20 (1985), 61–77. See also Eric F. Clarke, 'Structure and Expression in Rhythmic Performance' in *Musical Structure and Cognition*, ed. Peter Howell, Ian Cross, and Robert West (London: Academic Press, 1985), 209–36; Eric Clarke and Cathy Baker-Short, 'The Imitation of Perceived Rubato: A Preliminary Study', *Psychology of Music*, 15 (1987), 58–75; Caroline Palmer, 'Timing in Skilled Music Performance', Ph.D. diss. (Cornell Univ., 1988); and 'Mapping Musical Thought to Musical Performance', *Journal of Experimental Psychology: Human Perception and Performance*, 15 (1989), 331–46.

when it is not instructed to confine itself to values of a certain level, all sorts of tiny note values result, often arranged in strange groups. Machines can render the most complex rhythms effortlessly and with complete accuracy. When a human being performs, 'the intense struggle ... to play the rhythms correctly is inevitably heard.' The rhythm of Carter's music, for example, depends for its expressive effect not on the mechanical accuracy with which it is rendered, but 'on the intensity performers must invest to master it'.[17] Boulez has described the impossibility of producing rubato electronically, since, to be faithful to the concept, it would have to be different at each replay.[18] Recordings of performed music likewise present all nuances, including rubato, in a fixed state, even though they originate in live performance. From Boulez's point of view, such recordings are equally poor substitutes, presumably, for the uniqueness of an actual performance.

Each rubato, whether of the earlier or the later type, is indeed unique. It is such a personal device that it reflects the style of each performer: thus Chopin's rubato, for example, was different from Liszt's. It displays infinite variety and offers the performer extraordinary freedom. Thus one might identify innumerable categories and types. Some I have already named. The earlier rubato might be perceived, as we have seen, from different points of view, so that one might speak of a rubato of anticipation or of delay, a rubato of lengthening or shortening of note values, a rubato of displacement between parts or hands, or a rubato which steals from one note and gives to another. Both earlier and later types may be categorized by purpose: thus, a structural, expressive, or declamatory rubato. Concerning duration, we have already distinguished, especially for the later type, between a momentary, brief, or continuous rubato. The later type may vary in its essential nature, depending on whether it is relatively placid (a rubato of fluctuating tempo) or jerky and laboured (a kind which might be called a *stentato* type of rubato). The later type may be further categorized as an accelerating or retarding rubato, in distinction to an articulative rubato which coincides with *a tempo*. We have noted still other types and categories, and there are no doubt many, many more one might recognize in order to describe the infinite diversity inherent in the device as it passes through the hands of innumerable performers and composers during the course of its history.[19]

---

[17] Jonathan D. Kramer, *The Time of Music* (New York: Schirmer Books, 1988), 72–6.

[18] *Conversations with Célestin Deliège* (London: Eulenburg Books, 1976), 70–1.

[19] Musicians today seem to feel that there is complete agreement on what rubato means. Yet when I questioned 4 specific individuals, I got 4 quite different definitions: for a composition professor it was illustrated by a jerky, laboured rendition of a series of notes which would be

Equally striking is the extent of its presence both historically and geographically. As we have seen, both of the main types extend far back in Western music history. The later type certainly exists to some extent in all music; the earlier seems to reappear from time to time in composition or performance practice. We have noted both types in folk- and popular music as well as in art music. Furthermore, they both occur also in the music of many non-Western cultures. It is far outside the scope of this book, however, to trace the free rhythm in the preludial forms throughout the world or to study the results of heterophony, in which a fixed melody is accompanied simultaneously by a rhythmically free version that delays or anticipates some of the notes.[20]

Rubato in the West seems to survive continuously in one form or another regardless of the historical period and regardless of its chronological position within an historical cycle of ideas. Rubato will no doubt continue to adapt itself to whatever music the future may have to offer. The rubato at any particular time in the future may be of the earlier or later type, it may occur in newly composed works or in the performance of older music, and it may be included in the notated score or applied as performance practice. There is already abundant evidence, as we have seen, that the later rubato, as well as tempo flexibility in general, is increasingly desired by both composers and performers. The expressive intensity which will probably be required for new music during the second half of our modern period will no doubt be accompanied by a corresponding increase in the later type of rubato in all its forms. Furthermore, this increase will no

written presumably with equal duration; for a graduate piano major it was a compensated combination of retard and acceleration; for a musicology professor it was the expressive shaping of phrases; and for a graduate organ student it included also the modifications of tempo between sections in a continuous musical form.

[20] See Leonard B. Meyer, *Emotion and Meaning in Music* (UChP, 1956), 234–5 and 239–47. In Japan, metrical Noh singing may be accompanied by drums and flute in free rhythm (*New Grove*, ix, 519); in Nagauta music, the voice tends to 'sing out-of-time with the shamisen' and certain sections of music are in free rhythm: see William P. Malm, *Nagauta, The Heart of Kabuki Music* (Rutland, Vt.: Charles E. Tuttle, 1963), 50–1, 125–6, and 245. Free rhythm occurs in introductory flute solos in Gagaku (*New Grove*, ix, 515) and recitative-like passages in Bunraku (*New Grove*, ix, 521).

Curt Sachs, in *The Rise of Music in the Ancient World, East and West* (London: J. M. Dent & Sons, 1943), 145, describes heterophony in Oriental music: 'A singer's accompanist... is expected to follow behind by an irrationally small particle of time... His notes come in the correct... order, but are delayed when the voice unexpectedly restrains its ornaments and are ahead when the singer dwells upon a phrase.' In Indonesia, the rĕbab 'anticipates in rubato style the main notes' of the melody, and there are 'free-rhythmic preludes and postludes' (*New Grove*, ix, 175 and 178). There was a panel discussion at the national meeting of the Society for Ethnomusicology at Phoenix, Arizona, in Oct. 1988, entitled 'Flexible Rhythm, Elastic Meter, *Tempo rubato*: A Cross-Cultural Investigation'.

doubt be applied also in the performance of older music—and not only to nineteenth-century romantic works, but probably to almost all music from any period.

The works of some composers, such as Bach, Beethoven, or Chopin, endure without interruption through many centuries of changing moods and ideas. These works, however, enjoy a strange sort of immortality. During any particular period there will be stylistically good and bad performances, but during some periods most performances will be relatively close to the style prevailing at the time of composition and during others most will be more distant from that style. In either case, the performing style will be acclaimed by most listeners and preferred over any other, for it will match the spirit of the times—the same spirit that is embodied in the contemporary new music. During the first part of our modern period, for example, the neo-classical concern that performers keep strict time and avoid subjective interpretation was applied equally to new music and to the performance of old. Those in the Early Music Movement likewise performed in this manner and often deliberately selected music from the past that lent itself more or less to such a treatment. Thus, as any specific work moves through history, the expressive and rhythmic manner in which it is performed changes, sometimes matching, at other times deviating from the original style, adapting itself always to the way listeners want to hear their music. If Mozart and Haydn received more stylistically faithful performances during the first half of the modern period, then we might expect the same for Wagner and Brahms during the second half. And if the later type of rubato and tempo flexibility were restrained during neo-classicism, then we can no doubt expect the opposite in most music—new or old—during the period of the New Romanticism.[21]

The situation for the earlier type of rubato is somewhat more complex, since its existence depends upon an audible strict beat against which to operate. It seems to originate most naturally in solo vocal music, then spreads to various instrumental genres. First we find it added by the performer in the later, more subjective half of the Baroque period, where continuo-homophony provides in the aria—everywhere, that is, except the cadenzas—a continuous strict beat. From here the device moves to similar music for violin and later

---

[21] See Daniel Leech-Wilkinson's contribution to 'The Limits of Authenticity: A Discussion' in *EM* 12 (1984), 13–16; he states on pp. 14–15 that 'a work of music is far more adaptable than the people who listen to it', and shows, by a comparison of recordings, that early music performances reflect current taste. See also Richard Taruskin, 'The Pastness of the Present and the Presence of the Past' in *Authenticity and Early Music: A Symposium*, especially pp. 152–5 and 166.

for keyboard, and moves gradually from performance practice into notation. As this rubato continues into the Classic–Romantic period, new accompanimental figures provide the strict beat within the new homophonic style: first the Alberti-like basses with Mozart and Haydn, then the 'oom-pah-pah' accompaniments typical in nineteenth-century French and Italian opera and in many works by Chopin. The earlier rubato survives as long as simple, repetitive patterns in the accompaniment mark strict time—that is, in the relatively objective style of French and Italian opera, or in the still relatively strict rhythm of early romantic composers such as Spohr or Chopin. In the twentieth century, this type of rubato originates once again in solo vocal music—this time in jazz—then moves to instrumental jazz and later to the notated new concertato counterpoint of Carter and others.

Although the earlier rubato is basically an intensely expressive device, it can thus exist in both classic and romantic halves of a period. In itself, it is essentially a brief contrapuntal effect—to the extent that the melody and its accompaniment momentarily assert two independent approaches to rhythm. It thus becomes most conspicuous where contrapuntal devices are least expected, that is, in homophonic music, which reveals a special interest in making voices like one another. It may also occur, however, in the concertato counterpoint of the twentieth century, a style that displays the opposite attitude: a special concern for separate musical entities and for differentiating them when combined.[22] Some elements of concertato counterpoint still survive in the slashed and fanned beams of current music, which, when set against strict pulses in another voice, cause a brief contrapuntal effect similar to the earlier rubato. Perhaps such devices will constitute the extent to which this type of rubato will appear in the art music of the second half of the present period. And yet, within the modern style, new accompanimental figures may emerge which could mark the beat as strictly and clearly as the Alberti and waltz basses of the past.

A rubato more nearly like that of Tosi, Mozart, and Chopin might indeed arise again in popular music, especially that designed with a strict pulse for dancing. It could also very easily be revived in the

---

[22] It is difficult to know now whether later historians will see the present period as predominantly homophonic or polyphonic. The linear counterpoint of neo-classicism and the later concertato counterpoint seem to bear out the prophecy in Schoenberg's *Harmonielehre* (Universal, original edn. 1911, 3rd edn. 1922), 466: 'I believe that a further development of harmonic doctrine is not to be expected at present. Modern music . . . seems to be at a stage corresponding to the first stage of polyphonic music. . . . For it appears, and probably will appear more and more clearly, that we are turning toward a new epoch of polyphonic style'; trans. from William W. Austin, *Music in the 20th Century* (Norton, 1966), 204.

performance of older music, especially by singers, violinists, and keyboard players who perform the works discussed in this book. The cycles of Figure 1 show some historical parallels which suggest that this is a propitious moment for such a revival: 1690, the approximate date of North's first description of Tosi's rubato, corresponds three hundred years later to 1990 in our own cycle, and the years between 1828 and 1835, during which Chopin indicated rubato in his scores, correspond one hundred and fifty years later to 1978 and 1985. For those who feel uncomfortable with cyclic theories, it is perhaps sufficient to note that at the time of Chopin, as in our own time, art music had recently passed through an objective phase in which rhythm was generally very strict, and was just beginning to explore a more expressive and subjective approach.

Having been trained with the discipline of neo-classicism, many modern pianists already know how to play their left hand strictly like a conductor. Other factors also seem favourable at the moment for a rebirth of the earlier rubato, especially in performances of the works of Chopin. Modern performers have heard the device applied as a performance practice by jazz singers and instrumentalists, some of whom are still alive today. Some performers have heard it used within the complex new concertato style. Even the worst results of an unsuccessful rubato—the breaking of hands—is considered more sympathetically today as part of the performing style of late romanticism. In addition, there seems to be a crescendo of interest among scholars in Chopin's rubato and more uniformity of opinion that Chopin's instruction about the strict left hand should be taken at face value.

In spite of abundant scholarly attention, however, almost all authors dealing with the earlier rubato confine themselves to vague verbal statements and therefore offer the performer nothing concrete with which to approach the device. The only modern writer to offer more precise information to the performer, to my knowledge, is Paul Badura-Skoda, who presents an 'Essai de représentation graphique d'un rubato' in which selected notes of a melody appear earlier or later than those in the regular notation, printed directly below, of a six-measure passage from Chopin's B minor Piano Sonata. The displacement between the anticipated or delayed notes and the steady eighth-note arpeggiation in the left hand is thus clearly visible. He also depicts rubato by means of unusual groups, but, like many other authors we have already noted, selects none of his examples from passages where Chopin himself wrote the word *rubato*.[23]

---

[23] 'À propos de l'interprétation des oeuvres de Frédéric Chopin' in *Sur les traces de Frédéric Chopin*, ed. Danièle Pistone (Paris: Librairie Honoré Champion, 1984), 122–6 on 'Tempo: Regularité, Rubato et Agogique'. The music in his graphic representation in Ex. 33 on p. 125

Scholars, especially those who themselves are accomplished performers, may eventually offer explanations and justifications precise and convincing enough that performers in general will confidently put the earlier rubato to use. There may be at the present time a brief window of opportunity for this sort of rubato, for at some point in the future the increasing tempo fluctuation of the later rubato, in response to a desire for increasingly intense expression, may again deprive the melody, as it finally did in the later nineteenth century, of its relatively strict accompaniment. At the moment, however, there is probably enough sense of steady beat remaining from our recent neo-classical past to provide a strict background for the device. If the earlier rubato is indeed revived by performers, one word of warning is necessary, for the device is enormously habit-forming. The history of earlier rubato is filled, as we have seen, with complaints about its excessive or inappropriate use, even by artists of the first rank. It is a powerful effect, especially when applied within a style which is otherwise rhythmically restrained. Yet once a performer gets the knack of its execution, it is difficult to control and difficult to confine to a few appropriate spots. Furthermore, a revival of the earlier rubato would no doubt be accompanied eventually by all the mannerisms—the breaking of hands, the conspicuous asynchronization of chords, and so forth—generated partly by those who lack the ability to produce the real thing and partly by the confusing effect of increasing tempo flexibility in general. The mannerisms themselves might indeed be encouraged by those who, in the name of authenticity, attempt to revive the performing style of the late nineteenth and early twentieth centuries.

If the earlier rubato were generally restored in performances of Chopin, or in performances by singers or violinists, it would probably, due to its contagious nature, be applied by the same artists to the performance of new music as well. It might even lead eventually to notation by composers—probably, in order to retain for the word *rubato* its presently understood meaning, by marking *rubato* or *free* on one part and *a battuta* or *strict* on the others. In any event, of

comes from mm. 41–6 and 50 in the 1st movt. of the Piano Sonata Op. 58 and is thus far more extensive than even the *sempre rubato* in Chopin's Trio Op. 8 (see my Ex. 7.23a), which lasts for less than 2 measures. He seems to imply that it is appropriate for the performer to add this type of rubato during the entire course of a singing passage, for he precedes the example with the comment that this melody, 'inspirée par le bel canto italien . . . , requiert une grande indépendance rythmique entre le chant et l'accompagnement'. On pp. 123–4 he gives an example in which Chopin writes a sort of notated rubato: a descending figure which could have consisted of 12 even 32nd notes, but appears as 2 16th-note triplets and a sextuplet of 32nd notes. Badura-Skoda then applies such unusual groups, with the resulting displacement, to 2 descending passages which Chopin notated as even notes.

course, the earlier rubato will live again in the performance of the music from any period *only* if modern listeners *like* to hear it.

So far most pianists seem to feel that any conspicuous displacement between the hands in the performance of Chopin's works is outrageous. Chopin, of course, spoke of a strict left hand only in connection with those brief and infrequent spots where he marked the word *rubato*. However, tempo flexibility at that time in history, it seems to me, was probably very slight. There were, after all, other means of projecting expression, some so subtle that they would be overwhelmed and destroyed by perceptible fluctuations of the tempo. The following conditions thus seem to be required for the successful application of the earlier type of rubato: (1) a general restraint in rhythmic flexibility elsewhere in the music; (2) a deliberate strictness in the accompaniment during rubato; (3) some aspect in the text, the plot, or the unfolding music which motivates a level of expression almost too intense to control; and (4) the resulting displacement between two musical forces which *sound* as though they should, under normal circumstances, have been heard *together*. Clearly such an extraordinary procedure should occur for maximum effect only rarely. For those acquiring the technique, it might be a wise discipline, for example, to restrict it at first, as Liszt suggested to one of his pupils, to those places where Chopin himself marked the word. This might provide an insight and a feeling for the musical situations in which it is appropriate, so that it could later be applied, with taste and on rare occasions, also at other places.

Listeners of the future may know rubato—almost certainly the later type and perhaps also the earlier—from music newly composed. They may know it even better from the new ways in which old, familiar music is performed. Judging from the history of the past, the basic idea is tenacious and will surely continue in some form or other into the future. It may persist in ways familiar from the past. It may also, as it did in the new concertato counterpoint of the 1950s and 1960s, appear in new and unexpected ways—ways that we cannot even imagine at the present time.

Rubato in any form is an extraordinary device. It is seldom trivial, incidental, or commonplace. It is, after all, a violation of the orderly passing of time. Hence it bears such alarming names as *temps derobé*, *temps troublé*, *tempo perduto*, *tempo disturbato*, even *tempo indi-avolato*. Most commonly, however, it is *tempo rubato*: time that is robbed, stolen, broken, troubled, disturbed, devilish, or lost. It is a high-powered effect and must therefore be exercised with care, restraint, and artistic integrity. It is not for the beginner, the dabbler, or the timid. It is for those most sensitive to style and to the eloquent flow of musical thought. Indeed, it measures the capacity of a performer to comprehend the inner meaning of a work of art. For the listener it projects the sense of the most expressive moments in music, and instead of injuring the precious time continuum, acts ultimately to strengthen it, enhance it, and add to it special warmth and colour.

BIBLIOGRAPHY

Adam, Louis, *Méthode de piano du Conservatoire* (Paris, 1804/5).

Agricola, Johann Friedrich, *Anleitung zur Singkunst* (Berlin, 1757), repr. ed. Erwin R. Jacobi (Celle: Hermann Moeck, 1966). See also Tosi.

Alençon, J., *Korte aanmerkingen over de zangkonst, getrokken uit een itali-aansch boek, betyteld Osservazioni sopra il canto figurato di Pier Francesco Tosi* (Leyden, 1731). See also Tosi.

Anderson, Alice Jean, 'A Study of Tempo Rubato', M.Mus. thesis (Eastman School of Music, Univ. of Rochester, 1948; microfiche).

Atwood, William G., *Fryderyk Chopin: Pianist from Warsaw* (New York: Columbia University Press, 1987).

Bach, Carl Philipp Emanuel, *Versuch über die wahre Art das Clavier zu spielen*, 2 vols. (Berlin, 1753 and 1762; repr. Br. & H, 1969), rev. edn. (Berlin, 1787); both edns. trans. by William J. Mitchell as *Essay on the True Art of Playing Keyboard Instruments* (Norton, 1949).

Bacilly, Bénigne de, *Remarques curieuses sur l'art de bien chanter* (Paris, 1668; 2nd edn., 1679, repr. Minkoff, 1971); trans by Austin B. Caswell as *A Commentary upon the Art of Proper Singing (1668)* in *MTT* vii (1968).

Bacon, Richard Mackenzie, *Elements of Vocal Science; being a Philosophical Enquiry into some of the Principles of Singing* (London, 1824), new edn. by Edward Foreman in *MOS* i (1966).

Badura-Skoda, Paul, 'À propos de l'interprétation des oeuvres de Frédéric Chopin' in *Sur les traces de Frédéric Chopin*, ed. Danièle Pistone (Paris: Librairie Honoré Champion, 1984), 113–29.

Badura-Skoda, Eva and Paul, *Interpreting Mozart on the Keyboard*, trans. Leo Black (St Martin's, 1962).

Baillot, Pierre, *L'Art du violon: nouvelle méthode* (Paris, 1834); trans. by Louise Goldberg as *The Art of the Violin* (Evanston, Ill.: Northwestern University Press, 1991).

Barnett, Dene, 'New Light on Early Music', *The Canon*, 11 (1957/8), 33–6, 76–8, 99–101.

Barra, Donald, *The Dynamic Performance: A Performer's Guide to Musical Expression and Interpretation* (Prentice–Hall, 1983).

Barth, George Robert, 'The Fortepianist as Orator: Beethoven and the Trans-formation of the Declamatory Style', D.M.A. diss. (Cornell Univ., 1988; UM 88–21,213).

Bartók, Béla, *A magyar népdal* (Budapest: Rózsavölgyi és Társa, 1924). Trans. into German as *Das ungarische Volkslied*, in *Ungarische Bibliothek für das Ungarische Institut an der Universität Berlin, Erste Reihe*, xi (Berlin and Leipzig: Walter de Gruyter & Co., 1925); repr. of both the Hungarian and German edns. in *Béla Bartók, Ethnomusikologische Schriften, Faksimile–Nachdrucke*, i, ed. D. Dille (Schott, 1965). Trans. into English by M. D. Calvocoressi as *The Hungarian Folk Song*, ed. Benjamin Suchoff, with annotations by Zoltán Kodály, in *The New York Bartók Archive Studies in Musicology*, xiii (SUNYP, 1981).

—— *Béla Bartók Essays*, ed. Benjamin Suchoff (St Martin's, 1976).

—— and Lord, Albert B., *Yugoslav Folk Music*, i, in *Studies in Musicology*, vii (New York: Columbia University Press, 1951), new edn. by Benjamin Suchoff in *The New York Bartók Archive Studies in Musicology*, ix (SUNYP, 1978).

Bayley, Anselm, *A Practical Treatise on Singing and Playing* (London, 1771).

Bedenbaugh, Kay, 'Rubato in the Chopin Mazurkas', D.M.A. final project (Stanford Univ., 1986; UM 86–19,849).

Bellman, Jonathan, 'Chopin and the Cantabile Style', *Historical Performance: The Journal of Early Music America*, 2 (1989), 63–71.

Belotti, Gastone, 'Die rhythmische Asymmetrie in den Mazurken von Chopin', *Chopin-Jahrbuch* (1970), 109–37.

Bérard, Jean Antoine, *L'Art du chant* (Paris, 1755), repr. in *MMMLF*, 2nd ser., lxxv (1967); trans. Sidney Murray (Milwaukee: Pro Musica Press, 1969). See also Blanchet.

Berlioz, Hector, *Mémoires de Hector Berlioz* (Paris: Michel Lévy Frères, 1870; repr. Gregg, 1969), trans. David Cairns (Knopf, 1969).

—— *New Letters of Berlioz*, trans. Jacques Barzun, 2nd edn. (Greenwood, 1974).

Bernard, Jonathan W., 'The Evolution of Elliott Carter's Rhythmic Practice', *PNM* 26 (1988), 164–203.

Blake, Carl LeRoy, 'Tempo Rubato in the Eighteenth Century', D.M.A. thesis (Cornell Univ., 1988; UM 88–4,554).

Blanchet, Joseph, *L'Art ou les principes philosophiques du chant*, 2nd edn. (Paris, 1756). See also Bérard.

Boissier, Mme. Auguste, *Liszt pédagogue: Leçons de piano données par Liszt à Mlle. Valérie Boissier en 1832* (Paris, 1927); trans. in *The Liszt Studies*, ed. and trans. Elyse Mach (AMP, 1973).

Boulez, Pierre, *Notes of an Apprenticeship*, ed. Paule Thévenin, trans. Herbert Weinstock (Knopf, 1968).

—— *Boulez on Music Today*, trans. Susan Bradshaw and Richard Rodney Bennett (Faber, 1971).

—— *Conversations with Célestin Deliège*, original French version (Paris: Les Éditions du Seuil, 1975), English trans. (London: Eulenburg Books, 1976).

—— *Orientations: Collected Writings by Pierre Boulez*, ed. Jean-Jacques Nattiez, trans. Martin Cooper (HUP, 1986).

Brown, Howard Mayer, and Sadie, Stanley (eds.), *Performance Practice: Music before 1600* (Norton, 1989).

—— *Performance Practice: Music after 1600* (Norton, 1989).

Bruck, Boris, *Wandlungen des Begriffes Tempo rubato* (Berlin: Paul Funk, [1928]).

Burney, Charles, *The Present State of Music in Germany, the Netherlands, and the United Provinces*, 2nd edn. (London, 1775), repr. in *MMMLF*, 2nd ser., cxvii (1969).

—— *A General History of Music*, 2nd edn. (London, 1789).

Canon, Claudia von, '*Zwirnknäulerl*: A Note on the Performance of Johann Strauss *et al.*', *19th Century Music*, 2 (1978/9), 82–4.

Carter, Elliott, *The Writings of Elliott Carter: An American Composer Looks at Modern Music*, ed. Else and Kurt Stone (IUP, 1977).

Chopin, Frédéric, *Selected Correspondence of Fryderyk Chopin*, ed. Arthur Hedley (New York: William Heinemann, 1963; repr. Da Capo, 1979).

—— *Waltzes of Fryderyk Chopin: Sources*, i: *Waltzes Published during Chopin's Lifetime*, ed. Jan Bogdan Drath (Kingsville, Tex.: Texas A & I University, 1979).

—— *Œuvres pour piano, Fac-similé de l'exemplaire de Jane W. Stirling avec annotations et corrections de l'auteur (Ancienne collection Edouard Ganche); Introduction de Jean-Jacques Eigeldinger; Préface de Jean-Michel Nectoux* (BN, 1982).

Christiani, Adolph Friedrich, *The Principles of Expression in Pianoforte Playing* (New York: Harper & Bros., 1885).

Cinti-Damoreau, Laure Montalant, *Méthode de chant, composée pour ses classes du Conservatoire* (Paris, 1849).

Clarke, Eric, and Baker-Short, Cathy, 'The Imitation of Perceived Rubato: A Preliminary Study', *Psychology of Music*, 15 (1987), 58–75.

Clarke, J. William, 'Interpreting Rhythm in Bach: On the Expressive Aspects of Rhythm in the "Free Style" and *Tempo Rubato*', D.M.A. diss. (Univ. of Washington, 1971; UM 71–28,397).

Copland, Aaron, 'Jazz Structure and Influence', *Modern Music*, 4 (1926–7), 9–14.

—— *Music and Imagination* (HUP, 1952).

Corri, Domenico, *The Singers Preceptor* (London: Chappell & Co., 1810), repr. in *MOS* iii: *The Porpora Tradition*, ed. Edward Foreman (1968).

Corri, Philip Antony (Arthur Clifton), *L'Anima di musica* (London, 1810).

Couperin, François, *L'Art de toucher le clavecin*, 2nd edn. (Paris, 1717), modern edn. by Anna Linde, with English trans. by Mevanwy Roberts (Br. & H, 1933).

Cowell, Henry, *New Musical Resources* (Knopf, 1930).

Crutchfield, Will, 'Vocal Ornamentation in Verdi: The Phonographic Evidence', *19th Century Music*, 7 (1983), 3–54.

—— 'Brahms, by Those Who Knew Him', *Opus*, 2/5 (Aug. 1986), 12–21 and 60.

—— 'Gerald Moore and the Art of Accompaniment', *Opus*, 3/5 (Aug. 1987), 16–23 and 63.

—— 'An Open Ear', *Opera News*, 52/2 (Aug. 1987), 18–23.

—— 'Twin Glories II', *Opera News*, 52/12 (27 Feb. 1988), 11–14.

—— 'The Lure of History and Interpretive Power', '*The New York Times*, 25 Mar. 1990, Arts and Leisure Section, 31.

Culshaw, John, 'Tempo Mannerisms', *The Chesterian*, 27/172 (1952), 44–6.

Czerny, Carl, *Complete Theoretical and Practical Piano Forte School* (London, dedication dated 1839).

—— *On the Proper Performance of All Beethoven's Works for the Piano*, ed. Paul Badura-Skoda (Universal, 1970).

Dahlhaus, Carl, *Schoenberg and the New Music*, trans. Derrick Puffett and Alfred Clayton (CUP, 1987).

Debussy, Claude, *Lettres de Claude Debussy à son éditeur, publiées par Jacques Durand* (Paris: Durand, 1927).

Dolmetsch, Arnold, *The Interpretation of the Music of the XVII and XVIII Centuries*, 2nd edn. (Novello and OUP, 1946).

Donington, Robert, *The Interpretation of Early Music*, new edn. (Faber, 1975).

Dorian, Frederick, *The History of Music in Performance* (Norton, 1942; repr. 1966).

Duhaime, Ricky, 'An Historical Survey of *Tempo Rubato* as an Expressive Device', *National Association of College Wind and Percussion Instructors Journal*, 34 (1985), 4–9.

Edwards, Allen, *Flawed Words and Stubborn Sounds: A Conversation with Elliott Carter* (Norton, 1971).

Eigeldinger, Jean-Jacques, *Chopin vu par ses élèves* (Neuchâtel, Switzerland: Éditions de la Baconnière, 1970; 2nd edn., 1979; 3rd edn., 1988); trans. by Naomi Shohet with Krysia Osostowicz and Roy Howat as *Chopin: Pianist and Teacher*, ed. Roy Howat (CUP, 1986).

—— 'Chopin et l'héritage baroque', *Schweizer Beiträge zur Musikwissenschaft*, 2 (1974), 51–74.

Farrenc, Aristide and Louise (eds.), *Le Trésor des pianistes*, 23 vols. (Paris, 1861–72; repr. Da Capo, 1977).

Faure, Jean-Baptiste, *La Voix et le chant: traité pratique* (Paris: Henri Heugel, date on portrait 1886).

Ferrari, G. G., *A Concise Treatise on Italian Singing*, i (London, preface dated 1825).

Fielden, Thomas, 'Tempo Rubato', *Music and Letters*, 34 (1953), 150–2.

Finck, Henry T., *Chopin and Other Musical Essays* (New York, 1889), repr. in *Essay Index Reprint Series*, xi (Freeport, NY: Books for Libraries Press, 1972).

—— *Success in Music* (New York: Charles Scribner's Sons, 1927; copyright 1909).

—— *Musical Progress* (Presser, 1923).

Finson, Jon W., 'Performing Practice in the Late Nineteenth Century, with Special Reference to the Music of Brahms', *MQ* 70 (1984), 457–75.

Frigyesi, Judit, 'Between Rubato and Rigid Rhythm: A Particular Type of Rhythmical Asymmetry as Reflected in Bartók's Writings on Folk Music', *Studia musicologica*, 24 (1982), 327–37.

Fritz, Thomas, 'How did Chopin want his Ornament Signs Played?', *PQ* 29/113 (1981), 45–52.

Furtwängler, Wilhelm, *Gespräche über Musik* (Zurich: Atlantis-Verlag, 1948;

9th edn., Wiesbaden: F. A. Brockhaus, 1978); trans. by L. J. Lawrence as *Concerning Music* (B & H, 1953; repr. Greenwood, 1977).

—— *Ton und Wort: Aufsätze und Vorträge 1918 bis 1954* (Wiesbaden: F. A. Brockhaus, 1955).

Gable, David, 'Boulez's Two Cultures: The Post-War European Synthesis and Tradition', *JAMS* 43 (1990), 426–56.

Gagne, Cole, and Caras, Tracy, *Soundpieces: Interviews with American Composers* (Metuchen, NJ: Scarecrow Press, 1982).

Galkin, Elliott W., *A History of Orchestral Conducting in Theory and Practice* (New York: Pendragon Press, 1988).

Galliard, John Ernest, *Observations on the Florid Song*, 2nd edn. (London: J. Wilcox, 1743; repr., London: William Reeves, 1926). See also Tosi.

García, Manuel Patricio Rodríguez, *Traité complet de l'art du chant*, i (Paris, 1840 or 1841), and ii (1847), repr. of 2nd edn. of i (1847) and 1st edn. of ii (Minkoff, 1985); trans. by Donald V. Paschke in *A Complete Treatise on the Art of Singing*, i: *The Editions of 1841 and 1872* (Da Capo, 1984), and ii: *The Editions of 1847 and 1872* (Da Capo, 1975).

—— *García's New Treatise on the Art of Singing* (London, [1857]).

—— *Nouveau traité sommaire de l'art du chant* or *Neue summarische Abhandlungen über die Kunst des Gesanges* (Schott, [1859]).

—— *Hints on Singing* (London, [1894]), repr., with Introduction by Byron Cantrell (Canoga Park, Calif.: Summit Publishing Co., 1970).

Gatty, Reginald, 'Tempo Rubato', *The Musical Times*, 53 (1912), 160–2.

Goepfert, Robert H., 'Ambiguity in Chopin's Rhythmic Notation', D.M.A. diss. (Boston Univ., School for the Arts, 1981; UM 81–26,665).

Göllerich, August, *Franz Liszts Klavierunterricht von 1884–1886 dargestellt an den Tagebuch-aufzeichnungen von August Göllerich*, ed. Wilhelm Jerger in *Studien zur Musikgeschichte des 19. Jahrhunderts*, xxxix (Regensburg: Gustav Bosse, 1975).

Goodrich, A. J., *Theory of Interpretation Applied to Artistic Musical Performance* (Presser, 1899).

Gridley, Mark C., *Jazz Styles: History and Analysis*, 3rd edn. (Prentice–Hall, 1988).

Hanslick, Eduard, *Geschichte des Concertwesens in Wien*, ii: *Aus dem Concertsaal* (Vienna: Wilhelm Braumüller, 1870; repr. Gregg, 1971), 2nd edn. (Vienna and Leipzig: Wilhelm Braumüller, 1897); selections trans. by Henry Pleasants in *Vienna's Golden Years of Music 1850–1900* (S & S, 1950).

Heiles, William Hunter, 'Rhythmic Nuance in Chopin Performances Recorded by Moriz Rosenthal, Ignaz Friedman, and Ignaz Jan Paderewski', D.M.A. thesis (Univ. of Illinois, 1964; UM 65–832).

Herz, Henri, *Méthode complète de piano* (Paris, portrait dated 1837).

Higgins, Thomas, 'Chopin Interpretation: A Study of Performance Directions in Selected Autographs and Other Sources', Ph.D. diss. (Univ. of Iowa, 1966; UM 67–2,629).

—— 'Chopin's Practices', *PQ* 29/113 (1981), 38–41.

—— 'Whose Chopin?', *19th Century Music*, 5 (1981/2), 67–75.

Hiller, Johann Adam, *Anweisung zum musikalisch-zierlichen Gesange* (Leipzig, 1780; repr., Leipzig: Edition Peters, 1976); trans. by Suzanne Julia Beicken in 'Johann Adam Hiller's *Anweisung zum musikalisch-zierlichen Gesange, 1780*: A Translation and Commentary', Ph.D. diss. (Stanford Univ., 1980; UM 80–23,623).

—— *Anweisung zum musikalisch-richtigen Gesange*, 2nd edn. (Leipzig, 1798).

Hofmann, Josef, *Piano Questions* (Doubleday, 1909).

Houston, Robert E., 'Playing on Borrowed Time: A Brief History of Rubato', *American Music Teacher*, 38/1 (Sept.–Oct. 1988), 22–3 and 68.

Hummel, Johann Nepomuk, *A Complete Theoretical and Practical Course of Instructions on the Art of Playing the Piano Forte* (London, preface dated 1827).

—— *Méthode complète théorique et pratique pour le piano-forte* (Paris, 1838; repr. Minkoff, 1981).

Johnstone, J. Alfred, *Essentials in Piano-Playing and Other Musical Studies* (London: W. Reeves, [1914]).

—— *Rubato, or the Secret of Expression in Pianoforte Playing* (London: Joseph Williams, 1920).

Joseph, Charles M., *Stravinsky and the Piano*, in *Russian Music Studies*, viii (UMI, 1983).

Kalkbrenner, Christian, *Theorie der Tonkunst*, i (Berlin, 1789).

Kamienski, Lucian, 'Zum "Tempo rubato"', *Archiv für Musikwissenschaft*, 1 (1918), 108–26.

Karasowski, Maurycy, *Friedrich Chopin, Sein Leben, seine Werke und Briefe* (Dresden, 1877); trans. by Emily Hill as *Frederic Chopin: His Life and Letters*, 3rd edn. (London: William Reeves, 1938; repr. Greenwood, 1970).

Kenyon, Nicholas (ed.), *Authenticity and Early Music: A Symposium* (OUP, 1988).

Kernfeld, Barry (ed.), *The New Grove Dictionary of Jazz*, 2 vols. (London: Macmillan, 1988).

Kiorpes, George A., 'The Performance of Ornaments in the Works of Chopin', D.M.A. diss. (Boston Univ., 1975; UM 75–20,936).

—— 'Arpeggiation in Chopin: Interpreting the Ornament Notations', *PQ* 29/113 (1981), 53–62.

Klauwell, Otto, *Der Vortrag in der Musik* (Berlin and Leipzig, 1883); trans. as *On Musical Execution: An Attempt at a Systematic Exposition of the Same Primarily with Reference to Piano-Playing* (New York: G. Schirmer, 1890).

Kleczyński, Jan, *How to Play Chopin: The Works of Frederic Chopin, their Proper Interpretation*, trans. Alfred Whittingham, 6th edn. (London: William Reeves, [1913]).

Koch, Heinrich Christoph, 'Über den technischen Ausdruck: Tempo rubato', *AMZ* 10/33 (11 May 1808), 513–19.

Koczalski, Raoul, *Frederic Chopin: Betrachtungen, Skizzen, Analysen* (Cologne: Verlag Tischer & Jagenberg, 1936).

Kodály, Zoltán, *Ungarische Volksliedtypen* (Schott, 1964).

—— *Folk Music of Hungary*, enlarged edn. by Lajos Vargyas (Budapest: Corvina Press, 1971).

—— *The Selected Writings of Zoltán Kodály* (Budapest: Corvina Press, 1974).

Kravitt, Edward F., 'Tempo as an Expressive Element in the Late Romantic Lied', *MQ* 59 (1973), 497–518.

Kreutz, Alfred, 'Voraussetzung stilvoller Interpretation: Das Tempo rubato bei Chopin', *Das Musikleben*, 2/10 (1949), 260–4.

Kullak, Adolph, *Die Ästhetik des Klavierspiels* (Berlin, 1861); trans. of 3rd edn. (1889) by Theodore Baker as *The Aesthetics of Pianoforte-Playing* (New York, 1893; repr. Da Capo, 1972).

Kullak, Franz, *Der Vortrag in der Musik am Ende des 19. Jahrhunderts* (Leipzig: Verlag von F. E. C. Leuckart, 1898).

Lachmund, Carl V., *Mein Leben mit Liszt: Aus dem Tagebuch eines Liszt-Schülers* (Eschwege: G. E. Schroeder-Verlag, 1970).

Lampert, Vera, 'Quellenkatalog der Volksliedbearbeitungen von Bartók', *Documenta bartókiana*, Neue Folge, 6 (1981), 15–149.

Landowska, Wanda, *Musique ancienne* (Paris: Mercure de France, 1909).

—— 'Advice on the Interpretation of Chopin', *Etude Music Magazine*, 44 (1926), 107–8.

—— *Landowska on Music*, ed. and trans. by Denise Restout assisted by Robert Hawkins, 3rd impression (New York: Stein & Day, 1964).

Lange, Roderyk, 'Der Volkstanz in Polen', *Deutsches Jahrbuch für Volkskunde*, 12 (1966), 342–57.

—— 'The Traditional Dances of Poland', *Viltis*, 29/1 (May 1970), 4–14.

—— 'On Differences between the Rural and the Urban: Traditional Polish Peasant Dancing', *Yearbook of the International Folk Music Council*, 6 (1974), 44–51.

Lasser, Johann Baptist, *Vollständige Anleitung zur Singkunst*, 2nd edn. (Munich, 1805).

Leech-Wilkinson, Daniel, contribution to 'The Limits of Authenticity: A Discussion', *EM* 12 (1984), 13–16.

Lenz, Wilhelm von, 'Die grossen Pianoforte-Virtuosen unserer Zeit aus persönlicher Bekanntschaft: Liszt, Chopin, Tausig', *Neue Berliner Musikzeitung*, 22/37–9 (1868), 291–3, 299–302, and 307–10.

—— *Die grossen Pianoforte-Virtuosen unserer Zeit aus persönlicher Bekanntschaft: Liszt, Chopin, Tausig, Henselt* (Berlin: B. Behr's Buchhandlung [E. Bock], 1872); trans. by Madeleine Baker as *The Great Piano Virtuosos of our Time from Personal Acquaintance: Liszt, Chopin, Tausig, Henselt* (New York: G. Schirmer, 1899).

Liszt, Franz, *Frédéric Chopin* (Paris, 1852); trans. by John Broadhouse as *Life of Chopin*, 2nd edn. (London: William Reeves, [1913]), and by Edward N. Waters as *Frederic Chopin* (London: Free Press of Glencoe, 1963).

—— *Frédéric Chopin*, 2nd edn. (Br. & H, 1879).

—— *Des Bohémiens et de leur musique en Hongrie* (Paris, 1859), *nouvelle édition* (Br. & H, 1881; repr., Bologna: Forni, 1972); trans. by Edwin Evans as *The Gipsy in Music* (London: William Reeves, [1926]).

Long, Marguerite, *Au piano avec Claude Debussy* (Paris, 1960), trans. by Olive Senior-Ellis as *At the Piano with Debussy* (London: J. M. Dent & Sons, 1972).

Lussy, Mathis, *Traité de l'expression musicale* (Paris, 1874), trans. from 4th edn. by M. E. von Glehn as *Musical Expression, Accents, Nuances, and Tempo, in Vocal and Instrumental Music* (Novello, [1885]).

MacClintock, Carol (ed.), *Readings in the History of Music in Performance* (IUP, 1979).

Marpurg, Friedrich Wilhelm, *Anleitung zum Clavierspielen* (Berlin, 1755), 2nd edn. (Berlin, 1765; repr. Olms, 1970); trans. by Elizabeth Loretta Hays in 'F. W. Marpurg's *Anleitung zum Clavierspielen* (Berlin, 1755) and *Principes du clavecin* (Berlin, 1756): Translation and Commentary', Ph.D. diss. (Stanford Univ., 1977; UM 77–12,641).

—— *Principes du clavecin* (Berlin, 1756; repr. Minkoff, 1974); trans. Hays (see item above).

—— *Anleitung zur Musik überhaupt, und zur Singkunst* (Berlin, 1763).

Mason, William, *Memories of a Musical Life* (New York: Century Co., 1901).

Mathias, Georges, *Préface* to Isidore Philipp's *Exercises quotidiens tirés des œuvres de Chopin* (Paris: J. Hamelle, preface dated 1897).

Matthay, Tobias, *Musical Interpretation* (London, 1913; repr., Boston: Boston Music Co., 1980).

McEwen, John Blackwood, *Tempo Rubato, or Time-Variation in Musical Performance* (OUP, 1928).

Mozart, Leopold, *Gründliche Violinschule*, 3rd edn. (Augsburg, 1787; repr. Br. & H, 1983); trans. of 1st and 3rd edns. (1756 and 1787) by Editha Knocker in *A Treatise on the Fundamental Principles of Violin Playing*, 2nd edn. (OUP, 1951 and 1959).

Nathan, Isaac, *Musurgia vocales, An Essay on the History and Theory of Music, and on the Qualities, Capabilities, and Management of the Human Voice*, 2nd edn. (London: Fentum, 1836), repr. in *MOS* iii: *The Porpora Tradition*, ed. Edward Foreman (1968).

Neumann, Frederick, *Essays in Performance Practice*, in *SM* lviii (1982).

—— *Ornamentation in Baroque and Post-Baroque Music, with Special Emphasis on J. S. Bach*, 3rd impression with corrections (PUP, 1983).

—— *New Essays on Performance Practice*, in *SM* cviii (1989).

Newman, William S., *Performance Practices in Beethoven's Piano Sonatas: An Introduction* (Norton, 1971).

—— 'Liszt's Interpreting of Beethoven's Piano Sonatas', *MQ* 58 (1972), 185–209.

—— 'Das Tempo in Beethovens Instrumentalmusik—Tempowahl und Tempoflexibilität', *Die Musikforschung*, 33 (1980), 161–83.

—— 'Tempo in Beethoven's Instrumental Music: Its Choice and Its Flexibility', *PQ* 30/116 (1981–2), 22–9, and 30/117 (1982), 22–31.

—— *Beethoven on Beethoven: Playing His Piano Music His Way* (Norton, 1988).

Niecks, Frederick, 'Notes on Doubtful and Often Misunderstood Musical Terms', *MMR* 18/216 (1 Dec. 1888), 265–7.

—— *Frederick Chopin as a Man and Musician*, 3rd edn., 2 vols. (Novello, preface dated 1902).

—— 'Tempo Rubato', *MMR* 43 (1913), 29–31 and 58–9.

—— 'Tempo Rubato from the Aesthetic Point of View', *MMR* 43 (1913), 116–18.

—— 'Misconceptions Concerning Chopin', *MMR* 43 (1913), 145–6.

North, Roger, *Roger North on Music*, ed. John Wilson (Novello, 1959).

Palmer, Caroline, 'Mapping Musical Thought to Musical Performance', *Journal of Experimental Psychology: Human Perception and Performance*, 15 (1989), 331–46.

Perkins, Marion Louise, 'Changing Concepts of Rhythm in the Romantic Era: A Study of Rhythmic Structure, Theory, and Performance Practices Related to Piano Literature', Ph.D. diss. (Univ. of Southern California, 1961; UM 61–6,302).

Peters, Michael, 'Performance of a Rubato-like Task: When Two Things Cannot be Done at the Same Time', *Music Perception*, 2 (1985), 471–82.

Planté, Francis, 'Lettres sur Chopin', *Le Courrier musical*, 13/1 (1910), 36.

Pleasants, Henry, 'Afro-American Epoch—Emergence of a New Idiom', *Music Educators Journal*, 57/1 (Sept. 1970), 33–7.

—— *The Great American Popular Singers* (S & S, 1974).

Powell, Newman Wilson, 'Rhythmic Freedom in the Performance of French Music from 1650 to 1735', Ph.D. diss. (Stanford Univ., 1958; UM 59–3,719).

Quantz, Johann Joachim, *Versuch einer Anweisung die Flöte traversiere zu spielen* (Berlin, 1752), modern edn. by Arnold Schering (Leipzig: C. F. Kahnt, 1926); repr. of 3rd edn. (Berlin, 1789) in *DM* ii (1953); trans. by Edward R. Reilly as *On Playing the Flute*, 2nd edn. (New York: Schirmer Books, 1985).

—— 'Herrn Johann Joachim Quantzens Lebenslauf, von ihm selbst entworfen', in Friedrich Wilhelm Marpurg, *Historisch-Kritisch Beyträge zur Aufnahme der Musik*, i (Berlin, 1754), repr. in *Facsimiles of Early Biographies*, v: *Selbstbiographien deutscher Musiker des XVIII. Jahrhunderts*, ed. Willi Kahl (Cologne, 1948; repr., Amsterdam: Frits Knuf, 1972).

Ramann, Lina, *Liszt-Pädagogium: Klavier-Kompositionen Franz Liszt's nebst noch unedirten Veränderungen, Zusätzen und Kadenzen nach des Meisters Lehren pädagogisch glossirt* (Br. & H, 1901).

Read, Gardner, *Music Notation: A Manual of Modern Practice*, 2nd edn. (New York: Taplinger Publishing Co., 1969).

—— *Modern Rhythmic Notation* (IUP, 1978).

Reichardt, Johann Friedrich, *Briefe eines aufmerksamen Reisenden die Musik betreffend*, i (Frankfurt and Leipzig, 1774).

Robert, Walter, 'Chopin's *Tempo rubato* in Theory and Practice', *PQ* 29/113 (1981), 42–4.

Rosen, Charles, 'The Proper Study of Music', *PNM* 1/1 (1962), 80–8.

—— *The Classical Style* (Norton, 1972).

—— *Arnold Schoenberg* (New York: Viking Press, 1975).

—— *The Musical Languages of Elliott Carter* (Washington, DC: Library of Congress, 1984).

—— 'The New Sound of Liszt', review of Alan Walker's *Franz Liszt: The Virtuoso Years* (Knopf, 1983) in *The New York Review of Books*, 31/6 (12 Apr. 1984), 17–20.

—— 'Happy Birthday, Elliott Carter', *The New York Review of Books*, 35/19 (8 Dec. 1988), 24–5.

—— 'The Shock of the Old', review of *Authenticity and Early Music*, ed. Nicholas Kenyon (OUP, 1988) in *The New York Review of Books*, 37/12 (19 July 1990), 46–52.

Rosenak, Karen C., 'Eighteenth- and Nineteenth-Century Concepts of Tempo Rubato', D.M.A. Final Project (Stanford Univ., 1978).

Rosenblum, Sandra P., *Performance Practices in Classic Piano Music* (IUP, 1988).

Sachs, Curt, *Rhythm and Tempo: A Study in Music History* (Norton, 1953).

Saint-Saëns, Camille, 'Quelques mots sur l'exécution des œuvres de Chopin', *Le Courrier musical*, 13/10 (1910), 386–7.

Sanford, Sally Allis, 'Seventeenth and Eighteenth Century Vocal Style and Technique', D.M.A. diss. (Stanford Univ., 1979; UM 80–2,045).

Schenkman, Walter, 'Tempo Rubato: Sorting out the Confusion', *Clavier*, 13/5 (May–June 1974), 19–20 and 29–37.

Schiff, David, *The Music of Elliott Carter* (London: Eulenburg Books, and New York: Da Capo Press, 1983).

—— 'Elliott Carter at 80', *Encore* (Grove Music Society), 3/3 (1988), 1–3.

Schindler, Anton Felix, *Biographie von Ludwig van Beethoven* (Münster, 1840), trans. by Ignaz Moscheles as *The Life of Beethoven* (London, 1841; repr., Mattapan, Mass.: Gamut Music Co., 1966; another edn., Ditson, 1841?).

—— *Biographie von Ludwig van Beethoven*, 3rd edn. (Münster, 1860), trans. by Constance S. Jolly in *Beethoven as I Knew Him*, ed. Donald W. MacArdle (Faber, 1966).

Schmitz, Hans-Peter, 'Jazz und Alte-Musik', *Die Stimmen*, 18 (1949), 497–500.

—— *Prinzipien der Aufführungspraxis alter Musik* (Berlin-Dahlem: H. Knauer, 1950?).

—— *Die Kunst der Verzierung im 18. Jahrhundert* (Bärenreiter, 1955).

Schonberg, Harold C., *The Great Conductors* (S & S, 1967).

—— *The Great Pianists*, rev. edn. (S & S, 1987).

Schuneman, Robert A., 'Playing Around with Tempo', *The Diapason*, 62/6 (May 1970), 16–19, and 62/7 (June 1970), 16.

Schwartz, Elliott, and Childs, Barney (eds.), *Contemporary Composers on Contemporary Music* (New York: Rinehart & Winston, 1967).

Seashore, Carl E., *Psychology of Music* (New York: McGraw–Hill, 1938).

—— (ed.), *University of Iowa Studies in the Psychology of Music*, iv: *Objective Analysis of Musical Performance* (Iowa City, Ia.: University Press, 1936 on inner title page, 1937 on cover).

Shaffer, L. H., 'Performances of Chopin, Bach, and Bartók: Studies in Motor Programming', *Cognitive Psychology*, 13 (1981), 326–76.

Sobieski, Marian, and Sobieska, Judwiga, 'Das Tempo Rubato bei Chopin und in der polnischen Volksmusik', *The Book of the First International Musicological Congress Devoted to the Works of Frederick Chopin, Warsaw 1960* (Warsaw: Polish Scientific Publishers, 1963), 247–54.

Somfai, László, 'Über Bartók's Rubato-Stil', *Documenta bartókiana*, 5 (1977), 193–201.

—— 'Die *Allegro barbaro*—Aufnahme von Bartók textkritisch bewertet', *Documenta bartókiana*, 6 (1981), 259–75.

—— 'Liszt's Influence on Bartók Reconsidered', *New Hungarian Quarterly*, 27/102 (1986), 210–19.

Spohr, Louis, *Violinschule* (Vienna, 1832), trans. by C. Rudolphus as *Louis Spohr's Grand Violin School* (London, preface dated 1833).

Sternberg, Constantin von, 'Tempo Rubato', *The Musician*, 17 (1912), 524–5.

—— *Tempo Rubato and Other Essays* (New York: G. Schirmer, 1920).

Stowell, Robin, *Violin Technique and Performance Practice in the Late Eighteenth and Early Nineteenth Centuries* (CUP, 1985).

Strauss, John F., 'The Puzzle of Chopin's Tempo Rubato', *Clavier*, 22/5 (May–June 1983), 22–5.

Stravinsky, Igor, 'Chronological Progress in Musical Art: An Interview', *Etude Music Magazine*, 44 (1926), 559–60.

—— *Chronicle of My Life* (London: Victor Gollancz, 1936), published also as *An Autobiography* (Norton, 1962).

—— *The Poetics of Music* (HUP, 1947).

—— 'Stravinsky Reviews Three Rites of Spring', *HiFi/Stereo Review*, 14/2 (Feb. 1965), 60–3.

—— 'On Beethoven's Piano Sonatas', *Harper's Magazine*, 240/1440 (May 1970), 37–46.

—— *Selected Correspondence*, ed. Robert Craft (Knopf, 1982–5).

—— and Craft, Robert, *Conversations* (Doubleday, 1959).

—— *Expositions and Developments* (Doubleday, 1962).

—— *Dialogues and a Diary* (Doubleday, 1968).

—— *Themes and Conclusions* (Faber, 1972).

—— *Memories and Commentaries* (UCP, 1981).

Stravinsky, Soulima, Interview with Ben Johnston, *PNM* 9/2–10/1 (1971), 15–27.

Stravinsky, Vera, and Craft, Robert, *Stravinsky in Pictures and Documents* (S & S, 1978).

Strunk, Oliver (ed.), *Source Readings in Music History* (Norton, 1950).

Sulzer, Johann Georg, *Allgemeine Theorie der schönen Künste*, 2 vols. (Leipzig, 1771 and 1774).

Taylor, Franklin, *Technique and Expression in Pianoforte Playing* (Novello, [1897]).

Thomson, Virgil, *The Musical Scene* (Knopf, 1947).

—— *A Virgil Thomson Reader*, with Introduction by John Rockwell (Boston: Houghton Mifflin Co., 1981).

Todd, Neil, 'A Model of Expressive Timing in Tonal Music', *Music Perception*, 3/1 (1985), 33–57.

—— 'A Computational Model of Rubato', *Contemporary Music Review*, 3 (1989), 69–88.

—— 'Towards a Cognitive Theory of Expression: The Performance and Perception of Rubato', *Contemporary Music Review*, 4 (1989), 405–16.

Todd, Neil, Shaffer, L. H., and Clarke, E. F., 'Metre and Rhythm in Piano Playing', *Cognition*, 20 (1985), 61–77.

Tosi, Pier Francesco, *Opinioni de' cantori antichi e moderni, o sieno Osservazioni sopra il canto figurato* (Bologna, 1723); for trans., see Agricola (German), Alençon (Dutch), and Galliard (English).

Trechak, Andrew, 'Pianists and Agogic Play: Rhythmic Patterning in the Performance of Chopin's Music in the Early Twentieth Century', D.M.A. diss. (Univ. of Texas at Austin, 1988; UM 89–9,777).

Türk, Daniel Gottlob, *Sechs leichte Klaviersonaten* (Leipzig and Halle, 1783).

—— *Klavierschule, oder Anweisung zum Klavierspielen für Lehrer und Lernende* (Leipzig and Halle, 1789), repr. in *DM* xxiii (1967); trans. by Raymond H. Haggh as *School of Clavier Playing or Instructions in Playing the Clavier for Teachers and Students* (UNP, 1982).

—— *Klavierschule, oder Anweisung zum Klavierspielen für Lehrer und Lernende, neue vermehrte und verbesserte Ausgabe* (Leipzig and Halle, 1802).

—— *Treatise on the Art of Teaching and Practising the Piano Forte by D. G. Turk, Professor and Director of Music at the Royal Prussian University of Hall, with Explanatory Examples Translated from the German and Abridged by C. G. Naumburger* (London: Preston, preface dated 1804).

Volckmar, Wilhelm, *Orgelschule* (Br. & H, *Vorwort* dated 1858).

Wagner, Christoph, 'Experimentelle Untersuchungen über das Tempo', *Österreichische Musikzeitschrift*, 29 (1974), 589–604.

Wagner, Richard, *Über die Aufführung des Tannhäuser: Eine Mitteilung an die Dirigenten und Darsteller dieser Oper* (1852), trans. by William Ashton Ellis in *Richard Wagner's Prose Works*, iii (London, 1894; repr., New York: Broude Bros., 1966), 167–205.

—— *Über das Dirigieren* (Leipzig, 1869), trans. by Edward Dannreuther as *On Conducting: A Treatise on Style in the Execution of Classical Music* (London: William Reeves, 1887, 4th edn., 1940; repr., St Clair Shores, Mich.: Scholarly Press, 1976).

Walker, Alan, *Franz Liszt*, i: *The Virtuoso Years (1811–1847)* (Knopf, 1983), and ii: *The Weimar Years (1848–1861)* (Knopf, 1989).

Weingartner, Felix, *Über das Dirigieren* (Leipzig, 1895); 3rd edn. (1905) trans. by Ernest Newman as *On Conducting* (Scarsdale, NY: E. F. Kalmus, [1925]).

White, Eric Walter, *Stravinsky: The Composer and his Works*, 2nd edn. (Faber, 1979).

Williams, Adrian (ed.), *Portrait of Liszt by Himself and His Contemporaries* (OUP, 1990).

Wodehouse, Artis Stiffey, 'Evidence of Nineteenth-Century Piano Performance Practice Found in Recordings of Chopin's Nocturne, Op. 15 No. 2, Made by Pianists Born before 1900', M.F.A. Final Project (Stanford Univ., 1977).

Wolf, Ernst Wilhelm, *Eine Sonatine: Vier affectvolle Sonaten und ein dreyzehnmal variirtes Thema, welches sich mit einer kurzen und freien Fantasie anfängt und endiget* (Leipzig, 1785); trans. of the *Vorbericht (als eine Anleitung zum guten Vortrag beym Clavierspielen)* by Christopher Hogwood in 'A Supplement to C. P. E. Bach's Versuch: E. W. Wolf's Anleitung of 1785', *C. P. E. Bach Studies*, ed. Stephen L. Clark (OUP, 1988), 133–57.

—— *Musikalischer Unterricht* (Dresden, 1788).

Wolf, Georg Friedrich, *Unterricht im Klavierspielen*, 3rd edn. (Halle, 1789).

Yeston, Maury, 'Rubato and the Middleground', *Journal of Music Theory*, 19 (1975), 286–301.

DISCOGRAPHY

Bartók, Béla, *The Centenary Edition of Bartók's Records* (1981), i: *Bartók at the piano 1920–1945*, ed. László Somfai and Zoltán Kocsis, Hungaroton LPX 12326–33; and ii: *Bartók Record Archives: Bartók Plays and Talks 1912–1944*, ed. László Somfai, János Sebestyén, and Zoltán Kocsis, Hungaroton LPX 12334–8.

Beethoven, Ludwig van, *The Beethoven Society Edition of the Thirty-Two Piano Sonatas*, recorded by Artur Schnabel in London, 1932–5, released individually 1932–7 and as a set, RCA Victor (1956) and a later edn., RCA Victor Red Seal LCT 1109–10, 1154–5, LM 2151–8 (1957).

Carter, Elliott, *Piano Sonata*, performance and notes by Charles Rosen, Epic LC 3850.

—— *Variations for Orchestra*, New Philharmonia Orchestra, Frederik Prausnitz (conductor); and *Double Concerto for Harpsichord and Piano with Two Chamber Orchestras*, Paul Jacobs (harpsichord), Charles Rosen (piano), English Chamber Orchestra, Frederik Prausnitz (conductor); notes by Carter, Columbia MS7191 (1968).

—— *Sonata for Violoncello and Piano*, Joel Krosnick (cello), Paul Jacobs (piano); and *Sonata for Flute, Oboe, Cello, and Harpsichord*, Harvey Sollberger (flute), Charles Kuskin (oboe), Fred Sherry (cello), Paul Jacobs (harpsichord); notes by Carter, Nonesuch H–71234 (1969).

—— *Concerto for Orchestra*, New York Philharmonic, Leonard Bernstein (conductor); notes by Carter, Columbia M30112 (1970).

—— *String Quartet No. 2*, Composers Quartet, notes by Carter, Nonesuch H–71249 (1970).

—— *String Quartet No. 3*, Juilliard Quartet, notes by Carter, Columbia M32738 (1974).

—— *Duo for Violin and Piano*, Paul Zukofsky (violin), Gilbert Kalish (piano), notes by Carter, Nonesuch H–71314 (1975).

Chopin, Frédéric, *Everest Archive of Piano Music*, from Duo-Art and Ampico piano rolls 1916–25; X–902: *Ignace Jan Paderewski Plays Chopin*, X–904: *Josef Hofmann Plays Chopin*, X–911: *Harold Bauer Concert* (Chopin, Schubert, Weber, Saint-Saëns), X–919: *Ignaz Friedman Plays Chopin*, X–921: *Vladimir de Pachmann Plays Chopin*.

—— *Ignaz Friedman Plays 12 Mazurkas by Chopin*, recorded 1929?, Columbia Masterworks Set No. 159.

—— *Moriz Rosenthal Plays Chopin*, recorded 1935–6, Victor Set M–338.

Debussy, Claude, and others, *Welte-Mignon Digital: Berühmte Komponisten spielen eigene Werke*, Intercord INT 860.855 (1985).

Foss, Lukas, *Echoi*, Group for Contemporary Music at Columbia University: Charles Wuorinen (piano), Raymond Desroches (percussion), Arthur Bloom (clarinet), Robert Martin (cello); produced by Paul Myers and Lukas Foss, Epic Record LC 3886.

Gershwin, George, *Marni Nixon Sings Gershwin*, Reference Recordings LP RR19 and CD RR1900 (1985).

Grieg, Edvard, and others, *Famous Composers Play Their Own Compositions*, Allegro Records LEG–9021.

Orff, Carl, *Die Kluge*, Philharmonic Orchestra and Philharmonic Opera Company, Wolfgang Sawallisch (conductor), under supervision of the composer, Angel Records, Album 3551 B/L (35389–90).

Rakhmaninov, Sergey, *The Complete Rachmaninoff*, RCA Red Seal ARM3 0260–1, 0294–6 (1973).

Stravinsky, Igor, *Stravinsky Playing Piano*, Seraphim Mono 60183 (1972).

—— *Igor Stravinsky: The Man and his Music*, documentary radio series by Frederick Maroth and William Malloch, Educational Media Associates, Berkeley, Calif., EMA–103 (1977).

—— *Igor Stravinsky: The Recorded Legacy*, CBS Records GM31/LXX 36940 (1981).

Tatum, Art, *The Complete Art Tatum*, i, CD on Capitol C21K–92866; earlier LP: *Capitol Jazz Classics*, iii, M–11028.

—— *The Complete Art Tatum Piano Discoveries*, i and ii, French RCA T607–8; earlier in *Art Tatum Piano Discoveries*, i and ii, 20th Century Fox 3029 and 3033.

INDEX